A NATURAL HISTORY OF
INHACA ISLAND
MOZAMBIQUE

A NATURAL HISTORY OF
INHACA ISLAND
MOZAMBIQUE

Edited by
Margaret Kalk

Contributing Authors

J. de Koning, K. Balkwill, H. Feijen, C. Feijen, D. Broadley,
G. Ormel, A. van Bruggen, F. Hancock and F. Costa

Third Edition

WITWATERSRAND UNIVERSITY PRESS

Witwatersrand University Press
1 Jan Smuts Avenue
Johannesburg 2001
South Africa

© Witwatersrand University Press 1995

First edition   1958
Second edition 1969
Third edition   1995

ISBN 1 86814 208 6

All rights reserved. No part of this publication may be reproduced, stored in a retrieval system, or transmitted in any form or by any means, electronic, mechanical, photocopying, recording, or otherwise without the prior permission of the copyright owner.

Every effort has been made to contact the copyright holders for permission to reprint borrowed material where necessary. We regret any oversights that may have occurred and would be happy to recify them in future printings of the work.

Cover design by Wendy Matthews of Photo-Print, Cape Town

Typeset by Photo-Print, Cape Town

Printed in the Republic of South Africa by Creda Press (Pty) Ltd, Cape Town

Cover: A coral garden with hard coral, *Acropora* spp., soft coral, *Sarcophyton* sp. (lower centre) and a family of fish (goldies) (Robin Harris); Insets: Flamingoes amongst the mangrove trees; aerial view of the south bay and Ponta Torres strait.

Sponsors

Dr and Mrs A.J.M. Carnegie
Companhia do Pipeline Moçambique
Prof. R.M. Crewe
Dr J. Dhansay
Paul Dutton
Endangered Wildlife Trust
Ernest Oppenheimer Institute for Portuguese Studies
K.F. Gill
Laura Glyn-Thomas
Prof. S.A. Hanrahan
Dr David Hughes
Dr F. Kalk
Prof. J. Kalk
Dr John Ledger
Dr J.-P. Leger
Eckart H. Pfeifer
David Raulstone
SAFTAINER
Martie Sanders
Mary K. Seely
Dr and Mrs M. Strong
Dr H.H. Woermann

The publication of this book has been greatly assisted by a generous contribution by the Fundação Calouste Gulbenkian, Lisbon. Thanks to this support by the Gulbenkian Foundation we have been able to produce a high quality book at a price which should make it more easily accessible to students of this valuable and beautiful island off the coast of Mozambique.

Contents

Foreword   xiii
   Professor J.P.F. Sellschop, Deputy Vice-Chancellor (Research)
   University of the Witwatersrand, Johannesburg

Apresentação   xvi
   Professor Dr Narçiso Matos, Reitor, Universidade Eduardo Mondlane, Maputo

Preface   xvii

Acknowledgements   xix

Contributors   xxii

INTRODUCTION   1

PART I   THE ENVIRONMENTAL SETTING *Margaret Kalk*

1. GEOGRAPHY AND OCEANOGRAPHY   13
   Geographical Position of Inhaca Island; Topography of the Island;
   Geomorphology and the Origin of the Island; Climate; Tides; Water Movements
   as Determinants of Types of Substrata; Chemical Factors in the Environment;
   Penetration of Light and Turbidity of the Sea.

PART II   LIFE ON THE SHORES *Margaret Kalk*

2. ECOLOGICAL FEATURES OF THE SHORES   33
   Trophic Relationships; Interactions of Organisms with Physical Factors;
   Maintenance of Position on the Shores; Speciation on Tropical Shores.

3. SANDY SHORES ON THE SHELTERED AND EXPOSED COASTS   49
   Major Sandy Habitats; The Sandy Slope of the Sheltered Upper Shore on the
   West coast; The Flat Middle Shore; The Sand Flats of the Lower Shore; Sea
   Grasses on Sandy Mud of the Lower Shore and Associated Animals;
   Microbiota Between Sand Grains; The Sandy Shores Exposed to Strong Waves;
   Food Webs.

4. MANGROVE SWAMPS   90
   Mangrove Tree Species on Inhaca Island; Types of Mangrove Swamps on the
   Island; Environmental Variables in the Swamps; Adaptations of the Mangrove
   Tree Species; Distribution of Animal and Plant Associates; Adaptations to Zones
   in Mangrove Crabs; Preservation of the Mangroves.

5. ESTUARINE AND TROPICAL INFLUENCES IN THE SOUTHERN AND
   NORTHERN BAYS   125
   Conditions in the Northern, Southern and Western Bays; The Habitats of the
   Southern Bay; The Middle Reaches of the Southern Bay; The Mouth of the
   Southern Bay; The Habitats of the Northern Bay; The Sand-Bar Boundary.

6. SHELTERED AND VERY SHELTERED ROCKY SHORES   149
   Supratidal Fringe and Upper Shore; Middle and Lower Shore Rocks; The
   Subtidal Fringe on the Reef Flats; The Reef Flats and Coral Reef Associates.

7. ROCKS EXPOSED TO STRONG WAVE ACTION   179
   The Environmental Effects of Strong Wave Action; The Upper Shore on the Cliff;
   The Middle Platform; The Lower Platform; The Subtidal Fringe; Game Fishes of
   the Open Sea.

8. CORAL REEFS   211
   Distribution of Coral Reefs on Sheltered Shores; The Biology of Reef-Building
   Corals; Special Features of the Common Coral Genera; Phytoplankton and
   Zooplankton in the Sea Over the Coral Reef; Coral Reef Associates; Reef
   Destruction and Recovery at Inhaca Island; Biogeographical Perspectives

9. SEA MEADOWS AND SAND-BANKS   247
   Distribution of Sea Grasses on the West Shore; Biology of the Sea Grasses; Seaweeds
   in the Sea Meadows; Animals in the Sea Meadows; Sand-Banks on the Subtidal Fringe.

PART III LIFE ON LAND

10. TERRESTRIAL VEGETATION   281
    Jan de Koning and Kevin Balkwill
    Scrub Forest, Bushland and Thicket on the West Coast; The Eastern Dunes and
    the South-Eastern Peninsula; Freshwater Swamps; Agriculture in the Central
    Parkland; Natural Regeneration of Abandoned Cultivated Land.

11. BIRDS OF INHACA ISLAND   309
    Hans and Cobi Feijen
    Bird Habitats; Birds Among Trees of the Coastal Bush; The Low-Lying Central
    Area; Diurnal Birds of Prey; Birds on the Seashores; Seabirds.

12. AMPHIBIANS, REPTILES AND MAMMALS   318
    Donald G. Broadly and Margaret Kalk
    Amphibians and Reptiles; In and Around Freshwater Swamps; In the Grass,
    Among Trees and on Cultivated Ground; Evergreen Forest and Dense Coastal
    Bush; Marine Turtles in the Seas Around the Island; Mammals.

13. INSECTS   331
    Gert J. Ormel
    The Distribution of Insects on Inhaca Island; Odonata; Isoptera; Mantoidea;
    Orthoptera; Hemiptera; Neuroptera; Coleoptera; Diptera; Lepidoptera;
    Hymenoptera; Spiders of Inhaca Island *T. Steyn and A.S. Dippenaar-Schoeman*.

14. NON-MARINE MOLLUSCS   349
    A.C. van Bruggen and M. Kalk
    Snails and Slugs on Land; Snails in Freshwater Swamps.

15. CHANGES IN THE DIATOM FLORA IN THE FRESHWATER SWAMPS   353
   Florence Hancock

PART IV   THE FUTURE OF INHACA ISLAND

16. CONSERVATION AND DEVELOPMENT   361
   Margaret Kalk and Fernando Costa
     The Present Status of the People and their Problems; The Role of the Biologists in the Universidade Eduardo Mondlane; The Integrated Development Plan, 1990.

Copyright Acknowledgements   369

Aids to the Identification of Organisms at Inhaca Island   372

References   375

Index   384

FOREWORD

*Professor J.P.F. Sellschop, Deputy Vice-Chancellor (Research),
University of the Witwatersrand, Johannesburg*

It is a particular pleasure to join with Professor Dr Narciso Matos, the Rector of the Universidade Eduardo Mondlane, in this, the thirtieth anniversary year of the foundation of the University in Maputo, in presenting a foreword to this book on behalf of the University of the Witwatersrand, Johannesburg.

Members of the University of the Witwatersrand have enjoyed long scientific association with their colleagues in Mozambique. With the establishment of the Museu Alvaro de Castro in the then Lourenço Marques in 1911, the first visits of scientists to Inhaca Island took place. From 1922 onwards, such visits by academics and their students from South Africa became more frequent. However, from the mid 1930s the involvement of specifically the University of the Witwatersrand became firmly established, with a regular routine of winter vacation courses conducted on the island. So we see from ca. 1937 the appearance of scientific papers published in the international journals emanating from the biologists from Wits, as well as from Mozambique. The University of the Witwatersrand supported enthusiastically the erection of a building with research and teaching facilities and accommodation, opened in 1951.

The culmination of these historic research endeavours occurred with the joint congress of the Sociedade de Estudos de Moçambique and the South African Association for the Advancement of Science in 1958 (which I was fortunate enough to attend). This occasion was used to launch the first edition of the book *A Natural History of Inhaca Island, Mozambique*, edited by W. Macnae and M. Kalk. The impact of this book was profound and, not surprisingly, a second revised edition followed in 1969.

We now delight in the appearance of the third revised edition. The book has been

Neste ano do trigésimo aniversário da fundação da Universidade na cidade de Maputo, é um prazer singular colaborar com o Professor Doutor Narciso Matos, Reitor da Universidade Eduardo Mondlane, Para introduzir, em nome da Universidade do Witwatersrand, o presente livro.

Há longos anos que existe colaboração científica entre membros da Universidade do Witwatersrand e os seus colegas de Moçambique. Foi em 1911, aquando da fundação do Museu 'Alvaro de Castro na então cidade de Lourenço Marques, que cientistas da 'Africa do Sul visitaram pela primeira vez a Ilha da Inhaca. Estas visitas de docentes e estudantes sul-africanos a Moçambique tornaram-se mais frequentes a partir do ano de 1922. Foi, porém, em meados do anos 30 que Universidade do Witwatersrand, em particular, e Moçambique introduzincom cursos de férias anuais na ilha. Estes cursos realizavamse com regularidade e durante o Inverno. Foi assim que a partir de 1937, aproximadamente, biólogos das Universidades do Witwatersrand e de Moçambique começaram a publicar artigos em jornais científicos internacionais. Por sua vez, foi imiciada a construçao de um prédio em 1951, com facilidades que permitiriam a pesquisa científica e na qual os visitantes se poderiam hospedarse e que, foi entusiasticamente apoiada pela Universidade do Witwatersrand.

O ponto mais alto da história das relações cientificas entre os dois países foi o Congresso da Sociedade de Estudos de Moçambique realizado em conjunto com a South African Association for the Advancement of Science em 1958, congresso este a que tive o privilégio - e o prazer - de assistir. Foi precisamente na altura deste congresso que foi lançada a primeira edição do livro *A Natural History of Inhaca Island, Mozambique* editado por W. Macnae e M. Kalk. Tão profundo foi o impacto deste livro que, como seria de esperar, pouco tempo depois foi publicada uma segunda edição revista.

completely rewritten. It is now arranged according to habitats on the four shores, with the expansion of the single chapter on the general ecology of the shores, as in the first two versions, now into seven chapters. This changes the emphasis from anatomy, classification and keys to species, to one of adaptations to environment, including behaviour, physiology, ecology and development, drawing on the relevant literature of the last thirty years. The chronicling of the many changes on the island during this time, the decline and recovery of the coral reefs, the remarkable recovery of the bush and evergreen forests, for example, are of considerable interest.

The book in its third edition, completed in the 75th anniversary year of the Botany and Zoology Departments of the University of the Witwatersrand, was planned at Inhaca Island in September 1985, and is a fine example of scientific collaboration. The major credit must however go to Professor Margaret Kalk who has led, organised and persisted until the successful culmination of this worthy project and the appearance of this splendid book. I am proud to pay tribute to her outstanding scientific and managerial attributes. This book, the third of its title, will stand as a monument to her and as a beacon of the fruitful and ongoing collaboration of the University of the Witwatersrand and the Universidade Eduardo Mondlane.

Agora, temos imenso prazer em apresentar a terceira edição do livro. O texto foi completamente revisto, sendo, nesta nova edição organizado de acordo com os habitats dos quatro litorais da Ilha. Além disto, a ecologia geral das bargens, tratada num único capítulo nas duas primeiras versães, tem uma apresentação muito mais pormenorizada agora, ocupando sete capítulos ao todo. Assim sendo, a ênfase primordial deixa de ser a Anatomia, a classifiação e a distribuição das espécies, para incidir agora na adaptação ambiental, incluindo comportamento, fisiolgia, ecologia e evolução, e incorporando literatura mais relevante dos últimos 30 anos. De destaque particular é a narração cronológica das muitas transformações efectuadas durante as últimas três décadas, como por exemplo a deteriorização e subsequente restabelecimento dos recifes de coral , e outrossim, a recuperação notável da vegetação indígena.

Esta terceira edição do livro foi concebida e programada no Ilha da Inhaca em Setembro de 1985 e é um exemplo extraordinário de colaboração científica. A maior parte das honras e dos louvores têm que incidir sobre a Doutora Margaret Kalk que chefiou, organizou e, com persistência infatigável, realizou o notável projecto que agora culmina com a publicação desta magnífica obra. Tenho assim a honra de homenagear não só a categoria científica da Doutora Margaret Kalk, mas também a sua competência administrativa. Este livro, o terceiro do seu nome, permanecerá como um monumento às suas qualidades extraordinárias e um exemplo singular da assídua cooperação científica entre as Universidades do Witwatersrand e de Eduardo Mondlane.

Apresentação

Professor Dr Narçiso Matos,
Reitor da Universidade Eduardo Mondlane, Maputo

A Universidade Eduardo Mondlane mantém já há muito tempo um estreito relacionamento com a Ilha da Inhaca. A Estação de Biologia Marítima e as suas reservas marinhas e terrestres estão sob controlo directo da Faculdade de Biologia e a Estação tem sido palco de muitas visitas de estudo à Ilha. Pelo seu valor científico a Ilha da Inhaca tem sido visitada todos os anos por estudantes afim de adquirirem experiência prática nas diferentes disciplinas tais como biologia, ecologia, geografia e ciências marinhas.

Em anos recentes a Estação de Biologia Marítima tornou-se num centro internacionalmente reconhecido e providencia apoio à investigação científica feita na Ilha. Para além da importância educacional e científica a Ilha da Inhaca é considerada uma das mais belas ao longo da costa de Moçambique oferecendo reprouso e deleite a turistas do mundo inteiro. Por outro lado o ecossistema da Ilha da Inhaca é extremamente frágil e altamente vulnerável ao impacto do Homem. Assim o desafio aos cientistas e aos legisladores é compatibilizar as necessidades do Homem com a manutenção deste lindo e frágil ambiente.

Foi, portanto, com muito prazer que aceitei apresentar esta nova edição da *História Natural da Ilha da Inhaca*. A publicação do livro é oportuna e, tenho a certeza de que será extremamente prestável como guia para estudantes, cientistas e pessoas interessadas que tenham a oportunidade de visitar a Ilha da Inhaca.

The Universidade Eduardo Mondlane, Mozambique has maintained a close relationship with Inhaca Island for a long time. The Marine Biological Station and its marine and terrestrial reserves are under the direct supervision of the Faculty of Biology and the Station has been the site of numerous student visits to the island from many countries. Due to its scientific value Inhaca Island has been visited annually by our students in order to get practical experience in the field for various courses such as biology, ecology, geography and marine sciences.

In recent years, the Marine Biological Station has become an internationally recognised centre and provides support for all research undertaken on the island. Besides its educational and scientific importance, Inhaca Island is considered one of the most beautiful islands along the Mozambique coast, offering recreation for tourists of the whole world. On the other hand, the ecosystems of Inhaca Island are extremely delicate and highly vulnerable to human impact. Therefore the challenge to scientists and government is to counterbalance the realities of exploitation by man with the maintenance of this beautiful and delicate environment.

So it is with great pleasure that I accepted the invitation to introduce this new edition of *A Natural History of Inhaca Island, Mozambique*. The publication of this book is opportune and I am sure that it will be extremely useful as a guide to students, scientists and interested people who have the opportunity to visit Inhaca Island.

PREFACE

Inhaca Island is a small inhabited island in the Indian Ocean on the periphery of a large estuarine bay, on the fringe of the tropics. The four broad shores have a large tidal range, various degrees of shelter and salinity, different temperature maxima and several types of substratum; consequently communties are very diverse. There are easily accessible coral reefs, sea meadows and mangrove swamps as well as rocks and sand flats. The dunes also encompass a spectrum of habitats, some being exposed to erosion by wind or by sea. Large areas of the evergreen forest and bushland are unspoiled and others are in various stages of recovery. Since 1976 the land areas and some of the shores have been protected as nature reserves. All these habitats are within a days walk of the Biological Station (with laboratories and accomodation) that was built in 1951. It has become an international research centre in recent years. This facility has also given an impetus to the teaching of undergraduates.

The growing human population is facing the challenge of maintaining an equilibrium between human activities and both consequent and natural environmental changes. In the transition from a semi-subsistance economy to modern self-sufficiency, based on fisheries and tourism, the people are guided by the United Nations Development Plan of 1990. The Development Plan is based on recent research at the Unversidade Eduardo Mondlane, Maputo, Moçambique.

Two earlier editions of *A Natural History of Inhaca Island, Mozambique* have been used by students in Mozambique and other countries in East and Southern Africa. The previous editions have also have been popular with amateur naturalists. This completely revised edition was planned by biologists from both the universities. It seemed to us that a third edition with an integrated ecological theme, rather than a taxonomic one as in the earlier editions, would have condiderable educational potential. It would also help to cultivate an environmental awareness in support of development on Inhaca Island itself and on the other islands off the coast of Mozambique in which development will also depend on tourism. We agreed that the results of sixty years of study of the island's biota and their micro-environments would be described and reinforced by relevant information from South and East Africa and Madagascar. An analysis of the environmental complexity of the island and an outline of the ecological features of the tropical and sub-tropical shores precede the detailed accounts of life on the successive tidal levels, including the subtidal fringe. Adaptations to micro-habitats and the sharing of food resources and space are emphasised.

A striking phenomenon, touched upon in the texts, is the prevalence of commensalism on the lower shores, sometimes between the most unlikely animals. Competition between animals of the same species is usually expressed by token aggression, rarely resulting in death, and carnivores are species-specific in their choice of prey. Even among microphagous species sharing of planktonic food resources has evolved. In addition, in some genera a large number of species occur, that differ very slightly in morphology, but are physiologically adapted to different conditions. This edition also includes summaries of four recent and intensive studies on Inhaca Island of little known organisms that are universally present on the shores. These are the Harpacticoid copepods of the meiofauna (J.P. Wells, University of Aberdeen), the foraminifera of the benthic microfauna (A.R. Moura, Mozambique University), the flatworms fo the coral reefs and reef flats (S. Prudhoe, British Museum), and blue-green algae on rocks, sand, mangroves and plankton (S.M.F. Silva, U.E.M. and WITS).

Six new chapters have been added on both invertebrates and vertebrates, and on plants on land and in freshwater swamps. The trees, birds and insects have noticeably increased in number and diversity since the declaration of nature reserves.

In this edition many new diagrams have been included to facilitate recognition in the field. We have omitted keys to the identification of species that were present in earlier editions, since reliable keys are now available in a number of recent texts, listed at the end of the book. Classified lists of animals and plants present on the island (including about 3 500 species) recently updated by specialists are available from the Department of Zoology, University of the Witwatersrand.

Additional features of the book are the changes in human society on the island over the years and the growth of scientific endeavour. It includes an account of traditional conservation practices and ends with a description of present conservation practices on the land and shores and plans for the future.

Margaret Kalk

Acknowledgements

The initial stimulus for this revised edition arose from high level discussions between the two universities on forms of co-operation. I was encouraged to proceed with the prepartation of the book despite political difficulties at the time. Now it is my privilige to thank the present Rector of the Eduardo Mondlane Universidade (UEM), Professor Dr Narçiso Matos and Professor J.P.F. Sellschop, Deputy Vice Chancellor (Research) at the University of the Witwatersrand (WITS) for writing forewords to the book.

I wish to express my appreciation of the academic contributions from biologists at the UEM which enabled me to include chapters on the natural history of the land biota and to extend the chapters describing the shores of the island. Collaboration between Jan de Konig (UEM) and Kevin Balkwill (Wits) proved very fruitful in the interpretation of the ecological vegetation data.

I am very grateful to the University Research Committee (WITS) for the allocation of a grant to the editor, as Honary Research Fellow, to cover the cost of the production of the manuscript, including clerical and technical assistance. Facilities arranged by the Zoology and Botany Departments at WITS made the project feasible, and their encouragement has been heartening. Advice and critisism from colleagues at both universities have been most welcome. Garth James (WITS) to whom I am deeply indebted, patiently corrected the final manuscript. Shirley Hanrahan, Neville Passmore, Gert Ormel and Alan Critchley have been especially supportive.

I wish to thank Professor Alan Brown of the University of Cape Town for his invaluable pretinent and constructive scientific and stylistic comments on the manuscript.

This edition of the book has become a more comprehensive natural history by the welcome addition of authoritative chapters on amphibians and reptiles (Dr D.G. Broadley, The Natural History Museum, Bulawayo, Zimbabwe); non-marine molluscs (Dr A.C. van Bruggen, Leiden University, Netherlands), and freshwater diatoms, (Dr F.H. Hancock, WITS), all of which were based on earlier studies.

I have been fortunate in being able to use the diagrams form the earlier editions of the book, drawn by Haring Swart, Bill McNae, John Day, Edwyn Isaac and Jeanette Brandt. In the present edition, specialists illustrated some phyla: Alan Critchley (seaweeds), Barbara Pike (flowering plants) and A.C. van Bruggen (land snails). The rest of the illustrations were drawn by a succession of student illustrators: Ann Hanrahan, Ivan Jason, Emma Kalk, Thomas Kalk and Noeleen Murray. Many publishers granted copyright permission. Colour photographs kindly lent by old staff and students and scuba divers are acknowledged on the plates. Maps were taken from the 1990 United Nations Development Plan for Inhaca Island with the kind permission of the Instituto de Desenvolvimento Rural, Moçambique.

The onerous task of updating the specific names of the animals and plants was undertaken by many specialists[1] from museums, universities and research institutes to whom I am deeply indebted. The length of the list (3 500 species) prevented its inclusion in the main text. Warm thanks are also due to the many typists who handled the manuscripts so skilfully.

Finally, I must offer my warmest thanks to the Witwatersrand University Press editor, Jo Sandrock for her enthusiastic literary and technical advice, to the copy editor Lindsay Morton and to Lydia Weilert and Cheryl Brant for their patient final preparation of the manuscript in the face of many setbacks which have delayed publication.

NOTE

1. M.E. AIKEN, University of Natal, Pietermaritzburg – seaweeds; K. BALKWILL, C.E. Moss Herbarium, University of the Witwatersrand, Johannesburg – vascular plants; M. BERGGEN, Kristenbergs Marinbiologiska Station, Sweden – new shrimp species; G. BRANCH, University of Cape Town – polychaetes (ex proofs of the revised edition of the 1968 monograph by J.H.Day); D.G. BROADLEY, Natural History Museum, Bulawayo, Zimbabwe – amphibians and reptiles; R.K. BROOKE, Ornithology Institute, University of Cape Town – birds; A.J.P. CABRAL, Natural History Museum, Maputo – snakes; S. CHATER, Oceanographical Research Institute, Durban – fishes; B. COOKE, South African Museum, Cape Town – crustaceans; F. COSTA, Estaçao Biologia Maritima, Inhaca Island, – turtles, mammals; A.T. CRITCHLEY, University of the Witwatersrand, Johannesburg – seaweeds; A. DIPPENAAR-SCHOEMAN, Plant Protection Research Unit, Pretoria – spiders; G. de GRAAF, Transvaal Parks Board – mammals; J. de KONING, formerly Universidade Eduardo Mondlane, Maputo – vascular plants; J.J. and H.R. FEIJEN, formerly Universidade Eduardo Mondlane – birds; T.M. GOSLINER, California Academy of Sciences, San Francisco – sea slugs; F.D. HANCOCK, formerly University of the Witwatersrand, Johannesburg – diatoms; R.H. KILBURN, Natural History Museum, Pietermaritzburg – marine molluscs; F. MAPANGA, Estacao Biologia Maritima, Inhaca Island, Mozambique – vascular plants; N.A.H. MILLARD, formerly University of Cape Town – hydroids; A.J. MUIR, British Museum Natural History, London – polychaetes (from proofs of revised edition of the 1968 monograph by J.H.Day); G. ORMEL, formerly Universidade Eduardo Mondlane, Maputo – insects; M. PICKER, University of Cape Town – polychaets (from proofs of revised edition of 1968 monograph by J.H. Day); R.N. PIENAAR, University of the Witwatersrand, Johannesburg – dinoflagellates; G.L. PRINSLOO, Plant Protection Research Institute, Pretoria – insects; S. PRUDHOE, British Museum (Natural History) – flatworms; R.A. REDDY, C.E. Moss Herbarium, University of the Witwatersrand – vascular plants; M.H. SCHLEYER, Oceanographic Research Institute, Durban – corals; S.F.M. SILVA, University of the Witwatersrand, Johannesburg – blue-green algae; A.S. THANDAR, University of Durban-Westville – sea cucumbers; A.C. VAN BRUGGEN, Rijksmuseum van Natuurlijke Historie, Leiden, Nederland – non-marine molluscs; R. VAN DER ELST, Oceanographic Research Institute, Durban – fishes; J.H.C. WALENKAMP, formerly Universidade Eduardo Mondlane – starfish.

Contributors

Dr KEVIN BALKWILL, Director C.E. Moss Herbarium, Department of Botany, University of the Witwatersrand, Private Bag 3, WITS 2050, Johannesburg, South Africa.

Dr D.G. BROADLEY, Senior Curator of Herpetology, Natural History Museum, P.O. Box 240, Bulawayo, Zimbabwe.

F. COSTA, Conservationist at Inhaca Island, Faculdade de Biologia, Universidade Eduardo Mondlane, C.P. 257, Maputo, Moçambique.

Dr JAN de KONING (ex Universidade Eduardo Mondlane), Rijksherbarium/Hortus Botanicus, Rijksuniversiteit te Leiden, P.O. Box 9514, 2300 RA, Nederland.

Dr H.H. FEIJEN and J.J. FEIJEN (ex Universidade Eduardo Mondlane), United Nations Food and Agricultural Organisation, Bhutan.

Dr F.H. HANCOCK (ex University of the Witwatersrand), Department of Botany, University of the Witwatersrand, Private Bag 3, WITS 2050, Johannesburg, South Africa.

Dr MARGARET KALK (Professor Emeritus, University of Malawi), Department of Zoology, University of the Witwatersrand, Private Bag 3, WITS 2050, Johannesburg, South Africa.

Dr GERT J. ORMEL (ex Universidade Eduardo Mondlane), London, England.

Dr A.C. van BRUGGEN, Rijksmuseum van Natuurlijke Historie, Raamsteeg 2, Posbus 9517, 2300 RA, Leiden, Nederland.

INTRODUCTION

Inhaca Island stands at the entrance to a large calm bay, about 130 km south of the Limpopo River, which was marked on the first map of the east coast of Africa in 1502 as the Baia da Lagoa, the 'lake-like bay'. It was known to Arab traders in the fourteenth and fifteenth centuries and to early Portuguese explorers who sailed to India. The name was anglicised to Delagoa Bay by a British survey team in the last century. After independence was gained in 1976 it was renamed the Bay of Maputo (Fig. 1a) after the newly named capital Maputo (meaning the place where many waters meet).

A BRIEF HISTORY OF THE PEOPLE OF INHACA ISLAND: 1502-1990

Inhaca Island has a land area of about 40 km^2. It has been inhabited from early times by a small community of Ronga speaking people, whose chief, Nyaka, gave his name to the island, and a lineage bearing his name still lives on the northeast peninsula (Alberto 1959). Drinking water was readily available in the many swamps between the dunes. Tall reeds from these swamps and mangrove trees from intertidal, saline swamps provided materials for building huts and for making fish traps and small boats. The forests were a source of timber, fuel, fruits and medicines. The calm sea rising over extensive sand flats on the west coast brought an abundant supply of edible fish, caught by spearing and trapping. Crops of sorghum were grown around the periphery of swamps, and small stocks of poultry were customarily kept. Their homesteads were scattered in groups over the low-lying land behind the dunes, each surrounded by trees; many of them are still built in this traditional way (see Section 16.1).

In 1545 when Portuguese traders first started hunting elephant and taking ivory from the mainland for sale in Europe, Nyaka allowed them to establish a safely isolated trading post on a small nearby island, accessible on foot at low tide from Inhaca Island (Fig. 1c). This island became known as the Island for the Portuguese since it was used as trading headquarters throughout the sixteenth and seventeenth centuries, after which the elephant population was so decimated that the ivory trade was no longer profitable. The people of Inhaca Island have been mentioned in historical records for their hospitality to Portuguese, Dutch and British sailors (Wilson and Thompson 1969). Survivors from ships wrecked on the rocky shores of Southern Africa walked for many months to the north along the coast until they reached Inhaca. They were fed and housed and allowed to recuperate on the Island for the Portuguese from which they were ultimately rescued by trading vessels.

Towards the end of the eighteenth century the mainland was permanently occupied by the Portuguese, and their headquarters facing the Baia da Lagoa was fortified and called Lourenço Marques after the pioneer ivory trader of that name in 1545. The Island for the Portuguese was later evacuated. The British surveyors in the eighteenth century called the Island for the Portuguese Elephant Island, after the ill-famed trade.

Before the nineteenth century the kingdom of the Ronga people extended southwards for 150 km from Delagoa Bay and included Maputaland which is now part of KwaZulu in South Africa (Wilson and Thompson 1969). The kingdom was cut in half by the boundary drawn by the colonial powers in 1875 between Portuguese East Africa and British South Africa. There were several incursions by neighbouring peoples and the language spoken on Inhaca Island now bears traces of immigrants' languages such as Shangaan. There are now more than six clans on the island and until this century Nyaka was acknowledged chief (Wilson and Thompson 1969). A descendant still participates in the administration of the island. The original names given to villages, swamps and rocky headlands are in use today.

Towards the end of the nineteenth century Inhaca Island was claimed under the jurisdiction of the Portuguese on the mainland. An administrator known as the Chefe do Posto was appointed; he was housed in a semi-fort built on a cliff facing the bay. Cassava, sweet potatoes and ground-nuts were introduced; they grew better than cereals in the sandy soil of the island. A lighthouse was built at Cabo Inhaca near the north-east point in 1894 and a pilotage ship was anchored in a fairly deep channel just north of the small island, now known as Portuguese Island, to service trading ships from Europe. The pilot service continues to operate since the bay is dangerously shallow, and a single narrow, dredged channel must be followed by sea-going ships.

The economy of the people of Inhaca Island remained very largely at a subsistence level until about the beginning of the twentieth century when several Greek fisherman came to the island to trade in fish with the growing Portuguese population on the mainland (Alberto 1959). They employed men to fish and to build fishing boats with one large sail, big enough to carry ten fishermen. They sailed overnight to the capital with their cargo of fresh fish. The Greek fishermen introduced the use of heavy dragnets that were pulled along the bottom

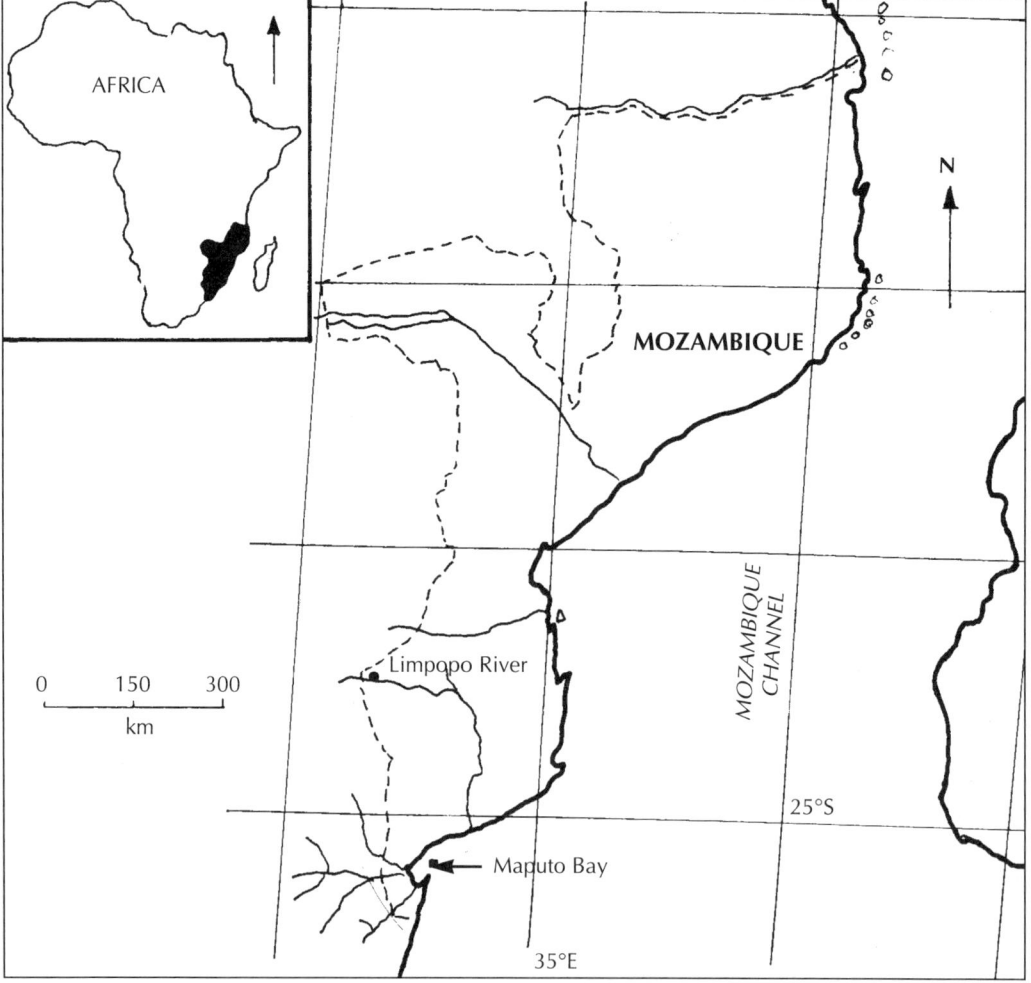

Fig. 1 Geographic position of Inhaca Island:
a. Mozambique in East Africa (Inset: Africa and position of Mozambique)

from the boat. This prosperous fishing industry declined and most of the Greeks had left the island by the mid-century; one who remained built a refrigerated packing shed and introduced a large motorised fishing vessel. Fish packed in ice could be transported within a few hours to the capital and then taken further afield for sale.

By the middle of the twentieth century most of the able-bodied men on the island had become migrant workers. Those who went to the sugar cane plantations in Mozambique and Natal in South Africa were able to return every summer to fish and work on the land. Those who worked in the South African gold mines on the Witwatersrand signed up for nine months to a year or longer. No family accommodation was provided for these migrant workers, thus the women remained on the island and the growing of crops was left entirely to them. Women and children supplemented the family diet by catching crabs, molluscs and small fishes on the seashore. A fish market and a small shop were built near the fishing harbour where rice, meal and bread could be bought for cash which the men brought back from their meagre wages.

Few other signs of Portuguese influence were to be found. A small clinic manned by a medical assistant was built on the west coast bluff near the Posto; falciparum malaria was meso-endemic and urinary bilharzia very common. A Catholic mission ran a small open-air primary school nearby. A long grove of coconut trees was planted near the fishing harbour and later a small hotel was built for tourists from South Africa who were attracted by the

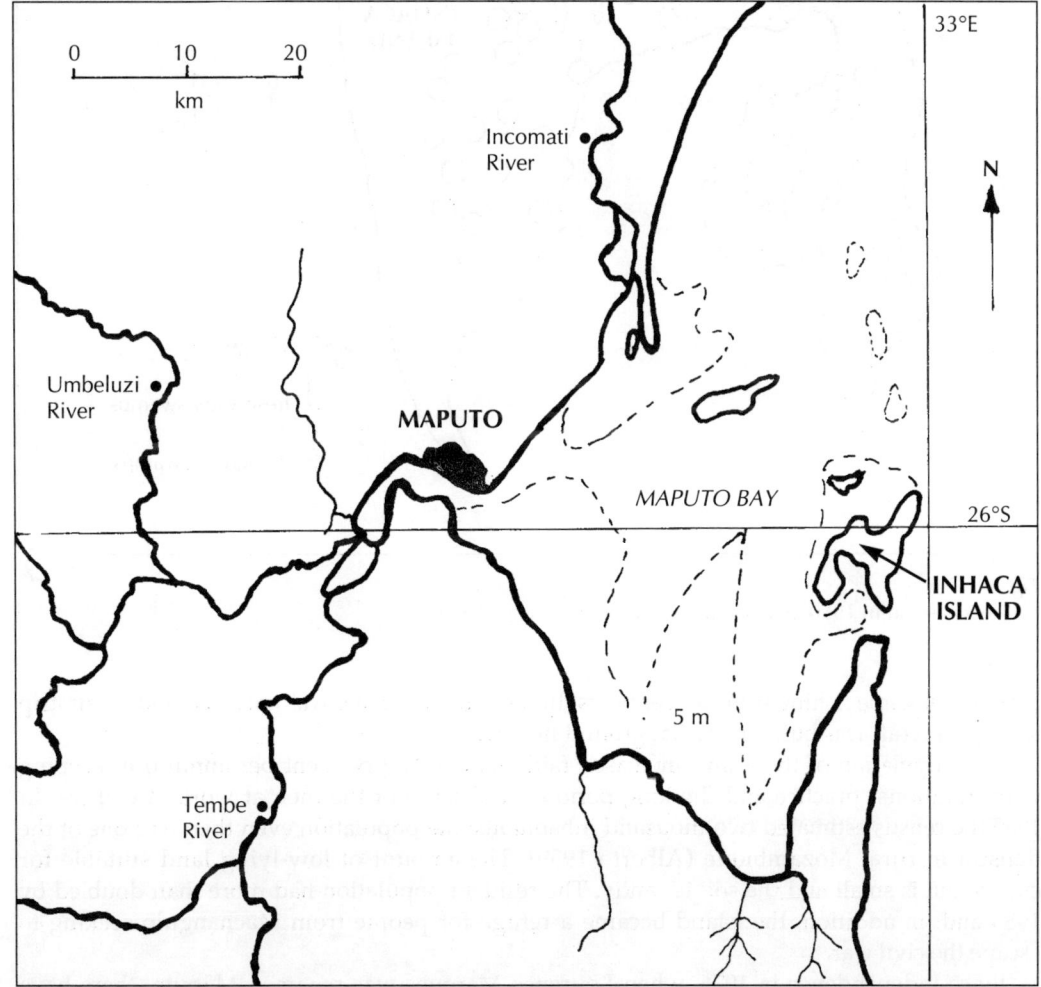

Fig. 1
b. Bay of Maputo with 5m depth limit and rivers (after A.J. de Freitas 1984).

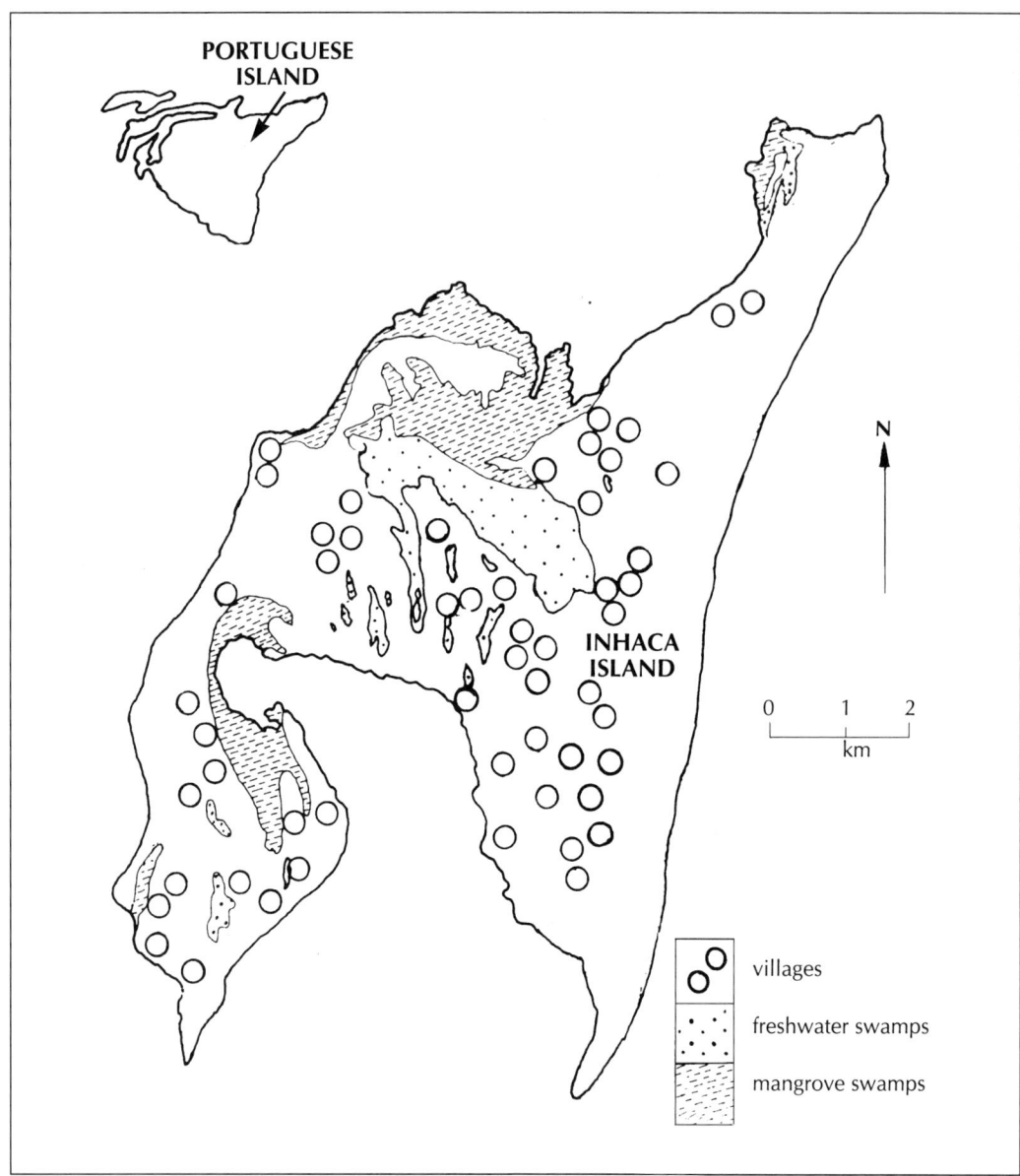

Fig. 1
c. Inhaca Island in 1950 (after Comodoro J. Moreira Rato, 1959).

prospect of game fishing from power boats. In the 1960s the hotel was enlarged and an airstrip for light aircraft was built on the flat ground nearby.

The population of the island increased fairly slowly (1,9 per cent per annum) in keeping with traditional practice and the long periods of absence of the men at work elsewhere. In 1950 the census estimated two thousand inhabitants; the population even then was one of the densest in rural Mozambique (Alberto 1959). The amount of low-lying land suitable for cultivation is small and the soil is sandy. The resident population had more than doubled by 1985 and, in addition, the island became a refuge for people from Machangulo seeking to escape the civil war.

Since Independence in 1976, when Lourenço Marques was renamed Maputo, there have been many changes in the lives of the people on Inhaca Island. There are large primary schools, vividly decorated with murals, in each of the three districts. The fishing industry has been organised as a state enterprise using the large, formerly privately-owned fishing boat and refrigeration plant at Portinho da Inhaca. There are also many smaller fishing boats of improved design which have been built under the direction of the Instituto de Investigaçao Pesqueira in Maputo. Small reed huts have been built for tourists and camping sites are available. The hotel has been enlarged and modernised and has become once again a focus for tourists, giving employment to more people on the island. At present a comprehensive plan of development is being implemented to improve the quality of life of the people, to diversify the economy and to stabilise the environment and its resources for posterity (Plano de Desenrolvimento Ihtegrada Ilha da Inhaca 1990). It is based on a scientific study of the problems and the potential of the island by scientists in Mozambique and aims to bring about an equilibrium between the people's activities and the survival of the environment (see Chapter 16).

SCIENTIFIC ENDEAVOUR: 1911 TO 1990

The Museu Alvaro de Castro was established in Lourenço Marques in 1911, and scientists started visiting Inhaca Island to study the fauna and flora of the shores and the forests. The island became known to biologists further afield as a source of tropical marine animals and plants. Biologists came to work on the island from Portugal, Britain and South Africa and from the Instituto de Investigaçao Cientifica de Moçambique, established in Lourenço Marques in 1955.

From 1922 onwards occasional study-visits were made from South African universities and museums. From the mid-1930s regular winter vacation courses were given by the staff of the University of the Witwatersrand under the leadership of the late Professor Van der Horst, an international authority on corals and enteropneusts (the acorn worms which exude huge wormcasts on the sandflats). Margaret Moss of the Botany Department pioneered a collection of plants which are stored in the herbaria at Inhaca Island and at the University of the Witwatersrand. In the early days the Chefe do Posto allowed the students to camp under the *Casuarina* trees on the bluff and lent a large room for use as a laboratory. Packing cases were used as benches and seats; paraffin lamps provided light for microscope work. One of the first publications on the land flora in 1937 was translated into Portuguese (Wolfowitz 1938) and others appeared in Portuguese, Moçambican and South African journals. These publications introduced the first phase of research. The importance of Inhaca Island to science was recognised by the Portuguese (Ferreira and Ferreira 1952).

The years of the Second World War interrupted these vacation classes. In 1947 Professor Aurelio Quintanilha, then head of the Cotton Research Institute of Mozambique and after Independence was appointed Head of the Biology Department at the University in Maputo, proposed that a marine biological station should be built on Inhaca Island for the study of biological sciences. The idea was warmly supported by Professor Van der Horst who promised that the University of the Witwatersrand would initially provide equipment for teaching, research and domestic use. A site was carefully chosen on the west shore in a sheltered cove about 5 km from the hotel at Portinha Inhaca in the District Ridjene, within easy reach of the coral reef, sea meadows and mangrove swamps. A modern concrete building was erected, equipped and opened in 1951 (Rato Morreira 1959). The buildings comprised a laboratory and museum, research rooms and a library, an electric generator house and comfortable accommodation for a large party of students and staff, including a dining room suitable for lectures. Donations of books formed the nucleus of a library.

For the next twenty-five years the Marine Biological Station was used for teaching biology during vacations by universities and teacher-training colleges from South Africa and Rhodesia, and even by some schools. When the University of Mozambique was opened in 1962 the Marine Biological Station was used by students and school pupils who visited the museum and studied the ecology of the island. Comprehensive collections of species from most phyla were sent to taxonomic specialists in museums and universities for identification and description (Kalk 1954). Students were required to preserve and mount specimens for the

Inhaca Museum and to add to the Moss Herbarium. Dr Piet Boshoff and members of the Underwater Research Group made a detailed study of the coral reefs over a period of forty years and left a unique collection of mounted corals at the museum on the island and a duplicate collection in the museum of the Oceanarium in Durban. Scientists in Mozambique studied many aspects of the biology and geology of the island.

At a joint congress of the Sociedade de Estudos de Moçambique and the South African Association for the Advancement of Science in Lourenço Marques in 1958, the main theme in the Biological Sections was life on Inhaca Island. The Congress launched the first edition of the book *A Natural History of Inhaca Island, Mozambique*, edited by W. Macnae and Margaret Kalk and published by the Witwatersrand University Press. The book attempted to present, in a taxonomic context, most of the information then available on the distribution of about a thousand species of animals and over five hundred species of plants on Inhaca Island.

The facilities provided for research on Inhaca Island became widely known to marine biologists. Later scientific papers concerning life on the island's shore and land are listed at the end of this edition; they post-date the publication of the first edition of the book.

Many of the names of species have since been changed and in this revised edition of *A Natural History* the names are updated by specialists according to more recent international literature.

Papers on other aspects of biology, published in international journals in the 1960s and later, included taxonomy, ecology, adaptive physiology, behaviour and genetics began. Meanwhile, climatic data were accumulated at the small meteorological station at the laboratory. In the 1970s geologists studied the rocks, the geomorphology of the island and the forces that shaped its shores.

Parts of the shores and dunes on Inhaca Island were declared nature reserves by the Mozambique government in 1965. It had become necessary to control excessive collecting by tourists and students in order to conserve typical habitats and to protect parts of the forests and mangroves from tree felling. At that time it was considered that sufficient resources of land for tilling and trees for fuel and building were left for use by the local people. The island was still under the control of the Maritime Department of Mozambique.

Following independence of the Republic of Mozambique, the Marine Biological Station at Inhaca Island has been administered since 1980 by the Faculdade da Biologia at the Universidade Eduardo Mondlane in Maputo. The University, established in 1962, was renamed in honour of the first leader of the independence movement. This strengthened the link with the University of the Witwatersrand in South Africa, since Eduardo Mondlane himself had been a student of Social Science there (Tobias 1978). He was refused permission to finish his studies and thus became one of the first victims of the apartheid policy imposed on universities in South Africa. The students and staff protested in vain and well remember his name. A biologist, resident on the island, with responsibility for conservation and the better utilisation of the natural potential of the land and sea, is closely associated with the university's faculties of biology and agriculture and with the research personnel in the forestry and fisheries institutes. He has a small staff of rangers who supervise the implementation of the conservation regulations in the Reserves. These have been recently been extended (see Chapter 16).

The island has become a well-used focus for education for Mozambican students and school pupils and for visiting students from Zimbabwe and other neighbouring countries.

In the last fifteen years, short-term research biologists from Holland, Sweden, Norway, Portugal and East Germany as well as from Mozambique, Zimbabwe and South Africa, have further extended specialist knowledge. Donor countries such as Sweden, Norway and Holland have generously added to the equipment in the laboratory, renovated the circulating sea water system for aquaria and contributed to the collections in the museum and herbarium. They have also provided motorised transport for investigations in water and on land and contributed funds for conservation and research.

Knowledge of organisms on tropical shores of the Indian Ocean has grown in recent years in universities and research institutes in newly independent countries of East Africa and in Malagasy, and much of it is applicable to organisms living on Inhaca Island's shores. The coastal universities and the Oceanographic Research Institute of South Africa have contributed substantially to marine, estuarine and dune ecology. They have published intensive studies of shores in Mozambique and Maputaland as well as around South Africa. This work has also served as a guide in the presentation of this study. The present extensive revision of *A Natural History of Inhaca Island*, is thus able to draw upon much wider resources than the earlier editions; these are acknowledged in the text.

The revised book has a new dimension: *temporal change*. Conservation of land in nature reserves for twenty-seven years has led to regeneration of soil and vegetation after shifting cultivation ceased. Experiments in dune reclamation fifty and fifteen years ago have shown positive results (see Chapter 10). Natural changes in currents have destroyed some coral reefs, rejuvenated others and have led to the growth of a new coral reef (see Chapter 8). Much of this aspect of the book may become useful for other coastal areas such as the Bazaruto Archipelago.

Research on the island has been more or less continuous over sixty years. *A Natural History of Inhaca Island* attracted nature lovers and holiday makers as well as biologists. At present scientific endeavour seeks to establish a basis for the control of many factors in the environment so that a change from a poor semi-subsistence economy to a partial cash economy can be sustained. The United Nations Development Plan of 1990 aims at exploiting the general interest in nature to attract tourists and so to help build the local economy by creating a market for local produce (Plano de Desenrolvimento Ihtegrada Ilha da Inhaca (1990), see Section 16.3).

In this new edition Part I describes the environment on the island in some detail, accounting for the variety of habitats. A general account of the principles underlying the complex relations of organisms on the tropical shores to each other and to their immediate microenvironments introduces Part II on the fauna and flora of the different kinds of seashore. Parts III and IV are new features of the book. A detailed description of the vegetation on land and freshwater swamps is followed by chapters on the land fauna: bird life, amphibians, reptiles and mammals, snails and also insects, on which so much of the plant and animal life depends. Changes in the diatom flora of the freshwater swamps indicates the need for future control of the Environment.

In Part IV, the new comprehensive development plan for welding together the ecology and the economy of Inhaca Island, under the guidance of the Marine Biological Station of the Universidade Eduardo Mondlane, is summarised. The conservation practices on land, shores and the coral reefs are described.

PART 1

The Environmental Setting

The main focus of this extensively revised edition of *A Natural History of Inhaca Island* is on the adaptations of species to their own special micro-environments. The seashore is one of the most stressful places on earth and every species in it has unique strategies for survival which are superimposed upon the phyletic structure, such as that of worm, crab, starfish, seaweed or flowering plant. An exceptionally large number of animals and plants live in the diverse habitats on the shores, in the shallow coral sea, the mangrove swamps, the freshwater swamps, and on the dunes (see Appendix A and B). This wealth of species reflects the complex assortment of environmental factors on the shores and dunes, which are exposed to different degrees of shelter and temperature resulting from the island's geographic position and topography. The majority of species are tropical in distribution and have extended their ranges from the north to similar habitats at Inhaca Island, although the full complement of the flora and fauna of the tropics is attenuated (Kalk 1959a). Many habitats and species on Inhaca Island occur on Bazaruto Island and the archipelagos of northern Mozambique; some of the organisms are characteristic of the Maputaland and Natal shores of South Africa to the south and, as on those shores, a subtropical component, some warm-temperate and also cosmopolitan species are present at Inhaca.

In Chapter 1 the geographical and oceanographical features, including the relevant, brief geological history of the island, are summarised. The resultant physical and chemical factors which affect life at different tidal levels are briefly outlined. These factors differ remarkably on the four shores, as others do on the western and eastern dunes (see Chapter 10). Taken together, they account for the variety of habitats in which populations of so many animal and plant species find their micro-environments.

1

GEOGRAPHY AND OCEANOGRAPHY

1.1 GEOGRAPHICAL POSITION OF INHACA ISLAND, BORDERING AN ESTUARINE BAY

Inhaca Island (lat. 26°S; long. 33°E) lies in the Indian Ocean 32 km due east of Maputo, the capital of Mozambique. Although the island lies outside the tropics, biogeographically the shores may be considered to be tropical since the majority of the organisms there are known from the Indo-West Pacific Region (Kalk 1959a): northern Mozambique (Kalk 1959b), Tanzania (Hartnoll 1976), Kenya (Lawson 1969), Somaliland (Vannini 1975, 1976) and Madagascar (Battistine and Richard-Vindard 1972). The reasons for this anomaly are outlined in this chapter.

Inhaca is an island by virtue of a strait, less than a kilometre wide, that separates it from the north-pointing Machangulo Peninsula which partly encloses the shallow Bay of Maputo (Fig. 1). This strait is called the Ponta Torres Strait. The Bay of Maputo has an area of 960 km^2 and is roughly horseshoe shaped, opening to the Indian Ocean to the north. The eastern boundary of the bay is prolonged beyond the peninsula and Inhaca Island by a line of subtidal shoals for about 36 km. Three shores of Inhaca Island are thus sheltered by the Bay of Maputo, and the east shore alone is exposed to the winds and waves of the Indian Ocean. The bay is shallow, only one fifth of its area being deeper than 10 m (measured at low tide), and then mainly to the north-west of the Island. Five large rivers, which arise in the mountains over 200 km inland, flow into the Bay of Maputo on its western boundary and reach the Indian Ocean northwards (Fig. 1b). The southern half of the bay is less than 5 m deep with many subtidal sand shoals occupying the greater part, through which smaller seasonal rivers from the Machangulo Peninsula cut channels and flow northwards. The Inhaca Channel, which runs parallel and close inshore on the west of Inhaca Island, is the deepest of these channels at present, 10-20 m deep. Movements and deposits of sand in the last thirty years have made some parts shallower and other parts deeper than they were at the time of the first edition of this book.

Inhaca Island forms the end of the eastern boundary of this large, complex, sheltered, estuarine bay; nevertheless the island's sheltered shores are also bathed by Indian Ocean waters with every incoming tide. Ocean water surges through the strait between the southern end of the Island and Machangulo Peninsula at 8 km per hour; a *greater volume* of ocean water sweeps in from the north around Portuguese Island and also breaches a sand bar connecting it to Inhaca Island and bathes the west shores. Thus the estuarine influence of the waters of the Bay of Maputo on the sheltered shores of the island is diminished.

1.2 TOPOGRAPHY OF THE ISLAND

Inhaca Island extends 12,5 km from its northern point, Ponta Mazondue, to the south-eastern point, Ponta Torres; it is about 7 km across the widest part in the central area. The island has a distorted H-shape because its north and south shores are deeply indented by bays including extensive intertidal sand flats (Fig 1.1). Thus the shores occupy an area of 60 km^2, an area larger than the land. The different aspects and slopes of the shores, with varying degrees of shelter, provide a large number of different habitats.

1.2.1 Land

The land is sandy but various altitudes, aspects and human occupation lend variety to the terrestrial flora and fauna. There are long dune ridges along the east and west coastlines and lower land in the interior of the island (Fig. 1.1). The eastern ridge is longer and higher; the highest point, Monte Inhaca (115 m), is about 3,5 km from the north-east point where the lighthouse stands 80 m above the sea. The outer face of the dune ridges is steep and bare except where special dune-binding plants withstand exposure to the salty winds of the Indian Ocean. Wind erosion occurs on the crest of the ridge. The gentler inner slopes of dunes facing the southern bay of the Island are covered with old evergreen forest or regenerating forest. On the western ridge the red cliff (Barreira Vermelha), about 3 km from the fishing village at the north-west point, is 61 m high and dominates the coastline. Low coastal bush clothes the slopes down to high tide mark facing the Bay of Maputo. The dune is topped by low trees which penetrate inland to some extent (see Section 10.1).

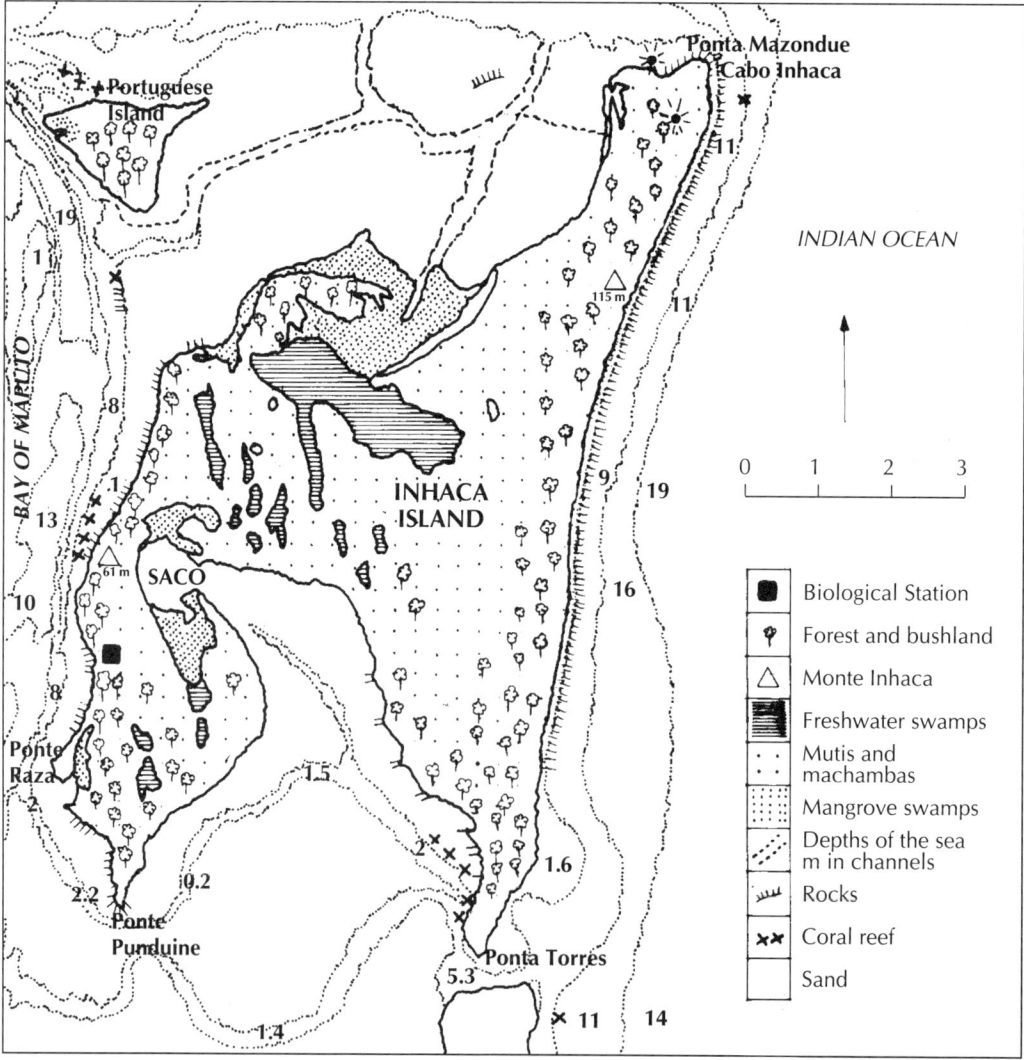

Fig. 1.1 Hydrography: map of Inhaca Island. Depths of the sea at low tide (from *United Nations Integrated Development Plan for Inhaca Island 1990*, (courtesy of Instituto de Desenvolvimento Rural, Moçambique).

The interior of the island consists of five interrupted lines of lower dunes lying about 500 m apart oriented in a roughly north-west/south-east direction. The slacks between them have freshwater swamps, some of which are still full of tall reeds and bulrushes (see Section 10.3 and Chapter 15). Other swamps are partly drained for agricultural use and the soil is darker and much less sandy than on the dunes. There is little humic material in the soils except in the old swamps and forests (see Section 10.4). There are two belts of red soil on high ground, one across the western dune ridge spreading inland north-east from the red cliff on the west coast and the other crossing the south-west peninsula. The people live in the lower-lying central area and on the south-west and north-east peninsulas in huts surrounded by large gardens in which indigenous fruit trees and trees of medicinal use are left standing. Much of the bush has been felled in these areas and grasses thrive among the scattered trees. The east and west dune ridges have been nature reserves since 1965 and the natural bush vegetation is regenerating well (see Section 10.5). Bird life is rich and varied, reptiles and many kinds of insects abound (see Chapters 11, 12 and 13). Dune forests, the inhabited areas (mutis and machambas) and freshwater swamps surrounded by fertile land are indicated on the map, Fig. 1.1.

There are no overt streams on the island. Rainwater seeps through the dunes to the freshwater swamps and to the mangrove swamps on the shores where small meandering channels occur. Seepage of fresh water can be detected below the dunes on the upper shores.

North-west of Inhaca Island, 2 km from the fishing village across the sand flats, lies Portuguese Island which is about 2,5 km^2 in area and uninhabited today; its vegetation is similar to that on the low dunes and mangroves of Inhaca Island. Aerial photographs taken annually since the 1950s show that Portuguese Island is changing in size and shape. A lagoon has been formed on the northern shore where land has been washed away, and now encloses a new coral reef (see Section 5.4.2).

1.2.2 Shores

The shores of the island are mainly sandy or muddy, but rocks occur on all the shores at various tidal levels as outcrops or as storm deposits. The four shores are markedly different in their major features.

The west coastline, facing the Bay of Maputo, curves gently into two shallow coves from the fishing harbour in the north to Ponta Raza in the south. Ponta Raza is a sandy spit which curves westwards for about 1 km to low tide, sheltering the cove where the Marine Biological Station stands. When the tide is out, outcrops of low rock are visible at intervals along the sandy slope of the upper shore. The lower shore is a vast expanse of sandy mud, exposed for about 500 m at low tide. It is far from being a monotonous stretch of sand flat; there are dry sandy areas, wet muddy areas and very shallow pools of standing water. Sea grasses grow on the lower shore, sparsely at first and more densely towards low tide levels (see Chapters 3 and 9). A considerable amount of coral rubble, consisting of large and small overturned slabs of rock, has been washed up from former coral reefs. These are scattered over the mud and lie in shallow pools giving the lower shore the character of a reef flat (see Section 6.4).

The sand on the lower shore lies over a metre deep on a bed of old coral rock. Wherever sand is removed on the fringe of the shore by rapidly flowing water in the Inhaca Channel, living coral grows up from 1-4 m below low tide and forms a confluent reef. From time to time parts are destroyed by sand movements whilst other parts are freed from sand and grow anew (see Section 8.6).

A narrow tidal creek, lined by mangrove trees, opens to the west shore south of Ponta Raza. It is about 1 km long and brackish in its upper reaches (see Chapter 4).

South of Ponta Raza beyond the opening of the mangrove creek there are two small coves lined by low cliffs of a different type of rock, that end in the south-west point, Ponta Punduine. The upper shore is foreshortened here and sea grasses grow near the cliff. Ponta Punduine, which separates the western shore from the south bay, is in deep shade for most of the day. A tidal current sweeps close to the point and at various times has eroded the cliff so that a number of boulders have tumbled out of it. These boulders extend from high tide to low tide in just over 50 m and end abruptly in sand (see Chapter 6).

The shore facing the Indian Ocean contrasts with the west shore in every respect. A very steep and uniform sandy slope reaches from the foot of the dunes to a continuous fringe of flat low rocks at low tide. On the north-east shore, at Ponta Mazondue, a high vertical rocky cliff is exposed at half-tide. The lower shore consists of wide rocky platforms extending to low tide and covered with seaweeds. Pools and gulleys rich in plant and animal life break the surface of the flat rocks (see Section 7.1).

Viewed from the top of a dune at high tide, the southern bay resembles a huge triangular lagoon fringed by mangrove trees, submerged up to their lower branches so that the foliage appears to be floating on the surface of the sea. It is not completely enclosed by land for the dunes on the horizon to the south are on the mainland of Machangulo Peninsula and the strait is concealed by high dunes of the island. After half-tide the mangroves can be explored, but the sand is still very soft and sticky. The more extensive mangrove associations are at the upper (north) end of the bay, known as the Saco da Inhaca (see Section 4.2).

When the tide recedes hard sand flats, which can be crossed on foot, are seen to extend for 7 km to the mouth of the bay which is 6 km wide. There is a permanent channel down the centre of the bay with its banks stabilised by sea grasses. On the east bank of the channel near the mouth of the bay there is a coral reef over 1 km long, in two sections. Isolated rocks occur on the east shore of the bay; some of them are like enormous mushrooms with oyster-covered tops and 'stipes' worn away by strong tidal currents (see Chapter 5).

The rocky cliffs at the south-east and south-west points are very different from the rocks on east and west shores, and also from each other. The south-east point, Ponta Torres, marks the boundary of the strait between the island and the mainland through which ocean water surges with the incoming tide and flows out at ebb tide. The cliff is vertical with a sheer drop into 15 m of clear water and is exposed to strong sunlight. The lower part is studded with corals, some of which are not reef corals. There is no wave action and no seaweed (see Section 5.5.4).

The intertidal area of the northern bay measures approximately 3 km by 6 km and is sheltered from the open sea on the west by the long northerly projection of the island and on the northern border by Portuguese Island and the sand-bar. The sand-bar extends from east to west, from Ponta Mazondue to Portuguese Island and beyond. The apparently wide entry to the northern bay, south of Portuguese Island, is blocked by a subtidal sandy shoal which is an obstacle to ships. (Many a time the tug which used to bring students and equipment to the island was delayed for several hours while it waited for the rising tide to lift it clear of the shoal when the narrow channel had been missed.) There is a rocky area on this channel (see Fig. 1.1) which has an incipient coral reef on its edge. South of the shoal the Inhaca Channel continues into the bay in diminished streams. Thus shelter from wave action is complete, but there is continuity with the Bay of Maputo even at low tide through the extension of the Inhaca Channel and by means of breaches of the sand-bar by one, two, or in some years, three, narrow winding channels. These lead into the sandy northern bay which may be about 1 m deep at low tide in some years. In 1991 there were two breaches of the sand-bar, the eastern one being negotiable by small boats (Fig. 1.1). The sand-bar has a large intertidal rock in only one area and consists of soft sand that has increased in height and length over the last fifty years. The sea has encroached on the old mangrove swamp on the western third of Portuguese Island, completely obliterating it. Between the extension of the sand-bar and this island the new, small coral reef has developed.

The landward edge of the northern bay of Inhaca Island is fringed by mangroves. An unusual mangrove parkland has developed at the south end of the bay where a veneer of clean sand covers the mangrove mud (see Chapter 4). A new mangrove swamp has developed in the last 20 years on the western edge of the peninsula. Another conspicuous feature of this warm northern bay is the large meadow of sea grasses which are more luxuriant here than elsewhere on the island's shores. The sea grasses are submerged in very shallow standing water and have a wealth of associated tropical algae and animals (see Chapter 5).

1.3 GEOMORPHOLOGY AND THE ORIGIN OF THE ISLAND

The physical conditions of Inhaca Island are not static. Erosion, deposition of sand and silt, the drying of freshwater swamps and possible changes in the depth of the water table present problems today. These are best studied from the perspective of geological events which led to the formation of the island, as well as in the present climatic context of the last fifty years. The geological formation of the rocks determines their chemical and physical composition, which influences the distribution of animals and plants that live on them (see Section 6.3 and 6.5).

The coast of Mozambique was initiated in the Mesozoic period when Gondwanaland fragmented and Antarctica, Australia and India, Africa and Madagascar drifted apart. Subsequently the coastal land underwent a series of tectonic changes so that several periods of submergence by the sea followed periods of emergence. The last few changes of sea level in the late Pleistocene, during which Inhaca Island was formed, coincided with glacial and interglacial periods in temperate regions and may have been influenced by them. The present coastal plain of Mozambique was formed when the sea level was higher than it is today. Consequently when the sea level fell new coastlines developed and rivers incised their valleys over the extended plains. Low sea levels permitted the deposition of wind-blown sand on land. The high sea levels that followed were accompanied by sedimentation under water and the drowning of river mouths. The events concerning Inhaca Island in the late Pleistocene (Hobday 1977) are outlined below.

1.3.1 Late Pleistocene rocks of different ages on sheltered and exposed shores

The visible foundation of Inhaca Island when it was a part of the mainland peninsula is calcareous sandstone. This was deposited on earlier rock at the time of the penultimate glaciation more than 140 000 years before the present (BP), when the sea level was *lower* than it is today. This basal layer, upon which other layers of rock were subsequently formed, is exposed on the upper shore of the west coast. The structure indicates that the deposition was aeolian (windblown); it accumulated in a dry climate during a regression of the sea (the 'Riss' low sea level). The rocks show coarse cross-bedding and are composed of 'well-rounded, medium sized quartz grains, embedded in finely abraded shell fragments'. They are similar in age and appearance to those in KwaZulu, South Africa. This calcareous sandstone under the island supports the water table. It was submerged during the subsequent Flandrian transgression of the sea (140 000 to 80 000 BP) and so became consolidated and impregnated with calcium salts forming a 'calcarenite'. A relatively thin surface layer of sediment (about 20 cm thick) was deposited horizontally on top of the cross-bedded layers and also became calcified (see Section 6.1.1).

When the sea had risen at least 30 m above the present sea level, beach (swash) deposits were laid down under water during strong wave action on the east coast. This became plane-bedded sandstone, formed on top of the earlier cross-bedded calcareous sandstone during the Flandrian transgression. Such sandstone rocks are now exposed at the north of the island, but on the east coast they are covered with sand, except at very low tide.

The rocks on the present north east shore emerged in three stages as the Eemian sea level rose and fell, as may be seen in the cove at Ponta Mazondue, where the cliff is 12-15 m high (see Fig. 7.1). After the first fall of sea level and during a temporary stable period, a notch was cut by wave action about 2 m above the present mean sea level. It is recognisable as a small shelf, supporting a pool under an overhang above the barnacles and oysters. About 2 m below this incipient terrace the rocks spread out to form a gently sloping 40 m wide platform. This terrace was wave-cut during a later and longer 'standstill' of sea level. It lies just below the present mid-tidal level and its surface has been worn smooth by the sand-scouring current that sweeps around the point, carrying sand to the sand-bar which joins this point to Portuguese Island. In recent times its surface has been lowered by solution and shallow round pools with rims protected by algae and barnacles are scattered over the surface; some of the pools have been deepened by churning pebbles and half-filled with coarse sand, some have brown algae similar to those on the next terrace.

About 50-75 cm below the level of the broad platform is another wave-cut terrace with deeper pools and inlets which spreads beyond the low tide of today and is visible only at extreme low spring tides. This is the level of the low rocks along the east shore (see Chapter 7).

During the same transgressions of the sea, sand was being deposited on the sheltered side of the peninsula in *quiet estuarine* water. It became consolidated under water. The subsequent regression of the sea gradually exposed the future 'small-scale, cross-bedded sandstone' of Ponta Torres and Ponta Punduini, but the southern bay which now lies between them had not yet been formed.

1.3.2 Isolation of the island and consequent coastline changes

When the sea level last rose in the final post-glacial transgression, the low coastal plain was flooded and the rivers drowned once more. The valleys of the adjacent coastal land had been deeply incised as rivers scoured into the sediments filling the paleo-valleys which had been formed in earlier Pleistocene regressions. The rivers extended their courses and eroded trenches, seen today as canals between shoals in the Bay of Maputo. The sea level rose rapidly between 17 000 BP and 9 000 BP to a level 25 m below the present level. It continued to rise more slowly and some time probably between 6 000 and 5 000 BP, the sea broke through one of the previously incised valleys and isolated the tip of the peninsula as an island, now called Inhaca Island. It also shaped the inner coast of the Machangulo Peninsula. The nearest trench to the island was deepened by inroads of the sea through the strait. This became the Inhaca Channel running parallel to the west shore, 1-3 km offshore.

The *ocean current* through the gap between the island and the mainland peninsula eroded the south coastline of the new island, and probably also invaded some previous fresh-water swamps, to form the southern bay. This process was accentuated by drainage from the dunes and the result is the deeply indentated bay with its upper limit stabilised by the growth of mangrove trees. The new strait and the southern bay gave access to the strong south-east wind to the centre of the island so that the central dunes with less plant cover have been lowered; storms and blow-outs at intervals have accelerated this process. The lesser north winds and wave action formed the northern bay which was, however, somewhat protected by the higher ground of Portuguese Island and later by the sand-bar which gradually accumulated between the northern point of Inhaca Island and Portuguese Island.

Near the mangrove creek outlet immediately south of Ponta Raza at mid-tidal level a flat slab of rock, a few metres long and a metre wide, is exposed. The covering of recent sand has been eroded away at this point by the current from the mangrove creek and highly fossiliferous pebbly rock is visible. Twenty-four species of gastropod molluscs and fifteen species of bivalves, shell-lined burrows and microscopic foraminifera have been identified. They are very similar to those found on the intertidal flats today, including the large white moon-snail and the white-beaked bivalve which litter the surface of the lower sandflats of the southern bay. This rock was formed in quiet water when the sea level was 3-4 m higher than at present, during the last Eemian transgression. Fragments of the older calcareous sandstone have been rolled into pebbles and embedded in this rock which is covered with sand except at this one site.

Towards low tide on the west shore beneath the sand, the bedrock continues into the Bay of Maputo and is made of coral skeletons of the same age, formed when the last interglacial sea was warm (Bosazza 1956). The limestone rocks strewn on the lower west and south-west shores are slabs of coral thrown up from the coral reef by infrequent cyclonic storms in recent times (see Chapter 6). The coral reef on part of the west coast grows on this older bedrock coral on the banks of the Inhaca channel (see Chapter 8).

1.3.3 Formation of sand dunes and sand-bar

During the last regressions coastal dunes were formed of wind-blown sand and the process continues today. On the east coast the stages of their formation have been concealed by the accumulation of more recent sand, but at the red cliff (Barreira Vermelha) on the west coast, contemporary erosion by sea has revealed a detailed profile of its gradual geological formation (Nestler *et al.* 1984). The dune has three layers: (i) a lower part about 40 m high, formed on the sandstone rocks below and composed of aeolian deposits during the last Eemian regression; (ii) a marine deposit about 70 cm thick above it, deposited under water during a short transgression of the sea, and (iii) an uppermost layer about 5 m thick composed of recently deposited wind-blown sand.

The lowest layer comprises partly consolidated 'beach rock' lightly calcified during its submergence in the *last* high levels of the sea. The submarine deposit above it contains microfossils of a different nature from those in the lower deposit. The uppermost sandy part of the dune is redder than the lower dune rock since it has been leached by fresh water percolating through the sand and releasing ferric oxide that coats the sand grains.

During the last ten thousand years (the Holocene Period) the climate has become warmer and more humid, providing conditions for vegetation to develop and cover the dunes except where they are exposed to the very high winds on the east shore. The west coast dunes have been thoroughly stabilised by vegetation along the sheltered shore. The bare cliff at the highest point at Barreira Vermelha is the only area where sea erosion has recently occurred on the west coast.

On the east coast of the island high south winds and strong oblique waves cause a north-flowing longshore drift which carries sand. When this current loses momentum as it joins the wide mouth of the Bay of Maputo, the sand is deposited as a sand-bar running due west. The sand-bar has grown in height and the top of it is exposed at mid-tide. To the north-west of the island, the longshore drift also adds to the build-up of the subtidal sand shoals in the bay, between which there are occasional deeper channels cut by outflowing river water from the mainland after coastal deluges. Typical intertidal and subtidal *sandbanks* were created by the ebb tidal currents through the northern sand-bar and through the southern strait, at Ponta Torres (see Fig. 1.1).

On the west coast the recurved sandy spit of Ponta Raza (see Fig. 1.1) has been formed by deposition of sand by the south-flowing drift arising from the small waves generated in the Bay of Maputo. This is met by an arm of the ocean current through Ponta Torres strait, and both intertidal and subtidal sand-banks have been deposited south of Ponta Raza spit. The Inhaca Channel has cut into the most recent substratum of semi-fossilised coral which is covered by sand; on its shoreward edge, wherever a solid substratum is exposed by currents, living coral grows confluently to form a coral reef (see Chapter 8).

1.3.4 Contemporary problems

Freshwater swamps occur in the slacks between the dunes. They are partly drained for agriculture. The largest swamp, which almost impinged on the northern bay in 1930 was about 5 km long from east to west, with one long arm (3 km) running south. The others were much smaller. The slightly darker soil around them indicates that their surface areas were greater in the recent past. In the last sixty years it has been recorded that some swamps and natural wells have dried up, and others are now dry at the surface in the winter months. The water table lies on the basal calcareous sandstone many metres below so that, theoretically, there should be an ample water supply for traditional irrigation of the crops and as drinking water for the Island's people. The growth of the population in recent years and the increased demand for water from boreholes by the hotel and the Marine Biological Station may be partly responsible for what appears to be a lowering of the water table. The impact of drought during a five-year period in the 1980s was severe but recovery of the water table may occur during years of good rains in the long-term cycle.

Inhaca Island has erosion problems. The outer eastern dune is being built higher and higher, and sand blows over the crest on to the forest. The red cliff (Barreira Vermelha) on the west shore is eroded at its base by infrequent storms and large boulders have fallen out of the cliff. Deposition of sand (derived from erosion of Portuguese Island) on the western flats raised the spring tide level to the base of the dune in recent years, almost impinging on the roots of the trees of the dune forest.

The dune below the lighthouse was threatened by the formation of a large enclave early this century, and in 1929 a successful conservation operation saved the foundations of the lighthouse (see Section 10.5.5). At Santa Maria on the mainland peninsula across the strait a wind-swept enclave had been developing for many years. The loose sand continuously threatened the growth of the Ponta Torres coral reef and in 1977 a similar attempt to stop the erosion by fencing and planting was made. The coral reef has continued growing successfully (see Sections 5.4.3 and 10.5.4). In 1992 *Casuarina* trees were planted at Ponta Torres for the same purpose.

Aerial photographs of the islands over the last 30 years show that Portuguese Island has been reduced in area by almost 40 per cent in about 40 years. A lagoon has formed behind the sand-bar to the north of the island, which shelters a coral reef. On the other hand, there are signs of natural reclamation of the shore from the sea in the north bay of Inhaca Island. In the curve of the north shore west of the lighthouse, the upper shore is being extended almost a kilometre seawards by accumulation of fine sand. New colonisation by seedlings of *Avicennia marina*, the pioneer tree of the mangrove fringes, has occurred over the last ten years (see Section 4.2).

1.4 CLIMATE

The composition of the flora on the *land* of Inhaca Island is determined mainly by air temperature, salty winds, rainfall and the nature of the soil. On the *shores*, however, water temperatures are much influenced by ocean currents from afar and by direct exposure to the sun of rock and sand between tides. Salinity and tides override the effects of rainfall; wave action and currents modify the substratum. At Inhaca Island the land climate favours sub-tropical evergreen forest and includes tropical species. The oceanographical features result in a larger tropical component in the flora and fauna of the shores.

1.4.1 Temperature and insolation

Inhaca Island lies in the region of transition from a warm temperate to a tropical climate with a pattern of hot, not very wet summers and drier winters. On the west coast of the island air temperatures in the Stevenson's Screen at the Marine Biological Station have been continuously recorded since 1951. The mean air temperature was 24,98°C. Sea temperatures have been taken at times on the west shore when inshore water readings were 3-7 degrees higher than in air, except in July.

The tendency of the sea to be warmer than the land at this latitude is the result of the southward flow of equatorial waters in the Mozambique current which travels parallel to the east shore of Inhaca Island along the edge of the continental shelf about 8 km offshore. This current arises from the South Equatorial Drift created by the South-east trade winds. The Drift flows westwards to the coast of Africa between latitudes 5°S and 15°S and is deflected both up and down the coast forming warm inshore currents. The effect of the Mozambique Current, which has a core temperature of 25°C or even more in winter, is felt nearest the shore at the latitude of the Limpopo Bight as the waters circulate in a mini-gyre towards the Bay of Maputo (Fig. 1.2). It is this inshore influence which supports the growth of tropical seaweeds and sea grasses on the island's shores and the subtidal growth of small tropical coral reefs with their hosts of associated animals.

Inshore waters, however, are raised to higher temperatures than those of the Mozambique Current. When the tide is out, the exposed sand and rock reaches over 45°C in summer, and

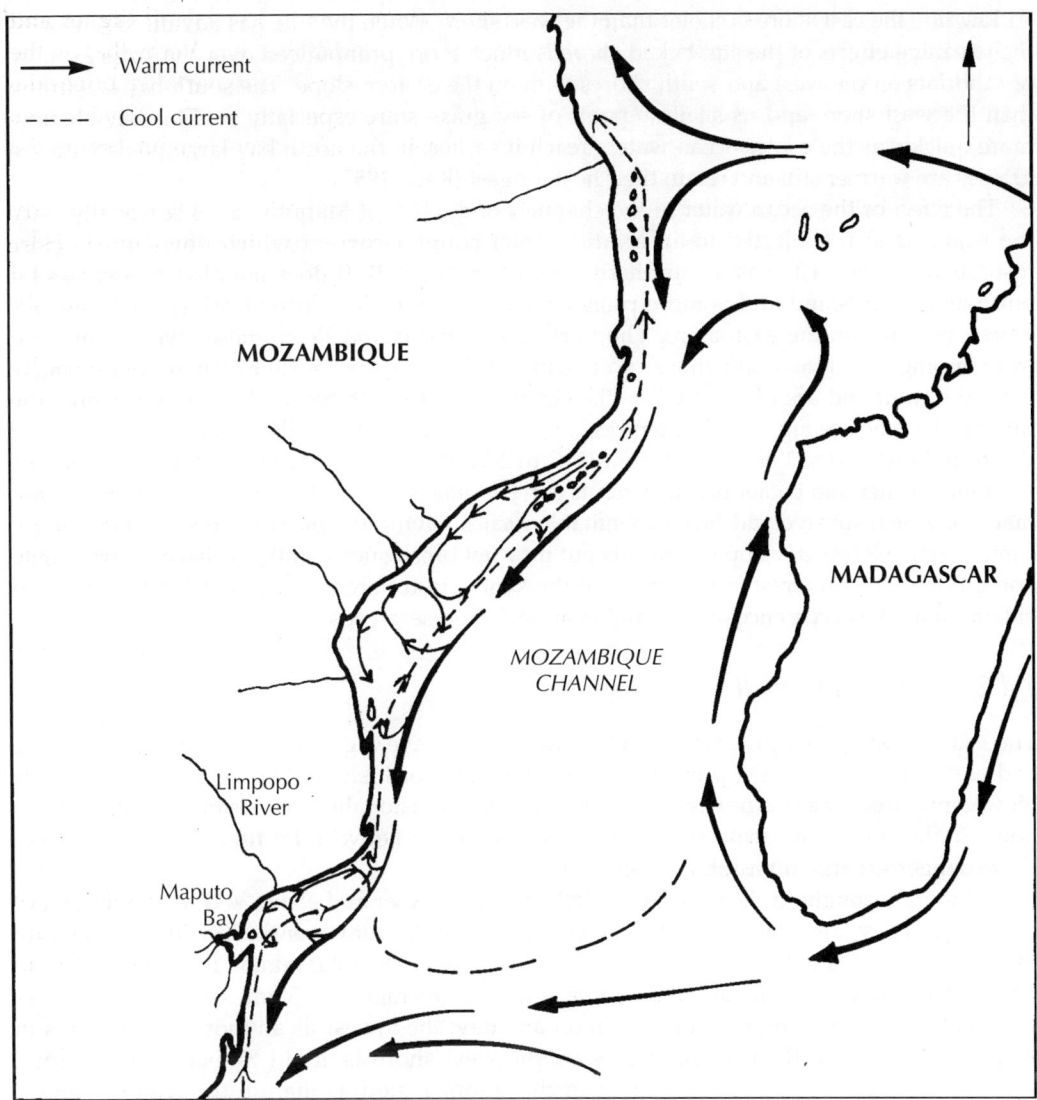

Fig. 1.2 Ocean currents in the Mozambique Channel: minigyres along the coast of Southern Mozambique, particularly in the Limpopo Bight, north of Maputo Bay (after A. J. de Freitas 1984).

when the shallow sea advances with the incoming tide the water becomes warmer. This lasts for a few hours around midday during a week of spring tides but is then diminished by the currents of cooler waters in the channels from the strait, when the tide flows in. In Table 1.1 a comparison of the water temperatures on three shores on a summer's day is given.

Table 1.1 Water temperatures (°C) on the shores, taken simultaneously. Low tide at 10.30 am, high tide at 4.20 pm.

	Eastern shore	Western shore	Southern coral reef	Southern sea grasses	Northern sea grasses
11:00 am	25,5	29	31,5	37,5	39
3:00 pm	27	35	27	33	36

At low tide the east shore is cooler than the west shore. When the tide has advanced half-way, the warming effects of the sun-baked shore is much more pronounced over the large expanse of sandflats on the west and south shores than on the eastern slope. The south bay is warmer than the west shore and its shallow pools of sea grasses are especially so. This bay changes more quickly as the cooler ocean waters reach it earlier. In the north bay large pools with sea grasses are warmer still and retain their heat longest (Kalk, 1957).

The effect of the ocean water in the channels of the Bay of Maputo varies seasonally since the island is also subjected to an erratic, cooler counter-current which flows north, close inshore along the east coast of southern Africa (see Fig. 1.2). It does not always reach as far north as Inhaca Island and is more pronounced in winter. This current brings cooler waters more especially to the east shore. The northern sand-bar and Portuguese Island prevent it from impinging on the north shore and to some extent on the west shore. These cooler waters enter the strait and are distributed by the channel in the south bay and across it towards the Inhaca Channel (see Fig. 1.1). The temperature in the water column above the coral reef off the Western shore during 1990 ranged from 18°C to 28°C (Gove and Cuambo 1990).

Both warmer and cooler ocean currents carry larvae, seeds and spores of shore organisms, many of which survive and breed in suitable niches, giving the biota its mixed geographical composition. Others develop into adults but may not breed successfully or their offspring may not survive. Thus interest in the study of the shores in different months of the year is often enhanced by the occurrence of uncommon animals and seaweeds.

1.4.2 Winds and rainfall

The island is subjected to strong ocean winds, which have a greater effect on the east shore and east dunes than on the protected west shore and north shore. On the west coast, winds blow gently from the south-west or north-east. Very occasionally severe storms, lasting a few hours in the afternoons, make inshore waters and the sea between Portuguese Island and the shore dangerous and impassable for small boats.

Rains fall throughout the year on over 80 days in summer and about 30 days in winter, but it is rarely heavy; the total rainfall for the year averaged 900 mm over the last 30 years. During each of the five years of relative drought in the early 1980s the total rainfall fell below 800 mm. The wettest months are January and February with an average of 145 mm rainfall, but in other months there is rarely more than 10 mm on any day; the lowest mean monthly rainfall is in August (24,85 mm). Relative humidity on the west shore is about 77 per cent and total evaporation is about 25 per cent higher in the summer months than in the winter months, reaching 105 mm in January, as measured in an evaporation pan. These are major factors in the lives of organisms on the shore and will be considered in several chapters in relation to adaptations to avoid desiccation and to adapt to varying salinity.

On land the dunes retain sufficient moisture to support a complete cover of vegetation, although at Ponta Torres only sand-binding plants can grow (see Section 10.2.1). The outer area of the east ridge is very steep and bare of vegetation.

1.5 TIDES

A knowledge of the tides at Inhaca Island is fundamental to the study of shore life. The rhythmic ebb and flow of tides exposes the shores and the organisms that live there to air and water alternately. This causes changes in temperature, salinity and illumination and affects the time available for feeding. Various animals and plants have adapted themselves to different lengths of time of submersion and emersion.

1.5.1 The causes of the tides

It is well known that the heights and times of tides change from day to day in a regular and mathematically predictable way. The tidal pattern is caused by the interactions of three forces in the solar system on the oceans around the earth. These forces are the gravitational pull of the moon and of the sun, that of the moon being about twice that of the sun (because it is much nearer, although its mass is less). The third force is the centrifugal force exerted on the oceans which is produced by the rotation of the earth and the moon about their hypothetical common axis. (This centrifugal force balances the gravitational attraction between the moon and the earth so that they remain at an equal distance from one another and orbit together around the sun.) These three forces have similar effects but act maximally at different times of the day, month and year. Thus:

- they produce high and low tides twice daily;
- a week of neap tides follows a week of spring tides;
- exaggerated equinoctial tides occur in March and September.

The tides are produced in the following ways: the forces of attraction lift the ocean water away from the solid earth since it is lighter; the earth is rotating and water slides easily, so the lift becomes a traction force pulling the water along from east to west towards the land masses, at about 80 km an hour. When the moon is overhead at any place the pull is greatest and the water rises. As the moon passes the water gradually subsides. This local rise and fall (about 50 cm in the open ocean) is the basis of the 'tidal wave' which is translated to the shores. (This 'tidal wave' is not to be confused with the catastrophic waves of hurricanes which are more properly called *'tsunamis'*.) When the long, low wave reaches shallow water its speed is checked by friction with the sea bed; its height is increased and it produces the rising tide. The trough behind each wave becomes the ebb tide (Pethick 1984). Wind driven waves are superimposed on the 'tidal wave' and have a different time scale.

At Inhaca Island the tides are semidiurnal and almost equal. The high tide at night is produced when the moon exerts its pull on the oceans more strongly on the area of the earth which faces it. Simultaneously on the opposite side of the earth in daylight the constant centrifugal force overcomes the moon's attraction which, at the greater distance, is stronger on the solid earth than on the water. Thus a high tide is produced there during the day. These two forces account for the succession of tides every day and night as the earth rotates.

Tides do not occur at the same time every day because the moon orbits around the earth more slowly than the earth rotates. The lunar day is 24 hours and 50 minutes so that each semidiurnal tide is *on average* 25 minutes later. Neither do tides reach the same height each time. The tide rises to its highest level and falls to its lowest level during 'spring' tides which occur around the times of the new moon and full moon. Then in each case the amplitude of the tide decreases gradually until, around the half moons in the first and third quarters, smaller tides occur called 'neap' tides. Neap tides have a reduced amplitude – lower high tides and higher low tides (see Fig. 1.3).

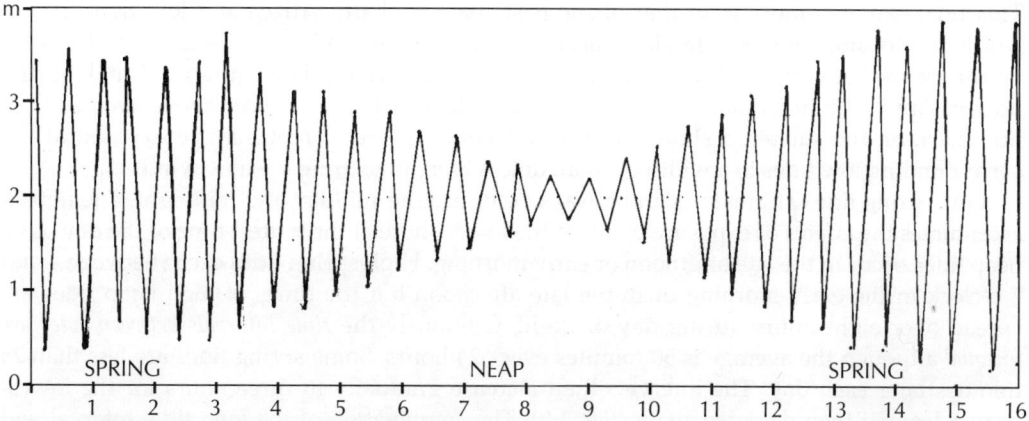

Fig. 1.3 Tides at Inhaca Island, September 1985, days 1-15 (from *Tabela de Marés* 1985, Maputo).

In a lunar month spring tides occur when the moon and the sun exert their gravitational pulls on the *same* side of the earth in a straight line (new moon). At full moon when the moon is on the *opposite* side of the earth to the sun, the sun's force more strongly augments the centrifugal force to produce the other spring tides. Neap tides occur when the pull of the moon and of the sun act at angles to one another and the resultant force is weaker.

The peak spring tides do not occur exactly on the night of the new moon or full moon, as would be expected, but one and a half days later. This lag is called the 'age of the tide' and is mainly the result of friction over the ocean bed which slows down the 'tidal wave' as it travels towards the shores.

When the sun is overhead at the equator during the equinoxes the earth, the sun and moon are almost in line and the strongest pull is exerted on the oceans creating the largest 'tidal waves'. These produce the extreme high and low spring tides in March and September, known as the equinoctial spring tides. The neap tide amplitudes are correspondingly smaller.

There are complications in this regular pattern because the oceans do not form a continuous shell around the solid earth. They are interrupted by irregular land masses (except in the Southern Oceans). The land masses distort the direction and the velocity of the 'tidal waves', varying the heights and times of the tides along a continental coast. On a smaller scale (and this affects Inhaca Island), bays and channels, promontories, estuaries and islands alter the heights and times of the tides to some degree. Daily tide data at ports are published annually in simple tide time-tables. Local tides near the ports can be calculated from information on the distance of the place from the port and the direction and degree of indentation of the coast. The data can be checked by observation.

1.5.2 Modified local tides on Inhaca Island's shores

The semidiurnal tides are about equal in range at Inhaca Island. From the *Tabela de Marés*, published annually in Maputo, the approximate times and heights of the local tides can be obtained for a particular day of the year. The exact data for Inhaca Island vary on the different shores because of the different degrees of shelter. The west shore heights and times are similar to those at Maputo Harbour, except that they are a few minutes earlier and some centimetres lower because of the island's position near the mouth of the Bay of Maputo. Tides are 30 minutes earlier on the east shore of the island and high spring tides appear to be about 2 m higher because this shore faces the open ocean and onshore winds produce wave splash and spray. High tides in the southern bay are delayed by about an hour because of the 7 km length of the bay.

Shores are usually characterised by the tidal range of spring tides. The maximum amplitude recorded in 1968 and predicted for 1985 at Maputo Harbour was 3,69 m. The mean spring amplitude has been calculated from the tide timetable as 3,5 m for Maputo; in the Inhaca Zone, it is given as 3,2 m – lower because of its greater proximity to the Indian Ocean. This tidal range is more than that of the east coast of South Africa and less than that in Northern Mozambique. The standard mean sea level in Maputo Harbour is 2,0 m and there is no difference between the level there and in the Inhaca Zone. The equinoctial high spring tides of April 1985 inundated 7-8 m up the sandy slope of the west shore above average high tide mark on the sheltered shores of Inhaca Island. The coral reef may be exposed at the corresponding low tides to a width of 10 m, unless there is a strong onshore wind.

Low spring tides at Inhaca Island always occur around midday and midnight. The former accentuates the effects of exposure to air on the lower shore at the hottest time of the day. Low neap tides occur in the late afternoon or early morning. High spring tides occur between 4 and 7 o'clock in the early morning or in the late afternoon but the times of high neap tides are spread over eight hours during day or night. Obviously the *time intervals between tides are unequal* although the average is 50 minutes every 24 hours. Some spring tides are less than 25 minutes later each day. The intervals then increase gradually to three hours for the lowest neap tides and then decrease again (Fig. 1.3). The combination of the long tidal interval and small tidal range of neap tides on a flat sandy shore results in the lower shore being continuously covered by shallow standing water at the lowest neaps, looking as though the tide had not gone out. A similar effect is seen during neap tides at the north-east rocks where the broad middle platform below the cliff is covered by the neap tide for many hours. The range of the least of the neap tides is only 40 cm.

1.6 WATER MOVEMENTS AS DETERMINANTS OF TYPES OF SUBSTRATA

Different degrees and directions of wave action have a formative effect on the substrata and on the survival of different kinds of plants and animals on the shores. Tidal currents carrying sediments of sand, silt, clay and organic matter, which are partly deposited on the shores as they lose momentum, are specially important on flat shores, where wave action is minimal.

1.6.1 Waves

The kind of waves that occur and the forces they exert on the shores depend on the wind speed and its duration, on the distance the wave has travelled over the surface of the sea and also on the slope of the beach. As waves travel towards the shore, the horizontal movement of water particles is slight; they move up and down in small orbits related to the height of the wave. When the waves reach shallow water they become higher and slower as friction with the sea bed increases. The horizontal velocity of the surface water particles then becomes greater than that of the waves so that the surface water tumbles over and waves 'break' on the shore.

On the east shore of Inhaca Island the prevailing strong south-east winds blowing over the vast expanse of the Indian Ocean induce waves which exceed 1 m or even 2 m in height. Their force is somewhat reduced by submerged rock a few metres below the surface near the shore, and by old submerged coral reefs, parallel to the shoreline, 5 km offshore. The east shore is very steep with a gradient of 1 in 10, so that the plunging waves break with considerable force on the sandy slope. Around the north-east point the impact of the waves becomes reduced on the wide rocky platforms around Ponta Mazondue. The pounding action erodes the rocks and sand. Coarse particles derived from the sandstone rocks remain to form the beach and some are deposited on the rocks leaving a light coating. Fine and medium particles are carried away in the longshore current around the north point to the sand-bar.

The sand on the east shore is coarse, well aerated and golden; the particles have a median size greater than 500 µm and the beach is thus very porous. Tidal water sinks to a water table over 2 m below the surface at the top of the intertidal slope. Consequently the sand is inhospitable to macroscopic organisms, but the surface layer supports a reduced community of minute animals of the meiofauna and microscopic algae. In contrast, the sandstone rocks support rich associations of animals and seaweeds. Millennia of wave erosion and wind have planed and creviced the rocks to make shallow and deep pools, with exposed and sheltered surfaces at all tidal levels and so created micro-habitats for many animals and for a rich covering of seaweeds below mid-tide level.

On the west shore, the gentle sea breezes rarely exceed a velocity of 15 km per hour. The waves are consequently low, less than 50 cm high – usually about 25 cm – and they break only on the slope of the upper shore and on the coral reef at lowest tides, where the tidal velocity is greatest. On the large stretch of flat shore tiny rippling waves occur. An occasional storm sometimes reaches gale proportions for a few hours and then waves over a metre high are generated. Normally the gentle waves result in sand accretion which repairs the erosion of infrequent storms, so that the shore is fairly stable. Above mid-tide level, waves have created a gentle slope about 20-30 m wide having a gradient of 1 in 30. The sand is clean and well sorted with a median particle size between 200 and 300 µm and there is less than 10 per cent fine sand. It is fairly porous, well oxygenated and is occupied by burrowing crabs and bivalves as well as by microscopic organisms. The water table is about 1 m below the surface in the sandy slope but the sand does not dry out completely because of capillary forces in the interstitial water between the grains (see Section 3.2 to 3.5).

Below mid-tide level wave action is minimal on the flats which extend over 500 m to low tide level. The proportion of fine sand increases down the shore from 20 per cent at the foot of the slope to 70 per cent in the clean sand bank in front of the coral reef (see Section 9.5). Storms have thrown up large and small slabs of dead coral which provide a hard substratum of limestone rock on the lower shore (see Section 6.4).

1.6.2 Tidal currents

The deposition of particles of different densities on various parts of the shore depends on the velocities of the incoming and outgoing tidal currents. On a broad open beach like the sand flats on the west shore of Inhaca Island, *minimal* velocity is around *mid-tide level*, where silt, clay and organic matter are deposited, producing a rich flat sandy mud. Cleaner sand is deposited at the head and foot of the beach where *maximum* velocities of the tide occur (see Chapter 3).

A different regime occurs in narrow bays and estuaries, such as the southern bay. There is partial reflection of the tidal wave because the tide has not fully receded before the next tidal wave arrives. Here the *minimal* velocities of the tidal currents are on the *upper* and *lower* shores. Silt, clay and organic material are deposited on the upper shore forming the soft mud where the mangrove association occurs. The mud becomes black due to biochemical processes described in section 1.7.2. Fine sand is also deposited around the mouth of the bay forming large sand banks (Fig. 1.1). The *maximal* velocity of the tidal current is on the middle shore where it has a strong scouring action leaving flat hard muddy sand (see Chapter 5).

The northern bay, having greater protection and a broad shape has been converted in fairly recent times into a shore of sand accretion.

1.6.3 Fresh water drainage

One other factor affects the substrata and the water table: fresh water draining from the dunes not only creates creeks and channels in the mangroves but also forms runnels on the sand and seepage below the surface, so that the water table is at the surface in some large areas of the flats. This causes a marked difference in the micro-environment from drier sand and is responsible for the occurrence of different associations of organisms (see Section 3.5.1 and 3.5.2).

1.7 CHEMICAL FACTORS IN THE ENVIRONMENT

The seashore environment differs from the terrestrial or that of fresh water in a number of significant chemical factors which contribute to the greater profusion of life.

1.7.1 Atmospheric gases

Aquatic plants and animals rely on the solubility of gases in water. The amount of oxygen dissolved in sea water at the temperature and salinity around Inhaca Island is about 6 ml per litre. Normally, it is not limiting because the oxygen used in respiration is replaced by day by means of the photosynthesis of macroscopic and microscopic plants. The content is doubled by day in rock pools with seaweeds and in sea grass pools on sand flats but depleted at night. Oxygen in the surface layers of clean sand is increased in the daylight by the micro-flora but the black mud of mangroves on the upper shore and that beneath the layer of sand on the lower shore sand flats is anoxic due to bacterial action. Many animals have adapted to the seashore by using various ventilating mechanisms in their burrows. In addition, most of them have respiratory pigments, which function at much lower oxygen tensions than the haemoglobin in land animals. The pigment may be used for a continuous supply of oxygen or as a store to prevent hypoxia when the tide is out (see Section 2.2.3).

Carbon dioxide is much more available in sea water than in air. It is much more soluble than oxygen, but its concentration is about 0,3 ml per litre, because it interacts with water to form bicarbonate and carbonate ions with which it is in equilibrium. This means that the concentration is maintained whatever the demand because the respiration of animals and plants ensures that the ions are inexhaustible. Some seaweeds, benthic micro-algae and phytoplankton can use bicarbonate as a source of carbon dioxide for photosynthesis; other marine plants such as sea grasses use dissolved carbon dioxide, as do land plants.

Nitrogen is more soluble than oxygen (10 ml per litre), and it is recycled by bacterial processes of nitrification and denitrification in interstitial waters of the sandy seashore, as it is in the soil. Nitrates and nitrites are thus made and used by the seashore plants. In addition, there is a significant increase in the turnover of atmospheric nitrogen because of the prevalence of nitrogen-fixing bluegreen 'algae' (Cyanophyta), which coat the rock and sand surfaces and live in close association with the roots of some of the sea grasses.

1.7.2 Organic compounds and nutrient salt cycles

The seashore is rich in decaying organic matter from both local (autochthonous) production and that imported by the tides (allochthonous). The former is derived from the dead shore plants and animals and is enhanced by the dense microflora, microfauna and the plentiful but minute meiofauna of the sheltered sandy shore. The meiofauna make a significant contribution to the nitrogen pool by excreting ammonia, augmenting that excreted by the macrofauna. The allochthonous production is enriched by the dead phytoplankton, the decaying sea grasses or seaweeds, and other material brought in from the subtidal zone.

In aerobic layers of the sandy mud, organic compounds are broken down by bacteria and fungi. Carbohydrates become lactate, acetate and pyruvate; proteins become amino acids and the sulphydryl (-SH) groups become sulphates. All these derivatives may be used by heterotrophic micro-organisms (bacteria, fungi, and protozoans). When oxygen is not available in the closely packed black layers of mud, anaerobic bacteria convert sulphate to sulphide. This reacts with the hydrated ferric oxide, which normally coats yellow sand, and reduces it to ferrous sulphide, giving the sand the black colour. When iron is limiting, hydrogen sulphide is liberated and the mud smells of rotten eggs, as happens in some parts of the mangroves.

The phosphorus cycle in black mud makes soluble phosphate available to plants so that sea grasses and mangrove trees grow well there. On the surface of the sand, aerobic bacterial decay of organic phosphate from detritus liberates inorganic phosphate, which reacts with the hydrated ferric oxide that coats the sand, to become insoluble ferric phosphate. This becomes buried in the anaerobic layers of the mud and is reduced to soluble ferrous phosphate. In this form phosphate can be absorbed by the roots of mangrove trees and sea grasses that penetrate the black mud. Some of the soluble phosphate moves up to the surface layers of mud with the rise in tidal water; surplus phosphate is again precipitated in oxidised form. The mud therefore acts as a nutrient phosphate *trap* and the cycle is repeated. The sea grasses release soluble phosphates from their leaves and roots which become available to the phytoplankton and heterotrophic micro-organisms in the tidal water. Thus the sea grasses act as a nutrient *pump* in the phosphate cycle.

1.7.3 Variations in salinity of the sea on different shores

There is yet another chemical environmental property in which the shores of Inhaca Island differ from one another. The salinity of the sea on the east shore is fairly constant, being the same as that of the Indian Ocean at that latitude, approximately 35,5g per litre, except for the tidal pools in which evaporation occurs. On the west shore, intertidal waters are slightly below this 'normal' value, whereas in the upper reaches of the north and south bays salinity changes twice daily to values above and then below the norms as a result of evaporation alternating with fresh water input (see Section 4.3.2).

Low tide water on the west shore has a value of 34g per litre, similar to that of the Bay of Maputo where input from the ocean is diluted by the rivers and completely mixed by the time the waters impinge on Inhaca Island shores. At high tide mark, however, freshwater seepage from the dunes reduces the salinity giving values around 30g per litre in a freshly dug pool. As the tide advances, saline water from the Inhaca Channel restores salinity of the inshore water to 34 g per litre (Macnae and Kalk 1962).

Much of the superficial fresh water of the island drains into the Saco at the upper end of the southern bay where freshly dug pools between tides may yield a value of only 5g per litre salinity. On the other hand, salt crystallises out on the surface of bare sand in a certain area of the landward fringe of the mangrove swamp during the heat of the day (see Section 4.5.1). The net result is the dilution of the water in the main channel to about 20 g per litre in its upper reaches at low tide. A fairly uniform salinity of 30 g per litre is restored in the bay at high tide by the ocean current which sweeps in through the southern strait. These variations in salinity are met by many adaptations in the specialised plants and animals of the mangrove swamps.

1.8 PENETRATION OF LIGHT AND THE TURBIDITY OF THE SEA

On tropical shores most seaweeds and sea grasses grow only on the lower tidal levels where they are submerged for most of the time during the weeks of neap tides. Although coastal waters are notoriously much more turbid than those of the open sea, photosynthesis will continue as long as the penetration of light is at least 1 per cent. An estimate of the limit of transparency needed for photosynthesis may be obtained by determining the depth at which a Secchi disc (a black and white disc, 30 cm in diameter) disappears from view when submerged.

On the west shore the penetration of light was on average about 3 m and since the tidal range is seldom much more, there is always sufficient light for plant photosynthesis on the shore. At the coral reef (where there may be less silt on occasion) the maximum penetration determined was 5 m on calm days, although 3 m was more usual, and even less when the wind was from the north. The growth of the coral reef depends on the endosymbiotic *algae* in the inner cells of the polyps, and thus illumination limits the depth to which they will grow (see Section 8.2). The corals on the east coast of the island grow at a depth of 12 m where the sea is clearer. The sea grasses grow to a depth of 5 m on the banks of the Inhaca Channel.

PART 2

Animal and Plant Life on the Shores

A general account of ecological interrelations of intertidal species with one another and with their environment is given in Chapter 2 to provide a framework on which to build the biological details in the following chapters. The common intertidal animals and plants are described in Chapters 3 to 9 in the order in which they occur on the sand or rocks from high tide to low tide. Their adaptations to life in particular habitats are discussed.

The four shores of the island exhibit different environmental features which determine the distribution of communities. The associations at successive tidal levels on soft substrata are discussed in Chapters 3, 4, 5 and 9, those on hard substrata are described in Chapters 6, 7 and 8.

The densely populated *sheltered* sandy west shore, adjacent to the Marine Biological Station, is compared in Chapter 3 with the sparsely populated east shore which is *exposed* to strong wind and wave action. The *very sheltered upper* shores which support mangrove associations are estuarine and exhibit a complete contrast; they are discussed in Chapter 4. The *very sheltered lower* shores in the southern and northern bays, described in Chapter 5, exhibit tropical features, including a coral reef and sea meadows.

Chapter 6 describes the flora and fauna of the rocky habitats of the sheltered west shore, including the 'reef flats' on which coral reef inhabitants spread up the shore in pools under coral debris. The contrast with life on rocks exposed to strong wave action is highlighted in Chapter 7.

The subtidal fringe of the sheltered shores supports a coral reef ecosystem, comprising the coral reefs themselves and the sea meadows. The associations overlap and are described in Chapters 8 and 9. The shifting banks of clean sand have a fauna of their own which is included in Chapter 9. The same species may occur on more than one shore. Physiological details of adaptations of species are then given either in the chapter where they are first encountered or where they are most significant. In both cases, they are cross-referenced. The less common species as well as those included in the text are listed in taxonomic order in Appendix A.

2

ECOLOGICAL FEATURES OF THE SHORES

The nature of life on the shores of Inhaca Island is profoundly influenced by the environmental features outlined in the first chapter. No less important are the inter-relationships of the organisms which sustain life in the various communities. The interdependence of plants and animals in regard to energy, food and space is discussed below; the relative importance of various parts of the patterns of life on the rocky and sandy shores is evaluated. Consideration is then given to the types of biological responses to environmental stresses which maintain the orderly distribution of organisms in the universal zonation on the shores. These general features give an ecological frame of reference for understanding the detailed study of organisms in the following chapters. Interrelations and responses of common species are described as well as their adaptations to their physical environment.

2.1 TROPHIC RELATIONSHIPS

It is well known that the energy supporting the familiar food chains in communities comes from the sun, whereas materials are recycled. Briefly, the organic material produced by the growth of green plants, termed 'primary production', depends on the trapping of light energy by means of chlorophyll. Animals obtain both their energy and their materials for growth and maintenance by feeding on plants or by consuming animals that have fed on plants. 'Decomposers', including the bacteria which replenish the mineral nutrients, are themselves part of the food chains. Respiration and photosynthesis recycle the carbon in complex pathways inside cells.

2.1.1 Primary producers on sandy and rocky shores; herbivores

On tropical shores the emphasis on each component in the food chains is somewhat different from that on land. Intense light from the tropical sun, the ready availability of nutrients and access to water twice a day from the tides ensure that green plants grow well. They cover all exposed surfaces, although the seaweeds on rocks are small in comparison with those on temperate shores and, in fact, the majority of algae are microscopic in size. Most conspicuous on sandy shores at Inhaca Island are flowering plants, growing in dense associations in three types of area: sea grasses on the lower sandy shores (see Section 9.2), mangrove trees on very sheltered upper shores (see Section 4.1) and the perennial succulent plants at high tidal level beyond the mangroves (see Sections 4.5). The gross primary production between tidemarks is very much greater than that of the sea, yet paradoxically, few animals attempt to eat the leaves of the prolific flowering plants *directly*. Some do graze on the living plants, but it is doubtful whether they digest the leaves; perhaps the thick coating of small algae and sedentary invertebrates on the leaves provide the greater sustenance.

Nevertheless these macrophytes do provide food for a large number of seashore animals *indirectly* when the dead leaves have been shed, then leached by the sea, fragmented by buffeting and decayed through the actions of fungi and bacteria. Leaves are shed continuously throughout the year because of senescence, desiccation and excessive salt load. The leaf litter is then converted into detritus before becoming a major source of food. The production of leaf litter in mangrove swamps has been estimated (as weight of carbon) to be higher than that of the litter mass in a tropical terrestrial forest (Dring 1982). On Inhaca Island the small areas of mangroves and smaller trees produce much less than in the true tropics. The formation of detritus and its consumption are outlined in Section 2.1.4 below.

Seaweeds, the primary producers on exposed shores, form an 'algal turf' on the rocks below mid-tide and line the rock-pools at Inhaca Island. Some small seaweeds occur on the leaves of sea grasses, on the aerial roots of mangrove trees, on dead bases of coral in the reefs and on the coral debris of the reef flat, but there are very few species on sand (see Sections 7.4.1, 7.5.1 and 3.5.2). Although the number of species of seaweeds is much greater on tropical rocky shores than that on temperate shores, the biomass is less since the large brown kelps and fucoids are absent. The chief algal grazers are members of the herbivorous families of molluscs, crabs, sea-urchins and fishes. Some scrape the surface of the rocks to obtain microscopic, encrusting and filamentous algae, whilst others in pools prefer the softer green algae to the thick-walled reds and browns (see Section 7.5.3). Much of the dead seaweed is carried out to sea and decays there, being brought in again by the tides as detritus.

The microscopic benthic flora is *directly* used by a large number of animal species as a source of food on both sandy and rocky shores. It lives on the surface or within the upper few millimetres of sand. Each grain of sand is coated with adhering diatoms, bacteria and organic particles in the process of decay (Fig. 4.22a and b). Between the grains in interstitial water are burrowing diatoms, flagellates, filamentous green algae, blue-green algae (Cyanophyta), and bacteria (see Section 3.2.2). This flora is grazed by large populations of small animals of many kinds, some of which use ingenious sorting methods to select diatoms or bacteria and to reject sand and silt (see Section 4.6.1).

Within the sand itself there are dense populations of meiofauna – that is animals between 50-500 μm (micrometres) in size, which live between the grains (see Section 3.6.2). Among them are herbivorous nematodes, copepods and ostracods which scrape the diatoms off the grains of sand. Some filter-feeding benthic animals have the means of selecting hard particles in the size range of 10-60 μm and choose diatoms rather than bacteria-covered detritus. For example, the wedge mussel selects large diatoms and oysters collect much smaller organisms (see Section 3.2.1). Algae and bacteria are also eaten by chance by the non-selective deposit feeders such as sea cucumbers when they swallow large amounts of surface sand (see Section 3.6.4).

A third source of primary production is the phytoplankton which floats in surface waters and consists mostly of diatoms, dinoflagellates and bacteria (see Section 8.4), 90 per cent of which is less than 60 μm in size and is known as the nanoplankton. This is consumed by the pelagic larvae of the majority of the intertidal animals during their temporary sojourn in the sea. There are also a large number of sedentary animals which are microphagous, such as worms, tunicates, sponges, bivalve molluscs, barnacles and other crustaceans which create currents of water from which suspended particles are filtered. The preferred size of particle is selected by the filter feeders by means of a number of incredibly complex mechanisms of interlocking cilia on mucus-covered surfaces or on mobile setose appendages (see Sections 3.3.2, 4.6.1, 6.4.4). Filter feeders are closely packed on a rocky shore exposed to strong wave action and their joint efforts must guide a veritable rain of mixed plankton and detritus towards themselves (see Section 7.5.5). Some of these mechanisms are described in the following chapters. In the open sea the main consumers of the phytoplankton are the permanent zooplankters, mainly copepods and other crustaceans (see Section 8.4).

Lastly, blue-green 'algae' (Cyanophyta) also contribute to the primary production on the shores. These are microscopic filaments or plates of cells in which the small cells contain chlorophyll but also have many structures in common with bacteria. The blue-greens at high tidal level form the dark coating on rocks and are directly grazed by some periwinkles. Those which form a mucilaginous film on mud, on sea grasses and on silted rocks of the lower shore are included in the diet of many animals but digested by only a few.

Some Cyanophyta play another role in increasing primary production on a tropical shore since they increase the nutrient salts available to green plants by 'fixing' atmospheric nitrogen. The process is photosynthetic: light energy splits water molecules, freeing hydrogen ions which reduce nitrogen in solution to ammonia. This is used in the synthesis of its protein by the blue-green alga and excess is converted into nutrient salts which enhance the growth of seaweeds and sea grasses. One genus of such blue-green algae, *Calothrix* (Fig. 8.7e), coats the roots of the pioneer sea grass *Thalassia* which is common in sea meadows.

2.1.2 Primary production in the coral reef: symbiotic algae

A special case of algal primary production supports the growth of corals in the reef. This production is the highest in the world and has been estimated to be about twenty times higher than that in the open sea. It may reach 5 000g C/m^2 per year in a mature reef, higher than that in cultivated fields of irrigated sugar cane, which holds the terrestrial record for gross production. There are several kinds of plants involved in primary production in a coral reef. A minor contribution comes from encrusting coralline red algae, known collectively as 'lithothamnion', which cements bases of old corals together. Another input is from a considerable growth of green algae attached to and penetrating into dead coral boulders on reef flats or in the reef itself. A larger (although indirect) source of food is the growth of associated sea grasses enhanced by the nitrogen-fixing blue-green algae (as mentioned above). Phytoplankton above coral reefs is denser than in the open sea or in other coastal areas because nutrients are rapidly recycled from the excretory products of the dense fauna living in the reef (see Section 8.4). The largest contribution to primary production, however, comes from the unicellular algae which live *within* the cells of the living coral polyps (see Section 8.2). The total weight of these symbiotic algae may exceed that of the living tissue of their animal hosts.

At Inhaca Island the primary production of the reef as a whole is not as high as that of a fully developed large coral reef ecosystem on the northern shores of Mozambique (e.g. off the Isle of Goa), largely because of the higher silt load in the tides and the lower temperature of the water.

2.1.3 Consumers: carnivores

On tropical shores a large number of species in the fauna are carnivores representing almost all phyla in the microfauna, meiofauna and macrofauna. Some fishes, swimming crabs, mantis shrimps and squids actively hunt their prey in the surface waters when the tide is high. Small carnivores such as snails and worms burrowing in the wet sand lie in wait for prey to approach, whilst corals and hydroids are sedentary. The latter catch single individuals in the zooplankton with their tentacles, first paralysing them with sting cells. Predators consume animals in all feeding categories using a number of different techniques and they display a remarkable specificity in their choice of prey. The number of individuals of each carnivorous species is, however, comparatively low.

2.1.4 The role of detritus as food: detrivores

The largest number of animal species appears to feed on the sand on the seashore. Some are very selective, using sensory mechanisms and adaptive motor structures to choose diatoms, meiofauna, bacteria or protozoans, as mentioned above, whilst others appear to be non-selective detritus feeders.

Detritus is formed by the fragmentation of plant litter into organic particles, less than 0,15 mm in size, that are coated with bacteria. Through the decay of mangrove leaves, sea grasses and seaweeds about 90% of the original dry mass of this comparatively indigestible material enters food chains either as minute particles or as dissolved organic molecules (Schleyer 1986). Conversion to detritus commences when senescent leaves, still attached to the parent plants, are attacked by fungi and bacteria which feed on the tissues. This paves the way for the cells to destroy themselves by autolysis when their own enzymes break down the cellular structure and large molecules. A few weeks after the leaves have been shed they are leached of all soluble matter and easily broken up by water movements and abrasion on the sand. Fragments are colonised by bacteria and fungi which feed partly on these fragments and partly on dissolved organic matter. With access to air, bacteria and fungi break down the leaf cell walls with cellulases, lignases and pectinases and thus expose the proteins to digestion. In mangrove swamps, fungi play the major role (see Section 4.3.1). A succession of bacterial populations specialise in different chemical reactions. During the formation of detritus the ratio of nitrogen to carbon increases due to the growth of the bacteria that synthesise proteins. The bacteria reproduce to densities of several million per gram of substratum and coat particles of organic matter adhering to sand grains or lying between them. An almost constant number of bacteria is maintained by protozoan grazing.

Ciliates and flagellates feed on bacteria in large numbers and by their predation stimulate a high rate of bacterial reproduction (see Section 3.6.1). The carnivorous organisms in the meiofauna, in the next order of size, feed on the protozoans (see Section 3.6.2), whilst shrimps and a few other animals select meiofauna, particularly in sheltered bays and estuaries. Some macroscopic detritus feeders consume the organic cores, the bacteria, protozoans and meiofauna as well as some diatoms and other members of the interstitial flora. Other detritus feeders among the crustaceans, such as some fiddler crabs (see Section 4.6.1), sift the sand with setose mouthparts and select the bacteria preferentially (as do those filter feeders extracting bacteria from the phytoplankton). In some areas aerobic bacteria are conserved by repeated turnover of sand by detritus feeders such as sea cucumbers and bubble crabs, which play a part in aerating the surface. It is notable that bacteria on land are mainly decomposers, but on the seashore they conserve plant food in their own cells and thus contribute substantially to the food of higher organisms.

Some detritus-feeding animals egest faeces in which organic particles have been stripped of all living material and only the indigestible organic core remains. These are excreted, recolonised by bacteria and become smaller still as the bacteria use them as food. The process is repeated again and again until the particles, reduced in size during each cycle, are less than 50 µm in size and are now readily suspended in water. The filter feeders which live in sand and mud such as bivalves, some snails, porcelain crabs and some hermit crabs and prawns exploit this resource, including the bacteria, more than they do the phytoplankton. Many animals in different phyla have evolved ways of extracting edible detritus from sand. They include some worms, snails, crustaceans, acorn worms, sea cucumbers, brittle stars, sea urchins and even fishes.

In contrast to the aerobic decay described above where carbon dioxide is an end product, the anaerobic action of bacteria in black mud is more akin to fermentation in the stomachs of ruminants since the end products are fatty acids. Some nematodes in the meiofauna feed on these bacteria and may utilise fatty acids (see Section 3.6.2). Some molluscs appear to maintain anaerobic bacteria in the intestine and absorb the fatty acids as a contribution to their nutrition. This is a less efficient method of energy production than the aerobic pathway of carbohydrate decomposition.

An outline of the food relationships on the sheltered shores of Inhaca Island is given in Fig. 2.1 showing the dominant role of detritus in the food web. Besides recycling mineral nutrients for *plants*, as on land, the bacteria in detritus on the seashore have a triple role in *animal* nutrition: they make the organic matter of seashore plants available as food; they are themselves direct contributors to the food chains; and they make dissolved organic matter available to the animals which can utilise it.

2.1.5 Commensals

Close relationships develop on tropical shores between two different species of animals; these depend on sharing food and shelter. This phenomenon is known as commensalism, meaning

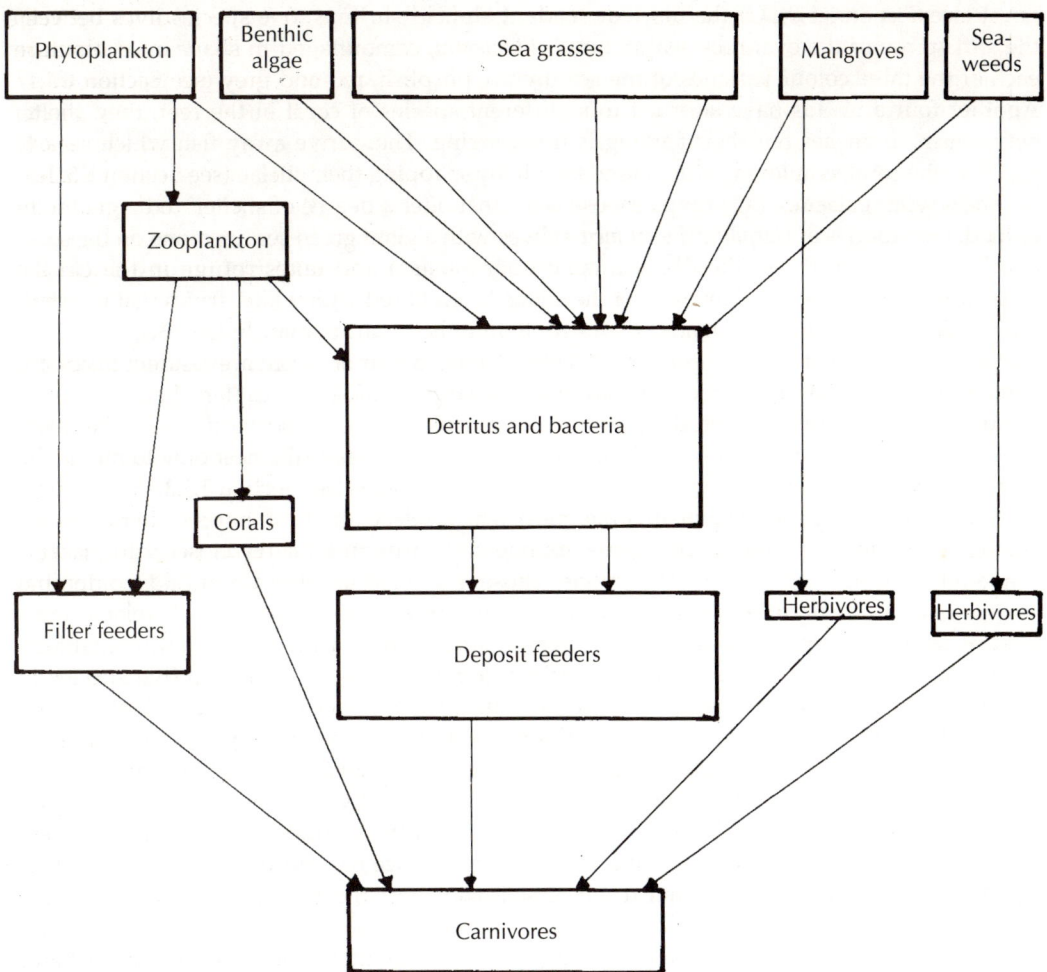

Fig. 2.1 The food web on the sheltered western shore of Inhaca Island. Blocks roughly proportional to population sizes.

literally 'eating at the same table', and it is probably a consequence of the close spatial associations of the very rich and diverse fauna of the coral reef ecosystem. The two animals concerned are usually of different size and phylum and it is typical that the smaller one, the guest, gains advantage from the larger one, the host, although reciprocity may be exhibited. The guest lives only with a specific host and differs in some adaptive way from the free members of its own family. The guest may receive more than food and shelter since the ventilation mechanisms of the host may increase the availability of oxygen for the guest. The host's mobility may sometimes substitute for the lack of motility in the guest. The loss of free movement in a guest may, however, complicate its reproduction and results in curious changes in size, structure and behaviour in both male and female. The relationship may become so intimate that the guest becomes a parasite!

Several hundred examples of commensal pairs are known on tropical shores and at Inhaca Island about 50 pairs of obligate commensals have already been observed. Commensalism is most common among the more highly evolved crustaceans, probably because their neurosensory systems and central nervous systems are most complex.

Some families of shrimps illustrate a scale of increasing dependence on a host. At the top of the scale are the pistol shrimps (alpheids) which are usually free-living and carnivorous. At Inhaca Island three species of them provide shelter in their burrows for three specific goby fishes; in this case each shrimp is the host which excavates the burrow and the goby is the guest (see Section 3.5.3). Of the more dependent alpheid shrimps, one species lives between the stiff spines of the oval rock sea urchin, *Echinometra*, camouflaged in shape and colour on each of the three colour varieties of the sea urchin, purplish, red and grey (see Section 6.4.1). Another four alpheids have adopted four different species of coral in the reef; they shelter between the branches but their feeding is free-ranging. They drive away fish which seek to graze on the coral, employing loud sounds made by snapping their chelae (see Section 8.5.2).

The smaller palaemonid shrimps choose hosts that offer a degree of shelter and a guarantee of food. One species of *Harpilius* (*Periclimenes*) lives with a giant green sea anemone on the sand; another lives with the snake-like sea cucumber *Synapta* and takes refuge in the cloacal respiratory chamber. A third member of the genus is bright red with white stripes that match its host, a large sea slug, *Hexabranchus*, and it shelters beneath its mantle (see Section 5.4.2). Dependence is carried to the extreme by *Anchistus custos*, a pair of which are resident inside the mantle cavity of almost every individual of the very large pinnid bivalves (fan-shells) in the sea meadows. The shrimps feed on the gleanings of mucus-collected food particles on the host's large gills, scraping them with specially modified chelae. They leave the host only to moult and mate and the female then returns to the shell to incubate her eggs (see Section 5.3.2).

In the cases mentioned above, it is the guest which selects the host, but in other cases the host chooses the guest. For instance, the sponge crabs (dromids) carry close-fitting jackets consisting of compound animals. One species choses a sponge, another a zoanthid (compound anemone) and others different species of compound tunicates (sea squirts). A related crab, *Dorippe* carries a purple sand dollar (sea urchin) supported by two pairs of abdominal legs, projecting dorsally like those of sponge crabs (see Section 9.4.2). There are five species of xanthid coral crabs which always live in pairs in different species of coral. They feed on detritus trapped in the mucus on the coral surface and in return they protect the coral by nipping the fins of fishes that graze on it, and thus drive them away (see Section 8.5.2).

Disparity in size between female and male guest is frequently seen and is exaggerated with increasing dependence. Females are bigger than males in the shrimps and pea crabs resident in bivalve shells. Tiny brittle stars cling to the under surface of sand dollars and the male is one third the size of the female. An extreme case is shown by the coral gall crabs (see Sections 9.5.1 and 8.5.2).

The mechanism for reciprocal tolerance of partners in commensalism has been a matter of much speculation. Some hosts are carnivorous and yet tolerate guests that may have been the right size for a meal when they first arrived in a post-larval stage. It has been shown that the fishes that are commensal with the large sand sea-anemone acquire a certain 'immunity' to the anemone's poison by adsorbing the host's mucus on their skin, so that the sting cells are not discharged on contact. One wonders whether the acquisition of chemical camouflage is a general phenomenon in commensalism.

A transition to parasitism appears to have taken place in two genera of snails which live on echinoderms. *Capula* has a little pointed shell and lives freely between the spines of the pencil urchin *Prionocidaris* in the sea meadows. Another, *Thyca*, is firmly attached under the narrow arm of the green starfish *Linckia*, which is common on sheltered rocks of the reef flat; it buries its proboscis in the soft tissues of the under side of an arm. The dwarf male snail is attached to the female's head underneath its limpet-like shell (see Section 6.5.1).

The strategy in these snails contrasts with the free association of the commensal scale worms (polynoids) which live with starfish, sea cucumbers and brittle stars or in burrows of other worms. Many more examples will be encountered in the study of the shores in the following chapters in Part II.

2.2 INTERACTION OF ORGANISMS WITH PHYSICAL FACTORS

2.2.1 Tidal zones

The seashore has the most variable physical conditions of all ecosystems in regard to daily, even hourly, availability of water and oxygen and sudden changes of temperature and salinity. The habitats are delimited largely, but not entirely, by the daily changes of tides, which create belts parallel to the tide line that are exposed to air and to water alternately for different lengths of time. The resident plants and animals possess a wide variety of strategies for coping with the consequences of these tidal changes. Every species has its preferred conditions, its own niche into which it fits and where it may affect the survival of other species. Organisms remain in their zones despite the contrary forces of waves and currents and despite the dispersal of their spores, seeds, eggs or larvae by the sea. In fact, species may be so strikingly limited in their vertical distribution that a tidal zone is known by the name of the dominant organism present.

On exposed rocky shores, for example, the upper area has a 'littorinid' or periwinkle zone and below it a 'balanoid' zone, named after the barnacles there, followed by an oyster belt. These zones are almost universally present on such shores. It is less easy to define dominants on lower shores because of the variety of habitats. There are shallower and deeper pools, sloping shelves of rock, overhangs and crevices, shaded or protected rock and vertical or horizontal faces which create conditions for different associations. In addition many animals which are more numerous subtidally find shelter among the seaweeds or sea grasses on the shore (see Chapter 7 and Chapter 9).

The usual subdivisions of the shores caused by tides are summarised in Table 2.1 which gives the vertical heights of spring and neap tides on the northeast rocks at Inhaca Island. The percentages of time that organisms are exposed to air in these zones during the tidal cycles of a lunar month are shown (Brehout 1982).

Table 2.1 Tidal zones and hours of exposure of north-east rocks at Inhaca Island

Zone	Tidal Limits	Range (m)	% Exposure to air
Supratidal fringe	EHS – MHS	>3,4	>95
Upper shore	MHS – MHN	3,4 – 2,5	95 – 70
Upper middle shore	MHN – MTL	2,5 – 2,0	70 – 50
Lower middlE shore	MTL – MLN	2,0 – 1,3	50 – 25
Subtidal fringe	MLS – ELS	0,5 – 0,2	<5

EHS (Extreme High Spring); M (Mean); N (Neap); MTL (Mean Tidal Level); MLN (Mean Low Neap); ELS (Extreme Low Spring).

On the reef flat of the lower shore, zonation is masked by the presence of seawater underneath each slab of dead coral in which organisms from lower zones live in shade and shelter, including many associates of coral, but not coral itself (see Section 6.5). On sand and mud flats tidal zonation is obscured by the variation in the depth of the water table below the surface which may depend on the freshwater drainage from the land. This effects a change on sandy shores from the horizontal zonation of exposed rocky shores to parallel areas at right angles to the tide line. This is seen, for instance, in the separation of the bubble crabs on dryish sand from the sentinel crabs on wet sandy mud on the west shore (see Section 3.4). In even more exaggerated form this phenomenon occurs in a zoned mangrove on the upper shores of the southern bay where the trees and the fiddler crab species associated with them are in belts parallel to the meanderings of the drainage channel rather than to the shore line (see Chapter 4).

2.2.2 Environmental stresses

The tidal pattern, which results in the various degrees of exposure to air down the shore from high tide to low tide levels (illustrated in Table 2.1), creates five problems for organisms

(Newell 1979). These confront both invertebrates and seaweeds that had their evolutionary origin in the sea and the flowering plants that evolved from families established on land.

i. Rock and sand temperatures may exceed 45°C on the shores of Inhaca Island when directly exposed to sun and air. This in itself is not lethal to tropical organisms exposed on the upper shores since water evaporates from them. Evaporation has a cooling effect but exacerbates desiccation. The organisms also undergo a sudden change in temperature when the tide rises.
ii. The supply of water is intermittent so that the partial daily loss of water between tides threatens the osmotic equilibrium of tissues.
iii. The osmotic pressure of the sea is fairly constant but water trapped in the substratum, where the majority of organisms on a sandy shore live, varies between tides. The salinity of mangrove mud falls when fresh water enters, whilst in pools on sand and rock it rises with evaporation, yet cells require a constant osmotic pressure. This problem would be acute for flowering plants which depend on absorption of water by the roots (and by leaves in sea grasses) and on a flow of nutrients in water between cells.
iv. In sea water the oxygen content is around 6 ml per litre and air has 200 ml oxygen per litre. Methods of oxygen uptake in water are not suitable for use in air and vice versa. There are also periods between tides when interstitial water in the sand becomes depleted of oxygen.
v. The tides interrupt the feeding of animals whether it takes place under water or when the animals are exposed to air. Neap and spring tides interfere differently according to the tidal level of the habitat.

2.2.3 Responses of organisms to variables induced by the tides

Survival in the face of these environmental problems is achieved by seashore organisms by a combination of four methods.

i. Species in the tropics may be genetically programmed to tolerate a wider range of temperature variability and higher temperature than temperate species of similar genera. This implies that cellular enzymes have different optima, if other features are similar.
ii. Animals may avoid the stress by evasive behaviour which shields them from disturbing forces.
iii. Specific structures may have evolved to counteract certain environmental variables.
iv. Physiological mechanisms may be brought into play to meet the changing circumstances. The internal environment of an organism may be so regulated that the potential effects of changes are excluded or remedied.

The position of any organism in a zone on the shore depends mainly on the interplay of its four types of adaptation to the abiotic variables. The immobile seaweeds, sea grasses and mangrove trees are confronted by gradients of change in all physical variables. The epifauna, consisting entirely of crustaceans and molluscs on the upper shores, is similarly exposed. The infauna and meiofauna and the inflora (diatoms and dinoflagellates) normally have smaller ranges of temperature and water loss with which to contend, but they may be located where salinity varies. The substratum of the lower shores may become anoxic when the tide is out and the organisms there will be in danger of oxygen depletion although other stresses are minimised.

Animals and plants on upper shores tolerate high temperatures better than those on the lower shores, although the former live closer to their upper lethal limit. A striking example of this genetic determination of adaptation to degrees of temperature stress is seen among the fiddler crabs of the mangroves at Inhaca Island: the five species of fiddler crabs occupy fairly narrow preferred areas from the inner to the outer tree fringes. These areas differ in shade and water content and hence in their temperature maxima (Edney 1960). It has been shown that the heat tolerances of the different species match the different maxima in their habitats (see Section 4.6.1). The converse, a lack of tolerance to heat and evaporation is shown by the broad-leaved sea grass, *Thalassodendron ciliata*, which grows best subtidally. When it is exposed to air and sun for a few hours in the subtidal fringe during extreme low spring tides in the summer months and the wind is offshore, a band of dead brown leaves appears. The terminal leaves die because the limit of tolerance has been exceeded. This reaction limits this seagrass to the subtidal and to intertidal pools (see Section 9.1).

Excessive heat is avoided by animals on the upper shores. Grapsid rock crabs at high tidal level migrate down the shore a little distance and hide in damp crevices. Land hermit crabs and ghost crabs spend the day in burrows and feed at night when the tide is out. Worms of several phyla, bivalves and smaller crustaceans live entirely in burrows. The only algae which survive the heat and evaporation of high tidal level are tiny red seaweeds of the *Bostrychia* association growing in the shade of overhanging cliffs and in crevices among rocks, which thus avoid this problem.

There are several structural devices for preventing that last increment of temperature that might kill. The lighter the colour of carapace or shell, the greater is the reflection of solar radiation. *Nerita* shells above high tide are white; the periwinkle at the highest level is pale pink and the ghost crabs of the high sandy slopes are pink or grey-green. Heat loss is slightly increased by convection from a sculptured surface in comparison with a smooth one. Thus the *Nerita* snail highest on the rocky cliff has not only a white shell but also many deep concentric grooves whereas the middle shore species are dark and smooth. Oysters above mid-tide are whitish whilst mussels exposed on the lower shores are black or brown (see Section 7.4.2). Fiddler crabs have chromatophores in the skin which enable them to change colour from pale in their burrows to dark in the light shade of mangrove trees. Fiddler crabs in sunlight on the lower shore remain pale longer.

Many sea snails utilise evaporation of water to cool themselves: those that live in exposed positions have shells in which the last whorl is inflated and filled with water before the tide recedes and they seal the shell with an operculum and mucus. Then, with no physiological effort, this water slowly evaporates and the animal cools without loss of water from the tissues. This method of coping with heat stress is seen in periwinkles and top shells. Neither the shell nor the operculum is completely impermeable to water, but loss from living tissue is avoided. Bivalves and barnacles close their shells with strong adductor muscles and only a little water evaporates from the mantle cavity (see Section 7.2.2).

Some crabs living on the upper sandy or muddy shores have the remarkable ability to *recycle* water from the gill cavity, passing it over the carapace and back again to the gills. They are thus cooled when they emerge for feeding between tides during the day. To compensate for the loss of water by evaporation special quill-like setae conduct water by capillary action in grooves from damp sand to the gill cavity. There are a number of specific variations of this method which are described in Part II (see Sections 3.4.1 and 4.6.2).

A parallel strategy exists in mangrove trees and salt marsh succulents. They have water storage tissue in their leaves which lose water in the heat of the day and regain it at night. The leaves have a high rate of transpiration during the day through open stomata on both upper and lower leaf surfaces, but at night the salt excreting glands secrete salt and these tissues are recharged (see Section 4.4.1). Red seaweeds at high tide levels, which appear to dry out between tides, lose water only from the mucilage *between* cells. None of these patterns of cooling allow water to be lost from essential tissues, and mechanisms exist for replacing it either by temporary access to water in the substratum or when the tide rises (see Section 7.3.3).

Most intertidal organisms are osmo-conformers, i.e. they are able to tolerate small, temporary changes in salinity in their body fluids which lose or gain a little water whilst the proportions of ions remain constant. A common method of overcoming such daily osmotic stress without altering ionic concentrations is the adjustment of the concentration of non-essential amino acids in the free amino acid pool in cells. Amino acids may be transported in either direction through cell membranes to offset changes in body fluids. In dilute media these amino acids may be deaminated and ammonia is removed by excretory organs (see Section 4.6).

Animals living on the upper shore may, as far as possible, avoid coming into contact with diluted seawater by remaining in air between tides as do some snails and some crabs in the mangroves. They may adjust the osmotic pressure of body fluids by producing either more dilute or more concentrated urine. As salts are lost they may be replaced by salt uptake in the gills and by drinking sea water. Tropical periwinkles save water by excreting solid uric acid, made by an enzyme system from carbon dioxide and ammonia (see Section 6.2.1). The various degrees of ability to withstand salinity changes are closely correlated with the natural distribution of organisms described in Part II.

There are two types of physiological adaptation to salinity stress in mangrove tree species at Inhaca Island. Three of the four species exclude salt physiologically at the roots and so minimise the problem. The other, *Avicennia*, absorbs salt but it does not accumulate salt in the sap beyond a tolerable level because special glands in the under surface of the leaves secrete salt at night. There are also some cellular mechanisms which protect the cells during the passage of sap through the tree (see Section 4.4). The seedlings are particularly protected since they start developing while still attached to the parent tree but are separated from its sap by salt-proof cells.

The sea grasses are able to avoid the salinity of the sea by actively pumping out salt from the epidermal cells of the roots, although the genera vary in their tolerance of dilution, a factor which partly determines their distribution (see Section 9.1).

The essentially aquatic seashore invertebrates are exposed to air for many hours unless they lie buried in wet sand. The simplest use of air for aquatic respiration is aerating the water which is normally used for ventilating the gills. Periwinkles and barnacles share the habit of sealing their shells and apparently exclude air when the tide is out, but each retains a bubble of air inside the shell aerating the water kept in the mantle cavity. Ocypodid crabs possess a similar strategy when they plug their burrows with a pellet of sand or mud before the advancing tide reaches them. As the water rises from the water table below, a pocket of air is trapped inside the closed entrance to the burrow (see Section 3.4.1).

Upper shore crabs use air for respiration directly by means of structural modifications of their respiratory organs. They have fewer gills than do totally submerged crabs and these are less permeable to water. In compensation, part of the respiratory cavity wall is strongly vascularised and acts as a lung; some crabs have air windows in the chitinous covering of their legs (see Section 3.4). The same principle is used by the mud-skipping goby which immediately blows up its gill cavity with air when it emerges from water to feed on land. The habit of recycling water is developed in sesarmid crabs to include aeration as the water flows over minute reticulations on the ventral carapace (see Section 4.6.2).

On the lower shores many of the burrowing worms make tubes of mucus that, being sticky, become reinforced with sand or mud. The worms live within the tubes, the ends of which remain open when the tide is out. The water is kept circulating by muscular contractions of the body wall or its appendages or by ciliated surfaces. Diffusion of air into the water in the burrows prevents complete oxygen depletion. The tubes are relatively impermeable and so insulate the animals from the black anoxic mud. Often the surface area of burrowers is increased by small gill-like appendages rich in blood capillaries through which gas exchange is augmented (see Sections 3.4.2 and 9.4.4).

The blood or coelomic fluid of burrowers contains respiratory pigments which in most cases are useful during oxygen shortage as well as during normal activity. The pigments (haemoglobin, haemocyanin or haemerythrin) load oxygen at fairly low ambient levels and unload it at extremely low oxygen tensions. In other words these respiratory pigments are biochemically adapted for working at the very low concentrations of oxygen encountered during low tides in muddy sand. They may act as stores of oxygen to be used in daily tidal predicaments.

All these precautions in burrowing animals for postponing the effects of a lack of environmental oxygen may be insufficient during low spring tides. Many worms live in collapsible burrows and bivalves close their shells completely. As a last resort, there are biochemical pathways which allow these animals to live temporarily without oxygen as do the muscles of terrestrial vertebrates when they engage in short-term intense activity. Seashore invertebrates do not use the same biochemical pathway as vertebrates, which results in panting during recovery to repay the oxygen debt incurred in the production of lactic acid anaerobically. Instead the carbon dioxide which accumulates in the tissues of invertebrates lowers the pH in cells and interrupts the tricarboxylic cycle of the respiratory pathway so that intermediate products accumulate. These may be excreted or, more usually, they are fed back into the normal cycle later.

Finally, periods of reduced activity imposed by the tides on all shore animals, either when submerged or between tides, reduce the demand for energy at just that time when oxygen is least available.

The flowering plants on shore, true to their terrestrial ancestry where oxygen was abundant when they evolved, maintain a flow of air through their organs both above and below ground. *Avicennia*, the mangrove tree which occurs on both the landward and seaward fringes of the mangrove associations has a host of pneumatophores, that is pencil-like air-breathing roots, which grow 20-30 cm vertically upwards in air from underground radial branch roots. Other mangrove tree species have prop roots or knee roots which serve a similar purpose. These are covered with lenticels from which leads a network of air capillaries. When water covers these modified roots the air pressure has been shown to drop until the tide begins to fall. Then once again air rushes in. The sea grasses also have capillary-like air passages throughout their tissues although they are less exposed to air. They obtain carbon dioxide from solution in the sea and oxygen produced by photosynthesis remains in these passages and is used for respiration (see Section 4.4).

The physiological mechanisms involved in all these responses to variable conditions can themselves vary. They may be upgraded or downgraded over periods of days or weeks by a process of slow adaptation called acclimation.

The tidal rhythm affects the times available for feeding since some animals feed when the shore is exposed to air and others only when covered by water. Some filter feeders have less time for feeding than individuals of the same species which live slightly lower on the shore; these have been shown to accelerate their feeding movements. The faster rate has been triggered in the laboratory in a simulation experiment by exposing the slower ones to air for a longer time before submersion. Some detritus feeders penetrate further down the shore if the detritus is richer there and then make do with a shorter feeding period. Molluscs and crabs which prefer hunting under water can be seen emerging from their superficial burrows as the tide rises and travelling inshore where prey may still be buried.

The distribution of seaweeds is limited by the necessity for exposure to light and by the absence of roots for absorbing nutrients. The first effects of the receding tide on seaweeds are a rise in temperature and an increase in the amount of light to which they are exposed. The response is an increase in the rate of metabolism, including the enhancement of photosynthesis, which can continue in air as long as there is a surface film of water in which carbon dioxide dissolves. This water is available for a limited time because of evaporation from the seaweed. The rate declines when the water loss exceeds 20% on average, and it is completely inhibited above 60% loss because of interference with electron transport in the chloroplasts. Thus growth is temporarily inhibited, but recovery from desiccation, i.e. loss of water *between* the tissues from mucoid substances, is complete when the tide returns. Similarly, since the general surface of seaweeds absorbs nutrients, uptake is impossible when they are exposed to air and are dried out. These two responses to tidal factors prevent seaweeds from succeeding on the upper shores *in the tropics*. The chloroplasts in both seaweeds and sea grasses are best able to absorb light for photosynthesis when covered by sea about one metre deep. Green, red and brown seaweeds have a variety of accessory photosynthetic pigments which facilitate light harvesting (see Section 7.3.3).

2.2.4 Periodicity

Many of the reactions of animals to stimuli on the sea shore, which appear to be direct responses, are, in fact, timed to respond at the right moment by genetic control mechanisms located in the central nervous systems (Brady 1979). The response appears to *anticipate* the stimulus; for example, the burrowing bubble crab closes its burrow *before* the tide reaches it (see Section 3.4); the carnivorous snail *Volema* starts hunting before other molluscs emerge on the sand flats. Moreover, when the animal is removed from the seashore and kept in a suitable constant environment, periodic behaviour is still displayed. Colour change in fiddler crabs, for example, is exhibited in synchrony with external factors for many days, sometimes for weeks, in the laboratory. The periodicity is controlled by a biological clock, a mechanism which is inherited and located in a centre in the nervous system in animals (see Section 4.6.1).

The simplest organisms on the sandy shores of Inhaca Island which exhibit periodicity are dinoflagellates (see Section 7.5.1). Biological clocks are evident in the flatworm, *Convoluta macnaei* (see Section 3.4.1) and in species from most phyla including molluscs, crustaceans, fishes and birds.

Not all responses are primed in this way. Sessile barnacles and mussels react to immediate stimuli such as being wet by the sea. It is said that even the sophisticated ghost crab emerges from its burrow when it senses the rise of the level of water in the exit tunnel of its burrow. Most organisms blend the direct response to a particular stimulus with neurally controlled periodicity. The fine tuning of clock-controlled behaviour involves reception of external stimuli and direct responses which modify the inherited stable periodic system. For example, the land crab *Cardisoma carnifex*, which burrows at the landward edge of the mangrove swamps, normally emerges to feed in darkness at night (see Section 4.5.1). If the morning is cloudy and overcast it will remain above ground and continue feeding, allowing the environment to slow down its diurnal biological clock. The rhythmic genetic control centres govern behaviour in general, in relation to light and dark, high and low tide, spring and neap tides. The clock mechanism thus contributes to the maintenance of specific intertidal positions on the shores to which the physiology and morphology of an animal is adapted.

Some of the responses of organisms to tidally induced abiotic factors on the shores, which determine survival and distribution, have been outlined here and will be amplified in the relevant chapters in Part II. The questions of how organisms find the correct zones, and manage to stay there, are discussed in outline below.

2.3 MAINTENANCE OF POSITION ON THE SHORES

Although the position of a particular species is governed primarily by responses to abiotic factors, its constant recurrence there involves a number of interacting patterns of behaviour which include reproduction, larval history and adult habits (Newell 1979). Competition between species plays a minor role.

2.3.1 Reproductive practices

It is rarely the direct responsibility of the parents of species on the shore to ensure that their offspring live in an optimum environment – a common practice on land, among insects for example. The majority of shore animals release their eggs and sperms into the sea in synchrony with seasonal temperature change or intensity of moonlight. Fertilisation is commonly external, depending on the specific recognition of eggs by sperms. Larval development usually takes place in the sea and the choice of habitat on shore is made by the larva before metamorphosis into an adult, in barnacles, worms and tunicates, for example.

Structures for internal fertilisation have evolved in some groups and among these are a few cases of incipient parental care of the young. False limpets, for instance, lay eggs in patches of jelly which adhere to rocks in pools where adults live and the larvae hatch there, some remaining to metamorphose. Some carnivorous snails lay yolky eggs in masses of impermeable capsules (see Section 3.5.3). Some dog whelks brood their young and so does the mangrove periwinkle for a short time. Female crabs 'in berry' carry eggs attached to abdominal appendages for some time. When the free larval life is shortened by hatching at a late stage the choice of site for settlement may be narrowed. Only rarely is the larval stage omitted, as in amphipods at high tidal level, in one species of brittle star (see Section 7.5.3) and in hydroids (see Section 9.4.5).

An exception to the relatively minor role played by adults in maintaining position on the shore is shown by a large number of 'colonial' animals, such as hydroids, zoanthids, corals, bryozoans and tunicates (see Section 7.5.2). These become compound animals by repeated asexual reproduction, that is through the growth of buds which remain attached and spread over the substratum. The mass of individuals produced is genetically related, structurally connected and physiologically co-ordinated, so that they are masters of territorial stability. They may also have free ranging larvae, which start colonies elsewhere when they metamorphose. Some tunicates retain larvae within a body cavity of the parent until they are ready to metamorphose into the adult form (see Section 6.4.3).

Surprisingly perhaps, most animals on the *upper* shore have numerous pelagic larvae, but these have a high mortality rate; there is, however, some compensation. The tactic is successful in this inhospitable environment as it allows time for mixing a large number of genetically variable larvae from individuals of one species over a wide area of inshore water. This ensures that at least some will be able to survive when they settle. On the *lower* shore, where microhabitats are more heterogeneous and conditions are less hazardous, there is a higher species diversity and populations are less dense. In this situation some species succeed by having smaller numbers but more protected offspring or asexual reproduction may be substituted for sexual reproduction.

Some degree of protection of young plants, which restricts dispersal, has evolved in one sea grass and three species of mangrove trees. Seedlings of *Thalassodendron ciliatum*, the sea grass at the lowest intertidal level, start developing whilst attached to the female plant. When released each is still supported by a floating parental bract which becomes entangled in other plants so that movement away from the habitat is restricted (see Section 9.2). Three mangrove tree species produce large torpedo-shaped seedlings (in which the hypocotyl has grown) while still on the tree. At abscission they drop vertically, pierce the soft mud below and almost immediately put out supporting rootlets so that they will grow in the same habitat as the parents (see Section 4.4). Similarly, seaweed spores often adhere to the substratum within minutes of release through the secretion of gum (from the golgi bodies) even before rhizoids are formed for attachment.

2.3.2 Larval adaptations for distribution and settlement

The larvae of many phyla of invertebrates are planktonic and constructed on a similar plan (see Section 8.4). They are microscopic in size and have means of staying afloat, such as bands of cilia or small setous appendages. They are equipped with sense organs sensitive to light, to the force of gravity and to variation in pressure. Movements are co-ordinated by small nervous systems, in which programmes have evolved to maintain the larva near the surface of the sea, but not at the surface. Larvae are positively attracted towards light and this response is reinforced by a negative reaction to gravity. The movements of larvae are therefore not random, since their swimming is directed upwards and they move towards the shore by taking advantage of the tidal currents or wave action to which they are willy-nilly subjected. The length of larval life varies in different species from 2-6 hours (when they do not feed) to weeks or months when they are adapted to feeding on plankton in the inshore waters. Some tropical planktonic larvae are reputed to be even longer lived. As the larvae grow their responses to light, gravity and pressure change so that they are brought inshore by the tide. They may sink or float according to the position of the future habitat underneath or on the surface of rocks or sand. They seek their appropriate micro-habitats by swimming or creeping, and their settlement is far from accidental for they develop new sense organs which detect features of their new habitats.

In many species that have been investigated there are three phases in the choice of habitat: temporary attachment, exploratory behaviour and settlement. At first, larvae in contact with a suitable substratum secrete a soft glue or fine sticky threads, which prevents them from being removed by the tide. Then each larva explores a limited area, inspecting and testing it with the new sense organs. The strongest stimulus for the settlement of barnacles, for instance, is a chemical secreted by their own species, which results in their crowding on to an existent zone (see Section 7.2.2). For worms the stimulus is the odour of the bacterial film on the detritus on which adults normally feed, rather than the size of the sand particles (see Section 3.4.2).

Molluscs may discern (by contact) the texture of the rock face. During the exploratory phase the larva may creep around without feeding, whilst the adult gut, sense organs and other structures are developing. Before finally settling on rocks, barnacle or oyster larvae turn around in all directions, as if estimating the potential space for growth unhindered by the proximity of others. If the larva is not satisfied that there is a suitable berth it will release the temporary hold and swim or creep away – unless the time programmed for exploration has expired. When the larvae have settled, metamorphosis takes place and adult means of fixation are used.

2.3.3 Adult behaviour

Some adult animals are as sedentary as plants, those with shells such as barnacles and oysters are permanently attached by a cement made of calcium carbonate and an organic adhesive. This method is also used by corals, serpulid worms in calcareous tubes and the strange worm-like mollusc *Dendropoma* on the middle levels of exposed rocky shores (see Section 7.3.4). The giant clam in the coral reef is at first held by byssus threads; as it grows it abrades the surrounding coral-limestone rock with the edges of its shells and thus embeds itself securely. It uses a rocking movement by alternately contracting the anterior and posterior pedal muscles (see Section 8.5.1). Sea urchins secure themselves in crevices by abrading the rock with teeth and spines. A tiny polychaete worm, *Polydora*, drills vertical burrows into the coral debris on the reef flat, curling up side down to use a row of strongly hooked setae on each side of the 'shoulder' segment (see Section 6.4.3). The 'date-stone bivalve' *Lithophaga* almost completely buries itself in coral limestone by using a chemical method: the mantle edge secretes acidic mucus which slowly dissolves more calcium carbonate as the animal grows (see Section 6.4.2).

The most elegant chemical method of attachment and burrowing is found in sponges. Amoebocytes (single cells) migrate to the external surface of the sponge and acid-etch minute chips of calcium carbonate from a rock or coral which are first outlined and then undercut. The chips are engulfed by the amoebocytes which then work their way to the inner layer of the animal where they are released into the exhalent current. A boring sponge will fill the cavity it makes in shell or rock with growing tissue (see Section 8.5.1).

A sedentary habit may be only a part-time occupation for some species when the tide is out or for others when the tide is in. Crabs may shelter in cracks and crevices when exposed to air, holding on to the rocks firmly with pointed dactyls on the ends of their legs. Snails that do not have an operculum, such as limpets and false limpets, graze on the rock coating of algae whilst it is still wet and then return to their home 'scars' to clamp themselves down as the surface becomes dry. The foot muscles become rigid and considerable surface tension is exerted through the very thin layer of mucus secreted between the surface of the foot and the flat rock. Periwinkles have an operculum and rely on drying mucus for attachment when the tide ebbs. They migrate some distance down the rocks for optimum feeding during neap tides and up again for spring tides (see Section 7.2.1).

The majority of animals on the sand flats and in mangrove swamps construct burrows and remain underground temporarily or permanently, almost all phyla that occur on the shores being represented. Soft-bodied worms of all kinds are almost cylindrical in cross section and have strong pointed ends, an ideal shape for burrowing in sand (see Section 3.3.2). Bivalves possess a spade-like foot and a wedge-shaped shell that cleaves the sand easily (see Section 3.2.1) and burrowing snails (see Section 3.7.1) have a specialised foot for digging. Burrowing crustaceans use a variety of adaptations of their articulated limbs, different species digging either forwards or backwards according to the consistency of the sand or mud (see Section 9.5.2).

All burrowers have three components in common: a penetrating organ, an anchorage system and a contractile procedure (Trueman and Brown 1992). Those without a hard skeleton for the attachment of muscles use the fluid enclosed in body cavities or blood spaces to give support for muscular action (see Section 3.7.1). Crustacea support their muscles by internal prolongations of the exoskeleton (apodemes) which give efficient leverage to their jointed legs. Echinoderms have evolved a hydraulic system of hundreds of tiny tube-feet which protrude through pores in the hard exoskeleton (see Section 9.4.1). These adaptations to burrowing and many more are described more fully in various chapters in Part II.

Flowering plants, as expected, are anchored by their roots. Those of some mangrove species are so designed that they stabilise the tree's growth in soft mud by means of extensive strong horizontal roots like cables, bearing small anchor roots not far beneath the surface mud. Seedlings and some mature mangrove tree species have buttress roots visible above the surface (see Section 4.4). Sea grasses are secured by many fine fibrous roots which grow from long rhizomes and become entangled in the sand.

Among seaweeds, as soon as sporelings have become stuck to rock, rhizoids, which produce mucilage, form attachments. Similarly, diatoms too produce mucilage for attachment to sand grains. The permanence of seashore communities depends therefore on every species knowing its 'place' in the ecosystem, using specific strategies to find its niche and to remain where it is best adapted.

2.3.4 Competition

On upper shores animals do not usually engage in interspecific competition in the harsher abiotic conditions to which they are adapted. The lower shore organisms cannot survive there even if they do, by chance, invade it. Mangrove trees do, however, compete for light. The pioneer tree is *Avicennia* which first stabilises the sand, as is happening now in the northern bay. When seedlings of *Bruguiera* grow in the new area they soon become taller than *Avicennia* and they produce shade. *Avicennia* seedlings cannot tolerate the shade of other trees, thus their adult distribution is limited to the margins of the mangrove association on both the landward and seaward fringes and the faster-growing *Bruguiera* occupies the interior (see Section 4.4).

Several examples of competition for space occur on the lower shores. Seaweeds on intertidal rocks have the advantage of being able to secure themselves on bare surfaces by means of rhizoids. When however the rocks become coated with a layer of sand the algae are abraded but the sea grass *Thalassodendron* can hold fast and grow, displacing the seaweeds. At Inhaca Island some pools which, thirty years ago, were rich in seaweeds are now populated by this sea grass. Sloping shelves of rock are gradually colonised by several species of zoanthids (compound anemones) which by repeated budding of the tough polyps can oust algae from rocks (see Section 7.5.2).

There are certain areas of rocks at mean low spring tides which are completely covered by a reef, often 30 cm thick, built entirely of coarse sand by the tubiculous polychaete worm, *Idanthyrsus pennatus*. Seaweeds do not usually germinate on these worm-tubes, perhaps because the feeding currents of the worms remove algal spores or the surface is too coarse for algae to settle (see Section 7.5.5).

Competition between deposit feeders and filter feeders is said to exist in sand and mud, since the detritus feeders remove the settling larvae of the suspension feeders. For example, the presence of a large population of the bubble crab, *Dotilla*, appears to exclude most animals through the daily disturbance of the surface of the sand. A concentration of this species appears to change the nature of the localised environment so that other species are excluded. This is one reason for the patchiness of distribution among organisms on Inhaca Island in which some species have excluded others (see Section 3.4.1).

In the stable communities on tropical seashores the large number of microhabitats is matched by the high species diversity. Animal species have very particular food requirements although they may overlap spatially. Specificity of prey for the predator, the choice of particle size for the filter-feeder, the division of detritus components among burrowing animals and

Fig. 2.2 Speciation in flatworms from the coral reefs and reef flats at Inhaca Island.
a. *Pseudoceros zebra*, Backround yellow, median band white, black bars, margin orange.
b. *P. dubius*, dorsal surface reddish brown, scattered yellow streaks, narrow black marginal band, ventral surface pinkish (approx x 1) (from Prudhoe 1989).

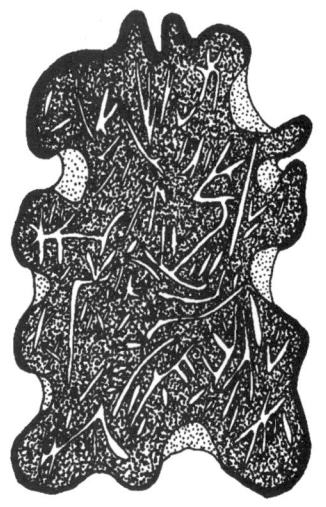

the preference for one plant species as food are shown to be correlated with the equipment of the animal concerned.

2.4 SPECIATION ON TROPICAL SHORES

The multiplicity of species in many tropical genera which inhabit the warm sands, the reef flats and coral reef is striking. A rapid scan of Appendix A, which lists species found at Inhaca Island in taxonomic order, will enable one to recognise the high degree of evolution in particular genera. Among coral reef fish for instance, there are 18 species of the butterfly fish *Chaetodon*. There are 6 species each of the carnivorous polychaete worms *Glycera* and *Eunice* as well as several species of the brittle stars *Ophiactis* and *Ophiocoma* among echinoderms. There are five species of the fiddler crab *Uca* in the mangroves. In the coral reef itself, *Acropora* exhibits much speciation.

A recent study of an earlier collection of flatworms from the reef flats and coral reefs, which is accompanied by skilful paintaings of live material, appears to reveal even more intensive speciation of one genus, *Pseudoceros*, than in any other phylum (Prudhoe 1989). Twenty-six species have been described which appear to differ mainly in colour and pattern (see, for instance Fig 2.2 a and b). The flatworms possess black, red, blue, yellow and possibly white pigments which may produce shades of purple, mauve, green, orange, brown and grey. Patterns are formed in median or peripheral bands, streaks and spots, large and small. Nine of these species have been reported elsewhere from the shores of the Indian Ocean, from the Red Sea to the Great Barrier Reef, including Sri Lanka and Indonesia. These have a similar *pattern* to some at Inhaca Island but they differ in shades of colour. What role does the colour or pattern play? If the animals in question were Birds of Paradise in New Guinea, a function in Specific Mate Recognition could be proposed. However, there is no evidence that flatworms can distinguish colours or patterns, nor is it known whether they can change colour.

These flatworms have many simple, cup-like eyes containing a small group of pigmented sense cells. The minute eyes in this genus are scattered over the tentacles (including their bases) and over the cerebral ganglia in species-specific patterns. Sense cells sensitive to colour have not been found, despite the existence of exotic colour combinations.

Is it perhaps conceivable that the definite arrangements of simple eyes in different species do recognise the *patterns* underlying the colours of their mates, but not the actual colours? There are very few, if any, other consistent differences between the 'species' of this genus, except perhaps in adult size. All are hermaphrodite and have internal fertilisation. The reproductive organs are relatively simple and do not differ within the genus. The flatworms may be carnivorous or scavengers; their sense of smell, located in anterior ciliated pits, is well developed. Feeding data of the Inhaca species have not yet been studied nor are their microhabitats defined. However, it is well known that among freshwater flatworms different species, co-existing in the same habitat, have specific diets. The flatworms are found underneath coral debris on the reef flats and in the living reef. This speciation presents an intriguing evolutionary problem.

In Part II, Chapters 3-9, some of the above ecological features are described in detail with regard to particular species. In other cases details of adaptations are not well known and challenge the reader to continue investigations.

3

SANDY SHORES ON THE SHELTERED AND EXPOSED COASTS

3.1 MAJOR SANDY HABITATS

Tropical sheltered, sandy shores are rich in both species and numbers of animals, most of which live in burrows that minimise the risks of desiccation and exposure to high temperatures when the tide is out. Some animals take advantage of the shelter of flowering plants such as the sea grasses on the lower shores. A few animals are active on the sand surface by day and many more may be seen in the shallow water as the tide ebbs and flows. Animals are more active at night and when the tide is high, but they leave signs of their activities above ground or underground, enabling the naturalist to discover their identities and modes of life.

The density of visible animals is greatly exceeded by those of the protists and meiofauna which live between the sand grains. The wealth of animal life as a whole depends on the microscopic interstitial flora and plankton, on the detritus formed from the sea grasses and very much less on the living macroscopic plants.

Tidal movements of water, together with wave action of varying degree, alter the nature of the substratum on successive levels on the shores. The sizes of particles and the mixtures of sand, silt and detritus resulting from water movements control the drainage and the level of the water table so that physical intertidal zones are distinctly defined. Accordingly, different communities of plants and animals occur in these zones and are adapted to the special conditions that prevail there.

The sandy shores of Inhaca Island differ in their degree of exposure to wave action: fully exposed, less exposed, sheltered and very sheltered. The first three types of shore are discussed in this chapter (excluding the subtidal fringe described in Chapter 9); the very sheltered shores are described in Chapters 4 and 5.

3.1.1 The sheltered west and south-west coasts

Sandy substrata predominate on the west and south-west shores, facing the Bay of Maputo but there are also hard substrata at various tidal levels, including upper shore rocks, reef flats and coral reef. For the location of habitats a convenient point of reference is the Marine Biological Station (Fig. 1.1). It is sited on the only flat foreshore on the island (apart from the fishing harbour), 50 m from high tide mark below a high dune. The station stands about halfway between the north-west rocky point at the fishing harbour and the south-west rocks at Ponta Punduine. A recurved sandy spit, Ponta Raza, 1,5 km south of the station extends to low tide mark and beyond to sandbanks on the subtidal fringe (see Chapter 9). A narrow mangrove creek hidden behind a low dune has its exit on the south-west shore immediately south of Ponta Raza.

The west shore is 6 km long and extends seawards for about 400 m in front of the station, the upper shore is a sandy slope 25-30 m wide and the middle and lower shores are almost flat. The shore ends about 50 m before a steep slope to the Inhaca Channel, which is 8-15 m deep. Several habitats may be easily distinguished. Below the upper sandy slope, a belt of sandy mud about 50-100 m wide spans mid-tide level; this soon becomes hard when the tide recedes, as the water table is 10 cm below the surface. On the next 200 m of sand flat down to mean low neap tide level, there are two alternative habitats: some areas, where the water table is just below the surface, are parallel to other areas where the sand is cleaner and drier, the water table being about 15 cm below the surface. These areas are occupied by different communities. Sea grasses cover the ground below mean low neap tide level, at first sparsely and then almost continuously beyond mean low spring tide level.

The south-west coast between Ponta Raza and Ponta Punduine is about 2 km long. Most of the upper shore is a rocky cliff; the lower shore is sandy and sea grasses may reach almost to the foot of the cliff.

The west shore facing the Bay of Maputo is moderately sheltered from wind and usually has low waves that break on the upper slope and to a lesser extent on the coral reef at extreme low tides; the sand flats have very little wave disturbance. The salinity of the sea varies daily from 30-34 g per litre and the water temperatures range from 18-35°C during the year.

3.1.2 The exposed east coast and less exposed north-west coast

The sandy shore on the east coast is 12 km long and, in contrast to the west coast, is an almost uniformly steep slope of coarse yellow sand, extending from the high dunes to rocks at low tide. This shore faces the Indian Ocean and is exposed to moderately strong wind and wave action. The salinity of the sea is about 35 g per litre and the temperature varies little from 25°C.

West of the northernmost rocky point, Ponta Mazondue, wave action decreases gradually as the north-west shore becomes increasingly sheltered by the large, subtidal sandbank that protects the northern bay (see Fig. 1.1). Below the dunes, the short steep upper shore has less coarse sand and the lower shore is rocky. The association of animals on this short strip of north coast sand differs from that on the east coast.

3.2 THE SANDY SLOPE OF THE SHELTERED UPPER SHORE ON THE WEST COAST

Along the whole of the west coast the upper sandy slope of the shore extends from the foot of the dunes almost to mid-tide level. It has a gradient of about 1 in 30 and analyses of sand particle size using a series of sieves showed 25% coarse sand, 65% medium grade and 10% fine sand. This composition makes the sand superficially dry between tides, but it is damp a few centimetres below the surface down to the water table, which is over a metre below the surface at the head of the slope. The sand is thus suitable for vigorous types of burrowing animals.

High spring tides in the early mornings and late afternoons now reach the foot of the dune forest, whereas up to about five years ago there was a band of sand between the highest drift line and the trees. This supratidal fringe has gradually decreased in width and is now almost obliterated. Deposition of sand has altered the slope of the beach; the sand-binding plants such as *Scaevola*, *Canavallea* and *Ipomoea* no longer grow there. Consequently the land hermit crabs which dug burrows in the sand under these plants for daytime shelter can no longer be found there. They can be seen at Ponta Torres, the south-east point at the foot of the dunes where these plants occur (see Section 5.5.1).

The drift line is barely marked except after storms when sea grass leaves, dead sea cucumbers and sand dollars are found.

3.2.1 Ghost crabs and wedge mussels

There is little sign of life on the sandy slope by day, but after sunset hundreds of holes leading to the burrows of the greyish-green ghost crab *Ocypode ceratophthalmus* become visible (Fig. 3.1a)

(Barras 1963). The crabs gradually emerge after dark, mend their burrows and forage at the driftline on the sandy shore for a time. They then migrate in hordes to the tidal flats of the lower shore where they hunt other crabs and prawns for a few hours. Adult crabs have carapaces 40-60 mm in length and the number on the west shore has been estimated as many thousands. The crabs return before the tide reaches the sandy slope and dig down almost vertically for 20-30 cm, using two legs on the side opposite the larger chela. They scrape sand over the top and close the burrow before the tide covers them. During the spring tide phase, burrows are dug higher on the slope but are always covered at high tide. In the neap tide phase burrows are dug lower and lower, leaving those of the spring tide phase open and empty. Juvenile ghost crabs tend to dig their burrows at higher levels, and some leave them during the day while they race down to the lower shore to hunt and to renew the water in the gill cavity. These crabs are seldom caught because their escape reaction is swift and their cryptic colouring enables them to vanish in depressions in the sand. On unfrequented shores in northern Mozambique and Tanzania the adults of this species hunt by day and a few that occur on the deserted east coast of Inhaca Island also do so.

Ghost crabs have a vaulted carapace that enlarges the air space above the gills in the respiratory cavity beneath it. Their large oval brown eyes are on long, tough eyestalks that may be held erect to extend their range of vision on the flats, or may be folded horizontally into orbits for protection while they burrow (Fig. 3.1b). Adults can perceive objects 100 m away and their near sight is acute. In the large family of Ocypodidae only *O. ceratophthalmus* has eye stalks which continue growing beyond their eyes as the crabs become adult. This suggested the specific name 'ceratophthalmus', meaning horny-eyed.

Ghost crabs are versatile in their feeding habits (Hughes 1966): they may scavenge along the driftline eating carcasses as large as sea cucumbers and they are fast enough to catch flies. When they hunt, they chase other crabs and pounce upon them, securing sentinel crabs (*Macrophthalmus*) or even carnivorous swimming crabs as well as the slower prawns. They dig up wedge mussels in the sandy slope (see the next sub-section) perhaps recognising the little depressions left in the wet sand by the burrowing mussels when the waves recede. Unerringly the crabs secure the shell in one swoop with two legs of one side or, if the shells are too deep, they use two legs on each side alternately. The shells are carried back to the crab's burrow entrance and crushed in the larger chela or, if too big, they are prised open after chipping the shell. The discarded shells accumulate there as 'kitchen middens' until washed away by spring tides. The ghost crabs supplement their diet by shovelling sand into their mouths with the minor chela and they probably digest the microflora and meiofauna, as their stomachs contain only sand. Adults may also be cannibalistic but the juveniles attain a greater speed than that of the heavy adults and few are about at night.

The top running speed in juvenile ghost crabs is, in fact, the highest known among crabs, reaching 2,1m per second (Burrows and Hoyle 1973). They have long legs relative to their weight and raise the body high off the ground so that their legs can move directly underneath the body instead of sprawling at the sides. Records show that each step may be longer than twice the length of a leg, suggesting that the young crab actually leaps, and photographs have confirmed that all legs are off the ground during the leap. They run sideways as do other crabs and when they run fast they do not use eight legs: the middle leg on the leading side alternates with legs 2 and 4 on the other side. At the highest speed the crab tilts its body up so that only two legs on one side alternate with each other in running. Their enemies are birds such as waders and carnivorous crabs, including their undiscriminating parents.

Ghost crabs can breathe in air, provided that the respiratory cavity is kept moist. They often interrupt their foraging on the lower shore by adopting a curious squatting position for a few seconds, choosing an area where there is still surface water left as the tide recedes (Hughes 1971). This posture enables them to refresh the moisture inside the respiratory cavity. Fine stiff hairs around the small apertures of the respiratory cavity at the bases of the third and fourth legs are used to suck up water by capillary action. High suction pressure is created by the vibrations of the scaphognathites (Fig. 3.1c). These are a pair of flexible fan-like appendages in the mouth cavity, each guarding a ventro-lateral aperture of the respiratory

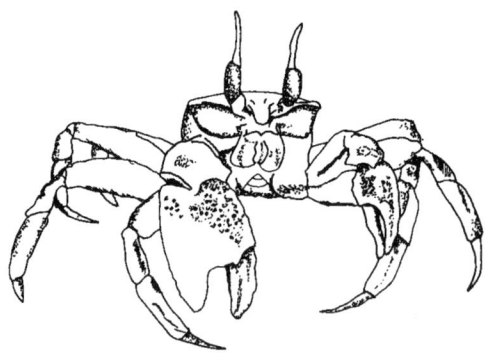

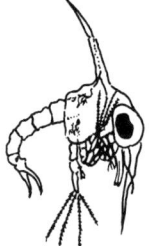

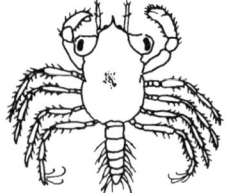

Fig. 3.1 Green ghost crab, *Ocypode ceratophthalmus* (x 0,5):
a. running (after Barras 1963);
b. dorsal carapace*, eyes resting;
c. respiratory fan, scaphognathite;
d. chela with stridulating organ (arrow)
e. larvae: zoea and
f. megalopa (x 20) (c-f after Barnard 1950).

cavity. The beat of the scaphognathites is reversible so that air may be drawn in anteriorly for aerial breathing. Inside the respiratory cavity there are fewer gills than in truly aquatic crabs and they are stiff and non-sticky so that they do not collapse in air. Richly vascular lung-like tufts grow out from the lining of the respiratory cavity and protrude from its roof, providing the large surface area required for aerial respiration.

In the breeding season males dig spiral burrows over one metre deep, marked by little pyramids of sand which have been considered mating signals (Hughes 1971). The burrow openings are 1-2 m distant from one another as a result of the threatening behaviour between two adult male crabs which happen to commence digging closer. The two crabs face each other standing vertically on tiptoe on two pairs of outstretched legs, with chelipeds extended forwards. They both perform paddling movements with the chelipeds: up, down and back, repeated rapidly at intervals. If they are the same size no clash occurs and they bypass each other, paddling their chelae. If unequal in size, a lightning clash may occur and the smaller one retires or is sometimes killed and eaten. When inside the burrow, a crab may signal occupation by tapping on the sand or by stridulating. Both sounds carry for about 10 m underground and have been tape-recorded. The territory is thus maintained without expending much energy in fighting, and partitioning of the sandy slope among the dense population is achieved.

When the males stridulate in the opening of their burrows the attention of females is caught. They make a low-pitched sound: krr – krr – krr – krr – krri – krr by rubbing the tubercles on the 'hand' of the chela (Fig 3.1d) against the ridge on the second segment of the same leg (Hughes 1971). Sounds are amplified by the flat sheets of thin cuticle on the two faces of the third segment (merus) on each leg. These vibrations are detected by the sensory cells in the myochordotonal organs. The sensory cells are situated on the very slender apodeme (internal cuticular skeleton) in the joint between the second and third segments of each leg on which the slender, accessory flexor muscle of the third joint is inserted. A female crab, encouraged by the paddling movements of a male standing in the opening of the burrow will follow him underground where copulation takes place. The female then lives in a side tunnel which the male has prepared, where fertilised eggs, attached to the setae on her abdominal appendages, are incubated. When the zoea larvae have two pairs of biramous thoracic legs as well as head appendages they are set free in intertidal water (Fig. 3.1e). After several moults the zoea becomes a megalopa larva (Fig. 3.1f). It swims near the surface of the sea for a time, then sinks and is deposited on the shore by waves.

The success of ghost crabs in the difficult habitat of the upper sandy shore depends on synchronising their burrowing and foraging behaviour with three sets of external factors: light and darkness, ebb and flow of the twice-daily tides and the weekly succession of spring and neap tides. Awareness of tides is considered to be the result of learned responses to the daily variations of a simple stimulus, namely the height of the water in the burrow, rather than by an inherent (nervous) rhythm. In addition, ghost crabs utilise complex social behaviour to maintain their species in the preferred zone. In other words, the behaviour is not governed by a simple biological clock as in fiddler crabs in the same family.

The most abundant bivalve mollusc on any shore of Inhaca Island is the small tropical wedge mussel, *Donax faba* (Fig. 3.2a) which may reach a density of over 100 per square metre on the upper sandy shore – an exceptional habitat for bivalves. They congregate around the level of high water of neap tides on the sandy slope in a narrow belt between 13 m and 15 m from the foot of the dunes. The mussels are buried 2-5 cm beneath the surface of the sand when the tide is out and may easily be found by hand in the dry sand. If a bucket of seawater is used to deluge the shells as they lie on the sand each will soon protrude a pointed mobile, muscular foot and burrow again (Fig. 3.2b). The shell of an adult, two to three years old, is about 22 mm long and 17 mm broad; the anterior end is wedge shaped and the blunt posterior end is often green with a superficial coating of the alga *Enteromorpha* which is present but, to the naked eye, invisible on the sand. *Donax faba* is unique on the east coast of Africa in having more than 30 different colour patterns.

This species of wedge mussel uses the standard bivalve methods of microphagous feeding and burrowing (summarised below) but in addition it has a number of modifications for living in the surf zone at high tidal levels in sand of mixed grade where there is little detritus. This mussel selects diatoms of size range 30-60 µm from the rich population of benthic diatoms, which are tossed up by the gentle waves, enriching the phytoplankton, and it rejects the stirred up sand. It makes the most of its very short feeding times by floating a few metres upshore for high spring tides and down again for high neap tides, although it has no locomotory organs. It can also endure the eight to nine hour interval of relative dryness when the tide is out by sealing off the shell.

Wedge mussels are best studied whilst they are active at the surface of the sand when the waves are advancing and retreating on the middle of the sandy slope. The mussel detects advancing waves by a special sense organ (a modified statocyst) which is enclosed in a capsule in the 'cruciform' muscle of the siphons; it is sensitive both to the sound of waves and to mechanical shock. In response to these stimuli the shell adductor muscles are relaxed and the shell opens by the pull of the elastic ligament outside the hinge. The animal thrusts out its foot against the sand beneath and is pushed up to the surface. As the sand is wetted by waves the siphons, only 5 mm long in *Donax faba*, are protruded from the posterior end until the openings are flush with the surface and so catch the sinking particles. These siphon openings are clearly visible at close range; the inhalent opening has a delicate lace-like sieve, which excludes particles of coarse sand, the exhalent opening has six petal-like processes which can open and close the aperture (Fig. 3.2a).

Fig. 3.2 Wedge mussel, *Donax faba*:
a. feeding (× 1,5) (arrows show direction of current);

b. burrowing (after Trueman 1975);

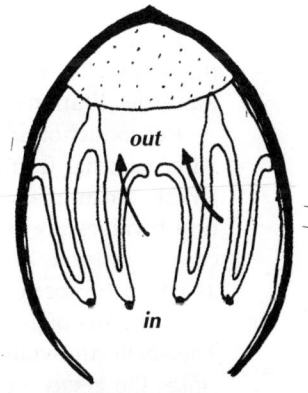

c. gills filter and collect food, arrow shows direction of water current in and out (after Newell 1979);

In the position described above, the mussels use the usual bivalve filtration mechanism of feeding. A current of water, created by the beating of cilia between the filaments of four large plate-like gills, is circulated through the mantle cavity (Fig. 3.2c); it is filtered through the pores of the gills from the inhalent to the exhalent cavity, particles being retained on the surface of the gills. The presence of particles stimulates the secretion of mucus and they become trapped. Surface cilia waft the food-laden mucus to the free edge of each gill where it becomes a mucus string lying in a groove. Large cilia in the groove convey the string to the palps around the mouth where the particles are sorted for size and density. The smaller particles lie on the ridges of the palps and are directed by cilia towards the mouth and swallowed. The larger ones (mainly fine sand) travel along the grooves of the palps, drop off into the mantle cavity and are expelled in the exhalent current. In *Donax* the gills participate in rejecting medium grains of sand by means of rows of large strong cilia on their surfaces.

The duration of feeding is three to four hours, twice in 24 hours, whilst the mussels are covered by the tide. The intensity of feeding is maximised by adjusting the positions of their burrows to the daily changes in the heights of tides, enabling the mussels to stay longer in the belt of breaking waves. With foot and siphons extended and shell partly open, the mussels appear to be temporarily floating in the swash or backwash, but this is not entirely a passive migration. The mussels are normally anchored in the sand by a swollen foot but they relax the hold by redistributing the blood from the foot sinus to the body sinuses. One wave will then carry them for a metre or so and they start digging in as soon as the wave retreats, burrowing being completed before the next wave comes, faster than other types of bivalve on the lower shore. During neap tides the mussels remain in the lower part of the slope and during spring tides they burrow in the middle part of the slope.

The wedge mussels dig a few centimetres into the sand as soon as it begins to dry after the last wave of the ebbing tide (Trueman 1975). The shell is half-closed and, as the siphons close simultaneously, jets of water squirt around the foot, softening the sand to ease penetration. The compression of the body by the closing of the shell drives the blood into the foot sinus so that it expands to form an anchor (Fig. 3.2b). The animal is pulled down by the muscles attached to the foot, the siphons are withdrawn and finally the foot is retracted and the shell completely closed. In this state the wedge mussel respires anaerobically between tides and, as in other burrowing bivalves, derivatives of the carboxylic acids in the respiratory cycle, rather than lactic acid from the glycolytic pathway, accumulate temporarily. Even the energy used in burrowing is supplied anaerobically, since there are few mitochondria in the foot muscles (Trueman and Brown 1987).

These adaptations – emergence, suspension feeding and rejection of sand, tidal migration, burrowing and anaerobiosis – are considered to be a series of responses to stimuli rather than parts of an endogenously controlled pattern of behaviour.

Donax faba does not occur on the south-west shore because the sandy habitat is replaced by the rocky cliff. Even where sand does occur at high tidal level immediately south of Ponta Raza, the dilution of the sea by the mangrove tidal stream prevents their survival.

Reproduction in wedge mussels shows no special adaptation to living at high tidal level. After external fertilisation and several weeks of floating at the surface of the sea as veliger larvae, the animals, having developed two shells, a foot and gills, sink to the bottom and the tiny bivalves are carried inshore by the waves. On the other hand, inheritance of complex colour patterns in *Donax faba* has been considered to play an adaptive role in evolution, assisting the species to occupy the hazardous upper shore.

Five pairs of genes control the main colour patterns of *D. faba* at Inhaca Island (Nolte 1954). The shells are white when the genes are in the recessive condition (Fig 3.3a). In homozygous states the genes are expressed as follows: completely purple shell, six purple radial stripes, three brown radial bands, three similar and overlapping rose-coloured bands and lastly, posterior dark edges of the shells (Fig. 3.3b-f). Over 60% of the shells are 'purple striped' and the other kinds are present in smaller numbers decreasing from 'purple' to 'brown' bands, to 'rose' bands to 'dark edges' and the multiple recessive 'white'. Smaller variations in the pattern are the result of various degrees of dominance of the genes and the presence or absence of a masking yellow cuticle. The different frequencies of the five colour genes have been found to be constant over the years, illustrating the genetic principle of 'balanced polymorphism', which is usually the result of some selective advantage for heterozygotes in the evolution of species. A connection between habitat, colour polymorphism and predation has therefore been sought, since not only *D. faba* on East African shores, but also *D. rugosus* in tropical West Africa and *D. variabilis* in America have many colour patterns, live on the upper shore in sand and are eaten by crabs and by wading birds.

To the human eye white shells and light-coloured heterozygotes are conspicuous on the wet sand, but when more than three genes for colour are expressed the shells are less obvious and juveniles particularly resemble sand pellets. On Inhaca Island, however, the predatory ghost crab does not see the colours of the shells before it digs them up, and birds are infrequent on the upper shore because of human disturbance. On Tanzanian shores (Smith 1975), an opportunity for studying predation and colour variation was presented since the green ghost crab emerges earlier and hunts by sight before dusk while the mussels are migrating over the sand in the waves. Ghost crabs see moving objects well and do have a limited colour vision. The colours of the shells that had been eaten could be determined by counting them in the middens which the crabs had accumulated outside their burrows.

It was found that all colour varieties were taken in numbers which were more or less proportional to their occurrence in the population, except that the light (heterozygous) shells predominated among the prey debris when the population was dense. When numbers had been reduced to one fifth in winter, the most numerous 'striped' were taken preferentially. In both seasons the rarer genes (purple, brown, pink), which increase the resemblance to wet sand, survived and thus colour polymorphism continues, protecting the species as a whole, when the habit of 'migration' to optimise feeding exposes them to predation.

3.2.2 The microflora of the sandy slope and sand flats

Microscopic unicellular algae, living in the upper 2-3 mm of the surface sand, are mainly diatoms and dinoflagellates. Diatoms have a rigid, silica-impregnated cell wall. Their chlorophyll *a* and *c* is masked by a xanthophyll pigment, fucoxanthin, so that they appear brown in colour. Photosynthesis is possible over a wide range of illumination, including extremely low light intensities, because xanthophyll has a light-harvesting function at low light intensity and a chlorophyll-protecting role in bright light. In this it resembles the action of xanthophyll in brown seaweeds but is unlike that in terrestrial plants.

Most of the benthic diatoms in sand are about 30-60 μm long with bilateral symmetry; the silica shell is beautifully embossed and engraved with intricate patterns, and there are

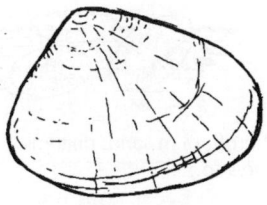
Fig. 3.3 Genetic patterns in *Donax faba*: a. white: multiple recessive; dominants:

b. purple;

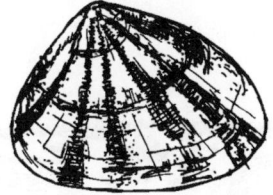
c. striped;

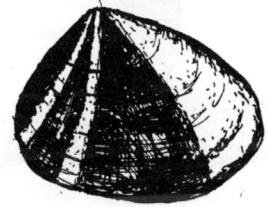
d. brown bands;

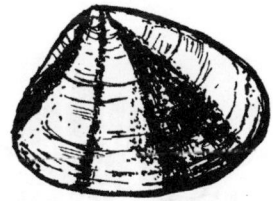
e. rose bands;

f. black edged.

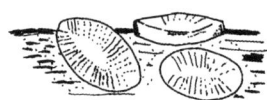

Fig. 3.4 Microflora in sand; diatoms:
a. *Cocconeis* sp.;

b. *Navicula* sp.;

c. *Nitzschia* sp. (all × 1000);

d. electronmicrograph T.S. *Nitzschia*: large chloroplasts; myo-filaments (arrows), raphe with mucoid rods (× 4 000);

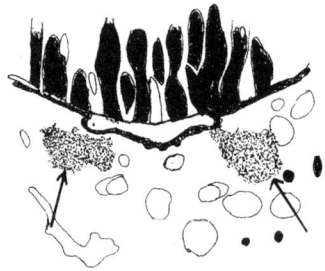

e. raphe enlarged, arrows to actin filaments (× 30 000). (c-e after Edgar and Pickett-Heaps 1983.)

pores which permit the entry of nutrients. There are two groups of benthic diatoms: epipsammic which are those attached to sand grains, and epipelic, which live freely in the interstitial water film between sand grains. The epipsammic diatoms have a single groove or raphe along one side by means of which they are attached to sand by the secretion of mucilage from the cell, e.g. *Cocconeis* sp. and *Navicula* spp. (Fig. 3.4a and b) (Pienaar 1965). Each grain of sand may have 20–40 diatoms adhering to it in little crevices. The epipelic diatoms, such as *Nitzschia* spp. (Fig. 3.4c) have a raphe on each side by means of which they are able to move between the sand grains. These movements are directional: the diatoms migrate to the surface of the sand during daylight hours when the tide has receded, enabling photosynthesis to occur; then they retreat a few millimetres before the tide returns and so avoid being dislodged by water turbulence.

A mechanism for the motility of diatoms has recently been deduced from observations of the ultrastructure of the cell (Edgar and Pickett-Heaps 1983). At each raphe the valves are held securely, yet loosely, by a tongue-in-groove joint, except at the central pore and at each end. Temporary adhesion to sand grains is provided by the discontinuous secretion of stiff strands of mucilage through the specialised plasma membrane beneath each raphe. The motive force is produced by contraction of pairs of bands of thin microfilaments, which lie close alongside each raphe (Fig. 3.4d and e). The microfilaments are made of the protein, *f-actin*, and resemble that in flagella and muscle fibrils. As the actin filaments contract the stiff strands of mucilage are displaced longitudinally, sliding along the raphe in the lipid component of the plasma membrane (at the ends where they are attached to the cell), whilst the other ends are still attached to sand grains. Thus the stiff strands move backwards and the diatom is pushed forward. The strands fall off at the end of the raphe, leaving a temporary mucous trail which is soon dissolved. Contractions of the actin filaments cease in response to a change in intensity of light falling on the diatom, but the light receptor has not yet been identified and is possibly in the chloroplast itself.

Although this migratory habit enables diatoms to maximise photosynthesis and yet retire to comparative safety when the tide approaches, a proportion of the free epipelic diatom population is dislodged by waves on the sandy slope and it contributes to the food chains on sandy shores, as described above for *Donax faba*. The attached epipsammic diatoms feature as food for herbivorous copepods and nematodes of the meiofauna (see Section 3.7.2) and for some ocypodid crabs. The assemblages of diatoms are larger on the sandy mud of the sand flats on the *lower* shore and they are more diverse. Species differ in their reaction to desiccation and illumination and their tolerance of organic material in mud. Large numbers are found in the stomachs of both deposit and suspension feeding animals as well as in detritus-feeders, such as crabs, molluscs and worms.

Dinoflagellates also inhabit damp sand, for example *Prorocentrum* and *Amphidinium* (Fig. 3.4f and g). These are unicellular algae with a cellulose wall made of several plates and with two flagella. The migratory habit to the outer surface from the interstices of the surface sand in tidal pools has been described in the case of *Scripsiella arenicola* (Fig. 7.18f) on the east shore of Inhaca Island (see Section 7.5).

In addition to the diatoms and organic matter on the uneven surfaces of grains of sand, clumps of bacterial cocci and minute groups of cells of blue-green algae (Cyanophyta) are retained. The blue-green algae are also known as Cyanobacteria since the cells are prokaryotic; they resemble bacteria in having no nuclear membrane and in the absence of membrane-limited organelles (mitochondria and an endoplasmic reticulum) which are characteristic of eukaryotic cells. The photosynthetic lamellae are, however, elaborate and carry chlorophyll *a*. The plasma membrane of each cell is surrounded by a cell wall of peptidoglycan and lipopolysaccharide. In many species several small cells are enclosed in a sticky mucous sheath of polysaccharide which attaches to sand grains; in other species filaments of a row of cells in a sheath are formed. Common species on sand grains at Inhaca Island are *Chroococcus minutus* and *Gomphosphaeria aponina* and *Lyngbia confervoides* (Fig. 3.4h, i, and j).

3.3 THE FLAT MIDDLE SHORE

The west shore in front of the Marine Biological Station becomes flat sand just above mean tide level. The first belt on the flats, about 50-100 m wide, supports an association of tube-building worms. Beyond this belt large wetter or drier areas, roughly parallel to each other, lie permanently side by side; they are riddled with crab holes and different species of crabs and animal associates occur in each, while both have a rich microfauna and meiofauna (see Section 3.6). Little macroscopic vegetation occurs below the crab zones until the level of mean low neap tides where sea grasses begin to cover the surface.

3.3.1 Green flatworms and their symbionts

While the tide is receding from the lower shore, the sandflats immediately below the sandy slope are often streaked with dark green furrows 30-70 cm long, where the drainage of freshwater lies near the surface. The green colour is the result of the congregation of millions of flatworms, *Convoluta (?) macnaei*, 3-4 mm in length and bright green in colour (Fig. 3.5a). The flatworms have incorporated between their cells many hundreds of individuals of symbiotic unicellular green algae, *Tetraselmis convoluta* (Fig. 3.5b). When these algae are free-living, they have a scaly cellulose wall, four flagella and an eyespot as do other members of the Prasinophyceae. *Tetraselmis convoluta* exhibits a circadian rhythm with regard to its cell cycle, dividing approximately once every 24 hours. During the division cycle the cells become non-motile and settle on sand grains and cocoons of the flatworm *Convoluta*. Once the colourless juvenile worms hatch they are voracious feeders and will ingest particulate matter including bacteria and non-motile *T. convoluta* cells. Once ingested, the algal cells undergo some marked changes, which include: loss of emergent flagella, degeneration of the theca and intraplastidial eyespot, and extensive lobing of the chloroplast. They also

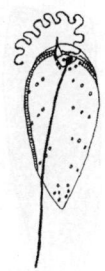

Fig. 3.4 Dinoflagellates in sand:
f. *Prorocentrum* sp.

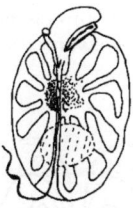

g. *Amphidinium* sp. (after Dodge 1982);

Cyanophyta:
h. *Chroococcus minutus*;

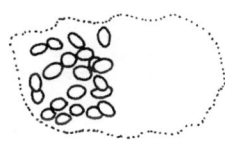

i. *Gomphosphaeria* sp.;

j. *Lyngbia confervoides* (all x 1 000) (h-j after S.F.M. Silva 1991).

Fig. 3.5 Burrowing worms on the middle shore:
a. Green acoelate flatworm, *Convoluta macnaei* (x 20);

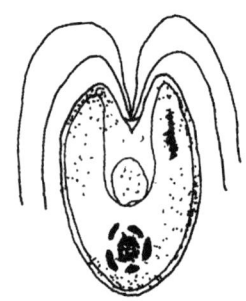

b. green algae in free-living stage, *Tetraselmis convoluta* (x 1 000);

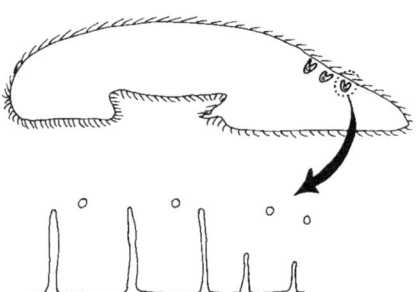

c. section of flatworm with ventral sucker, symbionts near dorsal surface of animal;

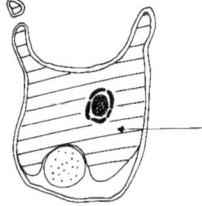

d. detail of a symbiont between host cells, presenting chloroplast (arrow) at the ciliated surface of the flatworm (a-d after R. N. Pienaar, unpublished).

migrate to the sub-epidermal region of the worm, where they would obtain maximum light when on the surface of the sand (Fig. 3.5c and d). It is generally accepted that the algal cells occur in an intercellular position, i.e. between the animal cells. The microalgae settle on the flatworm eggs which are laid on the surface sand; they are eaten by the young flatworms after they hatch and are absorbed by phagocytosis into gut cells. The algae shed their walls, flagella and eyespot, aided by intracellular enzymes of the flatworm cells, and find their way out of the cells (or are expelled) to lie between muscles, where they photosynthesise and reproduce by longitudinal fission.

This species of flatworm initially has a mouth, pharynx and internal digestive cells; its mouth closes permanently after engulfing algae. The flatworms depend on the algal symbionts to supply them with their excess metabolites (carboxylic acids, amino acids, fats and sterols) and do not feed themselves after the first few hours. The algae obtain their source of nitrogen from the excretory product of the flatworm which is in the form of uric acid crystals (visible with an electron microscope). The algae decompose the uric acid with the enzyme uricase and they utilise the ammonia formed. Photosynthesis is optimised by the migratory habit of the flatworm which rises to the surface in direct sunlight when the sand is exposed to air and sinks down a few millimetres before the tide returns, so that both host and guest are protected from the slight turbulence of the water. This rhythm is synchronised with the tidal movements and 'memorised' so that even when the flatworms are taken to the laboratory they still migrate up and down at the appropriate times. The worms move through interstitial water between the grains of sand with the aid of surface cilia and internal muscles and attach to sand grains by means of their two suckers. They may easily be seen if a little 'green sand' is scooped up in a small glass vial which is filled with seawater. The flatworms soon attach themselves to the sides of the tube and later, at the time of high tide, they will move down to the bottom of the tube.

3.3.2 Burrowing worms and their different feeding habits

In the first belt of the sandflats the proportions of fine sand, silt and organic matter to medium or coarse sand are higher than elsewhere on the west shore, since the tide slackens around mid-tide level and deposits fine particles. The sand retains sufficient water in the interstices to be hard underfoot when the tide is out. The water table is about 10cm below the surface where the grey-brown sand merges into black, anaerobic mud. The substratum is suitable for the support of the permanent burrows of several types of worms, although neither bivalves nor crustaceans are present.

Worms feed at different levels (Day 1967):
i. above the surface where the worm tubes protrude and phytoplankton is filtered from suspension;

ii. at the surface where benthic diatoms and detritus are selected by deposit feeders;
iii. within the sand where detritus is swallowed by submerged burrowers.
iv. near the surface and below, carnivorous polychaetes in temporary burrows where they may feed on other polychaetes and on meiofauna.

(i) A suspension-feeding polychaete worm, *Phyllochaetopterus elioti*

The first signs of this worm are the clusters of brown twig-like tubes sticking out of the sand for about 5 cm. The tubes are around 15 cm long and the lowest third is stained black from the anoxic mud below. The tubes are built by blue, luminescent chaetopterid worms, *Phyllochaetopterus elioti* (Fig. 3.5e) by means of a secretion of mucus from the ventral glands. This becomes coated with fine grains of sand, especially selected and affixed by the worm's lower lip. When the mucus dries it becomes a stiff horny tube with a diameter slightly wider than that of the worm and open at both ends. The worm lies in the mouth of the tube only while submerged by the tide and feeds on the smallest nanoplankton filtered from a ciliary current through a gossamer-fine funnel of mucus. Silt and larger particles are detected by cirri near the mouth and deflected by parapodia away from the funnel. Cilia on the parapodia of the middle region of the worm are responsible for the incoming current which causes a mucus sheet stretched across the tube to billow out into a funnel. At short intervals secretion of mucus stops and the mucus with entrapped food is rolled up into a ball by means of long cilia in the mid-dorsal food cup. The ball is carried forward towards the mouth and picked up by the muscular lower lip, then swallowed. This feeding mechanism is similar to that of the classic technique of the large cosmopolitan *Chaetopterus* (see Fig 9.8a, Section 9.11) in which muscular parapodia replace the more primitive ciliated parapodia of *Phyllochaetopterus*.

All the worms in the chaetopterid family have certain parapodia which, under the control of the nervous system, secrete granules that become luminescent, shining with a strong bluish-green light on contact with water (or perhaps on contact with the oxygen dissolved in it). One may speculate whether a cluster of points of light from a host of worm tubes would attract the phototropic dinoflagellates in the nanoplankton down towards the worms and so enrich the food laden currents?

(ii) Surface deposit-feeding terebellid worms

Little holes surrounded by fine radiating lines may be detected on hard sand while it is still wet from the receding tide. These holes lead into permanent vertical tubes of the small terebellids and the radii are impressions left by their large number of slender, ciliated, muscular tentacles. Terebellid worms are

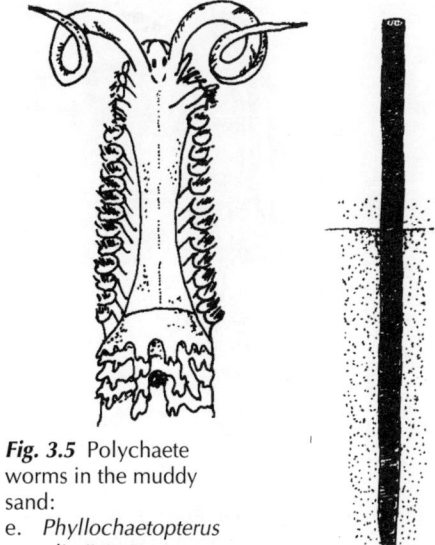

Fig. 3.5 Polychaete worms in the muddy sand:
e. *Phyllochaetopterus elioti** (x 3);
e'. black tube (x 0,5);

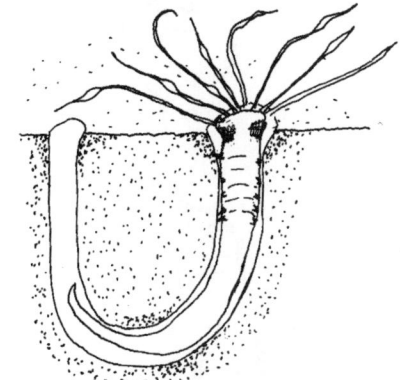

f. diagram of terebellid worm feeding in burrow;

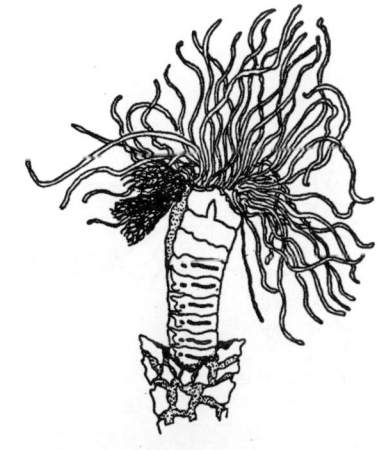

g. head of terebellid, *Pista brevibranchia** from coarse sand (x 3);

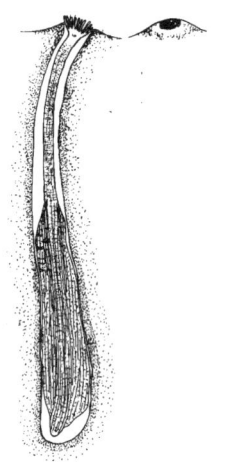

Fig. 3.5

h. Sipunculid worm, *Siphonosoma cumanensis* (x 0,5) in burrow and common conical exit/entrance.

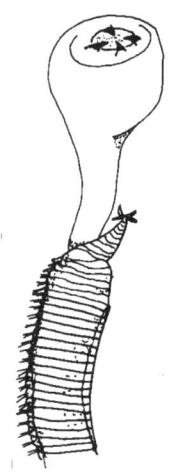

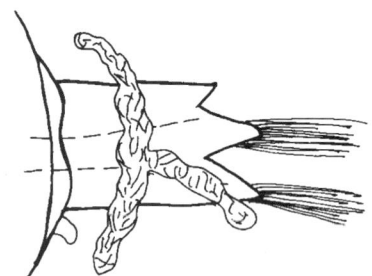

Fig. 3.6 Carnivorous polychaetes:
a. *Glycera subaena** with proboscis extended for feeding;

b. parapodium and gill;

tough and plump and have red gills on the anterior segments which protrude at the mouth of the tube among the tentacles (Fig. 3.5f and g). The tubes are made of mucus secreted by ventral pads and become parchment-like and fragile when dry, being coated with particles of sand. *Loimia medusa* makes wide tubes varying in colour from light to dark grey according to the aeration of the environment. *Thelepus plagiostoma* uses coarser sand, and *Pista brevibranchia* cements together fragments of gravel (Fig. 3.5g).

One can watch the action of red or yellow tentacles creeping out slowly by ciliary action over the surface sand, extending further to pick up particles, then rolling them into a dorsal groove on each tentacle to convey them to the mouth by ciliary action. The tentacles may also be suddenly withdrawn by muscular contraction. In both cases muscular lips receive the crudely collected material and sort it into edible detritus and sand suitable for tube-building or for rejection.

(iii) Detritus-feeding, burrowing sipunculid worms

Non-segmented sipunculid worms burrow a few centimetres beneath the surface of this organically enriched sand. Their holes in the sand are slightly bigger than those of the terebellids and each lies in the centre of a raised cone of loose sand which is left when egested sand is dispersed under water. The burrow has only sufficient mucus lining it to make its form permanent when supported by water and sand. *Siphonosoma cumanensis*, which is numerous at this level, has a tough pink cylindrical body without appendages or segmentation (Fig. 3.5h) and is turgid with body fluid held in a single, large coelomic cavity. The worm's blunt head may be retracted like a finger of a glove pulled off in haste. It is then protruded with the force of the contraction of body muscles which drives the body fluid into it; thus the head can be used as a battering ram to burrow in wet sand. The extended head has a number of sensory pits, tentacles and ciliated lips which are used in picking up detritus. Selection must be negligible since the gut is full of sand. The intestine is U-shaped and is three times as long as the worm, being looped and coiled, thus giving time for digestion. The anus opens just behind the head so that faeces are pushed out of the burrow through its single opening.

In the sipunculid type of burrow, which does not have a through flow of water like those of polychaete tubes, the worm is subjected to periodic low levels of oxygen between tides. It has no respiratory organs, except its skin, nor does it have a blood vascular circulation. A steady supply of oxygen is available to the muscles, however, since oxygen diffusing through the skin into the body cavity is absorbed by pigmented corpuscles. The pigment is haemerythrin, a protein like the haemocyanin of crustaceans, but with iron instead of copper. The pigment combines with oxygen when it is available and 'unloads' it when the oxygen tension falls, so that it serves as an oxygen store.

(iv) Carnivorous polychaete worms

In the surface sand around mid-tide level the most common carnivorous polychaetes are *Glycera* species. Glycerids are long, thin, red worms and are very active when extracted from the sand. They make superficial, branching, U-shaped galleries opening at the surface through which they travel with ease in a sinuous motion, their parapodia being very small. *G. subaena* (80mm) has three to five finger-like lobes of extensible gills on most parapodia (Fig. 3.6b), each red with haemoglobin. They lie in wait for very small prey and detect it with four minute tentacles on the small conical prostomium (Fig. 3.6a). A hunting glycerid rapidly extrudes a long and powerful proboscis, one third the length of the worm, which ends in four strong jaws with poison glands and is reinforced by skeletal supports. The proboscis is explosively protruded when the longitudinal muscles in the anterior third of the body suddenly contract and compress the free coelomic fluid which inflates the proboscis, turning it inside out. The technique of hunting in glycerids is reminiscent of the shooting tongue of chameleons. It uses the proboscis in a similar way for burrowing.

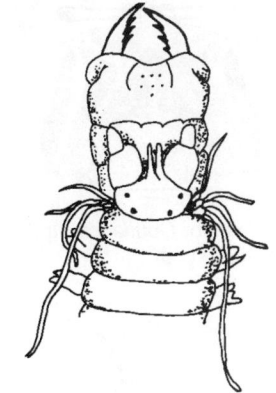

Fig. 3.6
c. *Ceratonereis erythraensis** head with proboscis and jaws extended for feeding;

Ceratonereis erythraensis (Fig. 3.6c) is a slender worm (100-160 mm) that burrows in sand by using the typical, undulating nereid manner of locomotion in which alternate contractions of sections of longitudinal and circular muscles of the body wall enable the worm to wriggle along using its strong mobile parapodia as levers (Fig. 3.6d). It has the typical short nereid proboscis with two jaws, which can be extruded for feeding

d. locomotory parapodium*.

(Fig. 3.6c), and well developed sense organs on the head: eyes, palps, tentacles and cirri, all of which suggest a carnivorous way of life, but like some other nereids it is a versatile feeder including diatoms and detritus in its diet.

3.4 THE SAND FLATS OF THE LOWER SHORE

3.4.1 Drier sand: bubble crabs; tube-building worms

In front of the Marine Biological Station and on parts of the lower shore where coral debris is absent, there are large tracts of sand stretching 100-200 m towards low tide where the surface drains well, the sand becoming fairly firm when the tide recedes. This sand is mainly of medium grade and the water table is about 15 cm below in black mud. When the tide first exposes these areas they are smooth and flat but after one and a half hours the surface becomes carpeted with tiny pellets of sand and shimmers with the movements of the small ocypodid crabs (with long eye-stalks), *Dotilla fenestrata* (Fig. 3.7a and b). The habit of forming glistening round pellets of sand has earned them the name of 'bubble crabs'. The generic name *Dotilla* refers to its globose shape and small size (7-10 mm) and 'fenestrata' describes the 'windows' or thin membranes on the ventral carapace and on the fourth segments of the legs. These membranes have a respiratory function like those in a closely related Australian genus, *Scopimera*, in which the vascular structure below the windows is visible and has been studied by the use of electron microscopy, whilst the function has been demonstrated experimentally. They have been called 'gas windows' which enable these crabs to breathe in air (Maitland 1986).

Bubble crabs spend most of their time in vertical burrows which do not reach the water table (unlike those of fiddler crabs in the mangroves); they emerge for three to four hours in

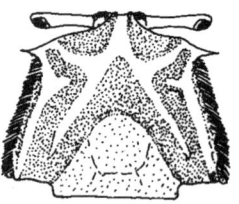

Fig. 3.7 Bubble crab, *Dotilla fenestrata*:
a. dorsal carapace (x 3)*;

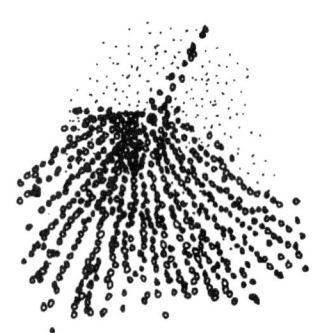

b. pseudo-faecal pellets radiating from burrow;

c. feeding 'spoon' from 2nd maxillipeds (after Vogel 1984);

d. capillary setae on abdomen (after Hartnoll 1973).

daylight and feed. The crab digs its burrow by excavating balls of darker sand, larger than itself, which soon fade in the sunlight as the black iron salts become oxidised again.

Dotilla feeds on the upper 1-2 mm of sand, utilising diatoms, meiofauna and organic detritus. A methodical system of working the sand is used in which the crab moves slowly forward and backward along successive radii of 10 cm or more from the burrow opening, picking up sand and taking pellets from the mouth from which food has been removed (Fig. 3.7b). About 30 per cent of the organic matter in the sand is said to be eaten every day by the dense population of bubble crabs (Fishelson 1983). Chelae are used alternately to pick up a pinch of sand at a rate of about 20-25 times a minute. The abdomen is meanwhile held clear of the ground and rejected pseudo-faecal pellets are deposited between the legs by each chela in turn. Edible material is sorted from the sand by the mouth appendages using a skilful 'brush and comb' action under two jets of water squirted from the respiratory apertures. This is achieved in the following manner: a scoop of sand is first pushed between the bases of the third maxillipeds, which tilt open, and is wiped off by the brush of setae on the outer surface of each first maxilliped. Combs of spatulate setae (called 'spoons') (Fig. 3.7c) on the inner edges of the second maxillipeds continuously work against the brushes on which two jets of water play. Food particles are washed into the mouth and the water is conserved by trickling back into the gill cavity through small openings at the bases of the maxillipeds. The loosened and scraped sand particles are worked into firm pellets and then removed by each chela in turn from the apex of the space between them. The pellets do not disintegrate immediately since they are bound by a slightly viscous secretion. This method is adapted to feeding in sand of medium sized particles by the simple nature of the 'spoons' (Vogel 1984). Fiddler crab species which feed similarly in sand, sandy mud, muddy sand or mud have more complex 'spoons' (see Section 4.6.1).

Aerial respiration is continuous through the 'gas windows' on the carapace and legs. Respiration is partly aquatic through the gills in the branchial cavity which occurs in spite of hours spent in the hot sun without apparent access to water. This is possible because every minute or so the crab momentarily ceases feeding to squat, resting its abdomen against the sand (Hartnoll 1973). Long capillary-like setae on the fourth abdominal segment suck up interstitial water which passes forward in the space between the abdomen and the sternum through openings on the margin of the fifth segment (Fig. 3.7d), and flows along well-defined grooves to the branchial cavity openings at the bases of the third and fourth legs. This water replaces that lost in the formation of sand pellets and keeps the gills moist.

In addition, the water can be used as a cooling device when some of it is directed from the exhalent apertures near the bases of the first antennae into the complex grooves on the dorsal surface of the carapace (Fig. 3.7a). The thin film of moisture evaporates so that the body temperature can be maintained as much as 10°C below that of the air. Squatting to absorb water may occur twice or three times a minute on a very hot day and once in five minutes on a cool day (Fishelson 1983).

About ten minutes before the rising tide reaches the crabs, they excavate their burrows, enter sideways and plug the openings with pellets of sand, either by dragging a heap towards the burrow over the opening or by plugging it from below. A bubble of air is retained in the burrow beneath the surface, serving to aerate the water which rises inside with the incoming tide.

The density of bubble crabs in the large patches where they occur at Inhaca Island is about 50 per square metre judging from counts taken while they are feeding, but probably only about 50 per cent emerge at any one time, since by digging an almost equal number is revealed. The surface of the sand is so disturbed and so much food is removed that not many other animals live there and those that do so are not deposit feeders.

The tubicolous green polychaete *Owenia fusiformis*, a suspension feeder, constructs a narrow, gelatinous tube covered with a minute mosaic of flat pieces of sand and shell fragments, elegantly dovetailed like the tiles of a roof. The tubes protrude a few centimetres above the sand and when covered by the tide the worms expose their heads to feed on suspended matter by means of a ciliated frilly crown (Fig. 3.8a, a' and b). Another worm *Mesochaetopterus minutus*, a very small chaetopterid, constructs clusters of very narrow, fragile, sandy tubes about 5cm long which just protrude above the sandy surface. The walls of the tubes are impermeable to oxygen which is therefore not lost to the almost anoxic interstitial water in the sandy mud. Thus a low level of oxygen is available to the worm when the tide is out. It uses the same kind of ciliary feeding as does the other member of the family, *Phyllochaetopterus* (in the muddy sand described in Section 3.3.2).

Mesochaetopterus (cf Fig. 3.5e) consolidates the sand at the edges of the *Dotilla* area and is there associated with crowds of the brittle stars, *Amphiura candida* (Fig. 3.8d). This brittle star is distinguished by its scaly yellow disc and small, prominent radial shields which are widely splayed. It is slightly larger than other sand-dwelling brittle stars (disc 6 mm in diameter) and its arms have four spines on each side of the arm segments.

A few widely scattered tubes of the scavenging eunicid worm, *Diopatra cuprea* are visible on the flats at this level down to low tide. The tube is remarkable for its impermeability and the way in which small fragments of shell are attached edge-on to the upper part, which protrudes 4-5cm above ground. When under water this worm, which has five long sensitive antennae on its head, can reach out of the tube (Fig. 3.8c) and with sharp jaws seize moribund amphipods or other small animals detected by antennae. It breathes under water by means of large pairs of bushy gills on the anterior segments; when the tide is out it retires to the bottom of the long tube and respires anaerobically.

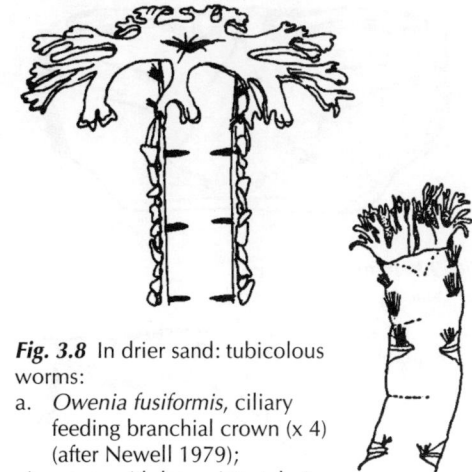

Fig. 3.8 In drier sand: tubicolous worms:
a. *Owenia fusiformis*, ciliary feeding branchial crown (x 4) (after Newell 1979);
a'. worm withdrawn into tube*;

b. tube;

c. tube of eunicid worm, *Diopatra cupraea* (x2);

d. brittle star, *Amphiura candida*, burrowing (x 5) (after J.B. Balinsky 1957).

3.4.2 Bare wet muddy sand: sentinel crabs; detritus feeding polychaetes; carnivorous and scavenging snails; hermit crabs

Large areas of wet muddy sand in which the water table is near the surface alternate with the drier *Dotilla* areas. These wetter areas extend from mean-tide level to beyond low neap tidal levels and may be invaded by a narrow-leaved marine angiosperm, *Halodule wrightii* (a sea grass) (Fig. 3.15 a-e). Although some parts of these wet areas lie nearer to mid-tidal level than do the drier areas, they are considered to be an extension of the lower horizons of the shore into which they merge, and here the animals are completely aquatic. The substratum is of mixed grades of sand with a higher proportion of fine sand (30%) and of silt and organic material than is the case in the drier areas. The black mud is nearer the surface, only about 5 cm below, and the sand is somewhat greyer. On the south-west shore, owing to the steeper slope of the shore at this level, the zone is colonised by sea grasses.

The most conspicuous animal is the ocypodid Sentinel crab *Macrophthalmus grandidieri*, and the number of associated animals is large. This crab is easily recognised from its shape, as the breadth of the carapace (32 mm) is twice the length (Fig. 3.9a); the proportions become more accentuated as it grows. The last joint of each cheliped is long and the narrow chelae have one median tooth on the lower joint. The crab makes permanent, almost horizontal burrows in the wet sand, into which it frequently sidles during a hot day with chelipeds folded beneath the body. Its habit of standing at the opening of the burrow, as though on the lookout, probably earned it the common name 'sentinel' crab. It feeds by day both by scavenging and by sifting sand as do *Dotilla* and other ocypodid crabs, but pseudofaecal pellets are rarely seen since the abundance of water available to the gill cavity and passing out over the mouth parts allows the sand to dribble out through a convenient lozenge-shaped opening between the third maxillipeds. *M. grandidieri* returns to its wet burrow frequently. The chelipeds, the first joints of the legs and the edges of the carapace bear thick clumps of hairs which retain water, thereby ensuring that water is available for respiration in the gill cavity and that the crab is kept cool. It resembles other ocypodids in communicating with the chelae, exhibiting aggressive behaviour in defence of its burrow and in courtship.

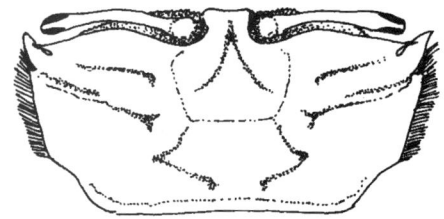

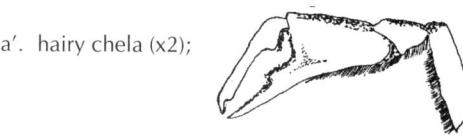

Fig. 3.9 In muddy sand:
a. *Macrophthalmus grandidieri**

a'. hairy chela (x2);

burrowing polychaetes (after Day 1974):
b. head of spionid worm, *Scolelepis squamata* (x 3);

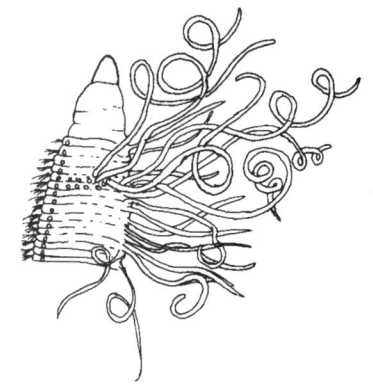

c. head of cirratulid, *Cirriformia tentaculata* (x 1).

The largest numbers of burrowing animals in the muddy sand are the polychaete worms. There are about 30 species here (Day 1967), but the density of each is not high; examples of the common ones are given below. Ten families are represented, and show various adaptations to different types of burrowing and to feeding on detritus either at the surface or within the sand, *dividing the food resources between them*. Parapodia are often much reduced and most of the species have a number of gills as fine outgrowths of the parapodia. Irrigation of the burrows may be effected by cilia, parapodia or bodily contractions. When the oxygen supply fails at low spring tides the worms cease activity and survive by means of anaerobic biochemical mechanisms of respiration.

Members of the following four families have tentacles on the head which are used to collect surface detritus. Terebellids (see Section 3.4.2) have numerous long tentacles which are both *ciliated* and *muscular*, spread over a circle of sand on the surface. Their ciliated lips provide a good sorting mechanism and much sand is rejected.

Spionid worms such as *Scolelepis (Nerine) squamata* (80 mm) (Fig. 3.9b) have two medium-long palps on the head, each with a ciliated groove for conveying the particles, gathered at the surface, to the mouth whilst the head is protruded from a fragile mucus-lined burrow. Sorting takes place on the lips to a limited extent by means of cilia.

Magelona cincta (30 mm) of the Magelonidae has a spade-like head and a distensible proboscis for burrowing; it has small complete parapodia but no gills. It lives and feeds just beneath the surface, collecting detritus with two fairly long palps which are drawn back to the mouth by *muscular* action after detritus has been picked up by numerous sticky papillae. It is thus not a very selective feeder and its alimentary canal is full of sand.

The Cirratulidae are stout worms with no appendages on the head. *Cirriformia (Audouina) tentaculata* (70 mm) (Fig. 3.9c) lies on the surface, usually on dark wet mud often under cover of a stone or plant debris. The prostomium is pointed for burrowing and its numerous very long feeding filaments, with *ciliated* grooves, arise in clusters behind the head. In addition it has characteristic long filamentous gills on almost every segment which stand out blood red on the dark mud.

The next group of three polychaete families, described below, burrow in order to eat and so they are non-selective feeders, operating at a lower level in the sand than do the members of the first group. They have no palps or tentacles on the head and the parapodia are much reduced in size and have numerous hooked setae.

The orbinid *Scoloplos madagascariensis* (120 mm) (Fig. 3.10a) has a pointed prostomium, soft protruding proboscis and flattened anterior segments which are useful for burrowing. The walls of the burrow are consolidated with mucus and remain open to well-aerated surface water between tides. The capitellid *Notomastus aberans* (60 mm) is a common species which protrudes a muscular proboscis for burrowing and swallowing sand and it burrows fairly deep. There are very small gills on the abdominal segments. Capitellids resemble earthworms in habit but secure themselves in sand by means of rows of hooks instead of setae.

Maldanids or bamboo worms are so-called because their segments are long and narrow, each with a slightly swollen glandular anterior end resembling nodes of a plant. They burrow downward and lie upside down with a frilly plate (pygidium) at the posterior end, plugging the opening. They feed on buried detritus washed out of the sides of the burrow by the peristaltic contractions of the body, which irrigate it; *Graviriella multiannulata* (80 mm) is a common maldanid worm (Fig. 3.10b). The blunt prostomium is contractile and it is covered with a fleshy shield which carries small sense organs.

Fig. 3.10 Polychaetes feeding on muddy sand:
a. orbinid, *Scoloplos madagascariensis*, head (x 3);

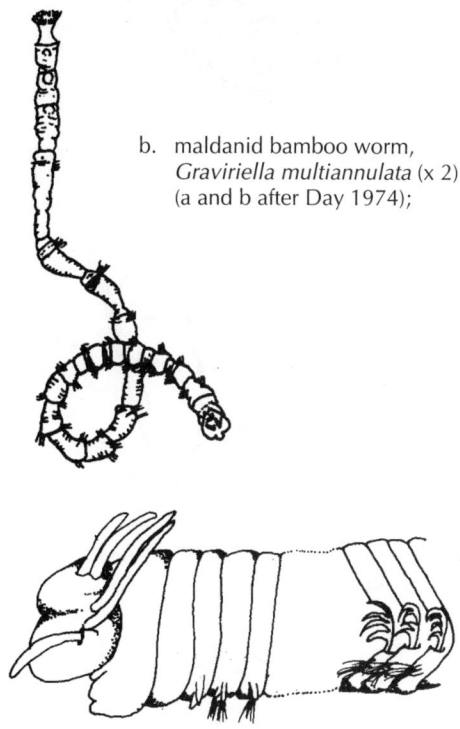

b. maldanid bamboo worm, *Graviriella multiannulata* (x 2) (a and b after Day 1974);

c. carnivorous eunicid, *Marphysa mackintoshi*, head* (x 1);

d. jaws of eunicid worm, maxillae (above) and mandibles (below) (x 5) (after Day 1951);

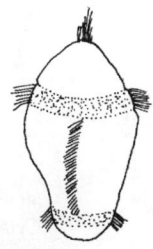

e. trochophore larva (x 200).

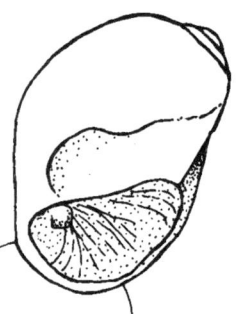

Fig. 3.11 Snails on the sand flats of the lower shore:
a. moon shell, *Polynices mammilla** (x 1).

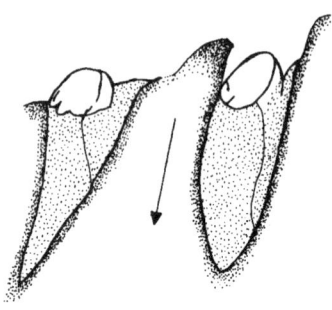

b. burrowing in sand (after E. R. Trueman and A.C. Brown 1992).

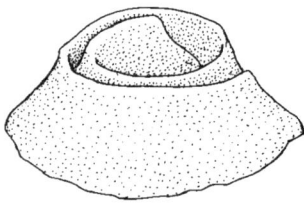

c. naticid collar of eggs (x 0,5) (a-c after Kilburn and Rippey 1982);

whelks:
d. *Nassarius coronatus** (x 1);

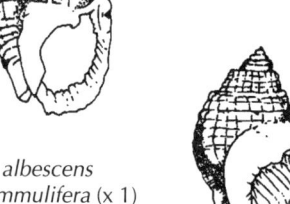

e. *N. albescens gemmulifera* (x 1)

f. attached hydroid, *Cytaeis nassa* (x 10) (e and f after Millard 1975).

The carnivorous polychaetes mentioned in Section 3.4.2 are present and there are more species of *Glycera*. In addition there are large members of the Eunicidae, such as *Marphysa mackintoshi* (200 mm) (Fig 3.10c) and *Lycidice corallis*. These worms have five short antennae provided with a chemical sense and well developed maxillae and mandibles on a protrusible pharynx (Fig 3.10d). Quite large gills of six filaments appear on the posterior parapodia. *Lumbrinereis papillifera* (100mm) is another eunicid worm which, in its adaptation to burrowing, resembles *Lumbricus*, an earthworm. It has a conical prostomium and lacks tentacles, palps and eyes, its parapodia are very small, but it still retains the characteristic of the family, the powerful and complex jaws.

Polychaete worms usually shed eggs and sperm into the sea and the fertilised eggs develop into planktonic trochophore larvae (Fig. 3.10e). It is shaped like a spinning top and has a girdle of ciliated cells for locomotion. There is an apical tuft of cilia above a sensory plate; the mouth is lateral and leads into a ciliated gut which terminates at the lowest point. The eye is anterior and the statocyst is posterior.

Carnivorous snails of the family Naticidae and scavenging snails of the family Nassariidae are common animals in the epifauna and they become more numerous toward low water. The largest naticid, the white moon shell, *Polinices mammilla* (Fig. 3.11a), which has an inflated rounded shell, may be discovered by following its trail. A winding ridge of surface sand is raised as it tunnels below out of sight in search of prey, ending after about 25-30 cm in a confused spiral where the snail may be found feeding on a bivalve mollusc. When caught it hastily withdraws the long siphon that conducts water to the olfactory organ (osphradium) in the mantle cavity, with which it detects prey; more slowly it retracts its over-large muscular gliding foot, which squirts out little fountains of water from the edges, and then it closes the shell with the yellowish horny operculum. The fully extended foot, three times the length of the shell, consists of a tapering propodium, the wide end of which is reflected over the shell anteriorly, and a blunt mesopodium which together partly enclose the shell (Fig. 3.11b). Burrowing is carried out by a series of alternate muscular contractions and expansions of these two parts of the foot. This mode of locomotion is called 'stepping', quite unlike that of a terrestrial snail. Penetration of the sand is effected by the extension of the propodium whilst the mesopodium expands as an anchor when its sinus fills with blood. The propodium then expands by the diversion of blood into its sinus and, exceptionally among snails, by absorption of water through lateral pores of the foot. The mesopodium then contracts bringing the animal and its shell underground (Trueman and Brown 1992). This 'stepping' motion continues until prey is found below the surface sand.

Polinices attacks bivalves and other snails by wrapping the large foot around them and smothering them in mucus. Then by means of the radula, a round hole is painstakingly bored in the shell of the prey by successive movements of the proboscis in small circles to right and left. The animal pauses at times to apply to the eroded area a gland on the upper lip with a non-acid enzyme secretion which softens the shell by dissolving the protein matrix. The feeding mechanism of the complex radula apparatus is explained in Section 6.2.1 and the species of bivalve prey such as clams and cockles are described in Section 3.5.3.

A smaller species, *Natica gualteriana*, which has a brownish, patterned shell and a calcareous operculum, lives in the same way. The naticid snails lay thousands of eggs enclosed in minute capsules in rows on wide gelatinous ribbons which become coated with sand and curled into the shape of a miniature 'shirt collar' about 5-7 cm in diameter (Fig. 3.11c). These may frequently be found on the surface of the sand.

The dog whelks (Nassariidae) are scavengers and have a characteristic form (Fig. 3.11d): a somewhat inflated shell with a prominent spire, spiral markings, a wide aperture and short siphon canal. There are four common species here, which vary in size, around 25 mm, and differ in the amount of callus thickening the boundary of the shell aperture and in markings on the shell (see Appendix A). These snails emerge from the sand as the water rises before the incoming tide and search for small carrion with the exploratory siphon raised. The shell of *Nassarius albescens gemmuliferus* often has a dark fluffy coating of the compound hydroid animal *Cytaeis nassa* (Fig. 3.11e and f), but the most common species *Nassarius coronatus* (Fig. 3.11d) does not carry a hydroid colony at Inhaca Island, although it does so on South African shores.

Another fairly common snail, *Volema pyrum* (57 mm), feeds on whelks but is both a carnivore and a scavenger. The large, brown pear-shaped shell is thick and heavy, fairly wide with a squat spire, and an oval aperture including a small siphon canal (Fig. 3.12a).

The empty shells of all the molluscs in this area may be inhabited by hermit crabs of the genera *Clibanarius* or *Diogenes* which may be distinguished by their antennae (Fig. 3.12b and c). *Clibanarius* has horny spoon-shaped finger and thumb on both chelae and feeds on detritus, whereas *Diogenes* is also a filter feeder, using its plumose antennae as a cast-net and removing the captured organisms caught on the setae with its mouth appendages. The uropods of *Clibanarius* are covered with bristles and bear spines which hold the crab firmly in the shell. Habits of hermit crabs are described in Section 5.5.3.

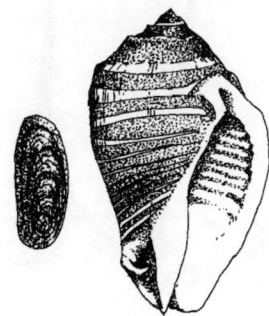

Fig. 3.12
a. carnivorous snail, pear shell *Volema pyrum* and operculum (x 1) (after Kilburn and Rippey

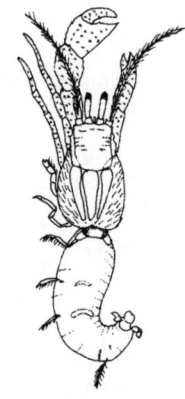

b. hermit crab, *Diogenes* sp. without shell (after Branch and Branch 1981)

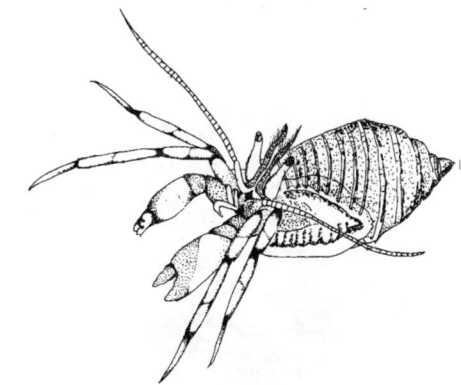

c. *Clibanarius* sp. in *Volema* shell

3.4.3 Minor inhabitants: brittle stars; brachiopods; nemertines; cerianthid anemones

A few brittle stars are always encountered when one digs in the sandy mud, usually the small cosmopolitan species, *Amphioplus integer* (Fig. 3.13a-d). The disc diameter is 2-4 mm and it has five pairs of closely set radial shields at the bases of the five jointed arms; the latter are about 20 mm long and bear small spines on the lateral edges of each little segment. The animal is greyish in colour, so that it is difficult to find, even when two or three arms are protruding from the burrow above the surface of the sand under water for feeding; at night it shines with a yellow luminescence. The arms of brittle stars move horizontally by means of longitudinal muscles between the numerous, segmental, calcareous ossicles. *Amphioplus* has three spines on each side of every segment of the arms, and ventral to each cluster of spines there are two tiny 'tentacle scales', each with a pore through which a mobile tube-foot (podium) is protruded when feeding (Fig. 3.13c). Mucus is secreted on the under surfaces of the arms, which forms a fine net over the spines. Detritus or plankton trapped on the mucus is compacted into a bolus and wiped on to the tentacle scales by the podia (Fig. 3.13d) (Pantreath 1970). Cilia, along the centre of the ventral surface of each arm, transmit these balls of food to the mouth where they enter the corners between the five jaws and pass straight into the stomach. The jaws are closed at intervals by the interlocking of the few simple teeth along their edges (Fig. 3.13c).

A tropical brachiopod, *Lingula* (Fig 3.14a), may be found by a diligent search of the surface of the muddy sand for a well-defined gaping slit about 5 mm wide, which is the opening of the burrow. *Lingula* has two shells about 20 mm long and 10 mm wide, which, unlike the calcium carbonate shells of bivalve molluscs, are made of chitin reinforced with calcium phosphate; they have no hinge but are held together by muscles. The animal has a tough contractile stalk about 50 mm long, which holds it upright in the burrow by internal hydrostatic pressure. When the tide covers the habitat the shells are protruded a few millimetres and stand agape so that the animal can feed on the plankton and detritus. A ciliary current is created by the lophophore (Fig 3.14b) which is a crown of tentacles surrounding the mouth and borne on two arms of a horseshoe-shaped support. These arms or 'brachia' give the name to the phylum Brachiopoda. The end of each arm is spirally coiled so that the number of tentacles is large and by means of their ciliated surfaces the current is filtered and the trapped particles are transferred to the mouth.

Lingula dates back to the Ordovician period (400 million years ago) and fossil species are far more numerous than the brachiopods living today. This genus has apparently changed little during the course of geological history. The sexes are separate in *Lingula* and there is external fertilisation. A pattern of early development resembling that of echinoderms and vertebrates occurs, in that the mesoderm (middle embryonic layer) develops as outgrowths of the embryonic gut instead of by repeated divisions of particular cells in the blastula (i.e. the first embryonic stage) as in other invertebrates. The free-swimming, feeding larva is not a trochophore but resembles a minute adult and there is no metamorphosis. The larval lophophore acts as a locomotory organ until the shell becomes heavier and the animal sinks to the bottom. The pedicel (stalk) uncoils and attaches to the sand and as it contracts the body of the animal is pulled down into the sand.

The most spectacular, but rare animal, encountered on the sand flat is the giant nemertine worm *Cerebratulus marginatus* (Fig. 3.14c). The worm is over a metre long, 25 mm wide and 8 mm thick and it may suddenly appear in the sand when one

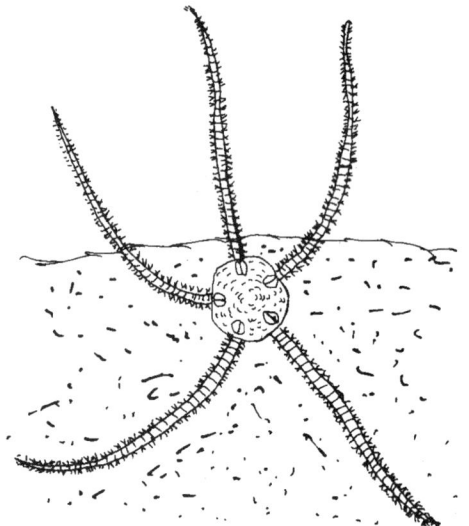

Fig. 3.13 Brittle star in muddy sand:
a. *Amphioplus integer* (x 2) feeding, arms above sand;

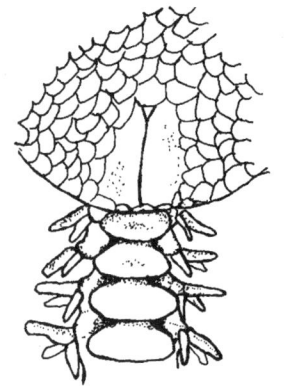

b. portion of dorsal surface of disc and arm*;

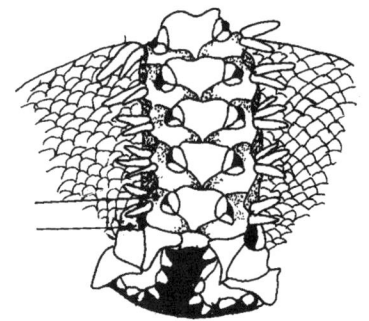

c. ventral, portion of disc and jaws* with oral papillae (arrow to hole in tentacle scale for podium);

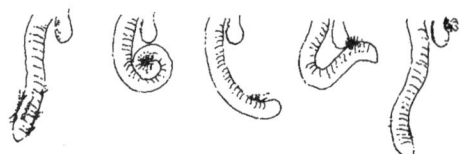

d. podia protruding, collecting food and transferring it to a tentacle scale (after Pantreath 1970).

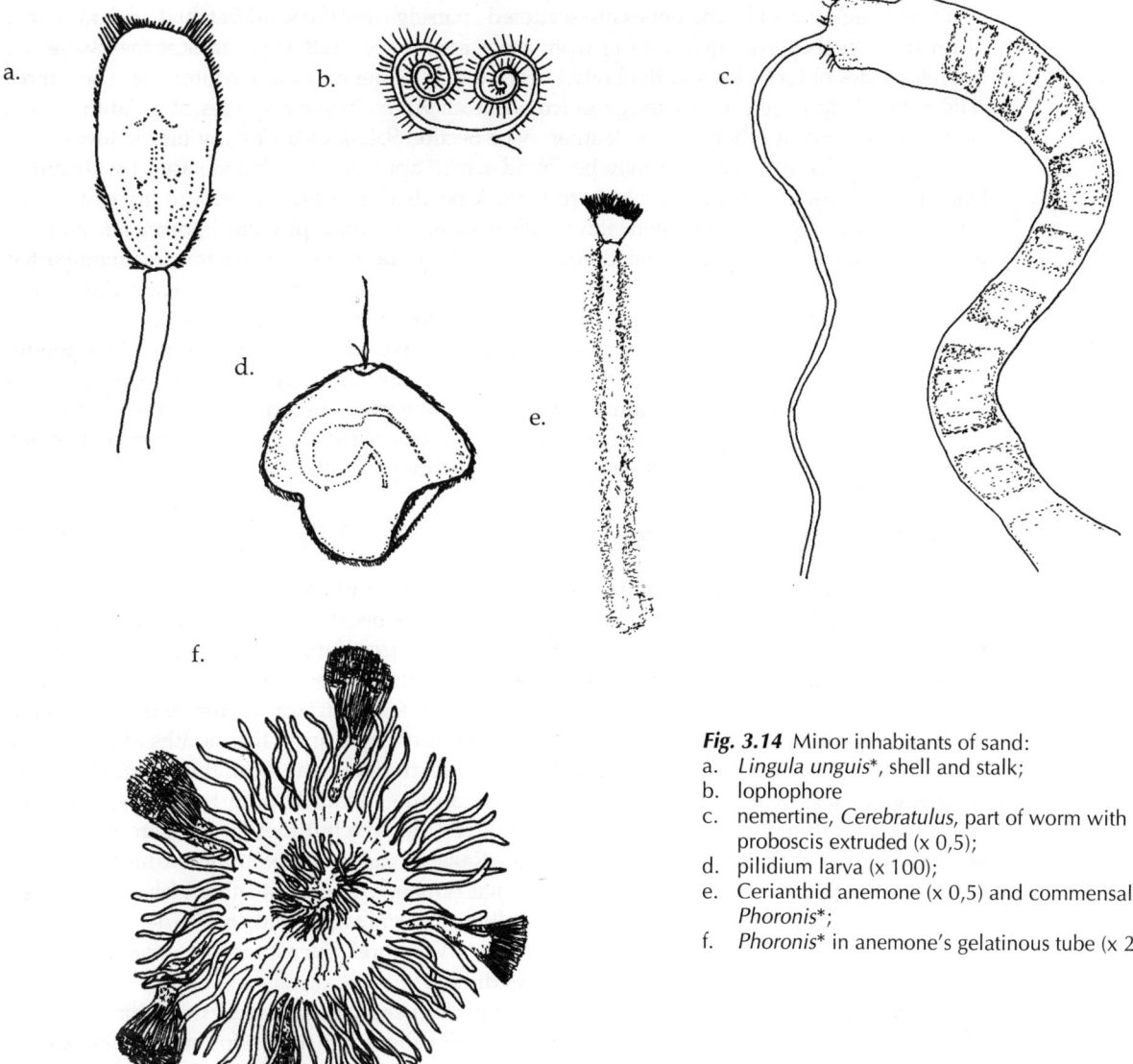

Fig. 3.14 Minor inhabitants of sand:
a. *Lingula unguis**, shell and stalk;
b. lophophore
c. nemertine, *Cerebratulus*, part of worm with proboscis extruded (x 0,5);
d. pilidium larva (x 100);
e. Cerianthid anemone (x 0,5) and commensal *Phoronis**;
f. *Phoronis** in anemone's gelatinous tube (x 2);

is digging. It is dirty white in colour with an arrow-shaped head and the body is flattened dorsoventrally. *Cerebratulus* turns on its side as it swims rapidly with sinuous motion about 10 cm beneath the surface and it cleaves through the muddy sand with a long rod-like proboscis which is suddenly extended, turning inside out as it is protruded. The hydrostatic pressure required for eversion of the proboscis is exerted when the fluid in the proboscis sac is squeezed by the anterior muscles of the body wall. It speeds through wet sandy mud (although there is no pathway) and when caught in a bucket of water, which one has strategically placed in a hole about a metre ahead, it continues swimming in the water like an eel. There is no bony tissue or other hard skeleton, but the muscles are supported by dense fibrillar tissue packed between the organs. The proboscis is also used to apprehend and paralyse prey which may be invertebrate or fish. The nemertines are unsegmented worms yet when captive in an aquarium this species will fragment into segments about 10 cm long, which are not known to regenerate as do those of other small nemertines. The larva of nemertine worms is a pilidium resembling the trochophore of a polychaete worm (Fig. 3.14d).

Burrowing cerianthid anemones are scattered sparsely over the sand flats but are common at lower tidal levels. Ceriantharia differ from other anemones (Actiniaria) in that they have two discrete circles of tentacles and the body is very much longer and more contractile. The surface cells of the body secrete a very tough mucous sheath which in some species of cerianthids may be gelatinous and in other species, leathery (see Section 9.4.5). Occasionally on the lower sand flats a pale cerianthid anemone may be found which appears to be surrounded by a circle of black flowers, each with numerous narrow black petals (Fig. 3.14e). These are the tentacles of several individuals of the peculiar animal *Phoronis* of the small phylum Phoronidea which is grouped with the Brachiopoda and Bryozoa as lophophorates. The phoronids are embedded in the gelatinous sheath of the cerianthid anemone and while feeding under water the ciliated tentacles of the lophophore are exposed (Fig 3.14f). They create a current of water which circles towards and then away from the animal thus serving two purposes: plankton and suspended detritus are carried to the mouth in the centre by a downward current and simultaneously the waste is washed away from the anus in the vortex. When disturbed, the 'worms' retract into their tubes and the cerianthid rapidly contracts out of sight. The young larva of *Phoronis* resembles the polychaete trochophore and soon develops pretentacular arms.

3.4.4 Birds of the intertidal zones

The small waders and plovers which feed on the worms and small crustaceans of the sandy shore of Inhaca Island do not usually appear numerous because they are dispersed. Most of the birds are migrants, flying about 15 000 km over a direct land route to or from the north. They breed in Canada, Greenland and the northern parts of Russia and return to the southern shores of Africa, where they spend up to six months fattening up for the return journey in August/September. Fairly large flocks may be seen roosting during the months of March and April before they leave Inhaca Island, but at other times of the year the resident species remain and those which, by chance, overwinter on the island. During the day on the west shore the migrating birds are widely scattered over the area but they congregate in flocks of several hundreds on the sandy spit at Ponta Raza in the early evenings at high tides – a spectacular sight! Nocturnal foraging by birds has not been studied at Inhaca Island.

The most common waders are the Sanderling, *Crocethia alba*, and the Curlew Sandpiper, *Calidris ferruginea*. Several others such as Curlew, Whimbrel, Turnstone and other sandpipers may also be seen (see Section 11.5.2 and Appendix A).

The Sanderling has a short, stout beak, pointed at the end. Some of these birds already have breeding plumage in April before they leave for the north. When foraging at the water's edge they characteristically run out after the water as the swash recedes, then they turn and run upshore as the next wave breaks and surges up the beach. They are very lively birds, quickly probing the wet sand for small crabs and worms between waves.

The Curlew Sandpiper is about the same size as the Sanderling and has a black, longish bill, curved down at the tip. Its non-breeding plumage is darker than that of the Sanderling, being mottled brownish-grey. Birds in breeding plumage are sometimes seen; the body is a dark chestnut colour with white around the bill and eye. When it forages on the muddy sand it walks along quickly probing the sand with its long beak and extracting polychaete worms.

Plovers are slightly heavier in build than sandpipers and have short, straight bills, slightly swollen at the tip, shorter legs, long pointed wings and a short tail. The migrant Grey Plover, *Squatarola squatarola*, and the resident Whitefronted Plover, *Charadrius marginatus*, are most common but others are sometimes seen, including the solitary Ringed Plover. The Grey Plover occurs in small flocks of 10-20 mingling with the waders but can be distinguished from them by the habit of running in short bursts with the body held horizontally. When it stops it pecks at the sand, not penetrating as deeply as the longer-billed waders. Some adopt breeding colours before migrating north, becoming spangled, black-and-white above and black below except underneath the white tail.

The Whitefronted Plover is a smaller, resident bird which nests on Inhaca Island in dry sand behind the dunes where a small hollow is scraped out and lined with fragments of shell.

They may be seen running very fast in pairs tucking the head into the shoulders as they forage along the water line. The plovers breed from August to December and the eggs are light in colour with few markings. The young are cared for by both parents.

3.5 SEA GRASSES ON SANDY MUD OF THE LOWER SHORE AND ASSOCIATED ANIMALS

On the sheltered west shore a zone of sea grasses is exposed during spring tides but remains very wet or covered by shallow water during neap tides. The layer of sand over the dark mud is somewhat thicker than in the previous bare zone. The sea grasses are sparse at first and then occur more densely in depressions, until at low spring-tide level the covering becomes continuous over large areas and penetrates to the subtidal fringe and to the shallow subtidal zone. A definite zonation of plant species occurs, that depends on the tidal levels and associated factors (Bandeira 1989).

3.5.1 Narrow-leaved and broad-leaved sea grasses

The sea grass *Halodule wrightii* (Zannichelliaceae or Cymodoceae) (Fig. 3.15a-e) occurs between low water of neap and spring tides; it merges with other species of sea grasses towards the lower levels. Its distribution is patchy in the *Macrophthalmus* areas (see Section 3.5.2) and even where it is denser it may be confined to slightly raised areas where sand is accumulating because of the plant's growth. *Halodule wrightii* is the pioneer amongst sea grasses in this type of tropical habitat, since it stabilises the substratum by means of slender, creeping rhizomes from which many roots and a short stem grow at each node, trapping sand from the receding tide. This species of sea grass is tolerant of fluctuations in salinity and so can flourish where there is fresh water drainage beneath the sand as on the west shore of the Island. *Halodule wrightii* can withstand several hours of exposure to air without damage, but when it grows in permanent pools the stems branch and grow much longer (see Section 5.3.3).

The leaf blades of *Halodule wrightii* are grass-like, usually 8-10 cm long and less than 1 mm in width. When examined through a hand lens the leaf tips may be seen to have two well-developed lateral teeth on either side of a conspicuous middle tooth protruding from the dark midrib (Fig. 3.15b). The leaves have three parallel veins.

Although this species is superficially similar to *Zostera capensis* (Zosteraceae) of South African estuaries, the latter is not common on the west coast of Inhaca Island; it is confined to the more sheltered southern and northern bays. The two genera of sea grasses can be distinguished by their leaf tips and veins since in *Zostera capensis* the leaf tips are rounded and have a central notch and the leaves are net-veined (see Fig. 5.5b).

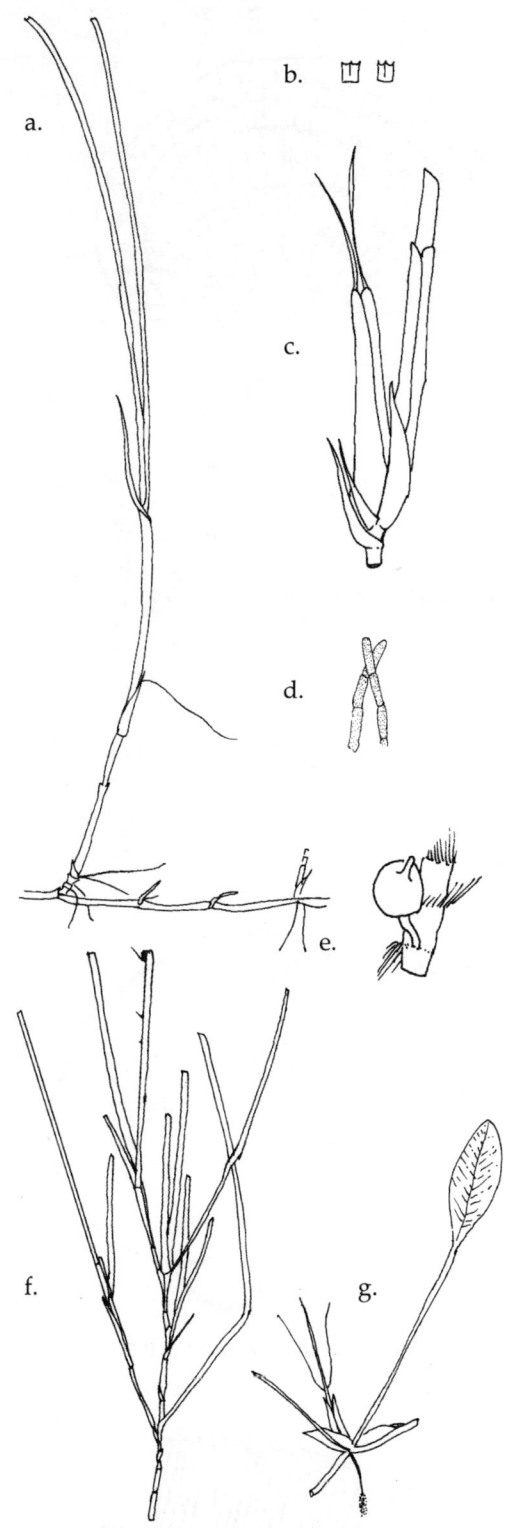

Fig. 3.15 Pioneer sea grass, *Halodule wrightii*
a. leaves and rhizome (x 1)
b. leaf tips,
c. female flowers,
d. pollen,
e. fruit. Associated sea grasses:
f. *Halodule uninervis* and
g. *Halophila ovalis* (a-g after F.M. Isaac 1968).

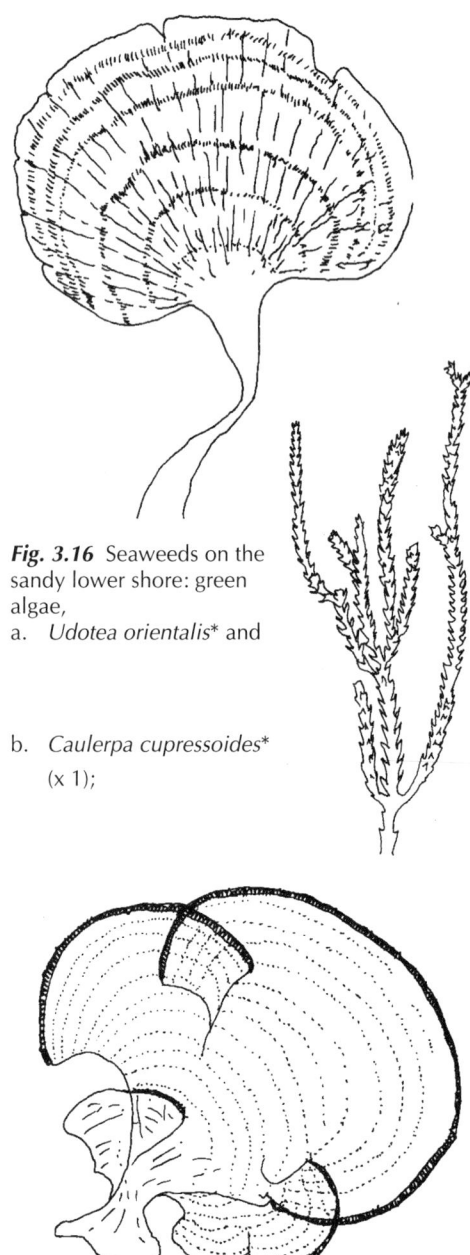

Fig. 3.16 Seaweeds on the sandy lower shore: green algae,
a. *Udotea orientalis** and

b. *Caulerpa cupressoides** (x 1);

c. brown alga, *Padina boryana** (x 0,5).

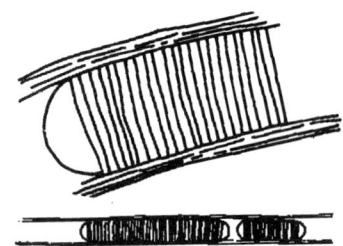

d. blue green alga, *Lyngbia majuscula* (x 1 000) (after S.F.M. Silva 1991).

Halodule wrightii flowers in summer, each flower being enclosed in a leaf sheath. The female flower is 1-5 cm long, has a small ovary and quite long styles (Fig 3.15c), the male flower is smaller and has two anthers, one longer than the other. The pollen is thread-like and floats on the water surface (Fig 3.15d). The fruit is black and spherical, about 2 mm in diameter, the base of the style persisting as a very small spike (Fig. 3.15 e).

A sibling species, *Halodule uninervis* (Fig. 3.15f) is sometimes associated with the pioneer species and is confined to depressions; it is a more delicate plant with slightly wider leaves (1-3 mm) in which the leaf tip has three equal, short teeth. A third (but less common), species in this area, *Halophila ovalis* (Fig. 3.15g), is a hardy plant in relation to temperature and salinity, but it is intolerant of being covered by sand and its growth keeps pace with sedimentation. This species has varieties of leaf shape: slender and almost grass-like and the broader, oval or raquet-shaped, to which the rank of subspecies was given (*linearis* and *ovalis*). A recent study (Bandeira 1989) has shown that the narrow and broad forms inter-grade and thus they belong to one species.

Much nearer low spring-tide mark a belt, about 30 m wide, of a broader-leaved species *Cymodocea serrulata* (of the same family) occurs in large depressions which retain water when the tide is out. It merges with the long-stemmed, broad-leaved sea grass, *Thalassodendron ciliatum* in the subtidal fringe. The latter continues into the subtidal zone (see Section 9.1; Fig. 9.1).

Cymodocea serrulata has fleshy rhizomes which become woody with age and are usually white and mottled purple, with one or more branching roots at the nodes. Leafy shoots with bright pink sheathing bases arise from the nodes and each leaf has a conspicuous ligule. The leaves are about 1 cm broad and 15 cm long; the leaf tips are rounded and bear tiny teeth (Fig. 9.2a and b). Flowers of this species are similar to those of *Thalassodendron ciliatum* (see Section 9.1).

The shore species of sea grasses, described above, have a higher photosynthetic rate than that of subtidal sea grass species and they maintain this advantage even when the tide has risen to cover them up to a metre deep. The rate of photosynthesis does not lessen when exposed to air in the shore species, whilst the more aquatic species exhibit a reduced rate.

3.5.2 Seaweeds on the sand

A firm substratum is necessary for motile or non-motile spores of seaweeds to attach themselves and germinate; thus macroscopic algae do not, in general, grow on sand or mud flats. Nevertheless some seaweeds may be found on the west shore sand. The green alga *Enteromorpha compressa* (mentioned in Section 3.2.1 as coating the living shell of *Donax faba*) covers any solid surface such as stones and small fragments of coral debris which are often half buried in the sand.

The large tropical group of green algae belonging to the Order Siphonales, is represented by two species, *Udotea orientalis* and *Caulerpa cupressoides* which grow in bare patches of sand between the narrow-leaved sea grasses below mean low neap tide. *Udotea orientalis* (Fig. 3.16a) has robust fan-like fronds (6 cm) which arise from a short 'stem' and spread on the surface sand. Along the surface of clean sand at low tide *Caulerpa cupressoides* (Fig. 3.16b) occasionally forms single runner-like axes which produce several moss-like branches 5-6 cm long.

Two brown seaweeds occur on the sand flats: *Padina boryana* and *Feldmannia irregularis*. *Padina* grows on sand that covers flat buried rocks. It has a delicate, pale greenish-brown, fan-like frond (6 cm) which grows from a short bunch of rhizoids forming a 'stalk' (Fig. 3.16c). The filamentous *Feldmannia* may cover parts of the muddy sand flats in summer near low tide.

A slimy blue-green alga (Cyanophyta), *Lyngbya majuscula* (Fig. 3.16d) covers the surface of large wet patches of sand-flats in summer. In addition to the algae mentioned above, the sand grains themselves support a varied microflora of diatoms (Fig. 3.4a) and also bacterial cocci which degrade the detritus (see Section 3.3).

3.5.3 Associated Animals: *acorn worms; sea cucumbers; molluscs; crustaceans; shrimps; commensal fishes*

One of the most striking features of the sand-flats around Inhaca Island is the number and size of 'wormcasts' on the surface. A few are scattered in the *Macrophthalmus* areas but there are many more among seagrasses, and in the sand between coral debris they occur in small groups. Small bare mounds of sand are visible when the tide has just receded, but before it rises again each of these has erupted into a 'wormcast' about 50-100 mm high in which the coiled sandy 'rope' is about 5-10mm in diameter. These are made by acorn worms (Van der Horst 1940), so-called because of the appearance of the anterior end of the 'worm', which has an oval proboscis arising from a thick collar, thus resembling an acorn (Fig. 3.17b). One can watch the wormcast growing as the animal feeds below (Fig. 3.17a).

The mouth is just in front of the collar, leading to a respiratory pharyngeal region in which microscopic pores open from the gut to the exterior, as do the much larger gills of fishes; this feature is the origin of the name 'enteropneust' (literally gut breathing). This region of the body has two flat 'genital wings' on each side which contain the gonads. A short 'liver' region follows with hepatic caecae visible externally as small ridges, and the acorn worm ends in a long 'abdomen' and terminal anus. Enteropneust species at Inhaca Island range in length from 80-300 mm, but one species over 800 mm long has been collected. Acorn worms are soft and flaccid when removed from their burrows and sheathed in a prolific secretion of thick, sticky mucus which, in some species, smells

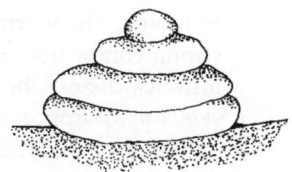

Fig. 3.17 Acorn worms, enteropneusts: *Balanoglossus studiosorum**
a. worm cast on surface;

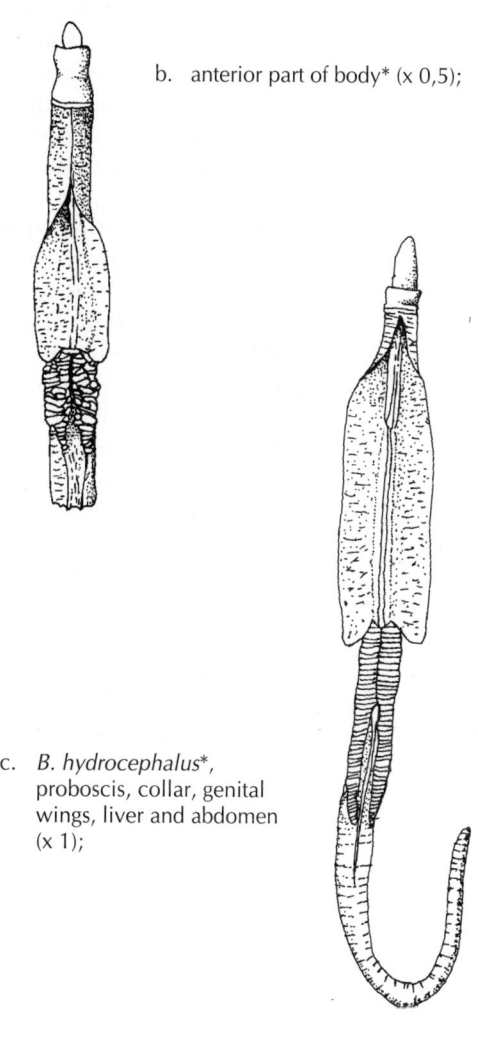

b. anterior part of body* (x 0,5);

c. *B. hydrocephalus**, proboscis, collar, genital wings, liver and abdomen (x 1);

d. skeleton of proboscis in *B. hydrocephalus* (after Van der Horst 1940).

of iodine. The worms become spirally tangled when removed from the burrow because the strong contraction of the longitudinal muscles and the weakness of the ventral and circular muscles distort the shape. There are neither hydrostatic forces of an internal fluid nor a skeleton to support the animal.

In order to extract this extraordinary acorn worm intact and to view it alive a special digging technique may be employed (Van der Horst 1940). Before the tide rises a large circle is drawn on the sand to include the exit of the U-shaped burrow (with the wormcast) and its anterior end, a hole in the sand surface within a metre away. Then a circular ditch is dug about 30 cm deep and the water that drains into it is rapidly bailed out by several workers. The 'island' left in the centre of the ditch begins to crack along the axis of the burrow and packs of muddy sand can be carefully removed until the animal is seen living in its mucus-lined burrow. If the abdomen is grasped it will break off, but the anterior end may be held in the hand while the rest is excavated. After the liver region is exposed and held, the rest may be safely pulled out without breaking the body.

Two of the six or seven species of enteropneusts on Inhaca shores are common on the west shore: *Balanoglossus studiosorum* occurs singly in fairly clean sand on the lower shore and *B. hydrocephalus* (Fig. 3.17c) is found in groups in the more calcareous sand among the coral debris of the reef flat. Both species are about 300 mm long but they differ in colour, in proportions and in the manner of burrowing as well as in habitat.

B. studiosorum is a sluggish animal with a small pale proboscis protruding a few millimetres from the darker collar which is elongated (20 mm), narrow and muscular. The proboscis is mobile and may help to loosen sand, but burrowing is really effected by the muscular contractions of the collar. The genital wings around the branchial region are folded dorsally and the liver region is orange anteriorly and green posteriorly; the abdomen is a dirty white. *B. studiosorum* was given its specific name in recognition of the many students who painstakingly aided Professor van der Horst in extracting whole specimens of the acorn worm in the early days of research on the island.

B. hydrocephalus (Fig. 3.17c) is more active than the former species and has a larger proboscis (15 mm) with a much stronger chitinous skeletal support (Fig. 3.17d), and this is used in burrowing. The name draws attention to the exceptionally large proboscis in this species. The collar, in contrast, is short and broad. The animal is more vividly coloured than *B. studiosorum*, with yellow proboscis, darker collar, brown to bright red genital wings, which are not folded together over the branchial region, a darker liver region and a greyish abdomen on which light and dark rings alternate.

The major force in burrowing (exerted by the collar in *B. studiosorum* and by the proboscis in the other species), is from weak peristaltic movements starting at the tip. A strong auxiliary force is supplied by the surface cilia of the body which beat backwards when the animal is pulled forward and vice versa in retreat. The worm is thus slowly propelled whilst floating in its own mucus secretion which lines the burrow. The mucus is reinforced by a secretion of protein which gives it a firm consistency and supplies sufficient support. The diameter of the burrow is determined by the width of the collar which fits snugly when the animal is at rest. The water of the respiratory current, created by the pharyngeal cilia, thus passes into the mouth and out through the gill pores, being then directed backwards, so irrigating the burrow.

When the acorn worm is feeding, the proboscis protrudes from the burrow and either lies flat or moves slowly to collect sand and detritus. Ciliary tracts convey particles to the collar where some sorting takes place. Mucus strings of the finer particles are formed and are passed backwards to the ciliary organ. This is a sensory depression between the collar and the mouth which is said to include an organ of taste; more sand may be rejected there. The mucus on the proboscis contains amylase, the enzyme of mammalian saliva, which initiates digestion. As the food passes down the alimentary canal digestion is continued through the action of 'liver' secretions. The movement of food and sand in the abdomen is carried out almost entirely by cilia since the gut has only very thin superficial circular and longitudinal layers of muscles.

The free-swimming larva of the acorn worm, known as the 'tornaria' has a complex winding band of cilia on its surface, which resembles that of the young echinoderm larva and suggests an ancestral relationship between the two phyla.

In spite of their appearance, lethargic movements and sand-eating habit, these animals are certainly not 'worms'. They are classified in the Phylum Hemichordata – as though a retrogressive relation of the Chordata. In fact they do have some of the hall-marks of chordates: a dorsal notochord, a dorsal nerve cord (as well as a ventral one in the abdomen), but only a peripheral network of nerve cells. There is a simple blood circulation but no heart. There is even a bizarre-shaped skeleton tailored to support the proboscis/collar complex (Fig. 3.17d). Much has been written about their evolutionary place among animals.

The large sea cucumber *Holothuria scabra* (30 x 10 cm) lies half-buried in the sand, sometimes as many as three or four per square metre (Fig. 3.18a). Its grey skin is thick, tough and fibrous and lined with a thin layer of circular muscle beneath which there are five separate strips of longitudinal muscle. All sea cucumbers have their skin reinforced with microscopic calcareous spicules of peculiar shapes: needles, hooks, anchors, buttons with button holes and spiny 'tables'. The spicules are useful in the identification of species. A fragment of skin that has been boiled in caustic soda will release its spicules (Fig. 3.18b) and these may be examined under the microscope and compared with pictures in a key to the Holothurioidea. The spicules in skin and tentacles vary in shape.

The power of locomotion in *H. scabra* is limited to slow muscular sliding under water using its longitudinal muscles; it has a few scattered tube feet on the lower surface or sole and it is very sluggish. When there is still a sheet of water over the sand the animal feeds at the surface on detritus, bacteria and meiofauna, by swallowing large quantities of sand unselectively. Twenty stubby tentacles with short branches, each ending in a suction pad (Fig. 3.18c), scoop up the surface sand and shovel it into the mouth. These tentacles are enlarged and much modified 'tube feet' which are characteristic of echinoderms (see Section 9.4.1). The detritus is digested and assimilated with about 50% efficiency; an enormous volume of sand passes through the gut and is voided as large greyish balls which accumulate behind the animal until the tide washes them away. Sometimes these balls of sand are the only sign that this cucumber is there, completely buried in sand (Fig 3.18d).

Quite a large proportion of *H. scabra* (30% in some areas) is parasitised externally by a small snail, *Mucronalia*, with a grey spiral shell and very small foot (Fig. 3.18e). These attach themselves permanently to the skin, mainly around the anus, by means of the proboscis, which is longer than the shell and penetrates deeply through the thick skin into the body cavity and into the respiratory trees. Enzymes are secreted through it, which dissolve the host tissues and the fluids formed are sucked into the parasitic mollusc's gut.

Two species of black sea cucumber are common on the surface of the sand and look alike. Each is 30 cm long and half

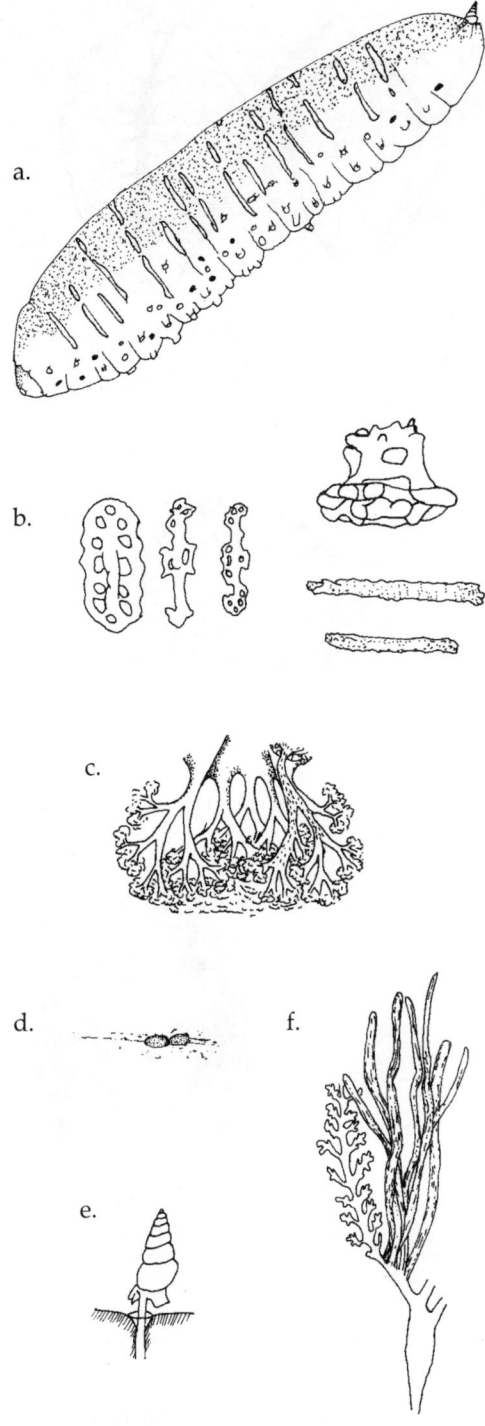

Fig. 3.18 Sea cucumber in muddy sand: *Holothuria scabra*
a. entire animal (x 0,3);
b. various spicules from the skin;
c. feeding tentacles;
d. faecal balls of sand (x 0,5);
e. snail parasite, *Mucronalia* sp.;
f. Cuvierian organs and respiratory tree.

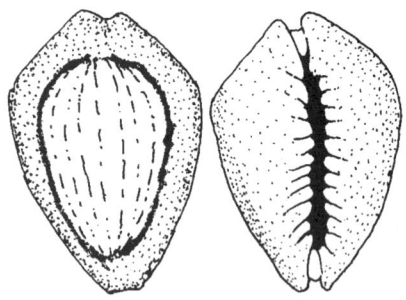

Fig. 3.19 Gastropod snails on the lower sandy shore:
a. Cowrie, *Cypraea annulus*, dorsal and ventral views (x 2);

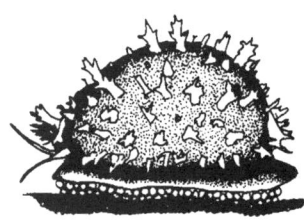

b. mantle expanded and eggs incubated;

c. young shell (x 2);

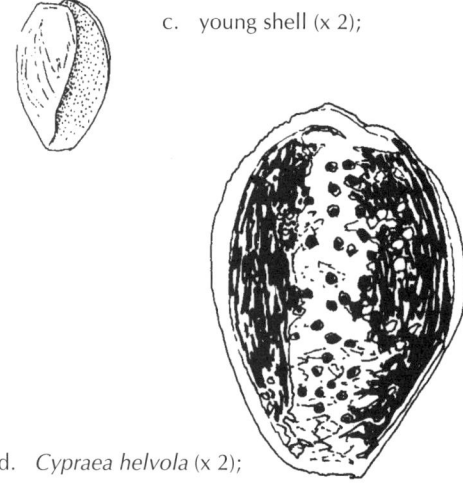

d. *Cypraea helvola* (x 2);

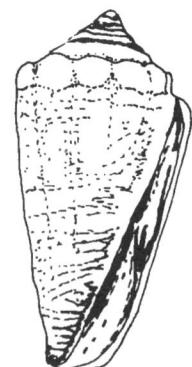

e. *Conus lividus* (x 1) and
f. poison tooth (x 10) (a-f after Kilburn and Rippey 1982).

the width of *H. scabra*, but can easily be distinguished since *Holothuria atra* is nearly always coated with adhering sand, whilst *H. leucospilota* is bare and shiny. The latter has the peculiar habit of expelling a mass of stiff white sticky strings from the cloaca at the hind end when disturbed (Endean 1957). These 'Cuvierian tubules' break off from the common stem of the respiratory trees attached to the cloaca (Fig. 3.18f), then lengthen and swell in sea-water, becoming extraordinarily adhesive; when they surround a potential predator it is completely incapacitated. Before expulsion the Cuvierian tubules are tightly coiled by means of longitudinal and spiral muscles, which surround a core of connective tissue containing muco-polysaccharide and collagen fibres similar to those in vertebrate cartilage. Ejection of these tubules is achieved by the forceful injection of water from the respiratory trees and relaxation of the cloacal muscles. The sudden stretching of the tubules bursts open the surface cells releasing two kinds of protein which absorb water and in so doing become intensely sticky, adhering to any objects except the sea cucumber's own surface skin. The strings continue to become increasingly sticky as though an adhesive and its primer were being mixed. The combination of tough, toxic skin and evisceration repels predators, although sea cucumbers are eaten by some fishes and by the large snail, *Tonna* (see Section 9.4.3).

Sea cucumbers usually have external fertilisation and pelagic larvae; the early larva, the 'auricularia', resembles that of the starfish in having extensive ciliary bands. At about 1mm in length it changes into a barrel-shaped 'doliolaria' larva with four girdles of cilia which show signs of bilateral symmetry (as in the adult) and some similarity to the larva of the acorn worm. Budding and transverse fission has been observed in *H. atra*.

Two species of cowries are the most abundant snails amongst the sea grass, *Halodule*, which are here unusually exposed to light; like most cowries they also occur under rocks as on the reef flat and coral reef. *Cypraea annulus* (20 cm) with a thick grey shell marked by a fine orange ring is most common (Fig. 3.19a). *C. moneta* (25 mm) the 'money cowrie' of the nineteenth-century slave trade has a thick yellow shell. The spire, present in the young shell (Fig. 3.19c), is concealed by overgrowth of the adult shell, and the wide aperture of the juvenile becomes slit-like with marked indentations. The surface of the shell is highly polished because when the animal is active overnight or under cover of a rock, the flaps of the mantle on each side cover the shell completely and continuously secrete an enamel-like layer of the shell. The mantle has many sensitive finger-like projections and a touch elicits its complete withdrawal. These processes resemble an algal growth and apparently act as camouflage whilst they serve for respiration (Fig 3.19b). The cowries browse on small compound animals, such as tunicates and also on sponges. They lay round clusters of eggs, each in a small white capsule, which adhere to the under-surfaces of rocks. The female broods the eggs by covering them with the sole of the foot until they hatch into late veliger larvae after 2-3 weeks (Fig. 3.20b). After the larvae settle and

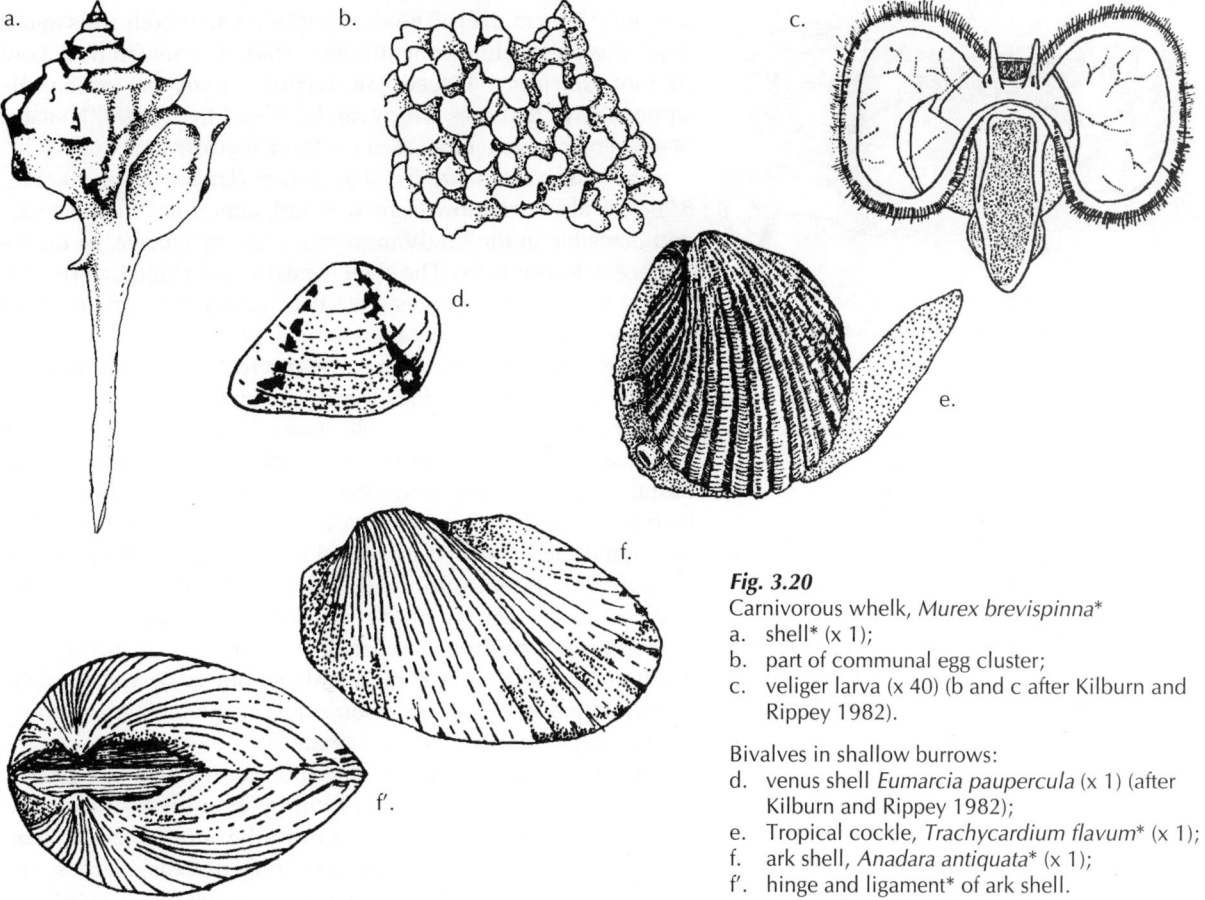

Fig. 3.20
Carnivorous whelk, *Murex brevispinna**
a. shell* (x 1);
b. part of communal egg cluster;
c. veliger larva (x 40) (b and c after Kilburn and Rippey 1982).

Bivalves in shallow burrows:
d. venus shell *Eumarcia paupercula* (x 1) (after Kilburn and Rippey 1982);
e. Tropical cockle, *Trachycardium flavum** (x 1);
f. ark shell, *Anadara antiquata** (x 1);
f'. hinge and ligament* of ark shell.

metamorphose the young shells are almost barrel-shaped, resembling bubble shells such as *Hydatina* although the shell is thicker and has a wide toothless aperture (Fig. 3.19c). *Cypraea helvola* which prefers a sheltered shore may be found under scattered coral debris. (Fig 3.19d).

The carnivorous and scavenger snails mentioned above in the *Macrophthalmus* area (Section 3.5.2) are still fairly numerous. In addition a few species of the most highly evolved carnivorous sea snails, the Neogastropoda, may occur in this zone although they are more common at the lower level (see Section 9.4.3). *Conus lividus* may be found here (Fig. 3.19e). The radula has only one to three pointed teeth (Fig 3.19f) in each row; they convey poisonous secretions which paralyse and then kill its prey. It feeds on acorn worms and polychaetes.

Here one of the smaller whelks, *Murex brevispinna* (Fig. 3.20a) (80 cm) is recognised by its swollen shell and low spire, a very long siphonal canal and many short, strong spines projecting around the shell. These spines are said to stabilise the shell in soft sand under moving water. The whelk buries itself between tides, with only the tip of the siphon canal protruding. It feeds on molluscs using a rasping radula and enzyme secretion of the foot gland in a manner similar to that of moon shells, described in Section 3.4.2. After fertilisation many females spawn communally and produce one large mass of stiff, parchment-like egg capsules linked together, to which several females cling for a time (Fig. 3.20b). The eggs hatch at a late veliger stage (Fig. 3.20c).

Most of the bivalves that burrow in the sandy mud are more often subtidal animals, but a few with thick shells extend their ranges in the intertidal narrow-leaved sea grass zone. The venus shell *Eumarcia paupercula* (*Chione ambigua*) (30 mm) (Fig 3.20d) is the most common here although its numbers are optimal only in estuarine conditions. It has a smooth convex shell with a buff ground colour and variable brownish radial markings, either zig zag lines or spots. When the bivalve becomes exposed on the surface it burrows actively using a hatchet-shaped foot, but

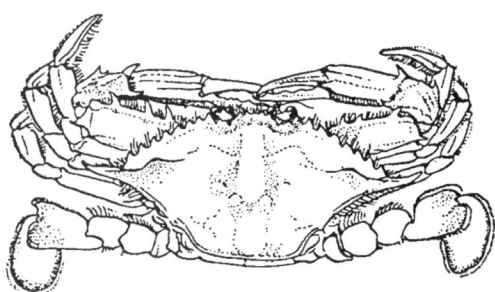

Fig. 3.21 Crustaceans on the lower shore:
a. swimming crab, *Portunus pelagicus** (x 0,5);

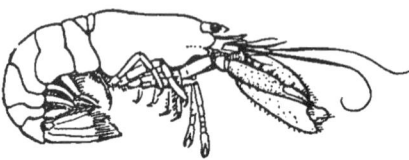

b. pistol shrimp, *Alpheus crassimanus** (x 2);

b'. snapping chela enlarged (arrow to the socket for trigger);

digs only 2-3 cm deep. The shell gapes open at both ends and if it is threatened while lying on the surface it is able to thrust out its foot and jump. It feeds on detritus in suspension and its siphons have tentacles which can be closed to prevent the entry of silt, a mechanism often seen in *Donax* species.

The tropical cockle *Trachycardium flavum* (60 mm) (Fig. 3.20e), is also a shallow burrower and sometimes the posterior end is visible in the sandy mud or it may, by chance, lie on the surface between tides. The shell is heavy and ridged, with little scales on the outer ridges, and it is sand-coloured; its short siphons end in tiny tentacles which bear eyes. Cockles can 'hop' by suddenly thrusting out the foot, a mechanism for evasion of predators and for limited migration on the shore.

A third bivalve, an ark shell, *Anadara antiquata* (70 mm) (Fig 3.20f and f') is another heavy, light-coloured shell with strong radial ribs, which may lie on the surface when the tide is out. It is distinguished from other bivalves by the long toothed hinge-line of the shell, which is trapezoid in shape. It is notable for the presence of haemoglobin in the blood corpuscles of the haemolymph, instead of the more usual haemocyanin. The pigment is not confined to specific muscles such as the heart or adductor muscle as in some bivalves, but is circulated. This bulky animal may thus gain some advantage in energy supply in the low oxygen level of this habitat.

The most conspicuous crustaceans amongst the sea grasses although not the most numerous, are the large carnivorous swimming crabs of the genera *Portunus* and *Charybdis*, with large eyes, which invade the area from the subtidal zone when the tide rises, but few burrow here. Swimming crabs have lightly built exoskeletons, the chelae being strenghthened by longitudinal ridges. The last segments of the fifth pair of legs are flattened and fringed with hairs for use as paddles, which requires exceptional mobility in the joints. *Portunus* (=*Lupa*) *pelagicus* (Fig. 3.21a) is the largest and fastest swimming crab, achieving a speed of over 1 m/sec (Hartnoll 1971). It is streamlined for swimming sideways: the carapace has short front and back edges and many lateral teeth, the last pair of which is produced into long spurs thus making the effective swimming 'length' twice that of the 'breadth'; the length is further exaggerated by the long tapering chelipeds which are held wide apart in a straight line whilst chasing prey. During swimming the forward paddle cleaves through the water with strong, bold, almost vertical strokes, giving both propulsion and lift and is 'feathered' on return. The paddle on the trailing side sculls to and fro whilst the walking legs make weaker movements concerned with steering. Most swimming crabs burrow sideways into the sand leaving an opening surrounded by shell debris. The chelae are toothed and narrow, with fast contracting muscle fibres, modified for catching shrimps, prawns or fish.

Shrimps and prawns may be found temporarily buried up to the eyes when the tide is out and may be caught by sweeping a net through the sea grasses at night when they are active under water; the numbers increase towards low tide and in shallow water subtidally. These are small swimming decapods, and, as in crabs, a carapace covers the head and thorax, concealing a gill chamber; the abdomen is extended, bearing biramous 'swimmerets' and ends in a pointed telson flanked by fan-like uropods (as in lobsters). Both shrimps and prawns have laterally compressed bodies and between the stalked compound eyes, typically, lies a rostrum, i.e. a spike with a number of 'teeth' which vary in number with the species. An obvious distinction between common shrimps (Caridea) and prawns (Penaeidea) is the 'hump-back' coupled with shorter swimmerets of the shrimp and the straighter abdomen of the prawns, when at rest. This distinction probably reflects the shrimp's habit of carrying eggs on the swimmerets until larvae emerge, whereas prawns migrate out to sea to breed and liberate their eggs. The prawn larvae return to estuaries and sheltered shores to feed and grow into adults. Although the prawns found intertidally are usually larger than the shrimps they are still juvenile, i.e. under a year old (see Section 4.5.3). Shrimps usually have the first three pairs of walking legs modified into small chelipeds, whereas prawns have only two pairs of chelipeds. The third pair of chelipeds is sometimes modified for a very particular diet or some other purpose in rock shrimps, coral reef shrimps and in two sand shrimps described below.

The common shrimps are the alpheid or pistol shrimps, each individual having one enormous, inflated chela. Near the hinge the small 'finger' of the chela bears a 'peg'; this may be 'cocked' like a pistol trigger and then snapped into a pit in the 'hand'. Under water a loud report is made, which is said to signal occupation of a burrow and may be used in hunting prey to stun it by means of the shock waves produced. Unlike most shrimps, alpheids have a very short pointed rostrum. *Alpheus crassimanus* (Fig. 3.21b and b') (the shrimp with the largest chela) is present but is more common in the mangrove of the south bay of Inhaca Island and there are other species, such as *A. rapax* and *A. rapacida*, both of which have setose chelae. These shrimps share their burrows with the gobies *Cryptocentrus cryptocentrus* and *Vanderhorstia delagoae* respectively (see Fig. 9.11a and b). Under cover of water the shrimps dig permanent slanting burrows and deepen them by repeatedly pushing out heaps of sand with the large chela. The sand is swept away by water movements leaving a scattering of grit and fine shell debris around the opening, by which one can recognise the likely presence of the occupants before they appear. The commensal goby perches at the entrance of the burrow while the male and female shrimps are at work; when disturbed it disappears into the burrow and the shrimps cease activity. The goby cautiously emerges after a time and apparently gives an 'all-clear' signal before burrowing is resumed. It is said that the fish communicates with its host by a flick of the tail on which the tip of a shrimp's antenna rests. These commensals may easily be studied while snorkeling in very shallow, calm water during a neap tide.

A pair of palaemonid shrimps, *Harpilius (Periclimenes) brevicarpalis* may often be found in association with the large green sea anemone, *Heteractis (Stoichactis) magnifica* (Fig. 3.21c), which burrows into the sand leaving its multi-tentacled disc exposed and flush with the surface sand (see Section 9.4.5). The shrimps (Fig. 3.21d) hover among the tentacles feeding on plankton and may disappear into the mouth or hide under the borders of the disc in the sand. These beautiful shrimps (30 mm) at Inhaca Island are completely transparent except for pure white patches on the third abdominal segment dorsally and on the anterior half of the telson; mature females also have two white thoracic spots. The tips of the uropods have orange spots ringed with blue; these apparently menacing caudal 'eyespots' are made more so by the continuous waving of the abdomen. It is said that fishes do not prey upon this species. Some species of *Periclimenes* with sharper chelae are known as 'cleaner-shrimps'; they remove particles between the teeth of some fishes or parasites from their gills. Sometimes pairs of damsel fishes, *Dascyllus trimaculatus* and *Amphiprion allardi* join the shrimps in association with the anemone. *D. trimaculatus* (35 mm)

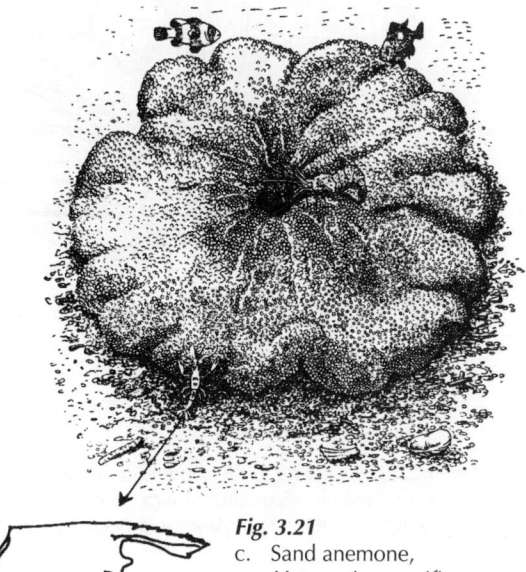

Fig. 3.21
c. Sand anemone, *Heteractis magnifica* (x 0,5) and commensal fishes and shrimp*;
d. arrow to shrimp (enlarged) carapace of *Harpilius (Periclimenes) brevicarpalis*;

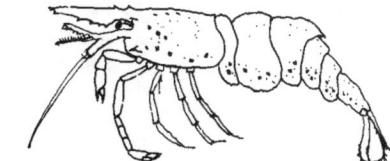

e. shrimp, *Hippolyte* sp. (x 1) (d and e after Kensley 1972);

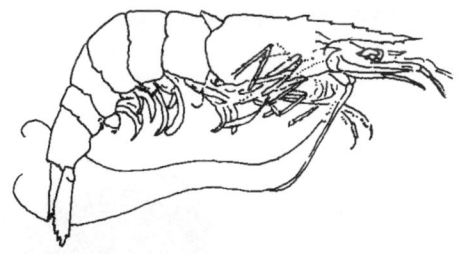

f. sand prawn, *Penaeus semisulcatus* juvenile (x 1); (after A.J. de Freitas 1984).

has three white spots on a jet-black background (Fig. 8.17b and c); *A. allardi* has a similar shape and broad vertical bands of colour. They all become protected from the nematocytes (poison darts) of the anemone by the adsorption of its mucus and so they are not recognised as alien by the anemone. (This anemone is a ciliary feeder on micro-organisms.)

Many other common scavenging shrimps occur amongst the sea grasses, such as the green-camouflaged *Hippolyte* sp. with the very flexible wrist of the second cheliped (Fig. 3.21e), the tiny *Latreutes pygmeus* and the fat, golden yellow *Processa aequimana* with equal chelae.

At least six species of prawns have been found at Inhaca Island, but four of them are restricted to the mangrove channel and sea grasses of the southern bay and are described in Section 4.5.3. *Penaeus semisulcatus* (Fig. 3.21f) prefers living among sea grasses on the sandy bottom of the west shore, whereas *P. japonicus* burrows in cleaner sand without vegetation. A study of their adaptations to the different habitats is reported in Section 4.5.3.

3.6 MICROBIOTA BETWEEN SAND GRAINS

A food web of organisms invisible to the naked eye exists on its own between the sand-grains on the sandy shores, the components of which are known as the 'interstitial flora and fauna'. The diatoms (discussed in Section 3.3) together with the bacteria causing decay are the primary producers in this food web. The consumers, usually less than 500 µm in length, are the unicellular protozoans, and the 'meiofauna' comprising many groups of minute multicellular animals. Both groups include herbivores, detritus feeders, predators and even suspension feeders. The criterion of size is somewhat arbitrary since most of the animals in both groups are long and thin and some protozoans are as large as the crustaceans, the nematodes or other groups of the meiofauna. The dimensions of the animals of the interstitial fauna vary according to the sizes of sand particles and the degree to which the sand has been consolidated, which together determine the amount of interstitial space. All of these animals have some kind of protective exoskeleton and most of them move freely between sand grains with some ability to adhere to them when disturbed by the tides.

3.6.1 Protists: foraminiferans and ciliates

Protozoans of predominantly two classes, foraminiferans and ciliates, live in marine sands. The former tend to occur in coarser sand and the latter in finer sand, rich in diatoms.

A foraminiferan has a simple amoeba-type cell structure but it secretes a shell with many perforations (foramina) through which specialised pseudopodia protrude (Fig. 3.22a). The shells have many compartments which are secreted in succession as the animal grows and all compartments are occupied by the animal. A very young foraminiferan, just formed by sexual or

Fig. 3.22 Microfauna in the sand: Foraminifera (all approximately × 20, after Moura 1965):
a. feeding mechanism in *Ammonia beccari*;

Shells of:

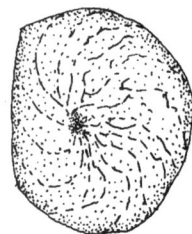

b. *Amphistegina lessoni*;

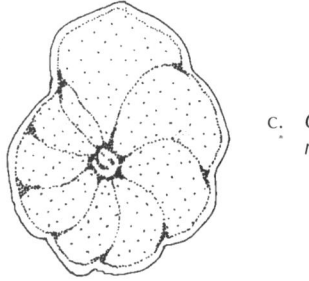

c. *Globorotalia menardi*;

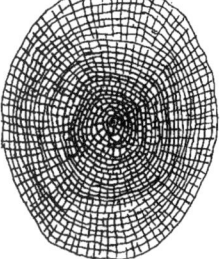

d. *Sorites marginalis*

e. *Peneroplis planatus*;

asexual reproduction, secretes a shell to fit itself, with many foramina and one larger aperture. The shell cannot increase in diameter so that when the protoplasm grows it bulges out through the large aperture of the shell and forms a lobe which secretes an adjacent larger compartment. This is repeated several times, and in different species the shell may become coiled, resembling a snail, or it may remain straight. The shell is formed of 'tectin', a mucoid substance, formed in Golgi bodies, that usually becomes reinforced with calcium carbonate by secretion through the plasma membrane.

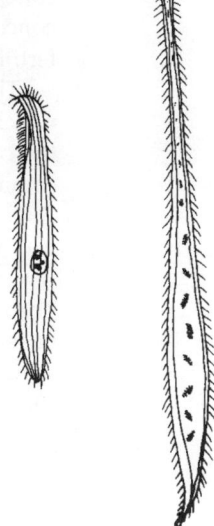

Fig. 3.22 ciliates (after Sleigh 1973):
f. *Trachelocerca* sp.

g. *Geleia* sp. (x 100).

The pseudopodia that protrude permanently through the foramina are long, thread-like and sticky, branched and interconnected. They spread out and float in the interstitial water of the sand and form a fine net which traps bacteria. Thus the Foraminifera are the suspension feeders of the interstitial fauna. Although the pseudopodial network (reticulopodia) is made of very fine threads, the individual branches have a two-way circulation of cytoplasm within them. Vacuoles containing digestive enzymes and bacteria flow towards the main body of the animal, since, as in amoeba, bacteria are absorbed by the formation of surface vesicles (phagocytosis) that fuse with the digestive vacuoles. Thus digestion is initiated in the reticulopodia outside the shell.

Some foraminiferan life cycles show an alternation of sexual and asexual generations as in lower plants; these are the only free-living protozoans that do so. At Inhaca Island over a hundred species of Foraminifera have been identified in samples of sand taken at low tide and in the subtidal fringe on various shores (Moura 1965). Of these species, 80 per cent were tropical in distribution and similar to those that occur on the shores of northern Mozambique. Many different species were found at all the sites investigated but five species appeared to be abundant and characteristic of certain habitats. *Ammonia becarii* (800 µm) (Fig. 3.22a), a planktonic form, occurred most frequently where the shore was exposed to ocean currents in the channels at Ponta Torres and Ponta Punduine in the southern bay. This species appeared to settle on the substratum when the tide recedes and may be resuspended when the tide flows in again. *Amphistegina lessoni* (1,2 mm) (Fig. 3.22b) occurred most commonly on the sand on the west coast, whereas on sea grasses *Globorotalia menardi* (1 mm) (Fig. 3.22c) was the most abundant. The shell compartments of the latter are much bigger than those of other species. *Peneroplis planatus* (1 mm) (Fig. 3.22d) has an asymmetrical shell and *Sorites marginalis* (1,3 mm) (Fig. 3.22e) has a flat disc; both were common in sand on the reef flat and at the Ponta Torres coral reef. The sizes of their compartments are much smaller than those of other species and the shell appears much denser. The density is said to be related to the greater availability of calcium carbonate in solution in their habitat where corals live.

Ciliates have the most complex structure of all protozoans. The surface of a ciliate bears rows of cilia, that form different patterns in genera which have different habits of feeding and locomotion. The cilia are remarkably co-ordinated by a system of microtubules and contractile microfibrils beneath them. The shape of an individual is maintained by a two-layered pellicle beneath the plasma membrane. This system of structures associated with the cilia allows for the bending, contraction and elongation of the animal's body as it glides between grains of sand in close contact with them or swims in interstitial water.

Ciliates that live in sand are usually flattened, long and narrow and almost ribbon-like. The cosmopolitan species, *Trachelocerca* sp. (Fig. 3.22f) and *Geleia* sp. (Fig. 3.22g) probably occur in fine sand. The former feeds on diatoms and bacteria and the latter is carnivorous. The food is absorbed through an oral funnel lined by longer cilia as in the familiar *Paramecium*. Ciliates cannot withstand wave action and are restricted to sheltered shores.

Ciliates differ from all other organisms in having two distinct nuclei in the cell. The larger one, the macronucleus, is derived from the original fertilised micronucleus of each generation by multiple divisions which do not initiate cell division. As in other animals this nucleus controls cellular differentiation and the synthesis of metabolic enzymes. The micronucleus is concerned with sexual reproduction only and undergoes meiosis prior to conjugation and fertilisation.

The protozoans and microflora provide a source of food for the meiofauna. The degree to which this food web is interlinked with the food web of the macrofauna varies according to the composition of the sand and is greatest on sheltered beaches, where detritus feeders are common.

3.6.2 Meiofauna: copepods and nematodes

The meiofauna are animals intermediate in size between that of the macrofauna, seen with the naked eye, and the microfauna. They are an integral part of the communities on sheltered sandy shores, living between sand grains in the interstitial water down to a depth of 20 cm or so, some penetrating into black anaerobic mud. Their numbers per unit area may be hundreds or even thousands of times higher than that of the macrofauna (McIntyre 1968). For example, on a tropical shore in the bay of Bengal in a *Dotilla* area comparable with that on the west shore of Inhaca Island, the number of individuals in the meiofauna was 300 times that of the macrofauna, although its biomass was about one twentieth that of the macrofauna. On another muddier part of the same sandy shore the ratio of the macrofauna to meiofauna was 1:15 000, and there the biomass of the meiofauna was as much as one third that of the macrofauna.

On a tropical shore the meiofauna consists mainly of nematode worms, benthic copepods and flatworms; and other minor phyla may be represented. All have evolved a long thin body form which enables them to move between sand grains. Their metabolic activity is about five times that of macrofauna, on account of their small size, so that their oxygen demand in the sand competes with that of the burrowing macrofauna. They make a significant contribution to the seashore communities, however, by their nitrogenous excretion which enriches the nutrient content of the sand and induces more rapid growth of the microflora. The small size of animals in the meiofauna demands that they have fewer cells in their tissues and thus fewer still are available for reproduction. Most of them, after internal fertilisation, produce a small number of eggs which are sticky and so remain in the same habitat. Development may take place in brood chambers or in egg capsules, or when larvae are produced they will remain in the local interstitial water column even though they are pelagic.

There is some diversity in the food consumed by the meiofauna and in the feeding mechanisms employed. Nematodes and copepods may be herbivores, grazing on diatoms, flagellates or bacteria which are either attached to sand grains or lying between them, or they may be carnivores that devour ciliates or other members of the meiofauna. The flatworms are scavengers and some will take only dead copepods. Some of the animals scrape dead organic particles from sand grains whilst some nematodes are reputed to absorb dissolved organic matter.

On sheltered shores in clean sand, sandy mud and mangrove mud, the meiofauna play a significant part in the food chain of the macrofauna, in that many detritus feeders select them, e.g. the bubble crab, *Dotilla*. Even larger numbers are consumed by undiscriminating detritus feeders such as sand swallowers among polychaete worms, sipunculids, sea cucumbers and acorn worms. On the sandy shores exposed to strong wave action, however, such as the east coast of Inhaca Island and the Southern African coasts, the meiofauna has not been found to enter the macrofaunal food chain.

The composition of the meiofauna varies numerically and taxonomically with tidal level, the slope of the beach and the degree of wave action and consequently with sand grain size. It is usual to find maximum densities and numbers of species occurring about 5-10 cm below the surface around mid-tide, decreasing toward both high tide and low tide levels. Numbers decrease with depth in the sand down to the water table but some nematodes occur even below it in the black mud.

An analysis of the taxonomic composition of the meiofauna on a tropical sandy shore in the Bay of Bengal shows that the numbers of nematodes and copepods vary reciprocally with sand grain size, the greater proportion of nematodes being in fine sand and in mud. Several studies on South African shores came to the same conclusion. At Inhaca Island where only copepods have been studied, numbers of species and individuals were much lower on the exposed shore than on the sand flats.

The usual method of collection of meiofauna from the substratum is to use tubes having an internal diameter of 3 cm and length of 20 or 30 cm; these are pressed into the sand until full and thus a measured core is obtained. The sand is washed out with 7% magnesium chloride so that the animals loosen their hold as their muscles relax. The sand is stirred and allowed to settle several times to disintegrate packed grains. After a final stirring the mixture is decanted through a sieve of mesh size 500 µm to remove the sand. The filtrate is poured on to a sieve of mesh size of 50 µm which retains the meiofauna. The animals are washed into 5% formalin and examined under a microscope. Copepods, nematodes and others then can be easily separated by hand under a binocular microscope.

Copepods are very small crustaceans, best known as filter feeding members of the zooplankton, which are called 'cyclopoid' when they have a median eye, a rounded head fused with the first of five broad segments known as the thorax, and a narrow, straight abdomen with five segments. These planktonic copepods use their antennae and mouth appendages to create and filter currents of water and pass the captured phytoplankton to the mouth. The last abdominal segment bears uropods, i.e. two branched appendages with many long setae, useful for floating. Very few cyclopoids are, however, adapted to living between sand grains. At Inhaca Island one species *Neocyclops affinis* (900 µm) (Fig. 3.23a), is found on the beach below the Marine Biological Station. It has a typical cyclopoid shape (but no median eye), and the mandible (Fig. 3.23a) and upper lip (labrum) have toothed scraping edges whilst the other mouthparts have small sharp setae which direct the scrapings to the mouth. The caudal appendages are short.

The majority of *benthic* copepods in the meiofauna, known as harpacticoids, crawl between sand grains in a wriggling manner. They differ from the cyclopoids in that the head and thorax are as narrow as the abdomen, giving them an almost cylindrical form. Their manner of feeding is also quite different from that of pelagic species. The benthic harpacticoids have cutting or scraping mandibles, their maxillae are much less feathery but have stout bristles and the maxillipeds often have grasping terminal segments; thus they are able to feed as herbivores and carnivores.

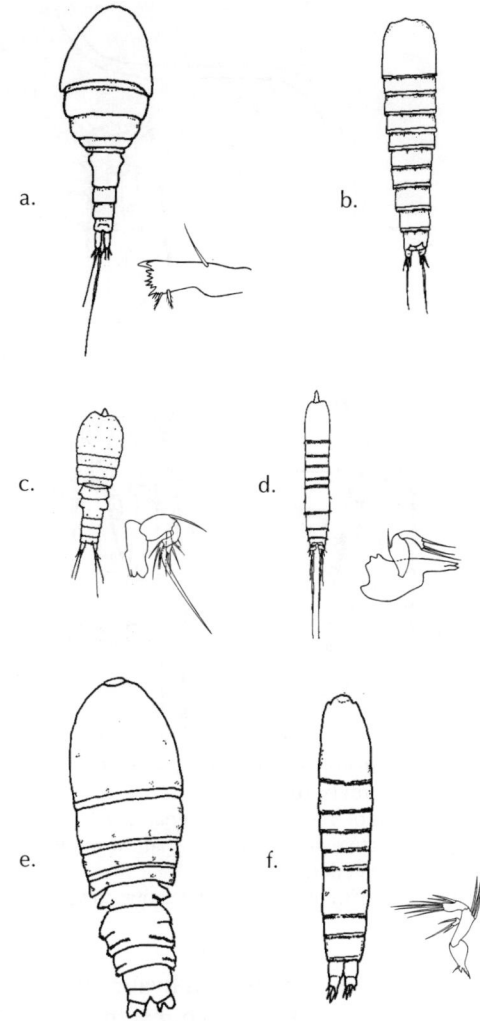

Fig. 3.23 Meiofauna in the sand: copepods (x 100 approx.) and their specialised mandibles (after Wells 1967):
a. Cyclopoid copepod, *Neocyclops affinis* harpacticoid copepods:
b. *Psammolaophante spinicauda*;
c. *Stenhalia unisetosa*;
d. *Karllangia psammophila*;
e. *Harpacticus* sp;
f. *Halectinosoma inhacae*.

The copepods at Inhaca Island have been studied intensively and the results published in a monumental work in which over four thousand individuals were observed; 135 species were identified of which 45 species were new to science. New and little known species were described in some detail (Wells 1967). About a third were cosmopolitan in distribution and others occurred elsewhere on tropical or warm temperate shores. The numbers of species which occurred in six different habitats illustrates the habitat preferences of these harpacticoid

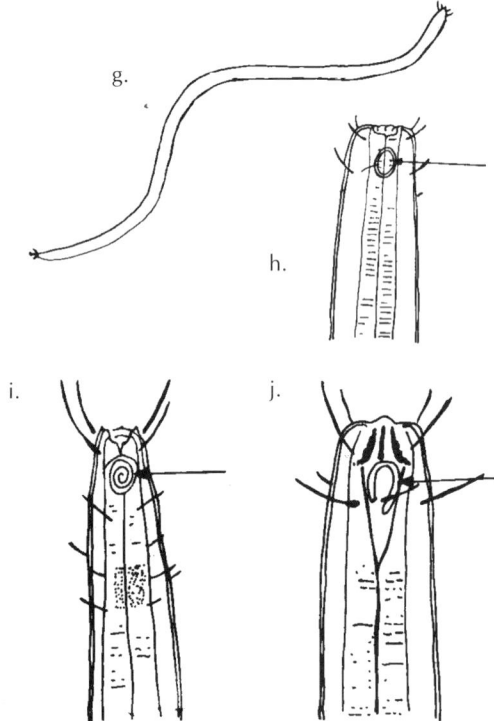

Fig. 3.23 nematodes (after Nicholas 1984):
g. entire worm (x 10);
anterior ends of 3 feeding types, enlarged:
h. non-selective detritus feeder, shallow mouth;
i. selective feeder;
j. predator with movable teeth (arrows to sense organs).

copepods. There were 80 species in sandy mud between coral debris and 55 in muddy sand between sea grasses, both areas that are rich in detritus. The clean sand, dominated by *Dotilla fenestrata*, had 49 species. On the surface of the leaves of seagrasses in the coating of silt there were 37 species, and a similar 34 species were found on seaweeds on the east coast. In the fine mud of the mangrove, which has less interstitial space available, there are only ten species of harpacticoids present and these were copepods that occurred in sandy mud on the west shore. Most of the species were represented by a few individuals in this survey, but in each habitat a few were common enough for 200-300 specimens to have been collected. The dominant harpacticoid species are briefly described below.

In the muddy sand between coral debris, *Psammolaophante spinicauda* (350 µm) is one of the dominants and it also occurs in clean sand. Its whole body is covered with minute hairs, except for the appendages, and it has sensillae (small sense organs) on the posterior edges of the segments (Fig. 3.23b). The mandibles have complex cutting edges and the maxillae have many spines, whilst the maxillipeds have prehensile claws; these are all structures that suggest a carnivorous habit.

Stenhalia unisetosa (460 µm) is one of the most common species on muddy sand between sea grasses and it also occurs in mud. It has many rows of sensillae on the dorsal surface of the segments. The mandibles have many sharp spines, some plumose setae occur on other mouthparts, whilst the maxillipeds are not prehensile. Presumably this species picks up detritus with its spines and brushes them into the mouth with the plumose setae (Fig. 3.23c).

The dominant species in clean sand is *Karllangia psammobia*, with an average length of 310 µm (Fig. 3.23d); it has a pointed rostrum and hyaline frills on the posterior edges of the segments of the body. The mouthparts resemble those of the carnivorous species in muddy sand, mentioned above.

The single dominant species on the surface of both sea grasses and algae is *Harpacticus gracilis* (550 µm), a cosmopolitan species that also occurs in sandy mud, but not in clean sand. This species of superficial habitat is somewhat broader than are the majority of burrowing harpacticoids (Fig. 3.23e).

Halectinosoma is one of the three genera that are co-dominant in fine mud. This genus illustrates the relation of intensive speciation and specialisation in different habitats among the meiofauna (see Appendix A). Five new species have been described at Inhaca Island, one of which occurs on detritus sand between coral debris, one on muddy sand between sea grasses and one species on both. Another species occurs only on sea grasses, while a fifth is present in all the habitats except clean sand. This is *H. langi*, which shows the widest tolerance of conditions including reduced salinity, and is dominant in mud of the mangrove. *H. inhacae*, which appears to inhabit only the sandy mud between coral debris, is illustrated in Fig. 3.23f.

The nematodes of Inhaca shores have not yet been studied in detail, but as on other shores, one would expect to find a variety of species and feeding habitats. Nematodes are very thin, smooth, unsegmented worms without appendages, tapering at each end (Fig. 3.23g). They have a thick protective cuticle that facilitates their wide distribution, which in turn depends on their tolerance of aerobic and anaerobic conditions and of desiccation or immersion in water of varying salinity. The nematodes move by writhing in regular waves since they have longitudinal muscles only and must have the support of a medium such as sand or fluid before they can move forward or backward. Attachment to the walls of the burrow is sometimes by means of adhesive glands on the tail.

The anterior mouth of nematodes may be adapted for a variety of diets by modifications of the six lips around the mouth, which may have lobes, plates, teeth and sense organs; in addition the muscular pharynx serves as a sucking organ. Three types of feeding are recognised (Nicholas 1984): those that select bacteria have small, simple, shallow buccal cavities; non-selective feeders have wide cup-like buccal cavities without teeth; others scrape diatoms from sand grains with small teeth. Predators swallow small prey using internal teeth to secure their prey. These types are illustrated in Fig. 3.23 h, i and j.

3.7 SANDY SHORES EXPOSED TO STRONG WAVES

Along the whole of the east coast from Ponta Torres in the south to Cabo Inhaca in the north a sandy shore, 12 km long, slopes steeply from high dunes to low rocks below low neap tidal level. This shore is exposed to moderately strong wave action and constant wind. The lower slopes of the dunes and the intertidal area are bare of vegetation. The sand is coarse except for an area at low tidal level at the extreme south of the island. There, fine sand is deposited on the lower shore by the tidal current which flows out through the Ponta Torres Strait from the Bay of Maputo.

3.7.1 The steep sandy slope of the east shore: pink ghost crabs; turtles; flatworms; plough whelks; mole crabs.

On the exposed shore the pink ghost crab, *Ocypode madagascariensis* (Fig 3.24a) digs its burrow at high tidal levels, replacing the green species of the sheltered shore (see Section 3.2.2). The pink crabs emerge by day to forage in the very shallow water at the edge of the waves when the tide reaches the sand of the lower shore. Although they run a short distance away from the rapidly advancing tide, when the water overtakes them they remain squatting and rooted to one spot while the wave rushes over them and the backwash tears at the sand. The pink ghost crab has stout, long eye-stalks but these do not protrude beyond the eyes as in the green species. The stridulating organ on the chela is composed of a short row of granules which, unlike that of the green ghost crab of the sheltered shore, is without hairs. Apparently the sound can be heard by the crabs in spite of the background noise of the breaking waves because its frequency differs from that of the waves.

Two species of turtle frequent the shallow waters of the east coast, the Loggerhead and the Leatherback (see Section 12.4.1) (Broadley 1990). Sometimes they may be seen, silhouetted against the sunshine filtered through the large waves as they rise before they break on the shore. *Caretta caretta* the Loggerhead Turtle (Fig.3.24b), which has a yellow plastron (undersurface), is the smaller of the two species, yet it exceeds

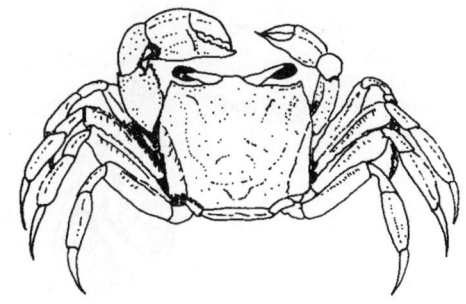

Fig. 3.24 On exposed sandy shores:
a. pink ghost crab, *Ocypode madagascariensis* (x 0,5) (after Crosnier 1965);

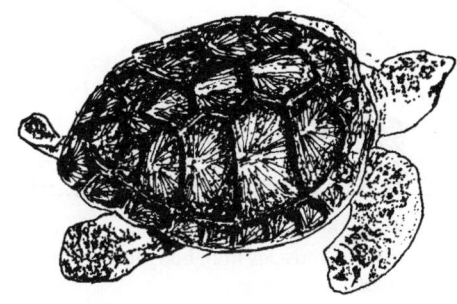

juvenile turtles:
b. Loggerhead, *Caretta caretta* and
c. Leatherback, *Dermochelys coriacea* (x 0,3);

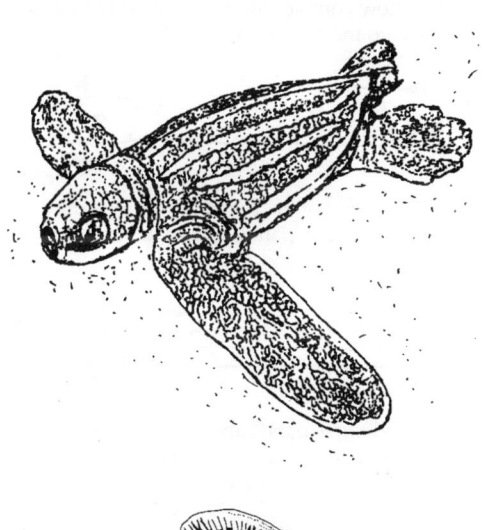

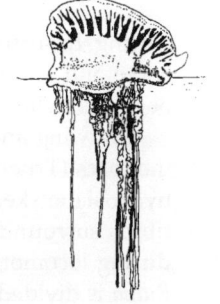

d. blue bottle, *Physalia* sp. (x 1).

Fig. 3.25 On the exposed sandy shore:
Bullia natalensis:
a. shell (after Kilburn and Rippey 1984);

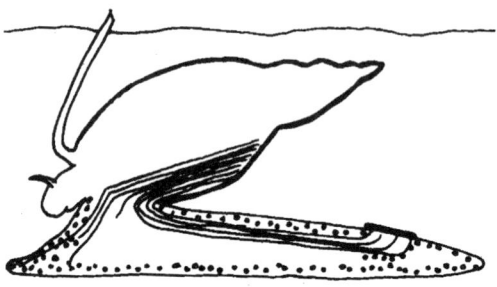

b. snail moving below surface sand, siphon protruding, muscles from body to operculum and foot;

L.S. foot, mechanism of burrowing: external arrows show direction of movement, internal arrows show contraction or expansion of foot muscles around blood sinus within:

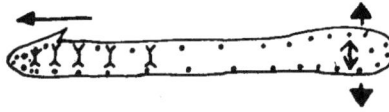

c. propodium contracts and moves forward, metapodium expands as terminal anchor;

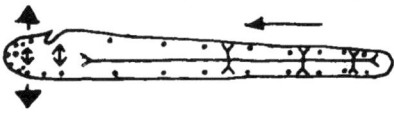

d. Propodium expands and anchors as metapodium contracts and moves forward (after Trueman and Brown 1992).

a metre in length; it feeds on crabs, molluscs and sea urchins. A larger turtle, which may be twice the size, the Leatherback, *Dermochelys coriacea* (Fig. 3.24c) is carnivorous, feeding on coelenterates, jelly fish and the stinging blue-bottles, *Physalia* sp. (Fig. 3.24d), which sometimes invade the shallow waters in hordes. The Leatherback is dark in colour, has no scales and is covered by a thick leathery skin with seven prominent longitudinal ridges dorsally and a pinkish colour beneath the throat. After mating in the sea from September to October the female turtles come ashore at night at ten-day intervals from October to January to prepare nests and lay their eggs at high tide level. A pit large enough to accommodate the body is first dug by great sweeps of the fore-limbs. A smaller deeper hole is then dug by the hind limbs to hold the eggs. Each egg is about 5-6 cm in diameter and 60-100 are laid at one time. The egg hole is then filled with sand and compacted by the hind limbs; the body-pit is smoothed over to leave little trace. Usually seven clutches of eggs are laid at intervals of 9-10 days in a season. They hatch at night from January to April after 62-72 days of incubation, when they have reached 5-6 cm in length. The baby turtles struggle slowly back to the sea down the sandy slope but many of them are preyed upon by the pink ghost crab, *Ocypode*.

At the level of high neap tides, runnels in the sand appear bright green when the tide has just receded. As on the sheltered shore, this colour is caused by dense populations of the tiny flatworm, *Convoluta macnaei* which contains green unicellular algae (see Section 3.3.1). If these flatworms become broken by the waves the fragments can regenerate whole worms, a process called architomy. The sexual reproductive organs appear to be degenerate.

On the middle level of the sandy slope the scavenging plough whelks, *Bullia natalensis*, emerge from their shallow burrows below the surface of the sand only when the incoming tide reaches them. The greyish shell is a tapering spiral which is smooth except for small axial pleats at the sutures (Fig. 3.25a). The extended foot is an oval disc only about 1 mm thick. The foot is not withdrawn when the animal is buried (Fig. 3.25b) or at rest on the surface, unless the whelk is attacked, when the foot is withdrawn into the shell by displacing water from the mantle cavity and the free space at the top of the shell; at the same time, blood is shunted from the foot into the visceral region. The foot is used in burrowing, in crawling over the surface and as an underwater sail for surfing. In surfing, or swash-riding, the animal turns onto its back at the edge of an advancing wave and allows itself to be swept ashore. Once stranded, it immediately rights itself and begins crawling or burrowing.

Crawling and burrowing in *Bullia* are faster and energetically more efficient than in other molluscs (Trueman and Brown 1992). The single, large blood sinus within the foot acts as a hydrostatic skeleton in which turgidity is maintained by the contraction of smooth muscle fibres surrounding it in a three-dimensional network. The thin, plate-like form of the foot during locomotion is the result of contraction of the dorsoventral muscle fibres. The foot of *Bullia* is divided functionally into an anterior propodium and a posterior metapodium, which contract alternately during crawling or burrowing, resulting in 'stepping'. The forward step is accomplished by muscular contraction thrusting the propodium forwards while the swollen metapodium acts as an anchor, preventing the whelk from being pushed backwards. Muscular contraction of the transverse and dorsoventral muscles of the metapodium now streamline its shape and push blood forwards into the relaxing propodium, which swells to become a terminal anchor. This holds its position in the sand while contraction of the longitudinal and collumella muscles draws the shell and metapodium forwards (see Fig 3.25c and d).

Many of the muscle fibres in the propodium are striated, contracting and relaxing faster than smooth muscle fibres and thus allowing a flexibility of movement denied to the metapodium; this flexibility is important in penetrating the sand and also in crawling while in search of food. Both types of muscle in *Bullia* are richly supplied with mitochondria, indicating a fundamentally aerobic mode of respiration, in contrast to the essentially anaerobic energy supply for locomotion in *Donax* (see Section 3.2.1).

Bullia has no eyes but the osphradium in the mantle cavity is particularly sensitive to methylamine, a derivative of the excretory product of fishes and of decay in invertebrates. The plough whelk consumes any carcass it can find from bluebottles (the small stinging jelly fish, *Physalia*, which may induce temporary paralysis in man) to small fishes. Before it can be covered by deep water of the advancing tide it burrows again. When the tide retreats, it emerges to feed again while in the surf, but buries itself before the sand dries. The whelk remains buried until the next incoming tide. This behaviour is reminiscent of that of *Donax faba*, the wedge mussel of the sheltered west shore (see Section 3.2.1).

The only other large animal encountered on the lower sandy shore is in the finer sand at the southern end of the island. This is the peculiar anomuran, barrel-shaped mole crab, *Emerita austro-africana* in which the abdomen has no segments and is reduced to a small flat triangular plate folded under the thorax (Fig. 3.26a). Four pairs of legs are flattened and have sharp terminal segments which are used for digging backwards into the sand until the animal is completely buried, except for the eyes at the end of long eye stalks protruding from the surface. The small fifth pair of legs can be inserted into the gill cavity to clean the gills, a very unusual adaptation, useful for living in surf, always laden with sand in suspension, which gets into crevices and cavities. The folded telson has a fringe of hairs to protect the openings to the gill chamber and it also serves as a brood chamber for the development of the eggs. Mole crabs resemble other anomurans, such as porcellanid crabs and some hermit crabs, in being adapted for filter feeding. The long antennae are fringed with a net-like array of branched setae which collect suspended food particles. The antennae are rolled up under the mouth when the mole crab emerges from the sand and the waves wash over it. The waves roll the animal up to the middle shore as the tide advances. There it stops to feed in the surf, taking advantage of the enriched source of food provided when the waves wash out particles from the sand, and lengthening the feeding time by keeping in the swash zone. Before the water becomes deep it burrows again and waits to be carried downshore with the receding tide, when it feeds in the surf again. The mole-crab, like *Donax faba* on the sheltered shore and *D. madagascariensis* on the north-west shore, exploits the enriched supply of suspended particles washed out of the sand by the waves. Its migration resembles that of *Bullia*.

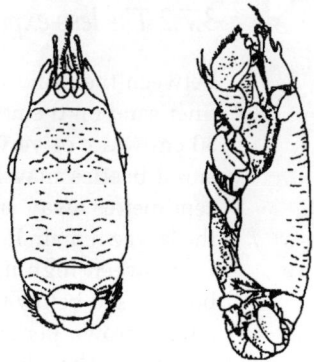

Fig. 3.26 On the exposed sandy shore
a. mole crab, *Emerita austroafricana** (x 1)

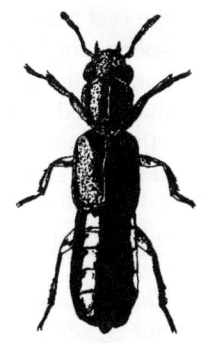

b. staphylinid beetle, *Bledius* sp. (x 5);

c. wedge mussel, *Donax madagascariensis* (x 1) (after Kilburn and Rippey 1982).

3.7.2 *The less exposed north-western sandy slope: burrowing beetles; wedge mussels.*

Between the dune vegetation and the rocks at mid-tidal level, there is only a short slope of finer sand on the northeastern shore. At the level of high neap tides a horizontal band, about 30 cm wide, of very narrow raised tunnels appears whilst the sand is wet. In these tunnels small beetles may be found (Fig. 3.26b), which when disturbed fly elusively only a few centimetres away and tunnel again. This staphylinid beetle, *Bledius* sp., lives in colonies and the larvae remain in the tunnels under parental care.

Between high neap and mid-tidal level a small wedge mussel, *Donax madagascariensis*, buries itself below the surface. The shell has conspicuous oblique ridges and is covered with a light brown periostracum which conceals its white colour (Fig. 3.26c). Internally it is white or purple. The edges of the valves are slightly dentate and fit each other closely. This bivalve mollusc feeds only in the surf zone and also exhibits migratory behaviour up and down the sandy slope and rapid burrowing similarly to *Donax faba* on the sheltered shore (see Section 3.2.1).

The protists and meiofauna are not numerous on the exposed shore in the clean sand. On the east coast one copepod species, *Karllangia psammophila* (Fig. 3.23d), is dominant among 40 species. In *K. psammophila* the rows of notched spinules between segments, the spiny telson and the thick spine on its head are adaptations to burrowing in clean sand. The maxillae and maxillipeds are prehensile, having sharp, finger-like claws which enable the copepod to seize other meiofauna as food. The meiofauna in the coarse sand burrow much deeper than on the sheltered shore to avoid desiccation and turbulence.

3.8 FOOD WEBS

The primary source of energy for biological production on seashores is solar as it is on land; but on seashores there is an auxiliary, predominantly lunar force, exercised by the tides. The tides irrigate the sand twice daily, distributing inorganic nutrients from the oceans as well as those produced by the nitrogenous excretion of the local macrofauna and meiofauna. Thus plant production is boosted. The tides also continuously replenish other supplies of animal food, bringing in plankton. Detritus is enhanced, being derived from subtidal sea grasses and seaweeds and added to the shore. Thus the energy wheel turns faster, providing conditions for the greater wealth and diversity of life on shore than on land.

At Inhaca Island the food web on the sandy west shore includes over 300 species of animals; these are recorded in Appendix A and only the common species (Macnae and Kalk 1962) have been described in this chapter. The exposed sandy shores, however, support less than ten species of macrofauna. An examination of the categories of food consumed by different kinds of animals suggests the reason for this difference.

The trophic relationships between plants, bacteria and animals have been described in general in Section 2.2. Bacteria, the heterotrophic producers living on the detritus of the substratum and in the plankton, play the major role in the food web on *sheltered* sandy shores which have been described as 'great digestive and incubating systems' in which organic particles, brought in by the tides, are converted into a bacterial biomass, much more concentrated than that in the sea. These are *lacking* on the *exposed* shore.

The value of phytoplankton should not be underestimated since it provides the main source of energy for the immense number of larvae, liberated into the sea by the majority of animal species on the seashore (see Section 8.4).

On the sheltered, sandy shores the three types of consumers – herbivores, detritivores and carnivores – are represented at all tidal levels with many representatives of each type: but there is a significant modification of the pattern on exposed sandy shores.

Detritus feeders and benthic microphagous algal feeders are few on the exposed shore since the turbulence of the sea on the steep sloping beach prevents the accumulation of detritus and reduces the amount of diatoms in the sand; there are no polychaetes, bivalves, small crabs, acorn worms or sea cucumbers. Consequently there are few carnivores.

This broad outline of the food webs on sandy shores leaves unsolved one large area of interaction: that between the meiofauna and the macrofauna. On exposed beaches this facet of the food web is minimal (Maclachlan 1972), but on sheltered beaches the density of the meiofauna and its ready availability suggest that it plays a part in the food web of the macrofauna as well as participating in a food web of its own with the microbiota.

4

THE MANGROVES

Mangroves, the trees that grow in the shallow tidal sea, occur on warm, very sheltered estuarine coasts, spreading over flat muddy sand on the upper shores. At Inhaca Island mangrove swamps cover 308 ha, equivalent to 7% of the total land area of the island. The trees provide an important resource for the people of the island since they are used for the construction of boats, in the building of houses and principally as firewood in household cooking and in the preparation of sea cucumbers for export, different species of tree being used for different purposes. These economic uses, however, endanger the survival of the mangroves which is essential for the following ecological reasons. The mangrove swamps form a natural barrier to the penetration of salt water from the sea into the adjacent agricultural land. They provide an ecological niche for the large edible crab until it reproduces in the sea and a nursery for the juvenile prawns which are later caught as adults in the Bay of Maputo. The mangroves retain sediments derived from the land and thus enhance the protection of corals. Sediments from the sea are also trapped and provide soil for the microflora and for a large number of animal species which represent a primary link in the food chain.

Current research aims at establishing an equilibrium between maintaining the ecological advantages of the mangrove trees and economic usage.

To explore the mangroves safely it is necessary to know the time of the incoming mid-tide when the reversal of flow occurs in the mangrove. It is better to start from the landward edge, where trees are less dense, and to follow a gully towards the main channel, deviating from it for short distances. Three hours after high tide the upper reaches can be traversed. Six hours later the ground water begins to rise again when the tide crosses the mid-tidal level. The sea flows inwards extraordinarily rapidly, faster than walking pace, since the ground is almost flat. At high spring tides all the tree trunks are submerged up to the lowest branches. High tides in sheltered bays usually occur later than stated in the official timetables. In the southern bay at Inhaca Island the tides are one hour later than on the west shore (see Section 1.5.2).

4.1 MANGROVE TREE SPECIES ON INHACA ISLAND

Four species of mangrove trees and an associated fauna, confined to the upper shore, extend from the highest spring tide to high-tide level of the lowest neap tide at Inhaca Island. Further north on tropical shores there are many more species of trees in the 'high' mangroves adjacent to the tropical rain forest; these grow 30 m tall and spread lower down the shore to mid-tidal level (Macnae 1968). In the drier coastal areas the number of species, the height of the trees and their density decrease with distance from equatorial regions. In the southern geographical region of 'low' mangroves, trees do not often reach 10 m in height and 3-5 m is more usual. Five species of trees are easily recognised in the field.

Avicennia marina (Fig 4.1a) is a spreading tree with small pale green leaves, grey and felt-like on the under surface. It is surrounded by hundreds of pencil-like aerial roots (pneumatophores) growing about 10-30 cm high vertically above the ground. *Ceriops tagal* (Fig

4.1b) is a bushy tree with yellowish-green, thick leathery leaves and 'buttress' roots protruding above ground level from the trunk and from underground roots. *Bruguiera gymnorhiza* (Fig 4.1c) is a taller, conical tree with large dark green leaves and more prominent knee roots around the trunk. These have emerged from the ground and then curved around to re-enter it. *Rhizophora mucronata* (Fig 4.1d) is usually a compact tree, which has stilt roots growing from the trunk above or below the branches and arching to the ground. The very large dark green leaves have a prominent, attenuated tip or mucron. A fifth, tropical species, *Lumnitzera racemosa*, is not common; it is a bush with bright green succulent, slightly notched leaves; it occurs at places on the landward margin of the swamps (Fig. 4.5g). These mangrove trees have air-breathing roots.

The mangrove associates (i.e. trees without air-breathing roots) which usually occur to landward of the true mangroves in estuaries in Natal (Berjak *et al.* 1986) and the east coast of Africa as far north as Kenya, are not well represented. On those coasts *Barringtonia racemosa*, *Hibiscus tiliaceus* and *Thespesia populnea* occur where fresh water accumulates. *Hibiscus tiliaceus* has increased remarkably in recent years in the Ponta Raza creek mangrove.

4.2 TYPES OF MANGROVE SWAMPS ON THE ISLAND

Fig. 4.1 Mangrove trees:
a. *Avicennia marina**;
b. *Ceriops tagal**;

There are seven areas of mangrove association at Inhaca Island, each one being somewhat different from the others in the pattern of distribution of the species of trees. Suitable shelter for mangroves, where saline and fresh water mix, occurs on the *upper shores* of the Saco of the southern bay, in the northern bay, in the creek behind the sand-spit at Ponta Raza and on Portuguese Island (see Fig. 1.1). The differences among them may be interpreted as representative of different stages in development of the association from the pioneer stage through maturity on Inhaca Island to partial destruction by coastline changes, as on Portuguese Island.

4.2.1 Unzoned trees: Pioneers and fringing trees

On the north-eastern shore of the northern bay the thick fringe of *Avicennia marina* is continuously extending on to the flat shore where silt and sand are accumulating because of the shelter afforded by the sand-bar between Ponta Mazondue and Portuguese Island. *Avicennia marina* seeds have germinated

year after year in the soft mud, growing into saplings which reach over 50 cm high, their root systems stabilising the substratum. The protruding pneumatophores trap more silt, thus extending the mangrove area which has consolidated the sand for about 200 m towards the sea in the last ten years. This area represents a pioneering stage of mangrove forest formation.

On the eroded parts of the shore of the southern bay there is a simple fringing mangrove swamp consisting of a belt of *Avicennia*, at most two or three trees wide, that lines the periphery on both east and west. It extends along the shores for about 3 km as far south as the freshwater influence spreads in the Saco (see Fig. 5.1). Germinating seedlings are not common

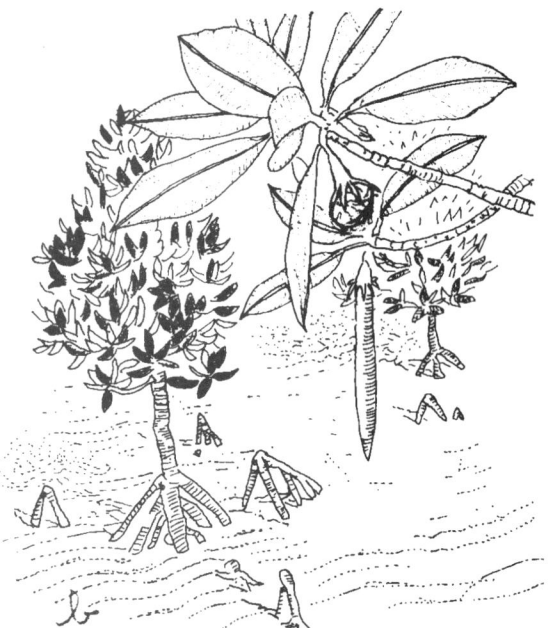

Fig. 4.1
c. *Bruguieria gymnorhiza**;
d. *Rhizophora mucronata**;

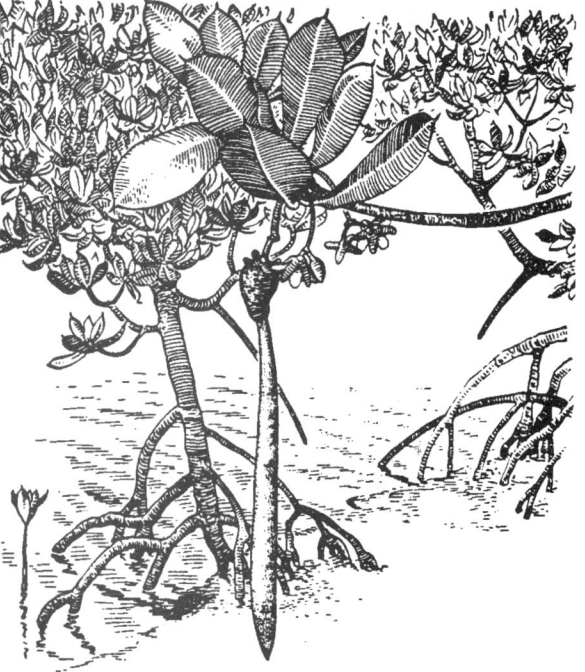

here since the seeds are washed out by the strong tidal current that comes in through the Ponta Torres strait at the mouth of the southern bay.

4.2.2 Zoned mangrove swamps: typical; creeks; parkland

Mangrove swamps that are fully zoned in relation to tidal channels occur in the southern and northern limbs of the Saco, in the central part of the swamp in the northern bay and in the creek mangrove at Ponta Rosa.

A pattern typical of tropical estuaries on east African coasts is best seen in the southern limb of the Saco (Kunwadlaniheni) (Fig. 4.2a). There is no river but drainage from the adjacent high dunes on the west is canalised into a few small 'tributaries' which enter a larger channel. The channel passes through the swamp and reaches the open shore at mid-tidal level; then it disappears, the drainage being subterranean until it reaches the main central channel in the southern bay.

Avicennia lines the landward fringe of the swamp; *Ceriops* grows in the drier, more sandy marginal areas of the central thicket of the swamp; *Bruguiera* grows in scattered groups in firm, fine black mud with water near the surface and forms a belt behind the next zone. The banks of the streams are lined by *Rhizophora* trees with roots arching into the water. The seaward fringe is formed by *Avicennia*. Narrow belts of salt marsh, with succulent plants and rushes typically line the sides of the mangrove swamp, where there are dunes on both sides.

The smaller mangrove swamp in the northern limb of the Saco is more complete than that in the southern limb since it is adjacent to flatter land at its head. It receives greater drainage from the freshwater swamps. Consequently there is an area of salt marsh at its *head* which is almost lacking in the southern limb; the zonation of trees is less well marked since the drainage inside the swamp is more diffuse. The landward edge of the swamp is lined by a hedge of the mangrove fern, *Acrostichum aureum* and the palm tree *Phoenix reclinata* grows just above the high water mark of spring tides.

The central part of the Tengeni mangrove swamp in the northern bay, which lies behind a low headland, is a well-developed zoned mangrove with a central channel (Fig. 4.2b). Freshwater swamps dominated by the reed *Phragmites* still exist on the south border of its head. Sixty years ago the whole of the landward border was in contact with the largest of the freshwater swamps which has been very slowly drying up. Nowadays the landward border of the mangrove swamp is separated from the freshwater swamp by a rough road to the lighthouse and by agricultural land. Thus very little salt marsh is left around the borders of this mangrove.

Another section of the northern mangrove (Sichwane) lies to the north of the small headland (Fig 4.2b). It appears to be a much older section that once had a good supply of freshwater since the *Avicennia* trees are tall, spreading and widely spaced, forming a beautiful 'parkland' bordered in places by a 'hedge' of *Lumnitzera*. Rushes and succulent herbs carpet some of the floor, indicating that the area still has a supply of fresh water. A thick layer of coarse sand lies on the peat beneath. This flat area extends seawards for about 200 m and is characterised by very few *Ceriops* and *Bruguiera* trees. *Rhizophora* is completely lacking. These trees may once have been present and have since died out. The height of the *Avicennia* suggests that they are old and that in their youth fresh water was plentiful. The excess of sand forming a layer above the mud indicates a shore of accretion, a mangrove invaded by sand. The flat area nearer to the hotel adjacent to this part of the mangrove association is land reclaimed from mangrove swamps on which 200 coconut palms were planted over a hundred years ago, as is the custom in Mozambique. It is highly probable that the Sichwane swamp is the remains of a former mature zoned mangrove.

Immediately south of Ponta Raza (at Xilthangalweni), another type, known as a 'creek' mangrove, lies in a narrow valley about 1km long running north-east/south-west between two dunes parallel to the west shore (Fig. 4.2c). The landward dune is about 20 m high and very steep; the seaward dune is lower and of later origin. The creek ends in a deep channel about 3 m wide which enters the small south-west bay, south of the Ponta Raza sand-spit. A shallow stream sweeps across from the channel exit on to the beach in the small bay and gradually disappears into the sand.

The distribution of trees is complex here as this valley appears to have two intermingled mangrove associations, an old and a new. The old mangrove stream had an exit on the west shore of the island about 500 m south of the Marine Biological Station, in a gap in the seaward dune. It is easily accessible from the west shore. The gap leads immediately to a large group of old *Avicennia* trees surrounding a group of old, gnarled *Rhizophora* trees, some dead and others dying, which mark the mouth of the original stream, long since dried up. When one penetrates further along the creek one finds a group of *Avicennia* surrounding a bare sandy patch, which is presumably the head of the old mangrove. The later mangrove swamp appears to commence about 200 m further down the creek. From a bare sandy patch, a single track leads to the southwest bay; it becomes wet, then wider and deeper until it forms a typical mangrove channel with the wide and deep exit described above. A comparatively recent dense growth of the very broad-leaved mangrove associate tree, *Hibiscus tiliaceus*, forms the boundary of the swamp along the foot of each dune, the long roots reaching down to less saline water. It is a small low-branching tree with large heart-shaped leaves on long petioles; both surfaces of the leaves are covered by short, soft hairs. The trumpet-shaped yellow flowers are conspicuous and as large as the leaves. The fruits are small (2 cm) and covered with short, yellowish hairs.

One may speculate that drainage of rain-water from the high landward dune was blocked by the seaward low dune that was built up by the long-shore current responsible for the formation of the Ponta Raza sand-spit. Fresh water surfaced in a swamp at the foot of the high dune which later became incorporated in the old mangrove through intrusion of the sea from the west shore. The drowned freshwater swamp is

Fig. 4.2 Distribution of trees in zoned mangrove swamps at Inhaca Island:

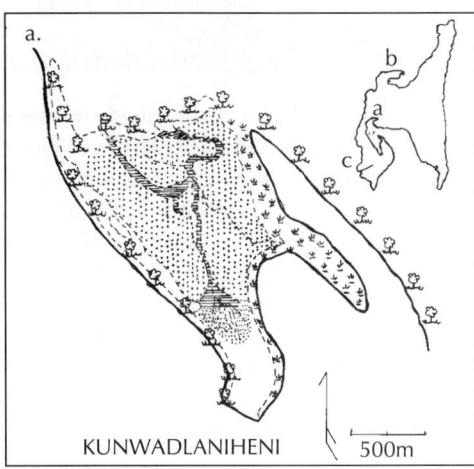

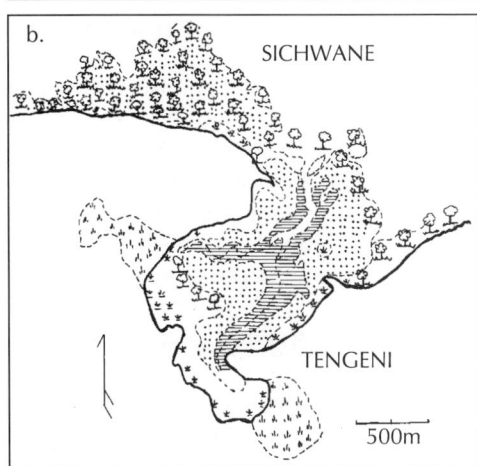

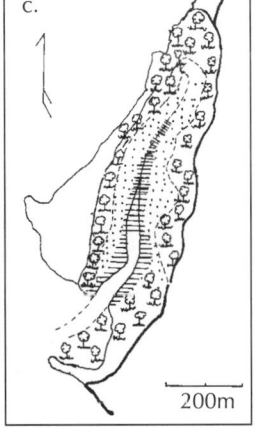

a. Kunwadlaniheni in the Saco:
b. Tengeni and Sichwane in the northern bay;
c. Xilthangalweni at Ponta Raza (a-c after Macnae and Kalk 1962).

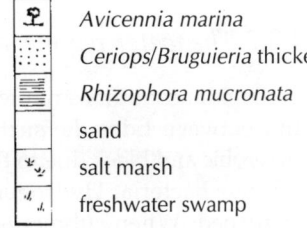

Avicennia marina
Ceriops/Bruguieria thicket
Rhizophora mucronata
sand
salt marsh
freshwater swamp

probably the historic pattern of mangrove swamp formation on the island, in contrast to the drowned river valleys on the mainland.

4.2.3 Partial destruction of mangroves on Portuguese Island

The area of mangrove swamp on Portuguese Island in 1958 was over 110 ha, distributed in several large patches. The western third of the island has been slowly inundated over the last thirty years and the land has been washed away; the last remaining mangrove swamp occupied about 26 ha in 1989. The mangrove channel was broadened by invading tidal currents and *Rhizophora* trees were washed away. The wide channel became blocked by new deposits of sand. It has become dry and saline, extensively colonised by halophytic, succulent herbs and the salt-marsh grass *Sporobolus virginicus*. The leaves of *Ceriops* bushes, once growing in the interior of the swamp and now exposed on the edges of the dry channel, have become a brilliant yellow and the trees are dying. *Bruguiera* and *Avicennia* are hardly affected and appear to have a wider tolerance of the changing salinity of the soil. The situation will be reversed if access to the sea occurs again naturally or by design (Hatton and Couto *in press*).

4.3 ENVIRONMENTAL VARIABLES IN THE SWAMPS

The high productivity of mangroves is made possible by the long history of accumulation and degradation of its decaying leaves. Detritus from nearby sea grasses and more distant seaweeds is also transported back to the shores by very gentle tides. The reduction in water velocity on the flat upper shore of a very sheltered bay results in the material in suspension being deposited. This contains powdered shell fragments as well as silt, clay and organic matter of mostly plant origin. These deposits undergo a number of biological changes before trees can grow on the shores.

4.3.1 Biological processes in the soil

Aerobic nitrifying bacteria in the surface layers of the mud deposits break down the proteins in organic matter into nitrates and nitrites. Anaerobic bacteria in the mud below the surface decompose the sulphur-containing organic matrix of shell material and in so doing form sulphates and sulphides; calcium ions are freed from the carbonate of shells to form soluble chlorides. In these processes of decay, phosphates are liberated (see Section 1.7.2).

Fungi play a major role in the decomposition of mangrove detritus (Jennings 1983). The species are fewer than found in forests on land on account of the lack of oxygen and the salinity of the water, although some are more abundant on the surface. An intensive study (Swart 1958) of fungi at Inhaca Island showed that many species of the Imperfect Fungi, *Aspergillus* and *Penicillium*, are present, whilst species of *Fusarium* and *Pestalotia* are abundant (Fig. 4.3a-d). Fungi that produce mushrooms and toadstools (Basidiomycetes) do not occur in the mangroves.

The stage is thus set for the activity of blue-green 'algae' (Cyanophyta) which are the first to colonise the soft, water-retaining mud and help to consolidate it. The habitat is thus prepared for microscopic green algae and diatoms which thrive in a solid, damp soil, rich in nutrient salts. Protozoans soon invade the mud, followed by nematodes and copepods of the meiofauna, which excrete ammonia and phosphate and so further enrich the soil, which is necessary for the growth of mangrove trees.

4.3.2 The water regime: aeration and salinity

In the tidal conditions of the mangroves the ground is inundated twice a day for some hours, and between tides the surface soil remains exposed to air. Below the surface the mud is anaerobic and black due to the hydrated ferrous sulphides produced by the reducing action of sulphur bacteria. Hydrogen sulphide may be liberated from waterlogged soil when it is disturbed. When submerged, water in the burrows of animals has thus a lower oxygen content than that of the sea, resembling the channel water, i.e. about 2-4 ml per litre. When the tide is out the burrows are filled with air since they are above the water table.

At Inhaca Island the fresh water component in the mangroves comes from local rainfall (average 900 mm per annum). The rain drains into a subsurface reservoir resting upon the calcareous sandstone base of the island and accumulates in the low-lying ground between the dunes as freshwater swamps which are adjacent to the very sheltered bays. The drainage of fresh water into the mangrove swamps is below the surface and becomes visible as bare wet tracks on the periphery of the mangrove when the tide is rising. These wet tracks become gulleys among the trees (by tidal erosion) which converge to form shallow channels in the densely wooded areas. Thus the mangrove soil and burrows contain diluted sea water.

The mixing of fresh and salt water is not uniform over the whole mangrove area and it varies from day to day, since the semidiurnal tides inundate the swamps to different heights on successive days of the lunar month. High spring tides completely cover the whole area, submerging all the trees up to their lowest branches and wetting the soil of the salt marsh plants on the supratidal fringe. The flooded area gradually declines in width as the spring tides give place to neap tides. At first only the salt marsh and the *Avicennia* fringe remain exposed at high tides; the soil of the *Ceriops* thickets receives less tidal water during neap tides until the lowest high neap tides reach only the *Avicennia* trees of the seaward fringe. The channels and gullies receive sea water at all high tides and it

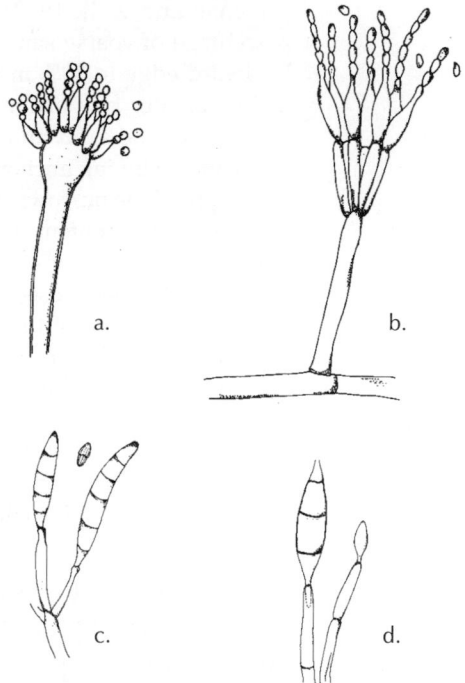

Fig. 4.3 Fungi with acrospores in mangrove mud (x 1 000):
a. *Aspergillus* sp.; b. *Penicillium* sp.;
c. *Fusarium* sp.; d. *Pestalotia* sp.

spreads to the adjacent black mud which remains waterlogged; this is the area where *Bruguiera* and *Rhizophora* grow. Thus a gradient of water in the soil is created by the changing tides from the seaward fringe through the mangrove to the landward edge. The water table is about 75 cm below the surface in the bare sandy patches at the head of the mangrove swamp but quite near the surface in the densely wooded areas of black mud. This determines the depth of the burrows made by the crabs.

The mixing of fresh and salt water (containing 35 g NaCl per litre in the mouth of the bay) produces variations in the salinity of the mangrove surface water. Evaporation of water from ground exposed to the heat of the sun accentuates the effect of infrequency of inundation of high tidal areas, increasing the salinity of the soil water. This is tempered by local drainage of fresh water, by rainfall and nocturnal dew. Mangrove water contains a higher proportion of calcium ions than the sea; the presence of these ions, derived from shell debris, diminishes in some degree *the absorption of sodium ions* by the tree roots.

The landward *Avicennia* may endure salinities twice that of the sea during neap tides, whilst *Ceriops* in well-drained sand is subjected to salinity little different from that of the sea. In the black mud, where *Bruguiera* and *Rhizophora* grow, the variation in salinity is below that of the sea and may be between 19 g and 30 g NaCl per litre. The seaward fringe of *Avicennia* is exposed to salt concentrations slightly lower than that of the sea because of drainage through the mangrove.

Thus, a permanent gradient in salt concentration cannot be measured from landward to seaward borders; the values for salt content vary daily and fortnightly with tides and seasons.

4.3.3 Gradients in texture and organic content of soils

As a consequence of tidal action, the sand grain size and silt content vary from the upper margin of the mangrove to the black mud in the interior, and from the surface to deeper layers

(Macnae and Kalk 1962). The yellow sand above mean high spring tides has a small percentage of coarse sand and 70% of medium grade; fine sand increases from 20% at the landward edge to 30% in the black mud. The grey soils around *Avicennia* and *Ceriops* trees have a greater proportion of medium-grained sand which facilitates drainage, whereas in black mud fine sand and clay are present. For trees, the composition of the soil presents problems involving anchorage, aeration and nutrient absorption; for crabs, it affects the form and depth of the burrows that they are able to make.

The organic content of mangrove soil varies from about 0,1-1% at the landward edge to about 6.4% in the interior of the swamp. A significant proportion of this organic content in the sandy soil that is exposed to direct sunlight consists of diatoms and dinoflagellates whilst the organic component of the black mud in the shade is due to bacteria, fungi and humus. Thus the food available for crabs and molluscs differs along the gradient of soil factors; this is reflected in the distribution of species (See Section 4.6.1).

To summarise the environmental factors: *the mangrove environment is nutrient-rich but inconstant, being variable in water content, salinity, oxygen content, consistency of the sand and organic content*. The plant and animal populations are dense and diverse since different species are specially adapted to conditions in the various zones, as discussed in Sections 4.5 and 4.6.

4.4 ADAPTATIONS OF THE MANGROVE TREE SPECIES: TRANSPORT, SECRETION AND EXCLUSION OF SALT; AERATION OF ROOTS; DEVELOPMENT

Trees that 'live in the sea' have roots, leaves, fruits and seedlings adapted in both structure and physiology to the peculiar environment of the tropical shore described in the last section. Mangrove trees on Inhaca Island belong to three different families that have independently evolved similar adaptations differing in detail, all of which, however, allow exploitation of the specific habitats.

4.4.1 The white mangrove tree

Avicennia marina is known locally as the 'white mangrove' because of the light colour of its bark and wood. It is the pioneer of the mangrove association, i.e. the first species to colonise the upper levels of very sheltered shores. The aerial roots of *Avicennia* trap more sand and silt and so the level of the ground is gradually raised and becomes fit for the germination of the seeds of other mangrove species.

Avicennia is the genus most tolerant of high and variable salinities, but it does have one handicap, a high light requirement; this explains its occurrence on both landward and seaward fringes and its absence in the body of the mangrove community. Although its seeds will germinate in the shade of other trees, the saplings cannot continue to grow in reduced light. When seeds of the other trees germinate in the shade cast by *Avicennia* their rapid growth outstrips that of the pioneer and in their shade *Avicennia* is ousted, remaining only on the fringes. The long hypocotyls of the other species raise their leaves above the seedlings of *A. marina*.

White mangrove trees may grow 5-10 m tall at Inhaca Island, the best growth occurring where salinity is slightly lower than that of the sea and where it is least variable, i.e. on the seaward fringe. On the very saline, bare sand at the head of zoned mangroves where a few widely spaced trees of *A. marina* may have germinated in slight depressions they remain *dwarfed*, barely 1 m high. Their branches are thicker and shorter than those of normal young trees, and there are fewer pneumatophores.

A. marina flowers in summer and is insect pollinated. The fruits ripen and fall in March, the time of equinoctial spring tides, which facilitates their distribution. The embryo undergoes some development before the fruit drops; the cotyledons are large and swollen, packed with nutrients, the hypocotyl has started growing and its distal tip has a heavy coating of bristle-like hairs which shelter the root primordia that protrude and provide them with a localised humid micro-environment. The meristems of the root primordia develop as soon as the fruit has been shed and the pericarp is sloughed off. The roots grow into the soil and anchor the seedling firmly. The hypocotyl continues to grow and raises the cotyledons clear of the ground. If the seeds are stranded on dry ground, the embryo loses water rapidly and irreversibly and will not develop.

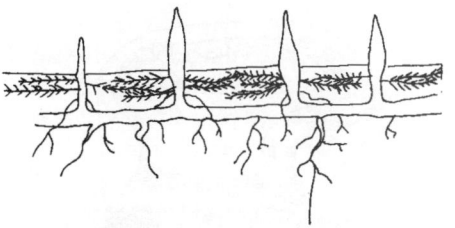

Fig. 4.4 Adaptations to mangrove conditions in *Avicennia marina*:
a. root system: cable root, pneumatophores and fibrous roots (x 0,2) (after Macnae 1968);

The root system of the white mangrove is subdivided for specialised functions (Fig. 4.4a). Anchorage is maintained by many long, strong, radial roots which act as cables and tie the tree down. A large number of fibrous and branch roots secure them in the black mud. There is no tap root. Lenticels are absent on these roots and so there is no danger of hydrogen sulphide being absorbed.

Gaseous exchange of oxygen and carbon dioxide for the root system is carried out by *vertical* roots growing upwards 10-30 cm *above ground* from the cable roots. These 'pneumatophores' are peppered with lenticels, pores which allow air exchange with the atmosphere, and have prominent cortical air spaces. From these, numerous passages connect with the air channels in all the other roots. When the tide covers the pneumatophores, air cannot enter and the pressure within the air passages drops, so that when the pneumatophores are uncovered again air is sucked in to restore an oxygen content of 10-18%.

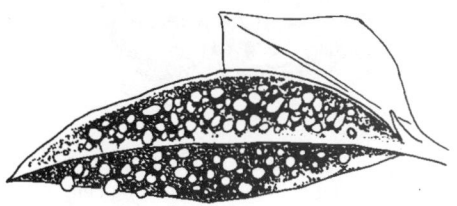

b. saline secretion on under side of leaf (after Berjak *et al.* 1986);

Absorption of nutrients is carried out by a feltwork of fibrous roots branching from the bases of the pneumatophores just below the surface of the soil. This position enables the absorptive roots to take advantage of the presence of inorganic nutrients provided by the action of aerobic bacteria near the soil surface. However, the roots do have the disadvantage of being in the layer of sand where evaporation may be high and salinity may increase above that of the sea or fall below it. Sodium chloride is absorbed and the sap in the xylem has about one hundred times more than in a land tree. The tree loses water by transpiration during the day, but the salt content of the tissues does not increase since at night the leaves excrete a salt solution which may be ten times more concentrated than that of the sap (Drennan and Berjak 1982). In the early morning droplets of salt exudate remain clinging to the under surfaces of mature leaves (Fig 4.4b). Salt is excreted by glands on the lower surface only. The leaves of *Avicennia* have water storage tissue between the chloroplast-containing cells and the upper surface (Fig. 4.4c). This may be

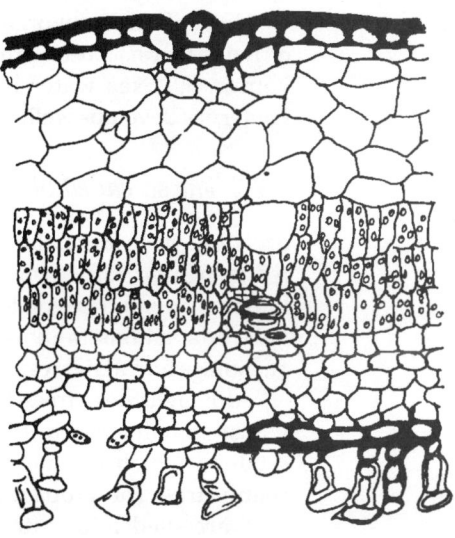

c. T.S. leaf with water storage tissues (x 50) (after Macnae 1968);

accessible to other cells as needed, and it plays a part in maintaining an even temperature in the leaf. Some cells in the leaves are rich in tannin bodies that sequester salt (Drennan and Berjak 1984). Although salt is not excluded at the roots, as is the case in other mangrove trees, it protects the cytoplasm of its cells from contact with sea water which would disrupt their organisation and inhibit the enzymes. Saline water is conveyed from the roots to the leaves via the apoplast, i.e. the non-living space between cells in the xylem (Drennan *et al.* 1987).

Evidence from the study of selectively stained ultrastructure of the excretory glands suggests that the salt solution is transported in *tubular endoplasmic reticulum* across cell walls to the salt-excreting cells on the lower surface of each leaf. The salt is then released by means of an active chloride ion pump accompanied by passive diffusion of sodium ions.

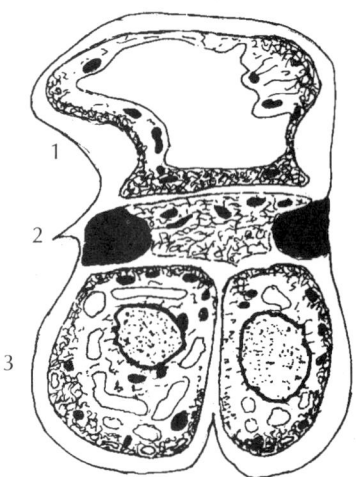

Fig. 4.4
d. salt gland, electronmicrograph (x 3 000):
1. collecting cell, 2. concentrating cell,
3. excretory cells. Note dense endoplasmic reticulum (after Drennan *et al.* 1987).

The excretory glands have a complex structure (Fig 4.4d). Each gland consists of two to four collecting cells, one stalk cell and eight to twelve excretory cells. The large vacuolated collecting cells adjacent to the endodermis of vascular bundles, receive sodium chloride in solution where the endoplasmic reticulum is in contact with the endodermal plasma membrane. The ions cross the cell wall within the tubular endoplasmic reticulum, contained in the plasmadesmata, and enter the stalk cell. The stalk cell is rich in endoplasmic reticulum and mitochondria near its boundary with the collecting cells. The boundary between the stalk cell and the excretory cells is similarly pierced by tubular endoplasmic reticulum and the salt solution is thereby conveyed to the excretory cells. Vacuoles are small in the stalk cells and the lateral walls are impermeable to water, being strongly cutinised.

The excretory cells on the other hand, are highly vacuolar. The boundary between the cells and the exterior has a thick layer of similar endoplasmic reticulum tubules, the outer ones being aligned with the cell membranes on the free edge. At these points the chloride ionic pump is driven by ATP (adenosine triphosphate), provided by the numerous mitochondria in this part of the excretory cell. This creates an electrochemical gradient which facilitates the passage of sodium ions to the external surface.

None of these cell organelles were developed to such proportions in the excretory glands of twigs that had been grown in salt-free solution. The experimental twigs studied had been kept in 50% sea water in which the excretory mechanism of the organelles in the leaf epidermis develops well.

4.4.2 The so-called black, red, and Indian mangrove trees

Bruguiera gymnorhiza, the black mangrove, grows in fine wet mud in the swamp interior, the red mangrove, *Rhizophora mucronata*, lines the banks of water courses, whilst *Ceriops tagal*, the Indian mangrove tree, colonises drier sandy soil, less often covered by neap tides. These trees have adaptations to living on intertidal flats that ensure reproduction, control of salinity, anchorage and aerial respiration of the roots, although they differ in detail.

Although these trees have been described as viviparous because seeds start developing whilst still attached to the parent plant, forming propagules, these are actually very large seeds in which the hypocotyl of the embryonic axis has extended to a length of many centimetres. The propagules are well adapted to implantation in wet mud and *immediately* they are shed into suitable conditions they germinate. The propagules are pendulous, torpedo-shaped structures, several times the length of the large leaves, and they grow on the tree for several months (Fig. 4.1b, c and d). Since the flowering season lasts the whole summer all the stages of development from flowers to ripe propagules may be found on the same tree. When the flowers have been pollinated by insects the calyx persists on the round fruit which contains only one seed (Fig. 4.5a).

The cotyledons of the embryo remain small and are concealed by the calyx and fruit wall. The hypocotyl, i.e. the axis between cotyledons and root, grows longer and thicker (Fig 4.5b). It is covered by a heavily cuticularised, green epidermis that is rich in tannin bodies (Drennan and Berjak 1984). It is thus resistant to immersion in seawater and also to microbial degradation for many months. During the three to four months of attachment to the tree the propagule receives much nourishment from the parent tree. The cells in the layer where parent and offspring are in contact are interdigitated to form a very large surface area through which food passes, as in a mammal's placenta. These cell membranes are, however, almost impermeable to salt so that the developing plant is partly protected from the salt load which may have accumulated in the branches. The salt that may penetrate is sequestered by the tannin bodies in the peripheral layers of the cotyledons and in the aerenchyma tissue of the thick cortex of the hypocotyl. A thin layer of chlorophyllous cells under the epidermis aids in assimilation. These cells form a thicker layer under the ridges which carry rows of lenticels through which the propagules are aerated (Fig. 4.5c).

When abscission from the parent tree takes place the propagules fall like arrows vertically, penetrating the soft wet soil when the tide is out. Development of seedlings is very rapid, roots appearing within hours of the propagules being implanted (Fig. 4.5d). Frequently trees are surrounded by growing seedlings of different ages. They germinate best in the shade of other trees and so cannot initiate a mangrove association. If the tide is high when the propagules fall they may be carried out to sea and back again to the same mangrove swamp or another, since they float easily, orientated almost vertically so that the delicate shoot is not immersed.

The differences between the propagules of the three species are small but are correlated with the conditions of the substrata in which they germinate. *Rhizophora mucronata*, on the banks of the channels, has the longest propagules that grow to 300 mm or more. At abscission the round fruit and four-sepal calyx are left behind on the tree (Fig. 4.5a) so that the weight of the propagule is concentrated in its lower half where it is somewhat swollen behind the tip. It sinks into the soft anoxic mud and the developing shoot is kept well above the surface.

Similarly the fruit with five of six sepals in *Ceriops tagal* remains on the tree when the propagule drops. As in *R. mucronata*, the first leaves begin to develop before the seedling drops and forms a collar-like extension below the fruit. The propagules are about 200 mm long and are very slender, curved and ridged, weighted towards the lower pointed end. They penetrate the more compact sandy mud when they fall after a spring tide has just receded.

The propagules of *Bruguiera gymnorhiza* are shorter, stouter, almost dagger-shaped and about 150 mm long (Fig. 4.5b). The fruit and calyx remain attached to the developing seedling at first, but are split off by the early emergence of the hypocotyl. The radicle grows strong rootlets immediately and anchors the seedling and the plumule enlarges (Fig. 4.5d). The tougher, heavier propagule of *B. gymnorhiza* penetrates the soft black mud in the interior of the swamp.

B. gymnorhiza and *C. tagal* have similar root systems. At their origin from the trunk the cable roots are flattened laterally into rugged buttresses, both above and below ground level (Fig. 4.5e). They also bend upwards at intervals along their length and protrude above the soil (Fig. 4.1b and c). At first they are slender and resemble 'elbows', then they thicken as the tree becomes older and form 'knee' roots. The exposed parts have lenticels and act as pneumatophores. Branching anchoring roots grow downwards from the cable roots and finer, more superficial roots for absorption remain near the surface.

Fig. 4.5 Germination in mangrove trees

a. Early fruit on tree, *Rhizophora mucronata* (x 0,5);

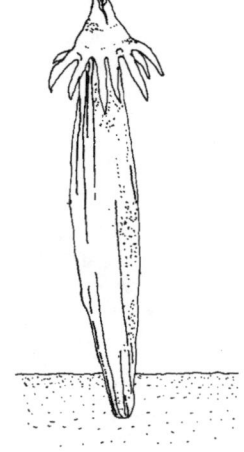

b. propagule penetrating mud, *Bruguiera gymnorhiza* (x 0,3).

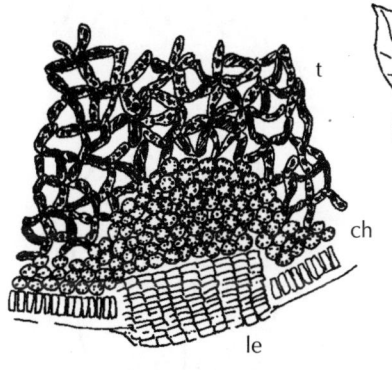

c. section of propagule, *Bruguiera* sp. (x 100): t tannin cells in aerenchyma; ch chlorophyll cells; le lenticel (after Cranson 1907);

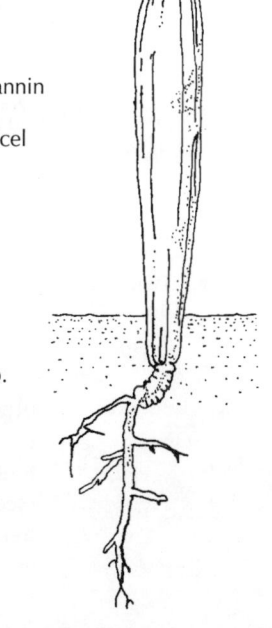

d. Seedling of *Bruguiera* sp. (x 0,3);

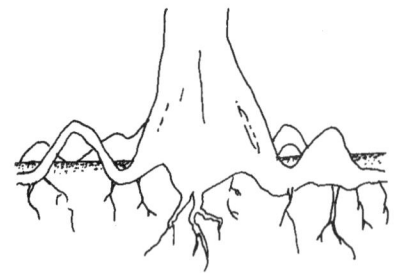

Fig. 4.5 Root Adaptations
e. Knee roots in *Ceriops tagal*;

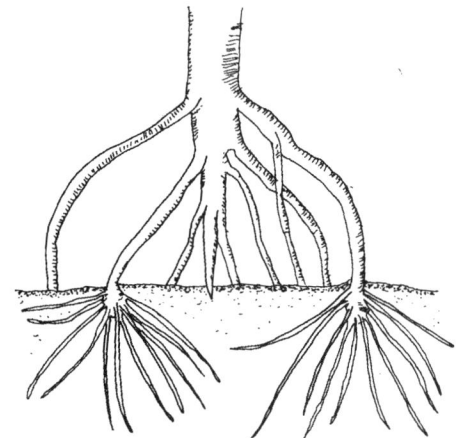

f. Prop roots in *Rhizophora mucronata*;

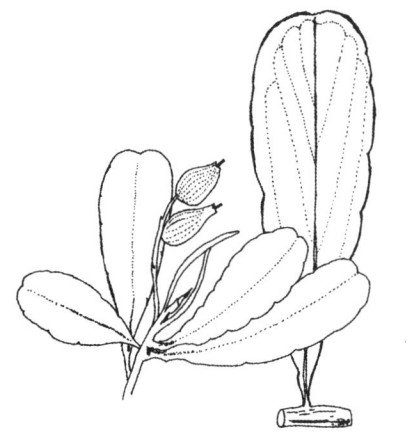

g. Twig of *Lumnitzera racemosa* (a, b, d-f after Berjak *et al.* 1986).

The root system of *R. mucronata* is adapted to the softer mud which is swept by the tidal current. The cable roots grow radially from the trunk from 5 cm to 2 m above ground depending on the size of the tree (Fig. 4.5f). They arch widely before growing down to the mud and so act as supportive props. Sometimes the prop roots branch above ground or may grow afresh from tree branches. Immediately below the soil surface they branch into nutritive roots and longer, fibrous anchoring roots occur at their ends. Lenticels on the portions of the roots above ground lead to a network of air passages as in the other trees.

The absorptive roots of the trees in this family have a subcellular exclusion mechanism involving active transport which prevents the absorption of salt, albeit incompletely. The osmotic pressure of the sap is variable and *lower* than that of sea water. *Bruguiera* grows best in dilute seawater and if the salinity exceeds 50% of normal sea water the rate of transpiration is reduced and less salt travels upwards to the leaves. Nevertheless, some salt does accumulate in the leaves, the termini of the transport system, since there are no salt glands. Salt is adsorbed on the tannin bodies present in the cells but as the salt level increases further the organisation of the cytoplasm in the cells of the leaves becomes deranged and they turn yellow and die. Leaves are shed throughout the year thus ridding the trees of the excessive salt load. Numerous leaf scars are visible along the branches, the functional leaves being crowded at the ends (see Fig. 4.1b and c).

4.4.3 The 'spring tide' mangrove

Lumnitzera racemosa is known as the 'spring tide mangrove' since its roots are inundated only by the highest of spring tides. The bush has light green, fleshy leaves with water storage tissue like those of other mangrove trees (Fig. 4.5g). Its flask-shaped fruits have one seed. The fruit becomes hard and woody, resistant to both salt water and to desiccation. It germinates on the sandy, landward edges of mangroves where such exist, usually in the shade of trees. The root system resembles that of *Bruguiera* in form but does not exclude salt; neither do the leaves have salt excreting glands. It has been described as a 'salt accumulator' storing salt in the bark and in the leaves until they are shed.

4.5 DISTRIBUTION OF ANIMAL AND PLANT ASSOCIATES

Over fifty animal species inhabit the mangrove swamps of Inhaca Island, about half of them being crustaceans; most of the others are snails or fishes. There are no burrowing bivalves nor polychaete worms. In contrast, the sandy slope between the same tidal levels on the less sheltered, west shore has about one tenth as many species; only the ocypodid ghost crabs and wedge mussels are numerous there (see Section 3.2). The high diversity and density of the animal populations in the swamps depend on the variety of the adaptations which enable them to be accommodated spatially along the physical gradients described in Section 4.3.

The animals that live in the mangrove swamps are influenced by those environmental factors which segregate the trees and thus appear to be associated with the trees. There are five species of fiddler (*Uca*) crabs, which retire into their burrows before the tide covers them; six species of leaf-eating (sesarmid) crabs emerge from burrows to feed at low tide, taking to high ground or to the trees when the tide comes in. The amphibious fish, the mud-skipping goby, builds nests for shelter from heat and predators, and for breeding, but it searches for food on drying mud and climbs trees when the tide is high. One species of shrimp feeds and breeds in its own burrow and never appears above ground. Two species of snail climb trees when the tide is high and others remain in or on the mud. These animals have evolved many kinds of structures, behaviour patterns and physiological mechanisms which fit each for a special niche in the swamp. Their distribution is described in this section and the adaptations of the crabs are discussed more fully in Section 4.6.

Fig. 4.6
a. Typical male fiddler crab, broad front (x 1)*;

b. *Uca inversa*, carapace (x 1,5);

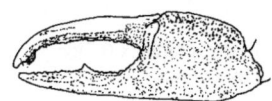

c. large male chela of *U. inversa* (x 1.5) (subterminal tooth on 'finger') (b and c after Crosnier 1965).

4.5.1 The supratidal fringe: salt marsh; land crab; grapsid crabs; fiddler crabs

There are three variants of the supratidal fringe in the mangrove associations of Inhaca Island which always include the outer *Avicennia* belt and the landward area as far as the influence of the highest spring tide extends. These are indicated in the maps in Fig. 4.2 a, b and c. The different areas are bare flats, salt marsh, and sand covered peat.

At the extreme southern end of the southern limb of the Saco mangrove a large area of sand, with no shade and almost devoid of vegetation, stretches from the dunes to the swamp. When this area has been covered by the highest spring tides, a sheet of sea water remains for some hours on the surface. High neap tides do not reach it, so that evaporation is accentuated and salt crystals glisten on the surface. This area is the most saline in the mangrove swamps and is the territory of the most robust of the fiddler crabs, *Uca inversa*, an ocypodid crab with the characteristic long eyestalks (Fig. 4.6b). It is one of three species of *Uca* with a broad 'front' between the eyes (Fig 4.6a). It is about 20 mm long and 27 mm broad; the dark carapace has a white speckled pattern and the strong legs are red. The male's major chela (Fig 4.6c), carried horizontally in front of it in the manner of all fiddler crabs, is a deep pink colour. The position of the tooth on the chela is 'inverted' i.e. this tooth is on the movable 'finger' instead of on the 'hand' of the claw as in other fiddler crabs.

Uca inversa is active only when the sand is moist from spring tide inundation, rain or dew. During spring tides these crabs feed from early morning to noon but during neap tide weeks feeding is confined to the very early morning. The flat is later deserted as the crabs remain in their closed burrows which are simple tunnels sloping down to the water table, dug out after spring tides when the sand is somewhat softer. The strong walking legs on one side of the body are used for scooping out balls of sand, about 10 mm in diameter, which are scattered near the hole but soon dry out and become dispersed; these balls are much larger than the sand pellets rejected whilst feeding.

During spring tides, depressions on the wet sand flat are colonised by blue-green 'algae' (Cyanophyta) which compact the sand grains into a thin crust when the tides are low. These are avoided by fiddler crabs. Low hummocks of sand occur here and there which are covered by a few salt marsh plants (described below); here the fiddler crab burrows are numerous. A few isolated dwarf *A. marina* trees occur towards the lower end of the flat. They have sturdy

Fig. 4.7 Salt marsh plants (x 0,5):
a. rush, *Juncus kraussii*;
grasses:
b. *Sporobolus virginicus* (a and b after Lubke and van Wyk 1988).
c. *Digitaria natalensis*;
fern:
d. *Acrostichum aureum*.

branches, but few pneumatophores and are only about 0,5-1 m high. There are very few normal *A. marina* trees between this type of flat and *Ceriops tagal* thickets, i.e. the lower part of the supratidal fringe is almost suppressed. Similar *U. inversa* territory occurs in the north-east corner in the northern sector of the Saco. This species has a tropical distribution.

At Inhaca Island the salt marsh type of supratidal fringe is best seen on the northern edge of the northern limb of the Saco (Kamakandi) where there is a transition from freshwater swamp vegetation to that of mangrove swamp. Gradually the *Phragmites* reeds become dwarfed and sparse in a wide meadow dominated by the rush *Juncus kraussii* (Fig. 4.7a). Several salt-tolerant grasses and sedges occur, the most conspicuous in the drier part being *Sporobolus virginicus* and in the wetter, more saline parts *Digitaria natalensis* (Fig. 4.7b and c). The mangrove fern *Acrostichum aureum* may be common at the landward edge forming a hedge (Fig. 4.7d). This fern grows about a metre high and has stiff fronds and pinnae. A few bushes of *Lumnitzera racemosa* occur here and there.

At the level of high water of ordinary spring tides the perennial herbaceous halophytes form a wide belt. *Chenolea diffusa* is followed by areas of *Sesuvium portulacastrum* and *Sarcocornia (Arthrocnemum) perenne*. The annual *Salicornia perrieri* invades the bare sand in summer. With the exception of *Sesuvium portulacastrum* these plants belong to the family Chenopodiaceae which have inconspicuous flowers. *Chenolea diffusa* has small, closely packed succulent leaves, grey-blue in colour and arranged in dense rows on the ends of vertical branches that arise from trailing stems (Fig. 4.8a). The leaves of *S. perenne* are very small, the stems are dark green, thick and fleshy and indented at the nodes so that they appear segmented. This plant has a creeping habit and forms low-lying green mats (Fig. 4.8b). *Salicornia perrieri* is similar to *Sarcocornia perenne* in stem and small leaves but it is a bushy plant growing about 15 cm high, and occurring in patches instead of spreading over large areas.

Sesuvium portulacastrum is quite distinct from the other plants since it has fleshy leaves that store water and grow in pairs at right angles from a prostrate stem (Fig. 4.8c). The leaves may be bright green, and in response to the high environmental salinity the stems become bright red. It belongs to the family Aizoaceae in which the flowers are not insignificant as in the other plants. Each flower occurs in the axil of a leaf and the calyx forms a pink star about 15 mm in diameter but there are no petals.

The halophytic herbaceous plants all have water storage tissue in stems or leaves, in the cells of which sodium chloride is confined to vacuoles and tannin granules. The osmotic balance between the salt-laden vacuoles of the storage cells and their living cytoplasm is maintained by the presence of relatively large amounts of organic solutes in the cytoplasm. The saltworts are immediately followed to seaward by a wide belt of both saplings and mature trees of *Avicennia marina*.

Occasional untidy pits about 20-30 cm across may be found among the outermost pneumatophores of *Avicennia* in the Kumakandi swamps in the north bay which are dug by a large land crab *Cardisoma carnifex* (Fig 4.9a). Each leads into a burrow about 10 cm in diameter which descends in a loose spiral to a larger chamber at the level of the water table. At night it comes out to scavenge for food. It very rarely appears above ground in daylight at Inhaca Island unless the sky is overcast. Its innate timetable can be changed by external stimuli. The carapace is dark, reddish brown and strongly convex, protecting a very large gill cavity with vascularised areas. The ventral area of the carapace is hairy and retains moisture. Juveniles have not yet been seen on the island. Where are they?

Three species of grapsid crab live in the *outer part* of the supratidal fringe among the rushes, grasses and saltworts. They belong to the sesarmid sub-family of grapsid crabs which burrow in the muddy banks of estuaries. The carapace is almost square, brown and smooth; the eyes are set well apart on short thick eye-stalks and the chelae are equal in size (Fig. 4.9b). They spend much of their time out of direct contact with water.

These three salt marsh sesarmid species are superficially similar in having no teeth on the sides of the carapace (Fig 4.9b). They may be distinguished by their chelae which have specific patterns of tubercles, hairs or toothed ridges, which are fitted in specific ways for cleaning sand and mud from various parts of the body which remain after digging burrows and feeding. *Sesarma catenata* (20 x 25 mm) has buff-coloured chelae with tufts of hair between finger and hand (Fig 4.9c). *S. eulimene* (15 x 20 mm) has purplish or rosy chelae without tufts of hair (Fig. 4.9d). *S. ortmanni* is slightly larger than the others and has orange chelae with large tubercles on the finger (Fig 4.9e). The differences in behaviour and feeding habits of these salt marsh species and the three mangrove species are discussed in Section 4.6.

Three more species of *Sesarma* occur in the area from the level where succulents invade the salt marsh down through the

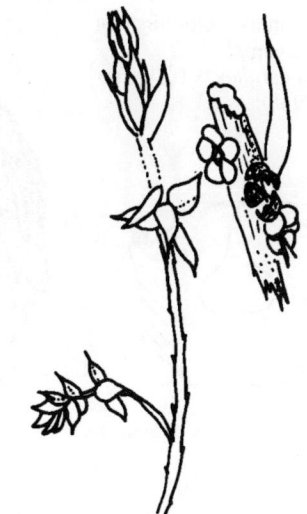

Fig. 4.8 Salt marsh plants (x 0,5):
a. *Chenolea diffusa;*

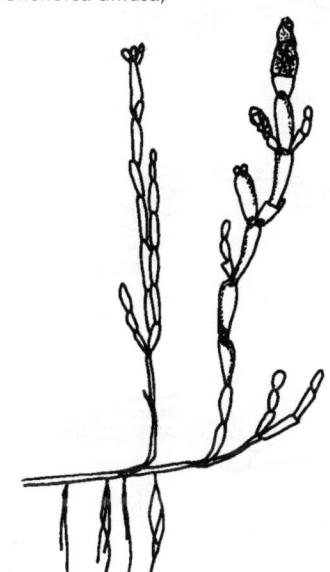

b. *Sarcocornia perenne* (a and b after Lubke and van Wyk 1988).

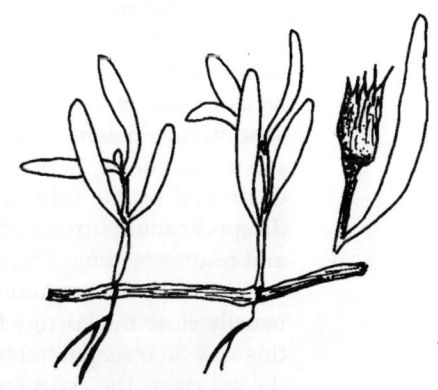

c. *Sesuvium portulacastrum.*

Fig. 4.9 Crabs on the outer fringe of the mangrove swamp:

a. *Cardisoma carnifex* (x 0,5) (after Macnae 1968);

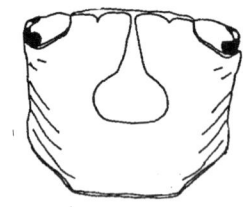

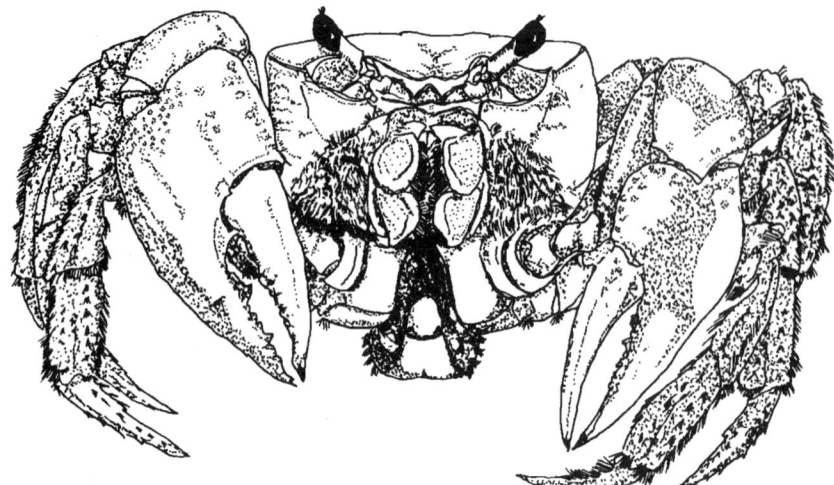

b. carapace of 3 species of *Sesarma* on fringe (x 1,5).

chelae (x 1):

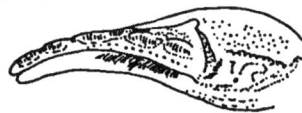

c. *S. catenata*;

d. *S. eulimene*;

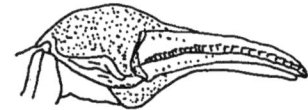

e. *S. ortmanni* (b-e after Crosnier 1965).

Avicennia belt and beyond among the *Bruguiera* trees where there is wet black mud. *S. meinerti*, *S. smithi* and *S. guttatum* are black and are larger than the other three species; they have one tooth on each side of the carapace (Fig 4.10a and b). They have been called 'leaf-eating' crabs because they pick up newly fallen leaves of *B. gymnorhiza* and take them to their burrows where they are broken up as food and consumed. Occasionally they will retrieve a carcass of a crab or other object, which is ejected after inspection if it is not palatable. The southern species *S. meinerti* is more numerous than the tropical *S. smithi* and *S. guttatum* and lives nearer the mangrove fringe. The chelae differ (Fig. 4.10 c, d and e). The legs of *S. meinerti* have furry extremities and retain water whilst those of *S. smithi* are bare. During low tides in the daytime these crabs spend most of their time at the entrance of their burrows, emerging to collect food now and again as leaves become available. The burrows of *S. meinerti* have hoods of mud sheltering the entrance, whereas those of *S. smithi* have small turrets. At high tide both these crabs remain *above* water level by climbing to higher ground, on to pneumatophores or the lowest branches of the trees. *S. guttatum*, only seen in the dark interior of the swamp, is described in Section 4.5.2.

The most numerous of all the mangrove crabs is the fiddler crab *Uca lactea annulipes* which lives beneath the *Avicennia* trees, avoiding deep shade. It has a black carapace (16 x 20 mm) with white markings, black legs and in the male a slender, smooth, major chela that varies in colour from almost white, through pale pink, to a deep salmon-pink (Fig. 4.11b). This species feeds for the greater part of the day between tides. Innumerable pseudo-faecal pellets, i.e. sand from which food has been extracted and which have been removed from the mouth by the minor chelae, litter the ground, giving the soil a crumb structure when the pellets are destroyed by the tide. *Uca* crabs are very sensitive to ground vibrations, and they will disappear into burrows at the fall of a footstep. After a few minutes they emerge cautiously and resume feeding. The crabs visit the burrows frequently during the day to fill up their gill cavities with water. About ten minutes before the tide covers the burrows they retire and usually close the burrow from within by moulding a soft sticky mud plug to fit the hole. In this way an essential bubble of air is retained in the burrow which may aerate its water when the sea rises. The crabs emerge and clean out large balls of sand to make the burrow usable when the tide has receded. Then each crab sits at the entrance and grooms itself, removing

Fig. 4.10 Sesarmid crabs in the interior of the swamp (after Crosnier 1965):

a. *Sesarma meinerti* (x 1) showing branchiostegites*;

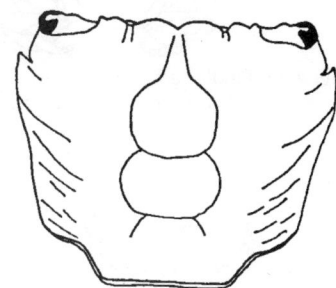

b. dorsal carapace of the three species (x 1);

chelae (x 1):

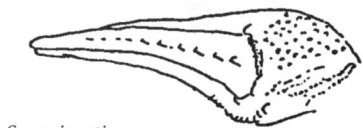

c. *S. meinerti*;

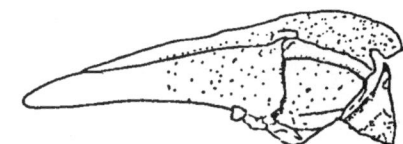

d. *S. smithi*;

e. *S. guttatum*.

mud from the eye stalks, body and major chela with the aid of the minor chela in the male whilst the female uses both chelae for the task.

In the summer this species has the most elaborate pattern of courtship and dancing of all the Inhaca *Uca* crabs. The biology of *Uca* and *Sesarma* crabs is discussed in Section 4.6.

The western part of the northern mangrove (Sichwane, Fig. 4.2b) has widely-spaced, large, spreading trees, and no saplings grow there now. The sandy soil between the trees is carpeted by *Juncus*, *Sarcocornia* and *Sesuvium*. The fiddler crab, *Uca lactea annulipes* is abundant between the innumerable pneumatophores but other species of *Uca* and *Sesarma* appear to be absent.

In the hidden sandy drainage paths lives another small ocypodid crab, *Macrophthalmus depressus* (9 x 13 mm) which may be found loosely buried when the tide is out; it has no formal burrow. This species is very similar in shape to its sibling *M. grandidieri* of the muddy flats of the west shore (see Section 3.5.2) but the mangrove species is darker in colour, being greyish-blue with brown spotted- legs (Fig. 4.12a). It feeds on detritus in sand when the sand is wet from the receding tides.

In the lower branches of *Avicennia marina* the largest species of periwinkle, *Littorina scabra* (Fig. 4.13a), rests when the tide is out during the day. These are also known from the sheltered rocks at Ponta Punduine (Section 6.2.2) but they are far more common on the trees of the seaward *Avicennia* fringe. At night when the tide goes out they descend to feed on the microscopic algal coating of the trunks of the trees and the pneumatophores, returning to the shade of the branches before the heat of the day. These periwinkles have a curious habit of suspending themselves on the end of a thick mucus string if dislodged from the tree; their hold on a leaf depends on the secretion of thick mucus.

Where the dunes are near to the mangrove swamp as on the western edge of the southern limb of the Saco. Kunwadlanheni mangrove, they are colonised by sand-binding plants such as *Scaevola plumieri*, *Ipomoea brasiliensis* and *Canavallia maritima* (see Section 10.5.1). Here the supratidal fringe is foreshortened and the succulents of the salt marsh form a very narrow belt in front of the *Avicennia* trees. The latter are well grown but do not cover a wide area. Between the dune and the mangrove may be seen the tracks of the land hermit crabs *Coenobita cavipes* and *C. rugosus* which are made during their nocturnal migrations from shelter under dune plants to the sea to feed and back (see Section 5.5.1 and Fig. 5.10d). There may be a drift line of

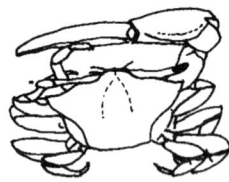

Fig. 4.11 Fiddler crabs:
a. carapace with broad front *Uca gaimardi**(x 1);

male chela:

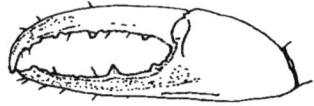

b. *U. annulipes*

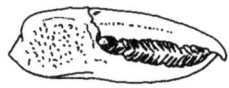

c. *U. gaimardi**(x 2);

d. carapace with narrow front (x 2)

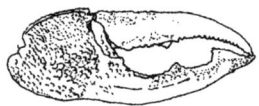

e. *U. urvillae*

f. *U. vocans* (x 2) (d-f after Crosnier 1965).

leaf debris sheltering the small sand amphipod, *Orchestia anomala*.

4.5.2 The interior of the swamp: crustaceans; snails; meiofauna; insects; mud-skipping goby; birds

Among the crowded trees in the interior thicket, crustaceans are abundant and there are a few specialised molluscs.

In the sandy mud in the shelter of *Ceriops tagal* at the edges of the thickets the smallest of the fiddler crabs, *Uca gaimardi* (c. 15 mm) digs its burrows (Fig. 4.11a and c). This species has a beautifully coloured carapace that varies from turquoise to a vibrant green, with red legs and major cheliped. Females are similar, but with muted carapace colours. The crabs spread out to feed in clear spaces but avoid both strong sunlight and deep shade. They do not venture into black mud but may be found feeding among *Avicennia* pneumatophores and *Ceriops* knee-roots.

In sunlit patches on the channel banks among *Rhizophora* trees, a fourth species of fiddler crab makes its burrows and feeds. This is *Uca urvillei* (25-30 mm), the males of which have a bright royal blue carapace. It is one of two species of fiddler crab with a narrow front between the eyes (Fig 4.11d). It has long, robust and toothed, golden-yellow major chelae (Fig. 4.11e); the females have a duller colour, brown to dark blue. In the mating season the colours become more vivid in both sexes.

The soft black mud in which *Bruguiera gymnorhiza* and *Rhizophora mucronata* trees grow is the territory of grapsid crabs. The smaller crabs include juveniles of a large species *Sesarma (Chiromanthes) guttatum* (21 x 26 mm). This species has one lateral tooth behind the post-orbital tooth on the dark carapace, and is distinguished by a large spine on the fourth joint of the chelipeds and distinctive chela (Fig. 4.10e). It makes a burrow rather wider than its body and often stands in the entrance by day.

Adult *S. meinerti* and *S. smithi* may also burrow in this area. The larger sesarmids may be active at night and spend most of the day in their burrows. When the tide rises they leave their burrows and climb on to exposed ground or the lower branches of the trees. This habit has earned them the name of 'water-haters'; in this behaviour they differ from fiddler crabs which retire into their burrows when the tide advances. Both these groups evade the predatory swimming crabs, *Scylla serrata* and fishes.

There are two representatives of another subfamily of grapsid crabs which do not burrow and most of which live on rocks at high tidal level. The smallest crab *Ilyograpsus padicula* (10 mm), is distinguished from juveniles of other species by the presence of four lateral teeth on the carapace (instead of one or two) (Fig. 4.12b). The larger brownish crab (20 x 25 mm), which has a speckled carapace and purple chelae, is the estuarine *Metopograpsus thukuhar* which may hide in other crabs' burrows (Fig. 4.12c). Both these species are carnivorous.

Occasionally a xanthid crab, *Eurycarcinus natalensis* (26 x 38 mm), which is bright purple in colour and has strongly toothed chelae, may be found under logs (Fig. 4.12d). This is the sole representative of the xanthid family of crabs which is usually confined to lower levels of rocky shores and is not adapted to exposure to air for as long as the grapsid and ocypodid crabs. When the tide is out it remains in shelter, under a log, completely inert.

The mud banks of the deeper channels are seen at low tide to be riddled with innumerable small holes. These lead into deep, narrow U-shaped burrows made by pairs of mud shrimps (thalassinids) *Upogebia africana* (c. 65 mm) (Fig. 4.12e). These pale, soft-bodied mud-shrimps never appear above ground and cannot easily be extracted by digging in the sticky mud. The burrows, which always retain water, have swellings at intervals enabling the shrimps to turn around. If strong pressure is firmly applied to the surface of the mud (by means of an inverted jam tin) some shrimps on the periphery will be forced out. They immediately start burrowing again with the first two pairs of legs. These legs are hairy and chelate and can be held together to form a basket-like scoop with which pellets of mud are removed. The third pair of legs is not chelate, and in this they resemble hermit crabs, to which they are related.

The mud-shrimp has a wide abdomen with fan-like appendages on each segment, uropods and a telson. It feeds by creating a current of water through the burrow by means of the abdominal appendages, from which the fringed setae on the first two pairs of legs strain out the plankton and detritus. The setous maxillipeds are used as combs to remove the food and transfer it to the mouthparts and thence to the mouth. This species of mud shrimp is capable of regulating the osmotic pressure of the body fluid over a wide range of dilution (Hill 1971). In the mangrove swamp where it is always wet, it occurs higher up the shore than it does on muddy beaches in South Africa, almost exclusively in estuaries and the most sheltered bays.

Three kinds of snail live on the mud of the swamp. The most conspicuous is *Cerithidium decollata* (30 mm) (Fig. 4.13b) which spends the days of spring tides securely attached to the shady side of trunks of trees, hanging with the apex pointing downwards. Trails of mucus lead to their 'perches' where they are made secure by dry sticky mucus around the aperture. There may be 20 to 50 individuals on one tree in a band 1–2 m high, beyond the reach of the spring tides. When the week of neap tides starts, the snails climb down on to the wet mud at night and feed, selecting mainly fungi from the surface; then they return to the trees (Cockroft and Forbes 1981). This semi-lunar feeding rhythm is unusual and may be triggered by a response to changes in humidity of the atmosphere since these snails are able to respire in air. The elongated, spirally and longitudinally ridged, grey shell is peculiar in that the apex is always broken off in young snails and sealed with a shelly plate, but young specimens smaller than 20 mm are rarely found. One wonders why?

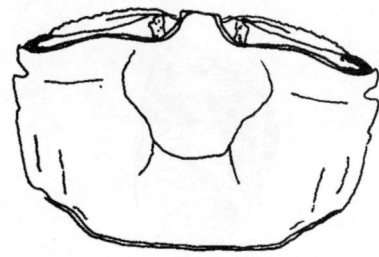

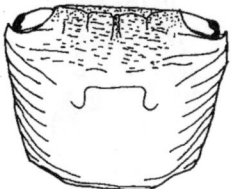
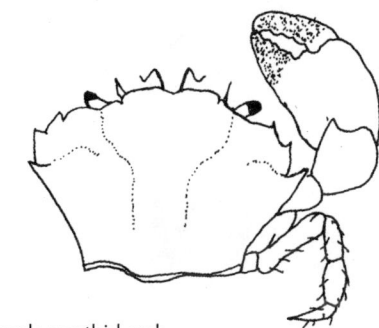
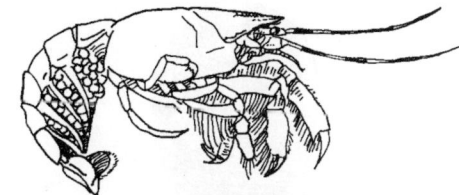

Fig. 4.12 More mangrove swamp crustaceans (all × 1):
a. Ocypodid crab, *Macrophthalmus depressus* (× 2)
b. *Ilyograpsus paludicola*;
c. *Metopograpsus thukuhar*;
d. purple xanthid crab, *Eurycarcinus natalensis* (a–d after Crosnier 1965);
e. mud prawn, *Upogebia africana**.

On the surface of black mud, particularly in the upper parts of the permanent channels, two species of pulmonate (air-breathing) snails may be found. *Melampus semiaratus* (6 mm) (Fig. 4.13c) has a black, barrel-shaped shell tapering towards both the apex and the lower end of the aperture. Below its midline there are a number of concentric, deeply incised lines. The aperture is long and almost slit-like and has a few widely spaced, small, blunt white teeth. Some individuals are found beneath the mud surface in small crevices or in crab holes. During

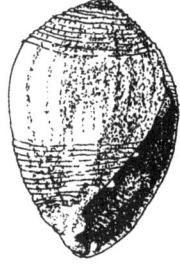

Fig. 4.13 Mangrove snails (x 1,5):
a. *Littoraria (Littorina) scabra**;
b. *Cerithidium decollata**;
c. *Melampus semiaratus*;
d. *Cassidula labrella* (c and d after D.S. Brown 1971).

spring tides some migrate to higher ground, but most are usually hidden, despite their aerial respiration.

The other pulmonate snail, *Cassidula labrella* (7 mm) is brownish in colour and slightly narrower. Its aperture is thickened and light in colour and has several large, blunt white teeth (Fig. 4.13d). Its surface is hairy so that mud sticks to it making it difficult to see although it does not seek shelter beneath the surface as does *Melampus*. Both these snails are said to be unselective feeders on the rich detritus of the black mud.

The meiofauna in the mangrove mud is denser than on the sandy shore, described in Chapter 3, because the shore is more sheltered and the detritus is richer. Nematodes are the dominant group (see Fig. 3.23g-j), but the harpacticoid copepods are also represented. The latter have been identified at Inhaca Island and no species peculiar to the mangrove swamp was found. Of the ten species (which also occur in detritus sand on the west shore) three are co-dominant: *Halectinosoma langi* (Fig. 3.23f), *Pseudostenhalia prima*, and *Stenhalia unisetosa* (Fig. 3.23c). The last-named is also very common in sandy mud, rich in detritus, on the west shore. The setose mouth parts and the non-prehensile maxilliped in each of these species suggests that they are not carnivorous. It has been suggested that the meiofauna in the mangrove are the major consumers of the micro-algae in the mangrove soils.

Among the foliage of *Ceriops* trees there are groups of nests of the tailor ant, *Oecophylla longinoda*, which are made by glueing together the crowded leaves on the ends of branches (Fig. 4.14a) (see Section 13.12.1) (Way 1954). While some worker ants hold the leaves together, others travel along the edges holding larvae in their mandibles and swaying from side to side, sewing the leaves together (Fig 4.14b). The larvae secrete sticky, silken threads which are attached to overlapping leaves to make a compact nest. These ants feed partly on the sugary secretion (honeydew) of insects living on the bark of the tree, that are relatively inconspicuous, being covered by a thin layer of transparent wax. The ants have a poisonous bite and they attack larger insects such as bees and flies. In fact, they do not mature unless their diet includes both 'honey dew' and protein-rich insects.

Several mosquito species inhabit the swamps. One species known in Kenya to be peculiar to mangrove swamps, is *Aedes pembaensis* (Fig. 4.14c). It may lay its eggs on the claws of the leaf-eating crab *Sesarma meinerti*, which takes them down to the end of its burrow where they hatch in water (Worth *et al.* 1961). The pale larvae and pupae develop in the permanent water in the crab burrows at the water table below. Clouds of adult mosquitoes are sometimes seen emerging from the burrows. *Anopheles gambiae*, a malaria-carrier, breeds in open areas behind the swamps and penetrates into the mangrove when adult (Fig. 4.14d). Other anophelines lay eggs in temporary pools in the interior of the swamp. It should, however, be noted that in contrast to tropical mangroves the mosquito population at Inhaca Island does not unduly worry daytime visitors to the mangroves, since most feed at dusk (See Section 13.10).

In the summer, bees visit the flowers of mangrove trees to gather nectar and pollinate them, but their nests are usually on land in trees (See Section 13.12). Tiny beetles may be found burrowing into the dead bark of mangrove trees and in the mud under fallen logs. Staphylinid beetle larvae in the mud are hunted by the mud-skipper, *Periophthalmus*.

Here and there in the swamp interior, especially along the banks of the streams in bright sunlight, an amphibious fish may be seen skipping along, resting on the surface or on roots, looking around with mobile eyes before pouncing on its prey (Stebbins and Kalk 1961). This is the mud-skipping goby, *Periophthalmus kalolo* (7 cm), which spends most of the day *out of water* hunting minute animals in the mud within a range of a few metres from its twin-turreted mud nest. In the summer months when the tide rises, these gobies disperse among the pneumatophores of the landward *Avicennia* trees, ahead of the tide, and then return as the tide ebbs.

This little fish has a large blunt head from the top of which two large eyes protrude from sockets, acting as conning towers as they swivel from side to side (Fig. 4.15a). Thus it has complete peripheral vision, as its generic name suggests, and can see over a distance of 10 m. It can withdraw its eyes deep into the sockets whilst under water and does so on land too from time to time, perhaps to moisten them with residual water in a depression below the eye ball. The head is held high when at rest (fig 4.15c) and is supported by the pectoral fins, which have an 'elbow' joint; spiny rays strengthen the fin fans which are used as a pair of crutches to 'walk' on land. The centre of gravity lies immediately behind the pectoral fins above the flat pelvic fins, each of which forms a semicircular part of the disc: the body is streamlined, tapering to the tail when the dorsal fin is folded. The tail is dragged over the surface mud whilst walking as indicated in the track recorded in Fig. 4.15b. The dark spotted and striped colouring of the skin matches the mud so that the fish is invisible unless it moves (Fig. 4.15c). The first leap it makes when disturbed is heard rather than seen. The projectile force comes from the tail which is curled and then pressed against the substratum as it straightens. This fish can swim under water like a normal fish, albeit slowly. More often in its escape reaction it ricochets like a fast-flung, flat stone that skims along the surface of the water. As it touches the water the tail of the fish curves and flicks, sending it flying through the air with dorsal fins raised like a sail to maintain balance.

The best time to observe the movements of these gobies is on the edge of the advancing tide when they flee helter-skelter over the mud away from the water and climb pneumatophores or lower branches of trees. They may also be watched on open sunny areas in the swamp at low tide where they feed, engage in the defence of their territories near their nests or in courtship in the summer, displaying the bright colours of the dorsal fin. One way to catch these elusive gobies, which nearly always

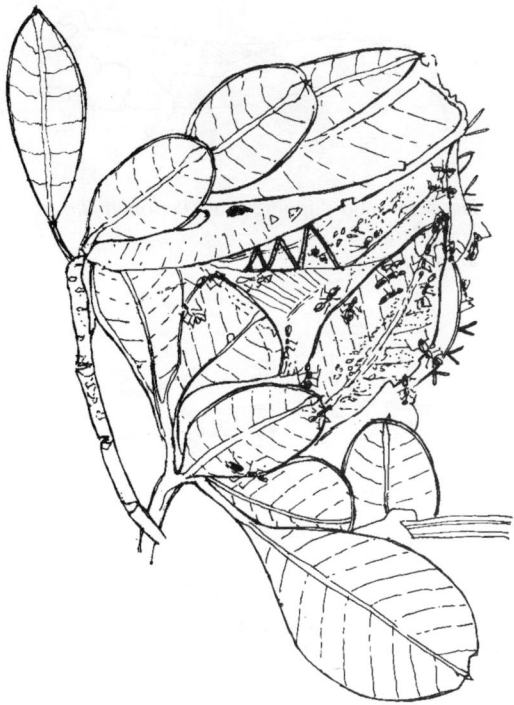

Fig. 4.14 Insects in the mangrove swamps
a. nest of tailor ants, *Oecophylla longinoda* (x 0,5) (from photograph by S. Hanrahan);

b. worker ant (x 3) (after Way 1954)

c. *Aedes* sp. (x 3)

d. *Anopheles* sp. (x 3) (cf *A. pembaensis*) (c and d after Skaife 1979).

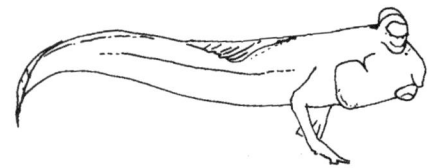

Fig. 4.15 Mud-skipping goby, *Periophthalmus kolreuteri* (x 0,7)
a. walking;

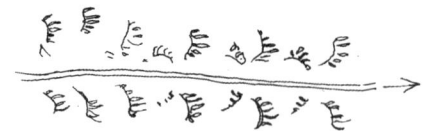

b. fin prints;

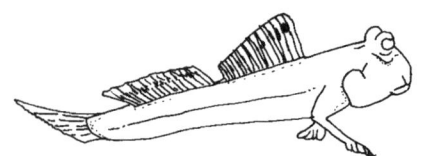

c. resting (x 0,8);

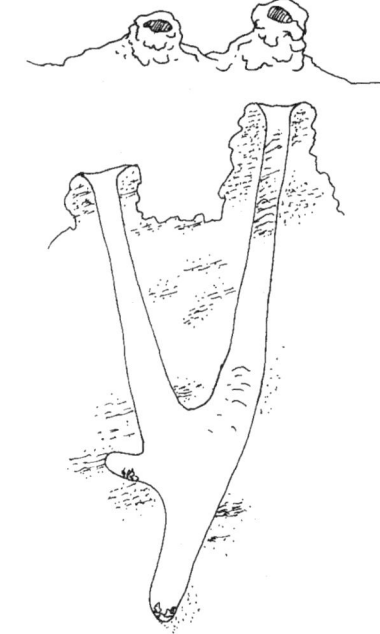

d. twin turrets and nest in section (x 0,25)
(a-d after Stebbins and Kalk 1961).

evade capture when chased, is to visit their nests (Fig. 4.15d) in the early morning at low tide before the sun shines on them. A sharp plunge of a large spade near the nest will compel the males to jump out into an awaiting handnet (see Section 4.6.2 below). A better way, which does not destroy the nest, is to pour seawater into the holes of the nest, which triggers the escape reaction to tidal flooding.

Periophthalmus hunts small animals it sees moving in the mud. The lens of its short-sighted eye is flattened on one part for clearer vision in air. The upper part of the retina, used when looking down, is very rich in rods but has few cones. Hence, with rods, it can easily detect shades of light and dark grey against the black background of the mud over a distance of 2-3 m. Very small crustaceans such as tanaids, young stages of small crabs and shrimps of various kinds, very small staphylinid beetles, larvae of ephydrid flies and nematodes have been found in the gut of these fishes. In fact, the tiny animals that live in the mud were first encountered at Inhaca by studying the diet of *Periophthalmus*.

Both male and female mud-skipping gobies construct well-spaced nests on the channel banks or among the trees, on wet mud in the shade. They dig out small pellets using the mouth and exude them on the surface around the hole in the way that toothpaste is squeezed out of a tube. The nests are hollow chambers about 25 mm in diameter reaching down 200 mm below ground level. Two pipes lead to the surface ending in turrets 10 and 30 mm high, on which fresh pellets can be seen drying. The fishes probably spend the night in the nest, since they are found there in the morning, and come out on land at low tides during the day, returning to the same burrow at night. They may visit it during the day, but when the tide rises, the fish waits in the turret until the sea almost reaches it and then jumps out and skips ahead of the fast-spreading water. The goby may reach the landward fringe and climb on to *Avicennia* pneumatophores or low branches to wait until the tide recedes. When several gobies are crowded on a low branch of *A. marina*, dorsal fin-flicking seems to express aggression in the competition for space.

In Inhaca mangroves the body temperatures of the gobies during the day range from 28-34°C and are similar to that of the mud on which they rest. They avoid mud warmer than 37°C, which is a few degrees lower than their upper lethal temperature. On cool mornings, the fishes remain in the nests until the sun warms up the turrets. They are lethargic on cooler days, as are all terrestrial poikilotherms, which depend on external heat for activity. This fish is capable of changing colour by means of skin chromatophores. In the cool shade the skin is pale beige with scattered dark markings and white flecks. In sunlight the skin becomes almost uniformly dark, the fins changing from yellow to black with a yellow border. Is it possible that the colour change is a means of thermoregulation – as in some reptiles?

Although *Periophthalmus* is amphibious, its mode of respiration is different from that of amphibians. *Periophthalmus* can survive indefinitely if artificially submerged in sea water, provided that it is sufficiently oxygenated; when the oxygen content falls, the fish rises to the surface to gulp air. Under water this goby respires by means of gills in the normal fish manner. On land it aerates its enclosed volume of water in the gill chambers, expanding the opercula. The small opercular apertures are sealed by a valve on each side, thereby enabling the gills to function normally for a time on land. The gills are smaller than is usual in fishes and the lining of the vacant space is highly vascularised. Thus, gaseous exchange can take place between the enclosed air and the lung-like lining. However, when the fish feeds on a tiny animal in the mud, both air and water are released explosively and audibly from the gill chambers so that it can swallow. The fish must then return to water to load up again with water, and then, on emergence, with air. These gobies often perform rolling movements from side to side on wet mud, presumably to keep the skin wet and possibly to mix air and water in the gill chambers. The skin between the scales is richly vascular (like that of a frog) and may be an accessory breathing organ.

Periophthalmus kalolo resembles amphibians in life history, in which aquatic larvae metamorphose into semi-terrestrial adults. The mud-skipping gobies mate in late summer after a display of courtship, when the male approaches a female flashing its dorsal fins, now black-and-white-striped and orange-spotted (Fig. 4.15c). It is notable that the lower portion of the retina of the eye, which faces upwards, is rich in cones so that the display of colour is appreciated. The female enters the male's burrow and fertilisation is internal. It is probably the female that adds a brood chamber in which she rests, neither feeding nor leaving the nest. *P. kalolo* is ovoviviparous and its eggs develop to the late cleavage stage before they are laid. The female climbs up the walls of the brood chamber depositing eggs which adhere to the mud as they are laid one by one (A.J. de Freitas, 1958, unpublished). The yolky eggs develop into small fishes in a few weeks. Once they have hatched they leave the nest on the next high tide when it is submerged. They escape into the open water, but many remain near the nest in the shallow streams. In late winter and early summer these larval fishes are abundant. After a few months they metamorphose into adults and live for much of the time on land out of water, constructing nests for themselves. When they become mature, females can be distinguished from males by the anal papilla which is elongated; it has a rounded end in males and is bilobed in females.

Periophthalmus kalolo is the only fish that breeds in the mangrove swamp. Its evolution from totally aquatic fish ancestors must have resembled that of amphibia. Mortality in isolated, temporary pools could be avoided by any ability to make use of the atmosphere for respiration. The biggest consequent advantage might then come from being able to feed on land.

The Mangrove Kingfisher, *Halcyon senegeloides*, which eats small crabs, has often been seen although it nests and breeds inland. Nests of the Lessermasked Weaver bird, *Ploceus intermedius*, may be built in the taller *Avicennia* trees near the coconut palms on the landward edge of the mangrove of the northern bay. The Chinspot Batis, the Purplebanded Sunbird and the Paradise Flycatcher feed on insects among the trees (See Section 11.2.2).

The little Greenbacked Heron, *Butoroides striatus*, nesting in the mangrove, is usually seen in the 'lagoon' of the southern bay. From September to December and again in March and April flocks of migrant waders, such as curlews, whimbrels, sandpipers and plovers congregate at high tide in the mangrove swamps and on rocks around the southern bay but they feed on the western sand flats. Flocks of the Greater Flamingo, *Phoenicopterus ruber*, feed on the flats adjacent to the mouths of the mangrove channels (see Sections 5.3.2 and 11.5.1).

4.5.3 Tidal Streams: oysters; shrimps; water-striders; fishes

The prop roots of *Rhizophora mucronata* on the banks of the streams are often colonised by the oyster *Crassostrea forskali* (Fig. 4.16a). This species is closely related to *C. cuccullata* which occurs at the same tidal level on the sheltered rocks of the west shore (see Section 6.3). The form of the shell differs from that of the rock-dwelling species in having pronounced spines.

In the tidal waters of the deeply shaded channels there appears to be very little plankton, but it has not yet been carefully studied here. The only shrimp found permanently in the

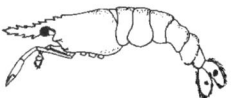

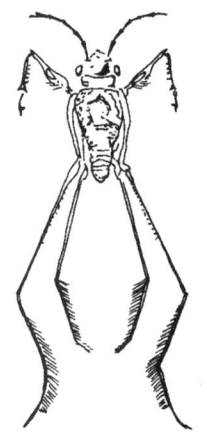

Fig. 4.16 Animals along the channels:
a. oyster, *Crassostrea forskali* (x 1) on *Rhizophora* roots;
b. shrimp, *Palaemon concinnus* (x 2) in channel (after Kensley 1972);
c. water strider, *Halobates* sp. (x 3).

channels, *Palaemon concinnus* (Fig. 4.16b), is distinguished by the presence of two lateral, post-orbital spines on the carapace and conspicuous statocysts in the uropods which are useful in a habitat of fluctuating density when salinity changes with the tides. Around the mouth of the channel in the Saco, the snapping of the chelae of the large alpheid shrimps *Alpheus crassimanus* (Fig. 3.21b) can be heard as they hunt their small prey in the water. They then return to their burrows in the banks of the channel.

On the surface of the sluggish mangrove streams where they emerge into sunlight, one of the few marine insects, the water-strider, *Halobates micans*, may be seen. These slender insects (Fig. 4.16c) are exquisitely adapted to living on the surface of quiet water, often near coral reefs. They do not get wet because the silvery ventral surface is covered by a close felt of scale-like, water-repellent hairs, which also cover the legs. They skim the surface of the water with two pairs of legs, supported by surface tension. Sensory areas on the feet pick up vibrations in the surface layer of water which occur when a stray insect falls into the water and floats. The water-strider skates rapidly towards its prey and seizes it with the first pair of legs, usually kept folded in readiness. The water-striders communicate with one another in courtship by making high frequency ripples on the water surface.

Juveniles of four species of prawns live in the mangrove channels at Inhaca Island, while one species is confined to the adjacent estuarine flats where the swamp effluent soaks into the sand on its way to the main channel down the centre of the bay. The four species are superficially similar (Fig. 4.17a-d and Fig. 3.13h) but they have evolved physiologically in different ways. They appear to have different responses to temperature, salinity, light, water depth and food (Hughes 1966, De Freitas 1986).

One species of prawn predominates in each of the three areas and a fourth species has a wider tolerance range and may be found throughout the mangrove channels.

The channels in the interior of the swamp are lined by the tree *Rhizophora mucronata*. In the deep shade of the trees, the temperature is fairly constant at about 19°C. The mud is soft, almost semi-liquid, making the water murky so that light penetration is low. The salinity of the water is similar to that in the bay, about 30g NaCl per litre, i.e. less than that in the open ocean. Here *Penaeus indicus* is dominant (Fig. 4.17d). It is probably concealed from predatory fishes by the darkness of the water; it never buries itself in mud as do the other species. There seems to be little temperature or salinity stress.

In areas of channels on the seaward side of the mangrove, lined by *Avicennia marina*, *Penaeus monodon* is characteristic of the shallow edges of the channels (Fig. 4.17a and b). It responds to the incoming tide by rising to the water surface and swimming with it; some may be found in the *Rhizophora*-lined channels. During the day when the tide is out it lies in shallow water, only partly buried, with eyes and sometimes parts of the abdomen above the sandy mud – but with uropods concealed. Its wide tolerance of salinity change (like that of *Avicennia*) allows it to occur in small numbers throughout the mangrove channels.

After the ebbing tide has retreated from the mangroves, water from the channels travels over the adjacent sand flats to the main channel in the centre of the south bay, which leads to

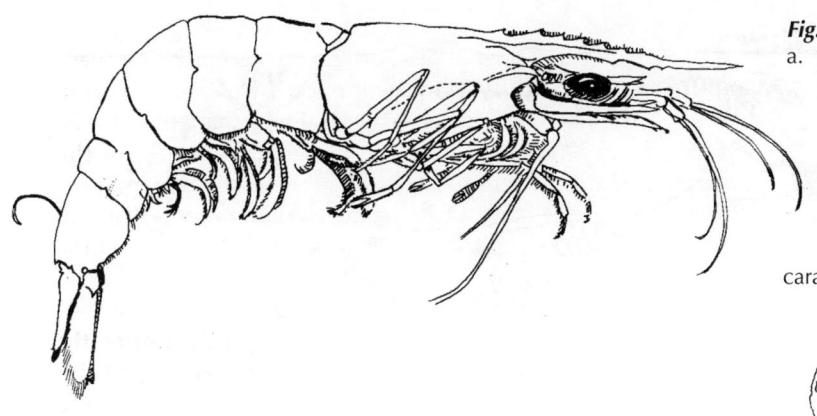

Fig. 4.17 Juvenile prawns:
a. *Penaeus monodon** (x 4);

carapaces (x 2):

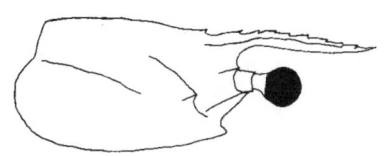

b. *P. monodon*;

c. *Metapenaeus monoceros*;

d. *P. indicus*;

the open sea. This effluent water drains slowly, leaving outlines of its course over the sand as clear water 5-10cm deep. In the absence of shade, the temperature of the sand and water may rise to 40°C and evaporation occurs so that salinity rises and approximates that of the sea in the main channel in the bay. This area of the sand flats is inhabited by *Penaeus semisulcatus* (Fig. 3.21f), which avoids the heat and light during the day by burying itself completely in the substratum. The greater intensity of light on this wet sandy area would enrich the diatom and dinoflagellate flora on which it probably feeds. The very young prawns of this species, when they enter the bay from the sea, are reported to cling to the undersides of leaves of sea grasses and to graze on the microflora (or meiofauna) attached to the leaves.

Small numbers of a fourth species, *Metapenaeus monoceros* (Fig. 4.17c) occur in the three mangrove habitats. It appears to have the widest tolerance range of salinity, including diluted sea water. It minimises exposure to high temperatures in the shallow water of the flats by burrowing and completely burying itself during the day.

A fifth species of prawn, *Penaeus japonicus*, not found in this mangrove association, occurs where there is slight wave action and the salinity is around 30 g NaCl per litre, on the west shore of the island in muddy sand. It too burrows completely during the day. It is probable that the composition of the intertidal food for prawns in the various habitats differs and that each species has a preferred diet (as do fiddler crabs). This question as well as the degree of osmoregulation and temperature preference await further investigation.

These juvenile prawns feed voraciously on small animals or microflora in the mud as well as on detritus and grow very rapidly, but do *not* mature in the mangroves. Their reactions to salinity and to tidal currents appear to change when they have grown near adult size and as they swim into the main stream of the Saco with the ebb tides they are carried out to sea. They spawn and eggs are fertilised in the deep sea in Maputo Bay, where they soon hatch into nauplius larvae with three pairs of appendages. After several moults they become protozoea larvae with antennae, stalked eyes, a wide thorax and long abdomen. After several more moults they become mysid larvae which are almost shrimp-like in form. These early larvae are planktonic in habit and are carried in currents and waves towards the shores. They moult yet again and become post-larvae that apparently are carried along gradients of decreasing salinity and reduced wave action which lead them into very sheltered bays and estuaries. Whether the migration is active or passive is not clear. The post-larvae grow into juvenile prawns with typical rostrum and appendages in their benthic habitats in and around the mangrove swamps (Fig. 4.17e).

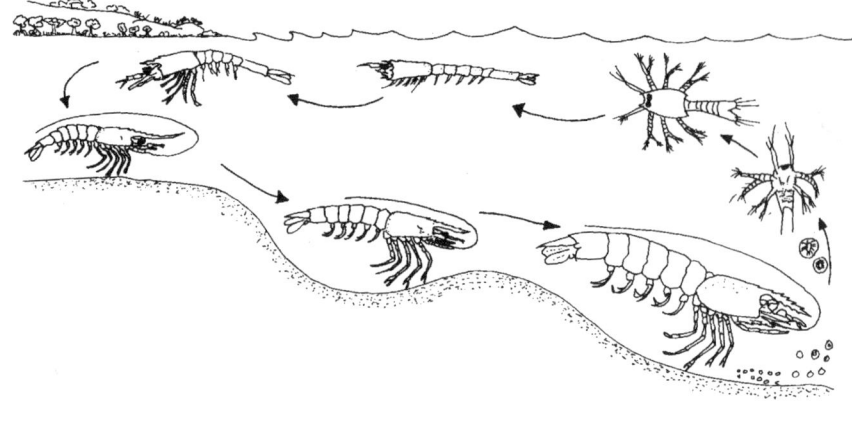

Fig. 4.17

e. life history of prawns: left, juveniles leaving mangrove swamps; right, adults breeding in the sea; top, larvae develop and return (b-e after A.J. de Freitas 1986).

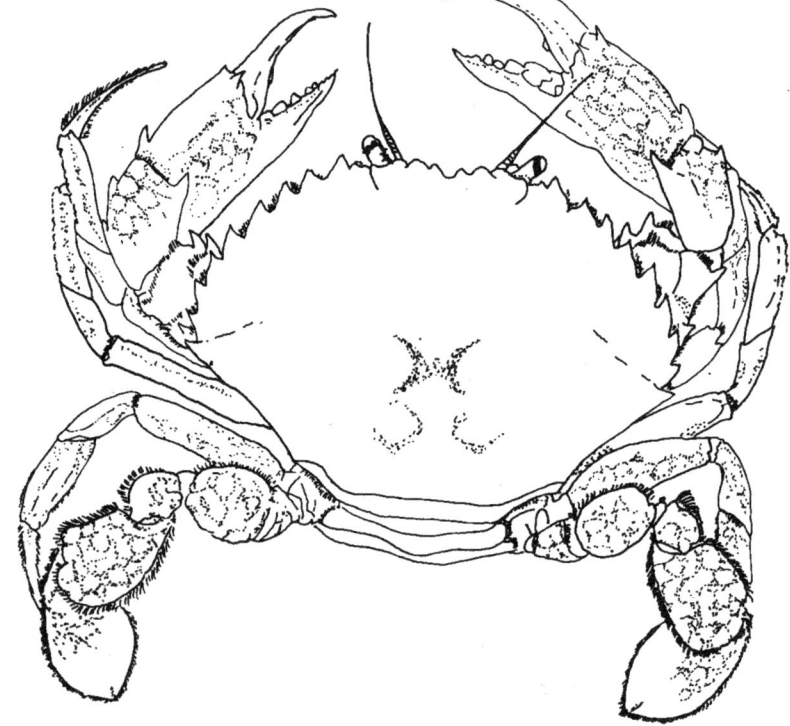

f. edible crab, *Scylla serrata* (x 0,5) (after Macnae 1968).

Among the roots of *Rhizophora mucronata* quite large elliptical holes are frequently seen in the muddy bank. These are made by the largest of the carnivorous swimming crabs, *Scylla serrata* (140 x 200 mm) (Fig. 4.17f). It has a fairly smooth carapace, except for three irregular horizontal lines of granules There are nine lateral teeth of equal size and many spines on the cheliped; the colour is mottled greenish brown. Individuals smaller than 20 mm are rarely found around the mouth of the channel but fishermen report that they live in the very narrow, shallow channels upstream. The males and females leave their burrows and migrate to the sea. They mate in the sea and the larval period is prolonged there before they return to the mangroves. This leads to their very wide distribution in mangrove streams over the whole Indo-Pacific coast, where they are collected for sale. They are also common on muddy shores of estuaries along the east coast of South Africa; they feed on small snails.

In the deeper water near the exits of the mangrove streams several species of fishes occur. The largest is the shadow goby, *Yongeichthyes nebulosus* (Fig. 4.18a) (13 cm) which is light grey in colour with five or six wide, dark cross-bands and black fin margins. It may hide in holes and

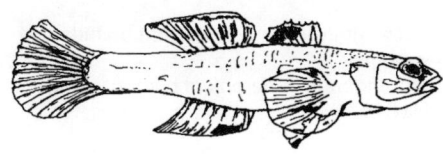
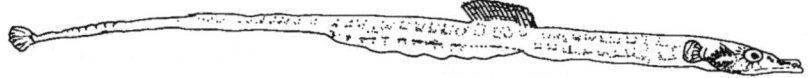
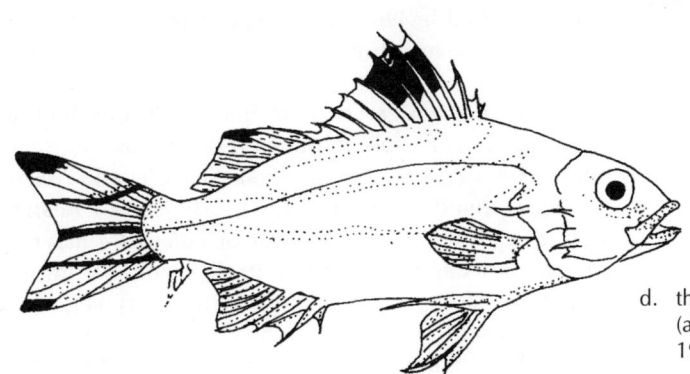

Fig. 4.18 Fishes in the mangrove channels: a and b. gobies:
a. *Yongeichthyes nebulosus* (x 0,5);
b. *Priolepis inhaca* (x 1);
c. pipefish, *Hippichthyes cyanospilus* (x 1),
d. thornfish, *Therapon jarbua* juvenile (x 1) (a-d after M.M. Smith and P.C. Heemstra 1986).

emerge to swim swiftly after fish prey. A much smaller fish, *Priolepis inhaca*, the meander goby (Fig. 4.18b) mixes with the juvenile mudskippers, but does not leave the water. It has diffuse, oblique dark cross-bars and the base of the tail fin is marked by a large, dark comma-shaped spot.

Pipefish may be seen, such as the very slender, graceful *Hippichthyes cyanospilus*, the blue-spotted pipefish (Fig. 4.18c). The body is protected by an armour of dermal plates and the mouth is at the end of a tube-like snout and has no teeth. The males incubate the eggs and carry the young in an abdominal pouch (see Section 8.5.6). The thornfish (Fig. 4.18d) is carnivorous.

4.5.4 *The seaward boundary: crabs; sea slug; whelk; barnacles*

The seaward fringe of *Avicennia marina* is populated by the fiddler crab, *Uca lactea annulipes*, which burrows among the pneumatophores. Here there are denser populations than on the landward fringe. In addition, in the sandy ridges in the soft mud seaward of the trees, the fifth species of fiddler crab *U. vocans* digs its burrows (Fig. 4.11d and f). It feeds on the surface material in the muddier parts. In the summer breeding season, *U. vocans* migrates to the *Avicennia* fringe where the substratum is sandier mud and there is some shade. Here it makes more permanent burrows and ventures back on to the mud for feeding.

Among the pneumatophores on the mud the air-breathing opisthobranch *Peronium peroni* occurs. It is dark green and has a tough, roundish slug-like body (Fig. 4.19a). The commonest mollusc is the large mud whelk, *Terebralia palustris*, which feeds on diatoms in the soft mud around the pneumatophores and especially in the bed of the channel outlet (Fig. 4.19b).

The pneumatophores of *Avicennia marina* on the seaward fringe are covered by many barnacles, *Balanus amphitrite* var. *denticulata*. They resemble the pink-striped *B. amphitrite*

Fig. 4.19 Molluscs on the seaward boundary of the swamps;
a. Opisthobranch, *Peronium peronii* (x 1) (after Day 1974);

b. gastropod, *Terebralia palustris** (x 1).

found at mid-tidal level on the very sheltered rocks at Ponta Punduine; but like the oysters, they too are more spiny in appearance and not as squat as the rock sub-species (Fig. 6.6e).

4.5.5 Epiphytic algae – the bostrychietum and blue-green algae

An association of very small maroon-red algae occurs in patches on the pneumatophores and bark of *Avicennia* trees on the seaward edge of the mangroves and on the trunks and arching aerial roots of *Rhizophora mucronata* trees. This is known as the bostrychietum after the dominant red alga *Bostrychia*. Eight or nine species of four genera are involved, most of which also occur in shady crevices on the upper shore rocks at Ponta Punduine. These algae colonise hard substrata at the same tidal level – namely high water of neap tides, which in the mangroves – are available on aerial roots in the shade of the trees.

None of the algae is more than 10-20 mm high and exact identification of a species depends on microscopic features. The four genera *Bostrychia*, *Caloglossa*, *Catenella* and *Murrayella* can be recognised in the field. The thallus of *Bostrychia* has cylindrical erect branches; that of *Caloglossa* has ribbon-like, short branches with long ventral rhizoids from each node. *Catenella* has a creeping habit with many segmented erect branches whilst *Murrayella* has a dorso-ventral habit with radially arranged, very fine, erect branches (Figs 4.20 a-f) (Silva and Cuamba 1991)

Several species of blue-green algae (Cyanophyta) grow on the different species of the Bostrychietum and also on the trunks of mangrove trees and prop roots. Common species are *Hydrococcus rivulatus*, *Chamaecalyx leibleiniae*, *Xenococcus acervatus* and *Microcoleus chthonoplastes* (Fig. 4.20g, h and i).

4.6 ADAPTATIONS TO ZONES IN MANGROVE CRABS

Many biologists have investigated the physiological responses to various stimuli in the different *Uca* spp. Does zonation depend on metabolic preferences for certain temperatures, levels of salinity, shelter and light, to the texture of the soils, the food availability and the microflora or meiofauna.

4.6.1 Fiddler crabs

The five species of fiddler crab appear to live in succession along a temperature gradient from the landward fringe of *Avicennia* to just beyond the seaward fringe in the following order *Uca inversa*, *U. annulipes*, *U. gaimardi*, *U. urvillei* and *U. vocans* (Edney 1960). It has been found that the upper lethal temperature of the crab species decreases in the same order from *U. inversa* (45°C) to *U. vocans* (42°C). The mud surface may be 2-3°C warmer than these lethal body temperatures and the crabs cool themselves by 'transpiration' (water loss) from the body surfaces, their bodies registering 5-10°C less than the surface mud. Loss of water in the first half hour of exposure is fastest in *U. inversa* from the habitat most exposed to heat. If the crabs are experimentally exposed to air for longer periods, the rate of 'transpiration' is controlled and water is conserved. Again *U. inversa* does this best and *U. vocans*, which lives in wetter

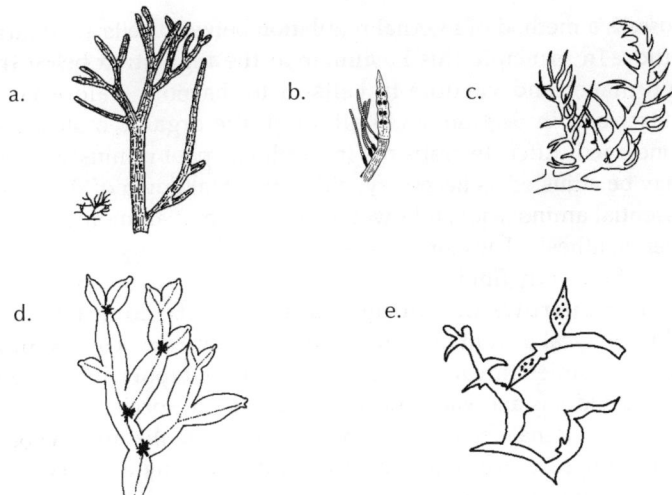

Fig. 4.20 Red algae of the Bostrychietum on tree roots: (natural size less than 10 mm).
a. *Bostrychia moritziana*;
b. *B. tenella*;
c. *B. binderi*;
d. *Caloglossa leprieurii*;
e. *Catenella nipae*;
f. *Murrayella periclados* (a-f after Jaasund 1976).

mud, conserves less successfully; the abilities of the other three species lie between the extremes. These temperatures are exceeded on more torpical shores, so the optima are labile.

The carapace of the genus *Uca* is the least permeable of those of all marine crabs. In all *Uca* species, desiccation is avoided in the field by frequent visits to their burrows where the temperatures are between 5°C and 15°C lower than that on the surface mud; in addition, there is access to water at the bottom of the burrows. During these visits to the burrows, water is sucked into the gill cavity through the apertures at the bases of the third and fourth pairs of legs by means of the usual pumping mechanism of the scaphognathites, the fan-like projections of the second pair of maxillae (Fig 3.1c). When the crabs are exposed to air, the reversal of the scaphognathite beat brings air into the gill cavity for aerial respiration. The lining of the walls of the gill chamber is folded into highly vascular tufts which carry out gaseous exchange in moist air. The gills are fewer in number than in aquatic crabs and are stiffened to prevent their collapse in air.

The water available to fiddler crabs varies in salinity daily and weekly according to the tides and in relation to the variable amount of fresh water drainage into the ground. These crabs avoid the direct drainage paths and so the water in the burrow varies less in salinity than does the surface water, yet there are changes to which the crabs respond. The osmotic pressure of their body fluids is normally less than that of sea water and the crabs are able to regulate it by raising or lowering the pressure in response to small external changes. In concentrated media, more water is excreted in the urine by the antennal glands, but this removes both water and salt. The loss of salt is made good by uptake by the gills through active transport. When crabs are exposed to more dilute media, sodium and chloride ions are excreted by the gills. Fiddler crabs can also drink sea water by spooning it up with the minor chelae; then the monovalent ions are absorbed by the gill epithelia but the divalent ones are excluded. The crabs also

Blue-green algae on the Bostrychietum:

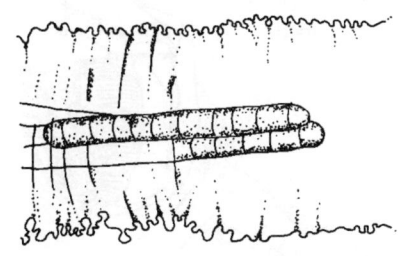

g. *Microcoleus chthonoplastes*;

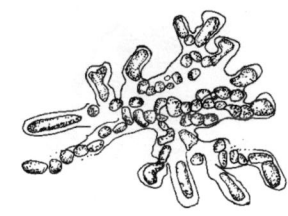

h. *Hydrococcus rivularia*

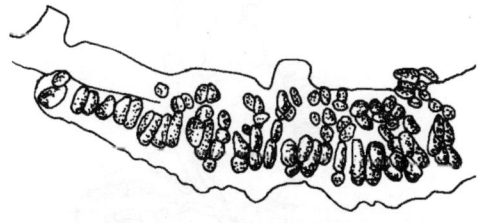

i. *Xenococcus acervatus* (after S.F.M. Silva 1991).

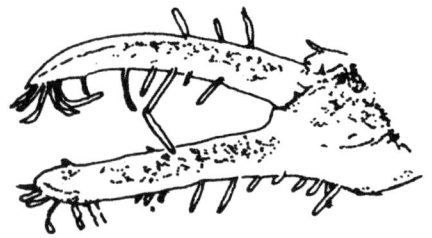
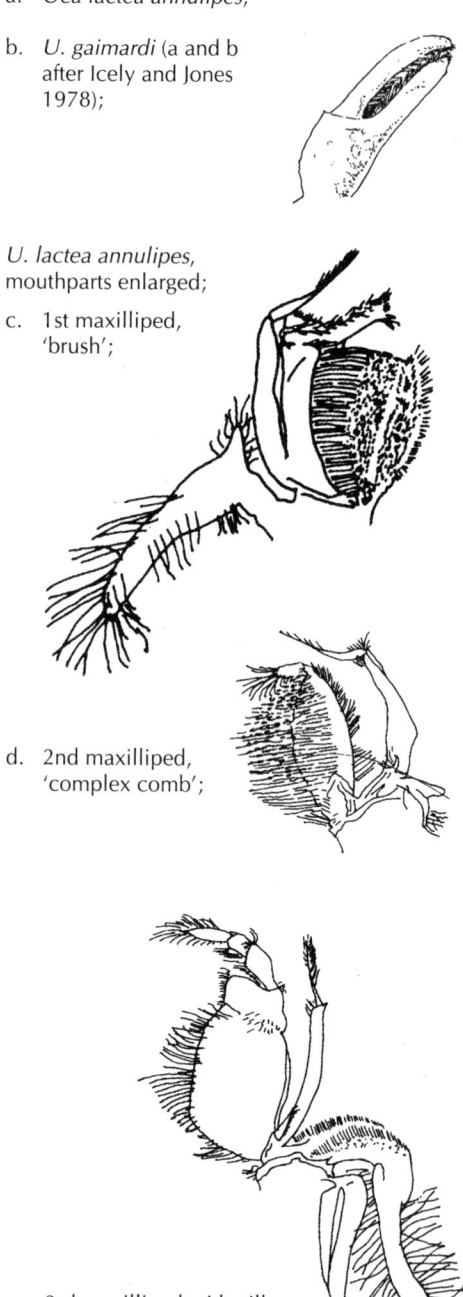

Fig. 4.21 Feeding mechanism in fiddler crabs: small chela (x 3)

a. *Uca lactea annulipes*;

b. *U. gaimardi* (a and b after Icely and Jones 1978);

U. lactea annulipes, mouthparts enlarged;

c. 1st maxilliped, 'brush';

d. 2nd maxilliped, 'complex comb';

e. 3rd maxilliped with gill and epipodite (c-e x 10, after Macnae 1968).

possess a method of internal regulation between cells and body fluids. In principle this is similar to the regulation between cytoplasm and vacuole in cells of herbaceous halophytes (described in Section 4.5.1) although the organic molecules concerned differ. In crabs the intracellular pool in muscle cells may be adjusted as necessary, either by breakdown of the non-essential amino acids, followed by excretion of ammonia, or by their synthesis. The osmotic pressure in cells is thus adjusted to that of the body fluids.

The burrows of fiddler crabs not only confer many physiological advantages, in addition they are a place of refuge when enemies threaten. The burrows are temporarily owned by individuals and vary in width according to species girth, the burrow of a male being larger to accommodate the major chela. Male burrows are defended from intruders by a display of aggressive waving of the major chela. Usually the threat suffices but sometimes this leads to a tussle in which one male seizes the chela of another and they both pull. The larger will toss the other aside as it releases its hold. Aggressive behaviour between species is rare.

Species distribution is dependent on another environmental factor that is probably decisive in separating the feeding territories of the five species of fiddler crabs from one another, namely the texture or particle size of the sand or mud and its consequent living organic content: micro-algae, protozoa, fungi and bacteria. The five species of crabs are adapted to handling different grades of particles by means of their minor chelae and *sifting the particles by means of the different kinds of combs on the second maxillipeds* (Icely and Jones 1978). The different kinds of masticatory structures in the stomach also play a part.

Uca inversa and *U. lactaea annulipes* pick up scrapings of wet sand with the minor chela in the male (and both chelae in the female), which are *forceps-like claws* bearing a few hairs at the ends (Fig. 4.21a). *U. urvillei, U. gaimardi* and *U. vocans* have minor chelae which deal better with mud. These have *spoon-shaped ends* and are more hairy, so are better able to retain wet mud. The 'finger and thumb' in these species are shorter, more especially in *U. gairmadi* (Fig. 4.21b).

The buccal cavity around the mouth, enclosed by the plate-like third pair of maxillipeds, has the six pairs of mouth appendages present in most crabs, but they have become adapted for holding small portions of sand or mud, sorting the living organisms from the grains and then propelling them to the mouth whilst afloat in water and finally, for rejecting pellets. It has been shown that the pseudo-faecal pellets contain much less organic matter than does the surface substratum before feeding in all zones.

All *Uca* crabs insert a pinch of surface sand or mud between the plates of the third maxillipeds when they move apart and open the buccal cavity for feeding. The load is wiped off on the brushes on the third joints of the *first* pair of maxillipeds and their endites, which face outwards (Fig. 4.21c). These are continuously sprayed by two strong jets of water from the gill cavity, which are pumped through the exhalent apertures by the reversed vibrations of the scaphognathites (see Fig. 3.1c). The setose combs on the *second* pair of maxillipeds (Fig. 4.21d), which face inwards, vibrate against the brushes on the first pair and so the trapped grains are scoured and removed. Fine organic particles are set free in suspension and wafted towards the mouth by the mobile, hairy parts of the maxillae and mandibles. The heavier inorganic particles fall to the base of the buccal cavity and are prevented from entering the gill cavity by the setose epipodites of the third maxillipeds which form a grid over the opening to the gill cavity (Fig 4.21e). The sandy particles are rolled along the hairy edges of the third maxillipeds; when sufficient sand has accumulated, it is removed by a minor chela and a pellet is deposited on the surface sand. Water in the buccal cavity is recycled to the gill cavity through openings at the base of the third maxillipeds.

Although this feeding pattern is common to the five species of fiddler crab the *sorting structures on the second maxillipeds* are modified according to the grain size and the food content of the substratum on which each species feeds (Dye and Lasiak, 1986). Sand grains bear adhering diatoms (Fig. 4.22a) and mud particles are coated with bacteria, blue-green algae and fungi (Fig. 4.22b). Between the grains there are members of the meiofauna: protozoans, harpacticoid copepods and nematodes (see Section 3.7).

In *Uca annulipes*, the second maxillipeds bear three different kinds of comb-like setae, separated spatially. These are short-stalked spoon-tipped setae, long-stalked spoon-shaped setae and plumose setae (Fig. 4.22c, d and e). Their distribution on the second maxilliped is shown in Fig. 4.22g (Icely and Jones 1978). The *spoon-tipped setae* scour the sand particles trapped on the first maxilliped. The area occupied by these setae is twice as great as in other species of fiddler crab. The *short-stalked setae*, which occur only in this species, have inflexible stems and are especially efficient in rasping away from the sand the hard particles which are *diatoms*. The *plumose setae* help retain the finer organic material such as bacteria, protozoans and blue-green algae. Thus *U. annulipes* can feed in both clean and muddy sand, i.e. on both the outer and inner fringes of the mangrove swamps.

U. vocans and *U. gaimardi*, having *no* stiff, short-stalked setae, cannot feed on sand. The second maxillipeds of *U. vocans* have long-stalked spoon-tipped setae and plumose setae with very small spoon-shaped tips as well as normal plumose setae (Fig. 4.22f). This species is known to feed mainly on *bacteria and protozoans* which are sorted from the mud it ingests (Dye and Lasiak 1986).

U. gaimardi has smaller second maxillipeds with long-stalked spoon-shaped setae set among the finer plumose setae (Fig. 4.22d). Feeding is restricted to the mud of the highest organic content, but it is not yet known which organisms it selects. None of the fiddler crabs has yet been found to utilise organic detritus. Further analysis of the food of each fiddler crab species is desirable for the study of the micro-evolution of these species.

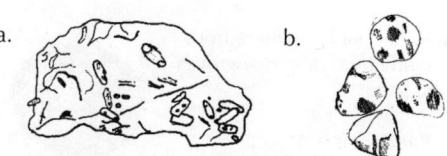

Fig. 4.22 Food selection in *Uca* species:
a. sand grain with diatoms (x 40);
b. mud particles with bacteria and blue green algae (X 100);

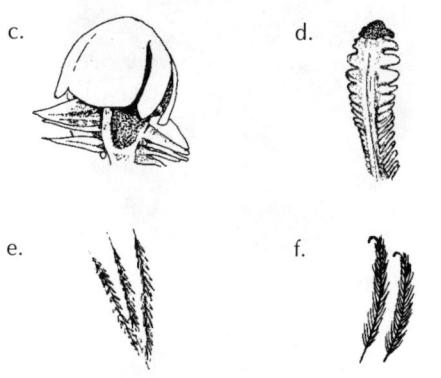

sorting setae on second maxillipeds:
c. short-stalked spoon (*U. lactea annulipes* only);
d. long-stalked spoon-tipped setae (*U. vocans* and *U. lactea annulipes*);
e. plumose setae (in 3 species);
f. plumose spoon-tipped setae (*U. vocans* only);

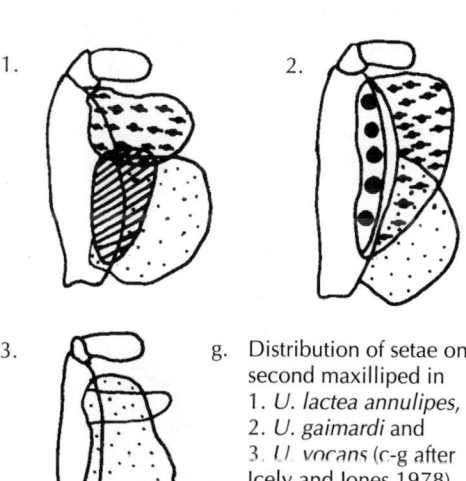

g. Distribution of setae on second maxilliped in
1. *U. lactea annulipes*,
2. *U. gaimardi* and
3. *U. vocans* (c-g after Icely and Jones 1978)

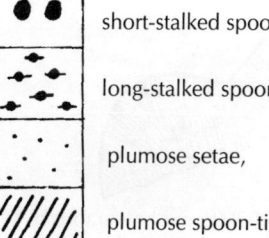

short-stalked spoons,

long-stalked spoons,

plumose setae,

plumose spoon-tipped setae.

Fig. 4.23 Diatom genera from the sandy mud near mangroves in the Saco;

a. *Raphoneis* (x 1 000);

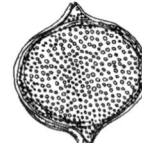

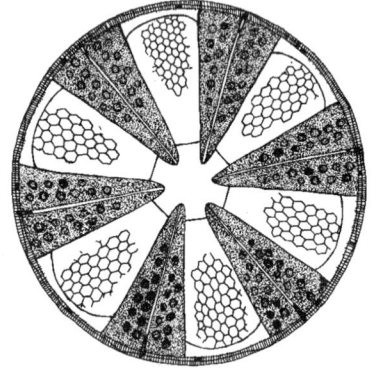

b. *Actinoptychus* (x 500);

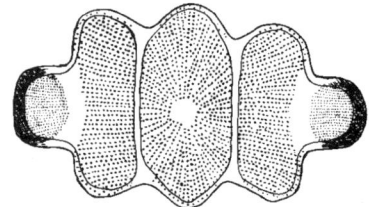

c. *Terpsinoe* (x 300);

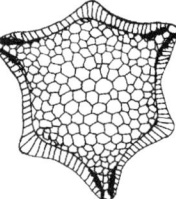

d. *Triceratium* (x 500);

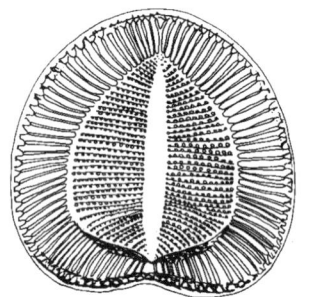

e. *Campylodiscus* (x 1 000);

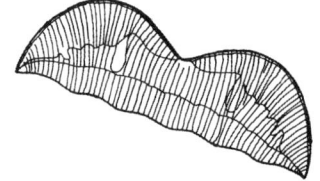

f. *Amphipora* (x 1 000) (from photographs by F.D. Hancock).

The stomach of most crabs has a gastric mill – consisting of a large median plate and two adjacent lateral plates – which grinds the food that has been swallowed. In *U. annulipes* the median plate has a coarse abrasive surface and the lateral plates have short, stout setae. These structures appear to be a suitable mechanism for breaking the hard siliceous frustules of diatoms. Six of the genera of estuarine diatoms that occur in the lower level feeding areas of *U. annulipes* on the seaward *Avicennia marina* fringe of the mangrove swamps in the Saco are illustrated in Fig. 4.23a-f. The substratum is sand, washed clean in the drainage from the swamp to the central channel.

U. vocans has a smaller median plate bearing hooked setae and on the lateral plates the setae are flexible. This arrangement seems not incompatible with the selection of protozoans and bacteria that live in fine mud (Dye and Lasiak 1986).

In *U. gaimardi*, the median plate is large and smooth and the lateral plates have short stout setae supplemented by several rows of small teeth. It is not yet known whether it selects blue-green algae, fungi, bacteria, protozoa or even nematodes. Harpacticoid copepods do not appear to be selected by any of these fiddler crabs.

The populations of the five species of fiddler crabs breed in the summer months. By means of hormonal action, their different bright colours become accentuated, more especially in the males, which engage in specific courting rituals that perpetuate zonation. The territories of *U. annulipes* abut on those of the others on both the landward and seaward fringes, and this crab has the most complex pattern of courtship dance (Gordon 1958). The male crab beckons with its major chela in the following way: the chela is swept out to the side from its bent horizontal position, then raised high and swept back to the bent position. At the same time the walking legs almost straighten and raise the body off the ground, the legs on the side of the chela even lifting off the ground. The minor chela stretches forwards and opens (Fig. 4.24a-c). After one or more signalling movements, the walking legs jump once or twice to one side or the other, as though stamping lightly on the ground. The male may circle around a female, the dance steps being repeated many times. More males emerge from burrows or stop feeding and join the dance. As the mating season progresses and even after some females have been fertilised, the dancing continues and intensifies in the bright sun of the early afternoons, mainly in January and February. One or two beckoning males will be joined by others that move in unison. A group of ten or so will dance rhythmically and move towards the same side when they stamp, regardless of whether they are facing the others or have their backs to them. They are joined by other small groups until a hundred or more are waving and side-stepping in synchrony. This delightful display will last 10-15 minutes and then one by one the males return to other occupations.

Females spend 3-4 weeks in intensive feeding and only when ovulation is imminent do they respond to the males

(Salmon and Zucker 1988). The proportion of mating females is thus small at any one time. Seldom is the female seen to respond while one is watching. In Japan, careful observers of the same species have recorded that, after displaying, a male walks to a female burrow and courts her by touching her with extended, vibrating walking legs and copulation may occur at the surface. Alternatively, a female may be lured to the male's burrow after his dance has been augmented by drumming on the ground with the major chela, kicking the legs sideways and curtseying. The male enters the burrow and the female follows, remaining inside the male's burrow during the next high tide. Next day the female returns to her own burrow and seals it. She incubates her large clutch of eggs beneath her abdomen and remains there without feeding until the larvae are about to hatch at new moon or full moon, when tides are high.

The other species of fiddler crabs at Inhaca are less demonstrative, the movements of the cheliped being smaller and different. *U. inversa* swings the chela outwards and in again in a horizontal plane when threatening other males, but stands motionless at the entrance to the burrow during courtship in late summer. *U. gaimardi* stands at the burrow opening until a female appears, then it swings the chela outwards and inwards repeatedly while walking around the raised rim of its burrow. They may mate above ground.

U. urvillei also swings the major chela horizontally and he chases a female to his burrow. At the opening of the burrow; the male may stroke or tap the female's carapace and eye-stalks with his minor cheliped. This species displays in the mornings for a short time when spring tides are low. When *U. vocans* migrates from the open mud of the flats next to the seaward *Avicennia* fringe on to the firmer sand under the trees, females build a rim around their burrows on which they stand for long periods. Males too spend some time plastering around the mouth of their burrows and then stand immobile. Both males and females raise and lower their walking legs on the same spot, a practice that has been dubbed 'drumming'. The raised rims of the burrows are said to act as devices for projecting the sound, emphasising the accoustic component of the ritual

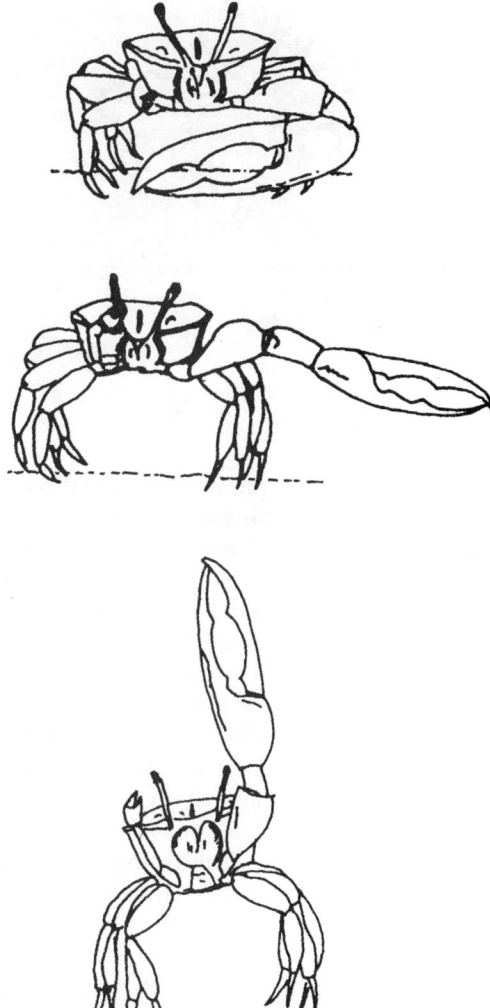

Fig. 4.24 Courtship dance — three positions in *Uca lactea annulipes* (x 0,5) (after Gordon 1958).

which is responsible for the specific name, *vocans*. In this species the females appear to play an active part in enticing males to their burrows. When displaying with the major chela in the morning a male dances around the female – a habit described as 'revolution', in which either the back or the front of the carapace is presented to her. In the case of every species the more often the observer spends time watching the fiddler crabs from day to day the more variations on the courtship theme can be discerned. The intense display is fairly brief in the two species which live lower down on the shore.

The size of the egg clutch is much smaller in these two narrow-fronted fiddler crabs and the eggs are protected by the abdominal segments. Females continue to emerge in their more humid environment to feed for one or two minutes at a time throughout the incubation period. The different reproductive habits of the broad-fronted and narrow-fronnted crabs are thus adapted to their different inter-tidal positions in the mangrove swamps. Mating behaviours ensures segregation of the species.

Displacement behaviour (Gordon 1955) has been observed in male fiddler crabs when, for example, the mating drive is thwarted by the disappearance of the female he was courting. He ceases his dance and makes very fast feeding movements with his minor chela without

Fig. 4.25 Mechanism of colour change: chromatophores (x 50)
a. pigment dispersed;

b. pigment concentrated.

picking up any sand whatsoever and no feeding pellets are formed. This behaviour ceases when the female reappears.

The activities of fiddler crabs are geared to the rhythms of day and night, to tidal cycles and to temperature changes as well as to mating seasons. The crabs are most active during low spring tides of summer days. They remain in their burrows in the dark and while tides cover their feeding grounds. They also change in colour rhythmically, a feature which has both camouflage and thermoregulatory functions (Fingerman 1968). These rhythmic functions are not *immediate* responses to the changes in the environments for they persist in the laboratory for many days in continuous light or in darkness and constant temperature, completely out of touch with tides. They are controlled by a 'biological clock' in the anterior nerve ganglia.

The fiddler crabs darken gradually during the day and become pale at night. This is accomplished by the dispersal of pigment granules into the fine branches of the black and red chromatophores in daylight and concentration in darkness (Fig. 4.25 a and b). Pigment granules in the white chromatophores respond in the opposite way. The changes in colour are indirectly under the control of the anterior ganglia of the nervous system that activate the neurosecretion of *peptide hormones*. There are three dispersing and three concentrating neurosecretions, one of each for each kind of chromatophore. These neurohumours are temporarily stored in the sinus glands in the middle of each eyestalk and in the post-commissural organs on each side of the oesophagus, that lie over blood sinuses into which the secretions are released. These are distributed by the blood circulation and thereby stimulate the chromatophores all over the external surface simultaneously according to the internal rhythm. When the environment is changed, as in the laboratory, the clock slowly runs down and is then reset according to imposed conditions.

This biological clock mechanism in the nervous system of fiddler crabs was one of the first among invertebrates to be investigated. It resembles, in principle, the persistent vertebrate rhythm of secretion of the adreno-cortical hormone releasing factor in the hypothalamus of the brain which, in man, is responsible for normal daily alertness. In man, persistence of the clock, which results in jet-lag after a trans-continental flight and in the difficulties of adjusting to working night shift, is possibly a parallel phenomenon. The biological clock mechanism on fiddler crabs perpetuates the segregation of the species.

4.6.2 Sesarmid crabs

The territories of *Uca* and *Sesarma* crabs do not overlap. Their habits contrast with each other in almost every respect. The species of sesarmids divide spatially into two groups of three species: those that remain in the salt marsh and those that can penetrate into the interior of the swamp.

Each species in the first group, consisting of *Sesarma catenata*, *S. eulimene* and *S. ortmanni*, has branching communal burrows that are not defended and several crabs may use one burrow for escape or to reach water when the tide is out. They spend the day sheltering in the shade of saltworts, rushes and grasses; at night they feed on detritus and patches of diatoms floating in pools. In the second group, *S. meinerti*, *S. smithi* and *S. guttatum* dig individual burrows in which they rest near the opening when the tide is out.

The temperature characteristics of sesarmid crabs have not been studied at Inhaca but their geographic distribution suggests that they differ (Hartnoll 1975). *S. catenata* occurs along the warm temperate coast of South Africa whilst *S. ortmanni* is tropical. *S. meinerti* has a wide distribution whereas *S. smithi* and *S. guttatum* are more tropical. Those that live in the interior of the swamp are subjected to lower temperatures than the salt marsh species which are in direct sunlight.

Access to water is provided at the bottom of burrows but, in addition, these crabs can take up water into the gill cavity from very shallow surface pools or even from films of water. The hairs around the inhalent openings at the bases of the cheliped and the third, fourth and fifth legs assist the uptake of water by capillary action. Sufficient water remains in the gill cavity for 2-3 hours isolation in *S. catenata*. Stored water lasts for 12 hours in *S. meinerti*. The gill cavity water remains oxygenated since while on land they use a special mechanism for continuously aerating the water in the gill chamber by *recycling* it externally over the ventral wall of the carapace beneath the gill chamber. This wall, composed of plates called branchiostegites, bears rows of minute, bent setae on ridges which increase the surface area and slow down the flow of water from the exit openings in front of the mouth to the re-entry openings at the bases of the chelipeds (see Fig. 4.10a'), so that the water is re-oxygenated.

The sesarmid crabs in the interior are efficient at absorbing water from thin films on the surface mud but they appear to abhor total immersion in water since they run ahead of the rising tide and even climb lower branches of trees or pneumatophores to avoid the incoming water, which brings in predators such as the crab, *Scylla serrata* and fishes from the bay.

In sesarmids water is lost by 'transpiration' twice as fast as in fiddler crabs because the permeability of the carapace is twice as high. Sesarmids, however, compensate for this handicap by confining themselves to the shade and being fairly inactive during the day. Sesarmids also differ from fiddler crabs in their mode of respiration, since they have no lung-like vascular areas in the branchial cavity and do not use aerial respiration. *S. catenata* will sometimes remain partly submerged in water for most of the day while it is exposed to the sun.

Tolerance of dilution of sea water is found to different degrees in all sesarmids. *S. eulimene* lives furthest up the estuaries of rivers (Boltt and Heeg 1975). *S. catenata* can tolerate 40% dilution of sea water. The salt marsh next to the freshwater swamp in the supratidal fringe is therefore well within their range. The mechanisms they use to osmo-regulate are similar to those in *Uca* crabs.

The food resources and feeding mechanisms of sesarmids and fiddler crabs are different. When a solid brown mass of diatoms occurs in a very shallow pool *S. eulimene* and *S. catenata* will scoop up the algae. They may pick up selected bits of mud which contain plant material and very small crustacea. *S. catenata* eats small fragments of saltworts; *S. eulimene* may eat the seeds of grasses; the sesarmids in the interior of the swamp feed on fallen mangrove leaves thus earning the name of leaf-eaters. *S. meinerti* waits in the entrance to the hooded burrow, which amplifies sound, and creeps out at the fall of a leaf to retrieve it. The crab dashes back to the burrow, tears up the leaf and eats it. Frequently a tug-of-war ensues when two crabs seize a leaf at opposite ends. It seems that the wide distribution of fallen leaves, the main food of hte sesarmid crabs in the interior, has not stimulated zonation of sesarmids in the swamp interior. In regard to communication, the sesarmids appear to have no equivalent to the courtship displays of the fiddler crabs but they raise their chelae high to signify defence or aggression.

The two genera do however resemble one another in being tropical in distribution. All the *Uca* species and seven of the eight *Sesarma* species occur in the mangroves in Tanzania (Hartnoll 1975) and in Morrumbene Bay, 400 km to the north on the coast of Mozambique (Day 1974).

4.7 PRESERVATION OF THE MANGROVES

The mangrove swamps are included in the nature reserves of Inhaca Island, especially the expanding Tivanini swamp on the north-west shore, the creek mangrove behind Ponta Raza in the south-west and also the degenerating mangrove on Portuguese Island. The degeneration on Portugues Island might be halted and recovery initiated if some official action were taken or a current with a new direction were to remove the sand blocking the exit to the sea.

There is always a certain natural mortality of trees in mangrove swamps as well as felling; simultaneously, trees are regenerated from germinating propagules. In a recent study (Hatton and Couto *in press*) it was found that regeneration in a mangrove reserve at Inhaca Island was about 20% for *Avicennia marina*, 27% for *Bruguiera gymnorhiza*, 36% for *Ceriops tagal* and 38% for *Rhizophora mucronata*. Although only dead trees are taken for firewood, excessive felling of trees has reversed the balance. The rate of tree-felling has recently been monitored by the Biology Department at the Universidade Eduardo Mondlane, Maputo, and planting of trees under the supervision of the Marine Biology Station on the island is encouraged. Consultations are taking place between the people of the island, the university and the Forestry Department and advice is being sought from other countries with similar problems in the conservation of mangrove swamps.

5

SOUTHERN AND NORTHERN BAYS: ESTUARINE AND MARINE INFLUENCES

5.1 CONDITIONS IN THE NORTHERN, SOUTHERN AND WESTERN BAYS

The southern and northern bays differ from the west shore in several respects. They are more sheltered from winds by the high eastern dunes and there is usually *no wave action.* Significantly more water drains from the freshwater mangrove swamps into the bays than on to the western shore, and the temperatures throughout the year are warm enough to support fully zoned mangrove associations. In the middle reaches the habitats of sand and muddy sand are similar to those on the west shore but they are enriched by species adapted to estuarine conditions. The more sheltered sea meadows in both bays are warmer than on the west shore (see Section 1.4.1); they have similar sea grass associations, which differ from those on the west shore. Coral reefs occur in the mouths of all the bays.

The northern and southern bays differ from each other in aspect and shape and are subjected to different kinds of tidal currents. A strong tidal current of ocean water flows through the Ponta Torres strait into the comparatively narrow southern bay. In contrast, a broad sweep of tidal water enters the northern bay from the sheltered Bay of Maputo in the west and in most years the sand-bar in the north is breached. Thus sand is carried into the bay; the northern bay is an area of sand accretion where mangrove seedlings are growing on the shore.

The mangrove associations on the upper shores of both bays are described in Chapter 4; in the present chapter the communities on the middle and lower shores are discussed. The long, narrower, southern bay has more small animal associations in restricted habitats than the other bays.

5.2 THE HABITATS OF THE SOUTHERN BAY

The southern bay is roughly triangular in shape, 10 km long from north to south and 6 km wide at the mouth. It opens into the Bay of Maputo opposite the Machangulo Peninsula on the mainland, which forms the southern boundary of the Ponta Torres strait, and hence to the Indian Ocean. The 'head' of the bay is expanded into a bilobed 'sac' about 3 km wide from north to south and is occupied by mangrove swamps which receive fresh water from the drainage of the dunes and from the freshwater swamps (see Section 4.2). The 'head' is connected to the widening part of the bay by a narrower 'neck' about 2,5 km long; the 'head' and 'neck' are known as the Saco da Inhaca (Fig. 5.1).

The wider triangular part of the bay is dominated by the ocean current that surges through the narrow strait between the south-east point of the island, Ponta Torres, and the mainland at Machangula. A branch of this ocean current enters the southern bay of the island and flows up the channel to meet the water draining from the mangrove. When the tide rises the volume of ocean water entering the bay is so much greater than the brackish water which leaves the swamps that mixing of brackish and saline waters occurs mainly in the 'Saco', which is thus more estuarine in nature.

In the middle reaches of the bay the channel from the ocean divides into two: one branch travelling northwards to the mangroves and the other leading south-west towards Ponta Punduine and the Bay of Maputo. The shores of the middle reaches have a wide variety of habitats: unzoned *Avicennia* trees line the shore on both sides of the bay, interrupted at intervals along the eastern shore at high tide mark by low-lying, roughly eroded rocks. South of the fork in the channel on the east bank, large buttress-shaped rocks divide sunny, protected coves on their northern sides from shaded sandy areas on their southern sides. The channel banks are covered by sea grass associations, which differ from those on the west shore (Fig. 5.1) (Bandeira 1989). A variety of animals are adapted to various clearly defined habitats (Macnae and Kalk 1962).

A large sand-bank covered by the South African sea grass *Zostera capensis* spans the mouth of the bay between the east and west channels. A coral reef grows along the east bank of the eastern channel for about 2 km from the rocky cliff at Ponta Torres. The animals near the mouth of the bay are coral reef associates (see Section 5.4.3). The mouth of the bay is bounded by the rocks at Ponta Torres and Ponta Punduine which differ in form and habitats. The tip of the peninsula at Ponta Torres is sandy, stabilised by dune-binding plants (see Sections 5.5.1 and 10.3).

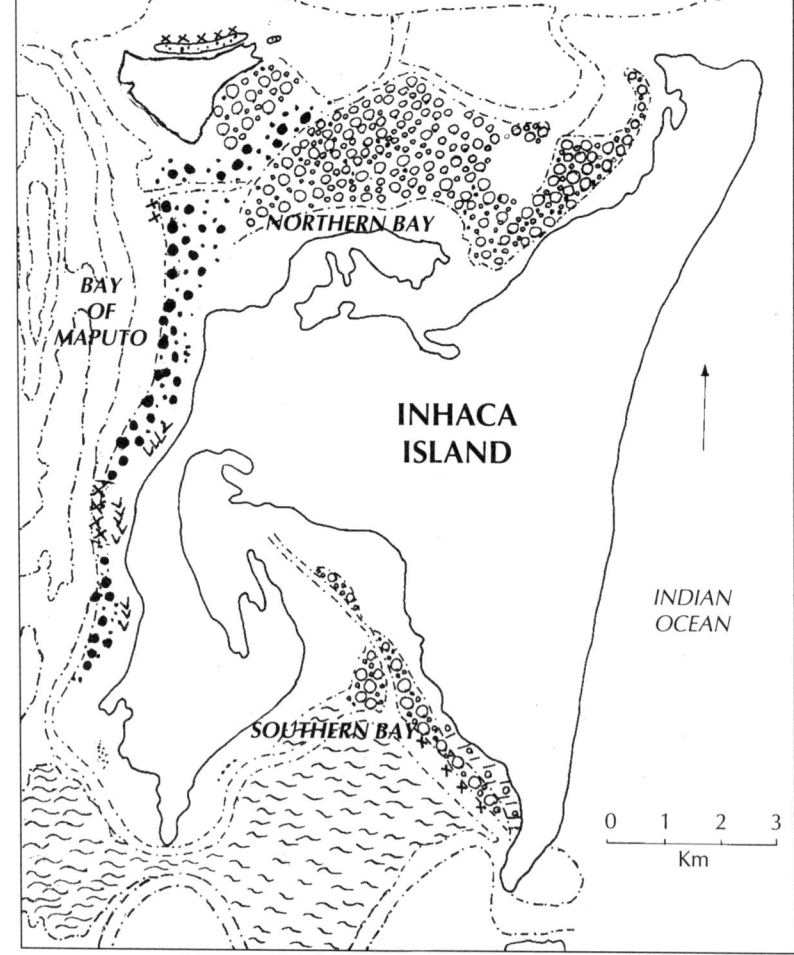

Fig. 5.1 Sea grass associations on the sheltered shores of Inhaca Island (after S.O. Bandeira 1990).

- *Halodule wrightii/Thalassia hemprichii*
- *Halodule wrightii/Cymodocea rotundata*
- *Thalassodendron ciliatum/ Cymodococea serrulata*
- *Zostera capensis*
- *Cymodocea serrulata*
- *Halophila ovalis*

Several habitats may be distinguished in the 'neck' of the Saco: sandy areas, more and less waterlogged muddy sand and a thick deposit of shell debris, partly buried in sandy mud.

5.2.1 Sandy areas on the seaward border of the mangrove swamp: Bubble crabs; prawns

Drier sand at mid-tidal level to landward of the drainage route in the Saco is occupied by *Dotilla fenestrata* (Fig. 3.7a) (see Section 3.5.1); there is no silt-covered tubicolous polychaete zone. The prawn *Penaeus semisulcatus* may be found buried in sand in the wetter areas (see Section 4.5.3).

5.2.2 Muddy sand flats: Greater Flamingo; crustaceans; commensal gobies; polychaetes; bivalves; diatoms

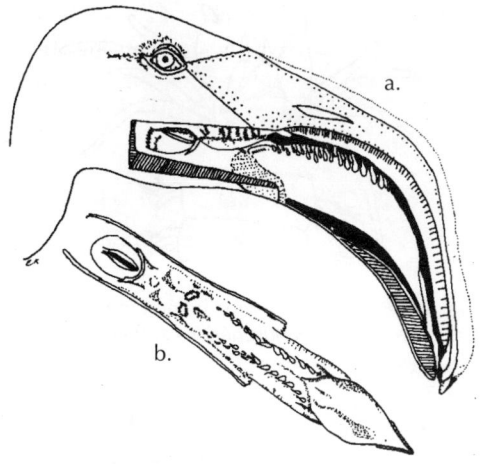

Fig. 5.2 Greater Flamingo, *Phoenicopterus ruber* head:
a. mouth in section;
b. lower jaw, inner surface (x 0,3).

A puzzling pattern of circular puddles with raised centres may be seen all over the 'neck' of the Saco when the ebbing tide first uncovers the surface. These are made by feeding flocks of the Greater Flamingo, *Phoenicopterus ruber*, when wading in shallow water. The birds stand over a metre high on pale pink, stilt-like legs and the white feathers over the body almost conceal the pinkish tinge of the wings beneath. Each bird feeds while stirring up sand as it rotates in a small circle, stamping its feet. The very long neck is bent down so that the head is upside down and partly under water. In this position, with slightly opened, bent beak, lined by fine grooves running inwards from the edges, it filters the water and loosened sand, retaining small invertebrates (Fig. 5.2). When the tide recedes and before the flats dry out, the birds fly together further down the bay. In flight the wings are conspicuously flame-red with black flight feathers. The Greater Flamingo occurs on inland waters as well as on shores; it is not known to breed on Inhaca Island.

Large areas of the 'neck' of the Saco are riddled with holes in the mud, made by two kinds of small crustaceans. The slanting burrows are inhabited by the sentinel crab *Macrophthalmus grandidieri* as on the west shore (see Section 3.4.2); the more vertical burrows are made by the tropical pistol shrimp, *Alpheus malabaricus* which shelters commensal gobies as do different species on the west shore (see Section 3.5.3). The shrimp (90 mm) is semi-transparent with reddish transverse bars on its carapace and abdomen. The uropods are deep blue and the chelae are greenish with brown dots. The dependent goby is also brilliantly coloured.

The muddy sand is inhabited by a large number of burrowing detritus-feeding polychaete worms of the spionid, orbinid and maldanid families. The species are similar to those on the west shore (see Section 3.3.2 and Appendix A). In the more waterlogged muddy areas near the mangrove swamp, oxygen is in short supply and the salinity of the water is more variable. Here two different beautiful, tropical nereid worms find their preferred estuarine conditions. *Dendronereides zululandica* has alternating red and yellow bands along its body and soft jaws on its proboscis (Fig. 5.3a and b). Finger-like gills protrude dorsally from about 20 pairs of anterior parapodia and the posterior parapodia diminish in size towards the tail, indicating a fairly inactive habit. The other estuarine nereid, *Dendronereis arborifera* prefers the blacker mud closer to the mangrove; it is pale in colour with pairs of bright red bipinnate gills on the anterior parapodia (Fig. 5.3c). Both species are adapted to the low oxygen content of the mud by their large gills and can also tolerate variation in salinity. The absence of strong jaws and teeth (characteristic of other members of the nereid family) and the presence of soft papillae on the proboscis in both species indicates a detritus-feeding habit.

Bivalves are poorly represented in both the sand and sandy mud in comparison with fully developed estuaries, although dead shells are numerous. Two small estuarine bivalves *Dosinia*

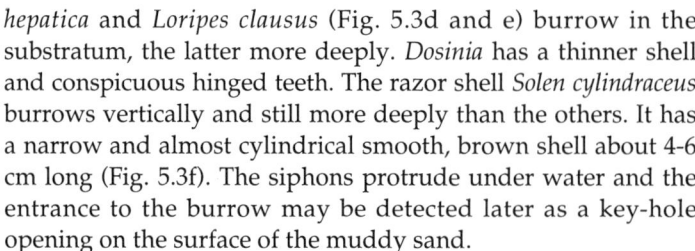

hepatica and *Loripes clausus* (Fig. 5.3d and e) burrow in the substratum, the latter more deeply. *Dosinia* has a thinner shell and conspicuous hinged teeth. The razor shell *Solen cylindraceus* burrows vertically and still more deeply than the others. It has a narrow and almost cylindrical smooth, brown shell about 4-6 cm long (Fig. 5.3f). The siphons protrude under water and the entrance to the burrow may be detected later as a key-hole opening on the surface of the muddy sand.

On the shores of South Africa and Mozambique, these three species of bivalves occur at much lower tidal levels in estuaries that have a continuous flow of fresh water from rivers (Day 1974). The attenuated population in the 'neck' of the southern bay at Inhaca Island appears to mark the limit of a significant influence of fresh water draining from the dunes and freshwater swamps.

The surface of the muddy sand is rich in estuarine diatoms (Fig. 4.23), which enrich the diet of some detritus-feeding animals. The rest of the bay towards the mouth is under the influence of the ocean water that enters the bay when the tide turns.

5.2.3 Shelly banks: worms and commensals

An extraordinary accumulation of very old, dead shells, mostly of *Loripes* and *Dosinia*, are buried 20-30 cm deep in muddy sand forming a wide band on the banks of the central channel near the point where the major arm turns westwards towards the Bay of Maputo. Many of these shells are encrusted with dead barnacles which resemble *Balanus amphitrite* seen on mangrove roots and rocks in estuaries. This thick deposit of dead shells of species that are not very common now suggests that it may be the result of a former catastrophe related to the ocean current.

No living bivalves or crustaceans are found beneath the shell deposit, but worms exist there and some such as *Owenia fusiformis* (Fig. 3.8a) make tubes between the shells. Small thin opheliid worms, such as *Armandia longicaudata* (30 mm), live upside down in the mud and ventilate their burrows by peristalsis to aerate their large and numerous gills.

Below the shell deposit the tropical sipunculid worm *Sipunculus nudus* makes permanent burrows in which small groups of the tiny bivalve *Montacuta* sp. live as commensals (Fig. 5.4a and b). They have very delicate brown shells and adhere temporarily to the burrow wall or to the worm itself, by means of a few delicate byssus threads. They are also capable of crawling by means of a long mobile foot. They are filter feeders and extract food from the water in the burrow. Another commensal in the sipunculid burrow is the flat, yellow polychaete *Ancistrosyllis falcata* (50 mm) (Fig. 5.4c and d). It has a very fragile body with small head appendages and a soft pharynx without jaws and is a detritus feeder. Many of its small parapodia have distinct dorsal hooks with which it anchors itself inside the burrow or to the sipunculid worm and thus is prevented from being washed out by the tide.

Fig. 5.3 Animals burrowing in mud in the Saco:
a. Polychaetes, *Dendronerides zululandica* head;

b. *Dendronereis arborifera* proboscis (a and b after Day 1974);

c. *D.arborifera* gill on parapodium*;

d. Venus shell, *Dosinia hepatica*;

e. Platter shell, *Loripes clausus*;

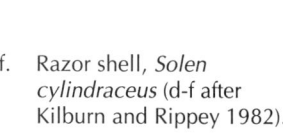

f. Razor shell, *Solen cylindraceus* (d-f after Kilburn and Rippey 1982).

5.3 THE MIDDLE REACHES OF THE SOUTHERN BAY

The sea grass *Halodule wrightii* (Fig. 3.15a-e) colonises the sandy mud near the channel in the middle reaches of the bay and spreads sparsely towards the fringing mangrove on the west and east. Another sea grass *Thalassia hemprichii* (Fig. 5.5) is associated with it down to the level of low spring tides and along the channel banks to Ponta Torres. (The west shore *Cymodocea serrulata*/*Thalassodendron ciliatum* association does not occur in the bay.) The *Halodule wrightii*/*Thalassia hemprichii* association covers a field 1-2 km wide and 5 km long and reinforces the banks of the channel by binding the dark muddy sand with their rhizomes and roots. *Thalassia hemprichii* (Hydrocharitaceae) in general appearance resembles *Cymodocea serrulata* except that the shoots are enclosed at the base by numerous shaggy old leaf bases. The flowers and fruit are quite different in the two families (see Section 9.1). The leaves (10 mm x 7 mm) appear to the naked eye to have smooth margins, but are actually very finely serrulate and, unlike those of *Cymodocea* spp., have no ligules. The creeping rhizomes are much branched and bear membranous scales at the nodes that persist for some time on the younger branches. The roots are short and unbranched and they are densely covered by masses of long fine hairs that bind the mud. *Calothrix* sp., a nitrogen-fixing blue-green alga, is associated with its roots.

5.3.1 Mud flats around the upper channel: brittle stars; molluscs

The sloping west bank of the channel appears to be richer in burrowing brittle stars than other flats on the island. Besides *Amphioplus integer* (Fig. 3.13a), which occurred on the west shore, new species have been described. *Amphiura (Fellaria) africana* has a brown disc, 5-10 mm in diameter, which is naked except for radial shields (Fig 5.6a, b and c). Another rarer and very fragile species, *Paracrocnida sacensis* (Fig. 5.6d, e and f), is slightly smaller and greyish-brown in colour. Its disc is covered with coarse scales. There are seven pairs of small spines on each segment, the middle ones being flattened. The jaws have very small dental papillae at the tips.

The food and feeding mechanisms of species of brittle stars at Inhaca Island have not been studied. A comparative study of feeding mechanisms in New Zealand (Pantreath 1970) suggests that those in mud may feed differently from those in the sand. The long, flat arm spines in *Amphiura africana* suggests that they may be used to support a network of mucous threads which would be formed by the rubbing of podia against them and would trap suspended matter, such as phytoplankton. The podia would collect it into a bolus to transmit to the mouth directly since there are no tentacle scales. The short arm spines and the presence of solid food may indicate that *Paracrocnida acensis* is a detritus feeder.

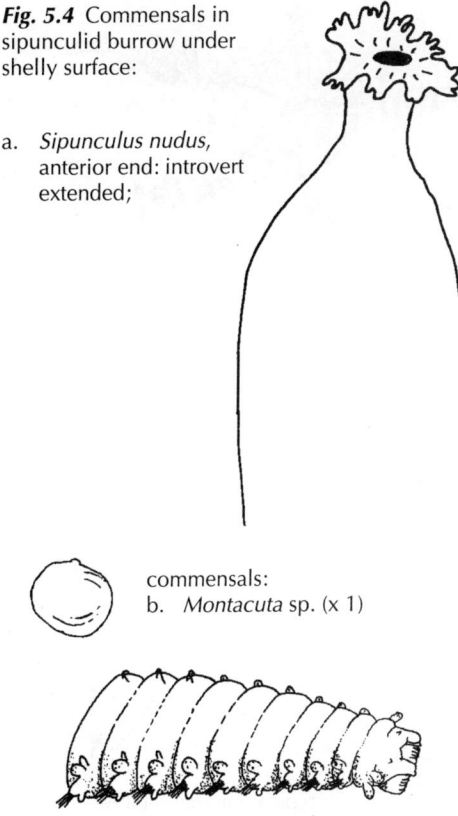

Fig. 5.4 Commensals in sipunculid burrow under shelly surface:

a. *Sipunculus nudus*, anterior end: introvert extended;

commensals:
b. *Montacuta* sp. (x 1)

c. polychaete, *Ancystrosyllis falcata*, anterior part of worm;

d. parapodium with hooks (x1) (c and d after Day, 1957).

Fig. 5.5 Sea grass, *Thallassia hemprichii* (x0.5) (after F.M. Isaac 1968).

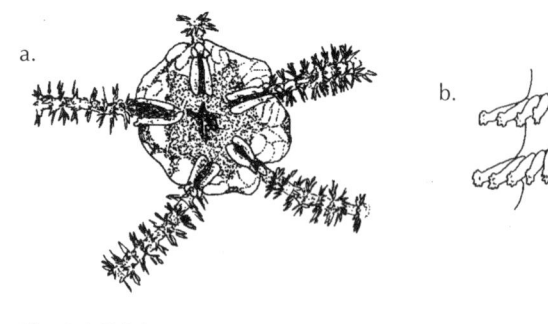
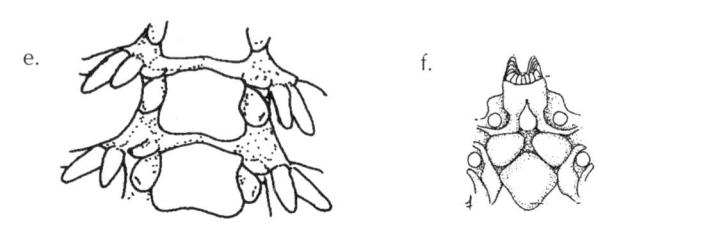

Fig. 5.6 Brittle stars in muddy sand in the Saco:
a. *Amphiura africana* disc (x 4);
b. under arm plates;
c. one jaw with few oral papillae (enlarged);
d. *Paracrocnida sacensis* disc (x 4);
e. under arm plates (a-e after Balinsky 1957);
f. one jaw* with many oral papillae (enlarged).

Fig. 5.7 Snails and slugs in the 'neck' of the Saco:
a. *Nassarius arcularius plicatus*;

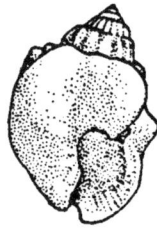

b. typical internal shell of tectibranch sea slug;

c. *Atys cylindrica** (x 2);

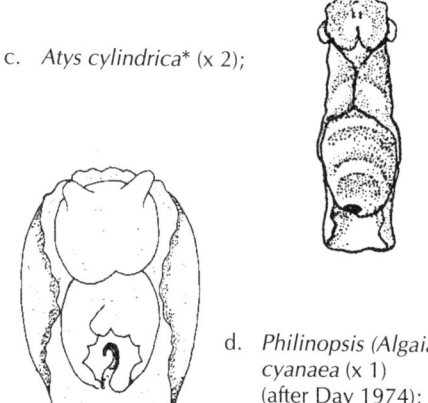

d. *Philinopsis (Aglaia) cyanaea* (x 1) (after Day 1974);

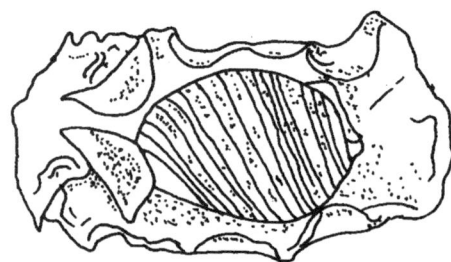

e. *Hydatina physis* (x 1,5) (after Kilburn and Rippey 1982).

About 40 species of ophiuroids occur on Inhaca shores, offering a fertile field for the study of their specific niches.

There are many scavenging snails as on the west shore (see Section 3.5.2) and, in particular, *Nassarius arcularius plicatus* (Fig. 5.7a). Its aperture is surrounded by a thick, shiny flattened 'callus'. The typical estuarine species *N. kraussiana* is apparently absent. The main enemy of these snails is the brown pear-shaped snail *Volema pyrum*, which is much more common than on the west shore (see Fig. 3.12a). Crowds of juveniles of this carnivorous snail emerge from the muddy sand ahead of the incoming tide and appear to skate along the surface. Some of the empty shells are inhabited by the hermit crabs, *Diogenes* spp. and *Clibanarius* spp. (Fig. 3.12b and c).

In the summer, small populations of tectibranch slugs may be found in shallow, sandy pools and runnels (Macnae 1962). These slugs have a thin, slightly coiled and inconspicuous shell (Fig. 5.7b); the gills are concealed in a posterior mantle cavity. Three species occur together and resemble one another in that the large foot is upturned as a shield over the head and side flaps cover the shell, so that they slide along the wet surface as though on a toboggan. They are distinguished by colour and size and they have specific food preferences. *Haminoea petersi* has a dark-spotted, yellowish-green body and its fragile sculptured shell (19 mm) has a golden cuticle. It feeds on surface diatoms which are raked into the mouth by hooked teeth on the radula. *Atys cylindrica* (Fig. 5.7c) has a translucent white body and a white shell with a yellow flush (25 mm). It possesses jaws as well as a radula and feeds on very young bivalves when they first settle in the upper few millimetres of sand. The third slug, *Philinopsis (Aglaia) cyanea*, is bigger, faster and carnivorous (Fig. 5.7d). It feeds on the other two species. It is brightly coloured, sometimes royal blue or black with purple-edged parapodia (side flaps of the foot); sometimes it has a pattern of brown and blue patches. A number of colour varieties *occur together* in one pool, breeding as *one species*. Another sea slug, not necessarily found with the other three, *Hydatina physis* is common in spring and summer. Its pale shell (25 mm) is almost round with sunken spire and is decorated by a large number of horizontal, thin brown lines (Fig. 5.7e). The animal protruding from the large aperture is beautifully tinted pinkish-violet edged with bright blue (see Section 9.4.3).

5.3.2 Among sea grasses around the middle channel: mantid shrimps; fan shells and commensals

As on the west shore, some of the bare patches of sand between sea grasses are small mounds, the hillocks left when worm casts of *Balanoglossus studiosorum* (Fig. 3.17b) are flattened by the tide. Some mounds are signs of several species of the sea cucumber, *Holothuria*, and are marked by round, sandy faeces. A squirt of water is ejected from the cloaca when the animal is disturbed (see Section 3.5.3).

The largest mounds are made as the result of burrowing by several species of mantid shrimps (stomatopods). A small depression in the top of the mound is the exit from a system of underground galleries. The entrance lies within a metre of the exit, hidden among sea grasses, and is a neat mud-lined hole leading downwards for 50-70 cm. The head and stout eye stalks of the mantid shrimp may be seen protruding from the hole when the entrance is just below water level.

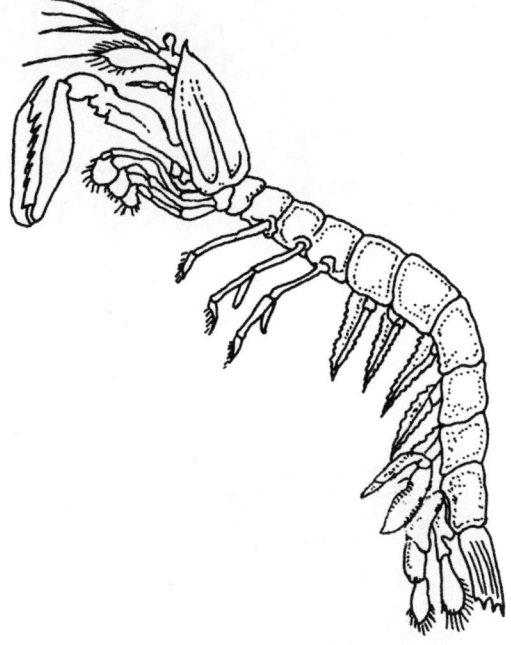
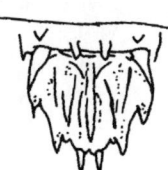

Fig. 5.8
a. Mantid shrimp, *Pseudosquilla ciliata* (x 1) (telson enlarged) (after Barnard 1950).

Stomatopods are not closely related to ordinary shrimps. The body is larger, stronger and dorsoventrally flattened. The straight abdomen ends in an extremely tough and spiny telson which may be used in fighting (Fig 5.8a). The stomatopods are so-called because eight pairs of appendages are modified as mouth parts instead of six pairs as in the Decapoda (shrimps and crabs). The most striking feature is the modification of the second pair of legs into formidable claws folded against the carapace. The claw is unlike that of any other crustacean in that the last joint, the dactyl, snaps shut like a jack-knife when catching prey, resembling the claws of the insect, praying mantis. The speed of the strike is one of the fastest animal movements known and takes about 8 milliseconds. A fish or crab may be impaled on the sharp spines of the dactyl and after descending into the burrow the mantid shrimp rips the prey apart with the third, fourth and fifth maxillipeds and its sharp, serrated mandibles. It swims very rapidly in pursuit of prey among the sea grasses and in the coral reef, using large swimmerets on the abdomen and the uropods on the tail. Three species have been caught, with some difficulty, since they can inflict deep wounds if handled. *Lysiosquilla maculata* (20-30 cm) has many broad black bands on the dorsal surface of the body. The dactyls of the claws have five teeth including the terminal one. *Squilla nepa* (15 cm) is light in colour with an orange-red abdomen and telson. Its raptorial claw has six teeth. *Pseudosquilla ciliata* (8-10 cm) is greenish and speckled and has a mid-dorsal grey stripe; it has three teeth on the claw (Fig. 5.8a). This small species may be dug out of the sand if its entrance hole can be found among the sea grasses. These three species represent the 'spearers' among stomatopods. Another group which lives among corals and coral debris is known as 'smashers' since they pound the prey with the 'elbow' of the chela until the carapace or shell is broken; their dactyls have no teeth (see Section 6.5.2).

The sea grasses become denser on the flat banks of the channel and the incoming marine current becomes stronger towards the mouth of the bay. *Syringodium isoetifolium* (Fig. 9.2c) and some *Zostera capensis* (Fig. 5.10c) occur with *Halodule* and *Thalassia*. The seagrasses almost conceal the dense aggregates of large tropical fan shells (pinnid bivalves). The 10-20 cm shells are horny and triangular; they stand upright in the sand with the narrow end pointing

Fig. 5.8. Fan shells and commensals in the sea meadow:

b. *Pinna muricata* shell (x 0,5) (after Kilburn and Rippey 1982);

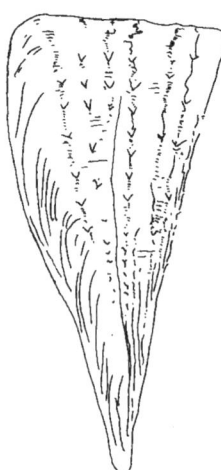

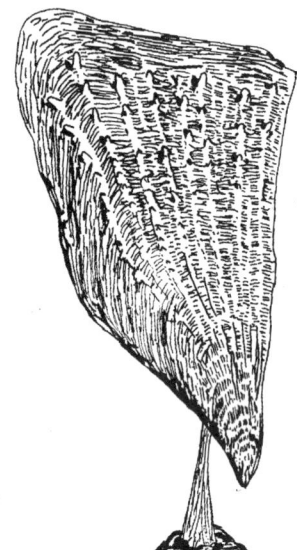

c. *Atrina pectinata** shell and byssus (x 0,5);

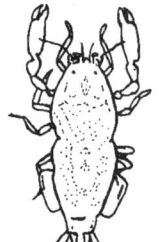

d. female commensal shrimp, *Anchistus custos**;

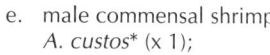

e. male commensal shrimp, *A. custos** (x 1);

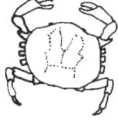

f. commensal pea crab, *Pinnotheres dofleini** (x 2).

downwards, attached to buried stones by strong byssus threads secreted by the foot glands. The fan shells lie almost completely buried with the broad gaping posterior edges exposed at the surface; these are razor sharp and placed so close together that it is impossible to walk between them. The 'dorsal' edges (i.e. on one side of the shell) are joined by an inelastic ligament along their length and the shells have a shiny nacreous lining. The shells break easily at their free edges since they are so little calcified but regeneration takes only a few hours. If pieces of debris fall inside the shell they are ejected by the 'pallial organ' a special finger-like extension of the mantle.

There are two common genera: *Pinna*, which has an internal longitudinal groove down the centre of the nacreous layer of one valve, and *Atrina* which has no groove. The most common fan shell is the tropical *Pinna muricata* (14 cm) which is buff in colour and has rows of scales on radial ribs. Its free edge is square cut (Fig. 5.8b). *Pinna bicolor* is smaller and darker with few radial ribs and is only partially buried. *Atrina pectinata* (20 cm) has more numerous ribs and bears finer scales (Fig. 5.8c). *A. vexillum* (20 cm) is occasionally found; the shell is smooth, heavier, reddish-brown and pear-shaped. These are the largest bivalves on the island, except for the giant clam (*Tridacna*) which occurs singly in the coral reef. The fan shells are filter feeders and the detritus and plankton which they collect is shared with two species of commensal crustaceans.

Nearly every shell of both genera, *Pinna* and *Atrina* contains a pair of commensal shrimps, *Anchistus custos* (Fig. 5.8d and e) and a female pea crab, *Pinnotheres* sp. (Fig. 5.8f). The shrimp has no spines on the rostrum, abdomen and legs, unlike the free-living members of their palaemonid family; the body is rounded instead of being stream-lined. These are obvious adaptations to its sedentary habit in a sheltered home. The chelae are expanded and spoon-like with long setae so modified that they can delicately remove the mucus and food collected on the gills of the fan shell without damaging them. The shrimps leave the shell to moult and mate and return to the shell where the female incubates her eggs. The female is the larger shrimp and she carries many more eggs than do free-living shrimp species. The post-larval stage is attracted to the fan shells and enters when very small.

The pea crabs are about 10 mm in diameter, usually soft-shelled and rotund with minute eyes and weak legs bearing setae used for clinging to the gills of the fan shell (Fig. 5.8f). After the moult preceding maturity the carapace hardens and the pea-crabs leave the shell to mate. They swim with the aid of *mobile setae* on their legs which stand up stiffly in the power stroke and fold flat on the recovery stroke (Hartnoll 1971). They mate in the sea and the female returns to a shell to incubate her eggs. The larvae are set free to develop and the last stage becomes more heavily chitinised before they seek fan shells and are attracted to them, probably chemically, to take up residence.

5.3.3 Warmer coves and shady nooks: algae among sea grasses; brittle stars; small acorn worms and parasitic molluscs; carnivorous tube worms; and burrowing anomuran shrimps

On the east shore of the southern bay a few rocky promontories, 2-3 m high, jut out from the upper shore into the sea grass beds, providing shelter from the south-easterly winds on their northern sides, which face the sun, and warm shallow pools of seagrasses remain when the tide is out. The temperature of the water reaches 37°C for a few hours each day in summer. In these coves the leaves of *Halodule wrightii* (Fig. 3.15a) grow more luxuriantly than on the west shore, reaching 20 cm in length; and it forms an association with *Cymodocea rotundata* (Fig. 5.9a) which occurs only in such warm pools. Each shoot has two or three leaves that have conspicuous sheaths at their bases, 20-40 mm long, some of which persist after the blades have been shed. The leaves may be 100-300 mm long and 3-6 mm broad; the margins are entire when mature and slightly toothed when young and the apex of the leaf is rounded. Flowers have seldom been seen at Inhaca Island, probably because the summer temperature is too low.

Clumps of tropical species of siphonous green algae are strewn over the flowering plants. *Boodlea composita* has the form of loose balls of thick, gelatinous, green string and is especially plentiful in summer, but rare in winter. *Valonia aegagropila* is loosely scattered over sea grass leaves as smooth green glassy lumps with small vesicles (cf Fig. 6.17d).

In every available bare spot of sand underneath and between plants hundreds of large brittle stars live in shallow holes. At extreme low spring tide when the area is drained for an hour, they are completely concealed. As the tide rises three arms protrude from every hole and wave very slowly through the water to feed; the disc and the other arms remain buried. The density of brittle stars is estimated to be about 100/m². Most of them are the beautiful *Ophiocoma scolopendrina* (Fig. 5.9b) which has a greyish granular disc 10-20 mm in diameter and arms up to 120 mm long. The arms have light purple markings and bear long spines and two pairs of tentacle scales on the lower surface of each segment. *O. valenciae*, another species with the same habit, is less common here and has strayed from the coral reef. The disc is nut-brown and feels like velvet; the arms are greenish with dark bands and each segment has a single pair of tentacle scales. The feeding mechanism is described in Section 3.4.3. These coves and their brittle stars are typical of sheltered estuarine bays in northern Mozambique, such as Nacala Bay.

In some parts of the warm pools where the sand lies thick on underlying rock, there is a large population of the yellow and green acorn-worm *Ptychodera flava* (10 cm) (Fig. 5.9c). The proboscis, which is longer than broad, has a strong collar. Wide genital wings cover the branchial region, which is perforated by large gill pores. Ten or more acorn worms may be picked up in a handful of sand. *Ptychodera* may also be found entwined among the clumps of the alga, *Boodlea*, lying on the sea grass leaves. This seems an abnormal place for a burrowing worm but this species uses the cilia on its proboscis and collar to collect detritus, rather than feeding by swallowing sand. The surface of the sand where the acorn worm lives has a slimy, crumb-like consistency owing to the presence of dispersed mucus-bound sand grains from

Fig. 5.9 Association in a warmer cove:

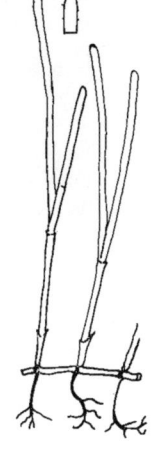

a. Sea grass, *Cymodocea rotundata* (x 0,5) (and leaf tip enlarged) (after F.M. Isaac 1968);

b. brittle star, *Ophiocoma scolopendrina* (x 1) (after Balinsky 1957);

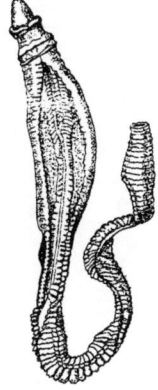

c. acorn worm, *Ptychodera flava** (x 1) (after van der Horst 1940);

d. parasitic snail, *Pyramidella dolobratus** (x 1).

potential worm casts. This dense population is reminiscent of similar shallow sandy pools on rocks on Mozambique Island.

Many individuals of *Pyramidella dolobratus* (Fig 5.9d) formerly considered to be a primitive shelled opisthobranch, live in the sand, feeding on acorn worms. This parasite attaches a small sucker at the end of a long proboscis to the acorn worm and inserts a sharp stylet into it. Tissue fluids are then sucked in by a pump-like muscular pharynx. The shell of *P. dolobratus* is about 5 cm long, slightly ridged spirally and tapering to an apex. The initial whorl is left-handed, and alternate light and dark bands mark the remaining whorls which turn to the right (see Section 9.5.3). The columella next to the aperture is pleated.

The rocky promontories, which shelter the sea grass pools on the northern side, cast shade on the accumulated sand on the southern side. In these areas there are no sea grasses and in one shady nook under an overhang, a dense population of carnivorous polychaete worms occurs. Tough narrow tubes of mucus and mud, coated with tiny flat pieces of shell debris, protrude like sprouting shoots from the sand. These are made by the eunicid worm, *Epidiopatra hupferiana hupferiana*, which is known from the subtidal zone on the coasts of the Cape and Natal. It is about 30 mm long and is said to be carnivorous, although it remains anchored in its tube. Its five tentacles have wide ringed bases and there are two small flat ovoid antennae (cf. Section 3.4.1).

In a different nook, coarse sand has accumulated (by some quirk of the ocean current) at the foot of a low sandstone cliff, and has covered the sandy mud beneath. Here the thalassinid shrimp, *Callianassa*, makes its burrows (McGinitie 1934). It is very common on the sheltered shores of estuaries in South Africa but this may be the only place where it is found on Inhaca Island and is probably at the northernmost limit of its distribution. *Callianassa kraussi* is a buff-coloured shrimp with a pinkish tinge, but related to hermit crabs (Anomura) rather than to true shrimps, from which it may be distinguished by the dorsoventral compression of the body and the large chelae (Fig. 5.10a). It digs its burrow with the chelae, one of which is bigger than the other, and it is completely subterranean, living *permanently* in its branching burrow. It feeds on detritus, loosened from the sides of the burrow and sifted through setae on the mouthparts. Compared with the size of the chelae and the abdominal swimmerets which create a through current of water, the thoracic legs are short and weak. They are adapted by means of their setae and tiny chelae to perform a number of functions such as loosening sand from the burrow walls for interstitial food, passing the food forward to the mouth, holding the shrimp securely in the burrow and grooming the body. One pair of small abdominal legs carries the eggs. So completely are these 'shrimps' confined to the burrow that mating, egg hatching and the non-planktonic larval stages complete their development there. These features enable a local population to build up and maintain itself in spite of its isolation.

5.4 THE MOUTH OF THE SOUTHERN BAY

5.4.1 Pioneer plant stabilisers at Ponta Torres; land hermit crabs; sea grasses

Wind-blown sand is being deposited along parts of the sandy shore of the bay and is being stabilised by dune-binding plants such as *Scaevola* sp. and *Ipomoea* sp. (see Section 10.2.1). Trails on the sand resembling snake tracks lead from the drift-line to the vegetation and back again. These are made by the land hermit crab, *Coenobita* (Fig. 5.10b) which rests by day at the foot of the dunes and returns to the shore at night (Vannini 1975). It carries a shell which leaves an imprint as it is bumped along. The shell is usually that of a small snail from the mangrove swamp when the hermit crab is young and that of the land snail, *Achatina* after it moults. The land hermit crab may be found during the day where the tracks end under dune vegetation, such as the succulents *Canavallea* and *Ipomoea*; they rest in shaded, conical burrows marked on the surface by a circle of loose sand. The crab digs in with large chelae sweeping downward together in a movement resembling the action of one's arms when swimming breaststroke. The tracks leading to the shore are made after dark when the land hermit crabs return to the lower shore to search for dead and decaying animals as food and to replenish their water supply. The migrations are synchronised with the tidal cycle so that they visit the shore only when the tide is out; they do not enter the sea. A study of *Coenobita cavipes* in Somaliland suggested that the daily migration is controlled by an endogenous tidal rhythm, i.e. a biological clock in the nervous system, and that the animal's responses are guided by offshore and onshore breezes and by the recognition of the different elevations of the horizons of dunes and sea. The site for burrowing is chosen in response to the greater dampness of the sand under the shelter of leaves.

Coenobita is pale cream to pink in colour, and has a laterally compressed carapace covering its head and thorax which is strongly calcified but not completely waterproof. The abdomen in hermit crabs is extended into a soft coil which fits into the snail's shell and is covered only by a thin cuticle. The abdominal legs (pleopods) on one side cling to the shell and are absent on the other side as in other hermit crabs. Several semi-terrestrial adaptations support the crab's migratory habit; the gill cavity has evolved into a lung-like structure with strong vascularisation of its floor and the gills are reduced in number and size. The 'lung' is kept moist by absorption of water on the shore at night. Its hairy chelae are dipped into damp sand or sheets of standing sea water, and the film of water in the shell is also replenished at night in these ways.

Two species of *Coenobita* occur on Inhaca Island, *C. cavipes* (40 mm) and *C. rugosus* (30 mm) which are similar in colouration even down to the red blotch on the larger left chelae, but besides differing in adult size, *C. rugosus* can stridulate and the other cannot. These sounds, associated with breeding, are made by rubbing the dactyls of the second and third legs on the left side against the tubercles on the left chela. Females in berry liberate their larvae as advanced zoea at the edge of the tide, thus ensuring a short larval life.

The much-eroded rocky promontories at Ponta Torres in the south-east and Ponta Punduine in the south-west mark the mouth of the bay. The ocean current through the strait is diverted to the north, west and south by two large sandbanks (see Fig. 1.1). A luxurious growth of the sea grass, *Zostera capensis* (Fig 5.10c) covers the sandbanks, stabilising them, and also lines the western bank of the channel that leads into the bay to the Saco. This is a northern outpost of this sea grass common in South African estuaries. It has been reported to occur in the larger ports of Mozambique, Tanzania and Kenya in similar situations, where it does not appear to flower (Isaac 1968).

Zostera capensis (Zosteraceae) has a slender creeping rhizome from which many roots grow at the nodes. The net-veined leaves of the shoot form a sheath at the base and are 8-20 cm long. The ligules are short and the leaf tips are rounded with a shallow notch in the middle. Bisexual flowers are borne on an inflorescence; the seeds are cylindrical and have about 20 shallow grooves.

These sand banks are almost inaccessible, although exposed at low spring tides, because of the strength of the current through the strait.

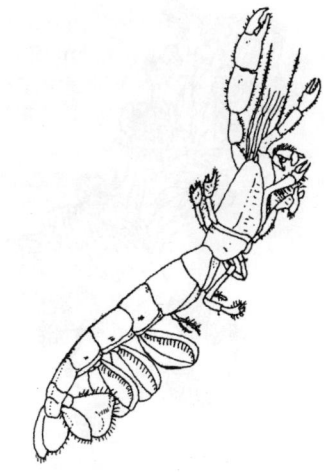

Fig. 5.10 In shady, sandy coves:
a. Thalassinid shrimp, *Callianassa kraussi* (x 1) (after Branch and Branch 1981);

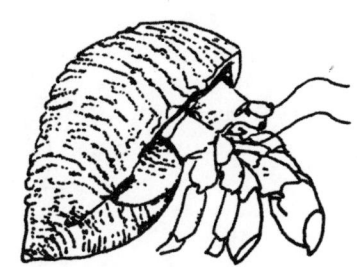

b. Land hermit crab on supratidal sand, *Coenobita* sp. in young *Terebralia* shell (x 0,5) (M. Vaninni 1975);

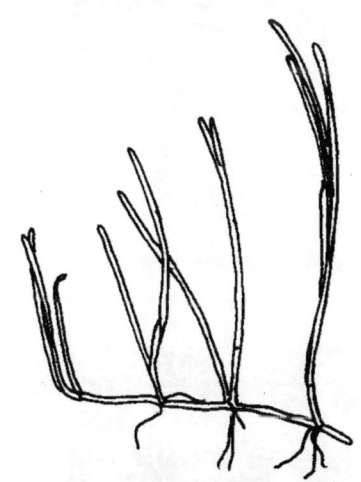

c. Sea grass, *Zostera capensis* (x 0,5) (after F.M. Isaac 1968).

Fig. 5.11 Corals at Ponta Torres reef:
a. *Tubastrea (Dendrophyllia) micranthus* (x 1) at base of cliff;

b. *G. savignyi* polyp (x 4)

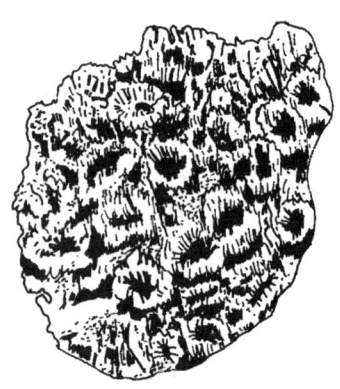

c. *Goniopora savignyi* skeleton (x 1);

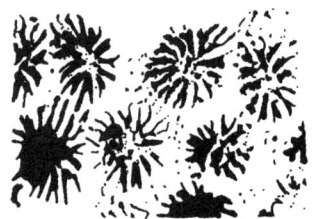

d. *G. savignyi* calyces enlarged;

5.4.2 The vertical cliff at Ponta Torres: ahermatypic coral; tunicate and commensal shrimps

The rocks at the mouth of the southern bay at Ponta Torres are subjected to a very strong tidal current through the strait from the Indian Ocean, carrying a heavy load of abrasive sand. At the foot of the dunes at high tide level, the rocks form a horizontal platform and then a vertical cliff descends into the channel. Ledges at 5 m, 8 m and 15 m sometimes support a meagre growth of corals, but these are short-lived because they may become buried in sand. The depth of the channel has varied over the years as sand accumulated or was swept away. In 1975 sand-binding dune plants were planted on dunes at Santa Maria on the opposite side of the strait in order to reduce the sand load (see Section 10.5.4). Coral growths have since improved along the whole reef of 1-2 km in the channel. One of the characteristic corals on the cliff is the *ahermatypic* jet black coral *Tubastrea (Dendrophyllia) microanthus* (Fig. 5.11a) It has no symbiotic algae and so does not depend on clear water for its slow growth; its red polyps are extended during the day for feeding.

Large tunicates (*Pyura stolonifera*) occur here at a depth of 5 m on a submerged ledge of rock as they do on the northeast coast (see Section 7.5.5). The branchial cavity shelters a pair of commensal shrimps, *Dasella herdmaniae* (Berggren 1990) which have not yet been found in this host on the northeast rocks.

5.4.3 The coral reef and associates: echiuroid worm; echinoderms; hermit crabs; slugs and their commensals; swimming crabs

Along the east bank of the channel that enters the bay from the Ponta Torres strait two areas of coral reef occur, one just before the channel branches south-westward towards the Bay of Maputo and the other beyond the fork along the east bank of the channel that leads to the Saco. The latter reef is less stressed by sand and silt since the strength of the current is less and a large proportion of suspended matter is diverted westward. This more northerly reef is more stable and the species are more diverse. In 1976 a stabilising programme of planting dune plants which fixed the sand was begun (see Section 10.5.5) (Salm 1976). The reef was monitored in 1982 and was found to be no longer in a 'juvenile' stage of small coral colonies. Large boulders of *Porites*, 4 m in diameter are surrounded by foliaceous colonies of species of *Pocillopora* and *Stylophora*, massive forms of *Favia, Goniopora, Platygyra, Echinopora* and *Montipora* (Nestler *et al.* 1984). These genera are described in Section 8.3.

This coral reef is within 50 m of high tide mark. Another unusual feature of the reef is the exposure of a band of corals 10 m wide, on the gently sloping upper bank of the channel for an hour or two at low water of spring tides, where there are large patches of a few genera. The most striking species is *Goniopora savignyi* (Fig. 5.11 b-d) in which pale flesh-coloured polyps are extended 2-3 cm from the skeletal calyces during the *day* – the only species at Inhaca Island to do so. This species forms luxuriant arborescent colonies which appear to be about 30 cm broad when the polyps are fully expanded, but when disturbed each colony contracts into a small 'stone' about 10 cm in diameter. Another genus, *Pavona* has three species here, *P. cactus* being common (Fig. 5.11e and f). It has small flat, prickly branches arranged in a cluster like leaves, on which very small orifices for the polyps are flush with the surface and connected with one another by rows of ridges. This structure seems to facilitate cleaning away sand. *Galaxea fascicularis* (Fig. 8.5d) is conspicuous since it has pale green calyces about 1 cm long spaced out on an erect branching skeleton that stands high above the few small, pink bushy skeletons of *Pocillopora* species (Fig. 8.4c). Small, stony mounds of *Porites*, with very small pores as calyces, occur on the fringe and larger boulders may be seen in deeper water (Fig. 8.4d). *Porites* is the most resistant of the coral genera to erosion by mechanical forces and is also a pioneer in the new reefs that form after being smothered by sand, being able to colonise the soft substratum. The more fragile species of *Acropora*, which have many long branches, occur mainly in the deeper water of the west coast (see Fig. 8.4a), and are not successful here.

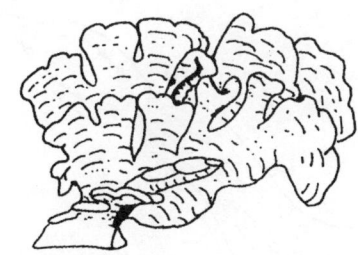

Fig. 5.11 Corals at Ponta Torres reef:
e. *Pavona cactus*, skeleton (x 0,2);

f. calyces enlarged.

A number of animal species occur in the Ponta Torres reef and the sea grass association of *Halodule wrightii* and *Thalassia hemprichii* in close proximity to the reef, which are not seen on the west shore, for reasons not yet analysed. Maybe the influence of the salinity of the Indian Ocean is greater; temperatures are slightly higher between tides, shelter from wave action is complete, the current is strong.

The beautiful echiurid worm, *Ochaetostoma zanzibaricus* (Fig. 5.12a) is found in large numbers in the semi-liquid muddy sand under flat rocks on the inshore edge of the steep part of the shore next to the first coral reef. It has a purple sac-like body with a large tongue-like, yellow proboscis, both tinged with green stippling. This is almost the only place on the island where it has been possible to collect echiurans (tongue worms), although the probosces of others have been seen on the sand flats, laid flat on the surface for the ciliary collection of detritus. It contracts very rapidly into the sand and disappears if the sand nearby is disturbed. Under the rocks there is, however, no retreat when the protective rock is raised. The echiuran worms have a single body cavity like that of sipunculid worms, but they may be more closely related to polychaete worms since the larva is a trochophore and becomes segmented when it develops, like that of polychaetes. At metamorphosis the segmentation disappears, i.e. the organs are re-absorbed (Barnes 1974).

The sea urchins of the sea meadows of the west shore (see Section 9.4.1) are well represented here. In addition, a red sea urchin with a partly collapsible, soft lightly calcified test, *Astropyga radiata*, is present. Three species of sea urchin with quill-like sharp, pointed spines are numerous on the reef itself. The smallest one, *Echinothrix calamaris* has black and white banded spines (Fig. 5.12b). These are much shorter than those of the large, purplish-black sea urchins, *Diadema setosum* and *Diadema savignyi*, in which spines reach about 20 cm in length at Inhaca Island. The two species of *Diadema* have irridescent blue spots on the test and are distinguished by the conspicuously raised anal cone which is black in *D. savignyi* and surrounded by a red ring in *D. setosum* (Fig. 5.12c). Very young individuals of *Diadema* spp. have banded spines like those of *Echinothrix* but the latter is distinguished by the rough texture of the spines when stroked towards the body. During the day the sea urchins tend to

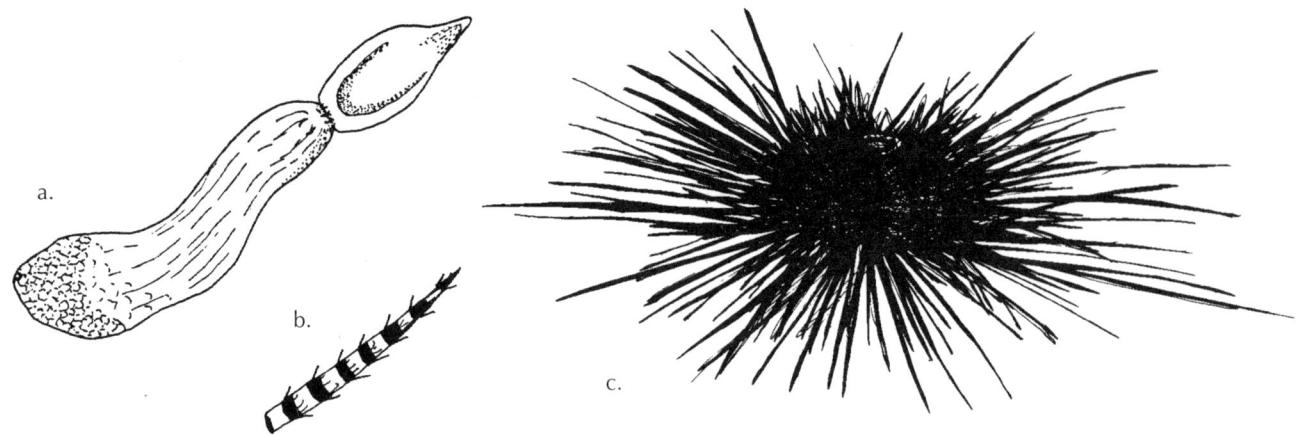

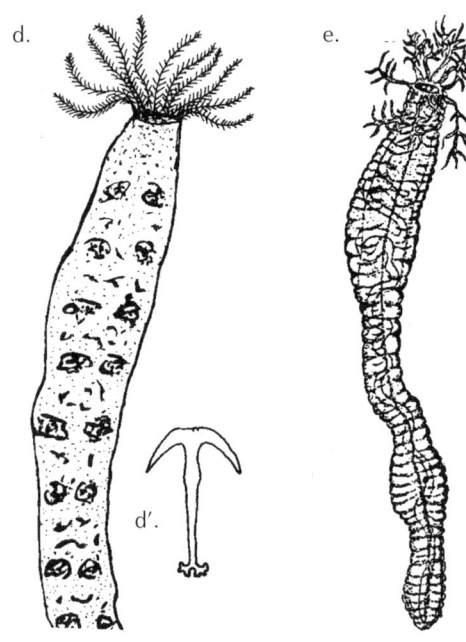

Fig. 5.12 Ponta Torres coral reef associates:
a. Echiurid worm, *Ochaetostoma zanzibarica** (x 1);
b. sea urchin, *Echinothrix calamaris*, banded spine;
c. *Diadema setosum* (x 0,3);
d. Sea cucumber, *Synapta maculata* anterior end (x 0,5);
d'. anchor-like spicule;
e. sea cucumber, *Ophiodesoma mauritiae** (x 0,5).

hide between rocks or corals with spines scarcely visible, but at night they walk on their ventral spines as though on stilts and take up exposed positions to graze on rock surfaces. Their spines are hollow and fragile and contain a poisonous fluid which is intensely irritant if, by chance, broken spines penetrate the skin. In spite of this armour these sea urchins are eaten by helmet shells (*Cassis* spp.), spiny lobsters and many species of fishes.

Two species of synaptid sea cucumbers (Apoda) may sometimes be found stranded on sand when the tide is out, although they normally live subtidally in sea meadows. These sea cucumbers lack tube feet. *Synapta maculata* (Fig. 5.12d) is almost a metre long and snake-like in appearance, having brown-spotted skin on a buff background. It is soft to the touch but the surface has the texture of sandpaper owing to the anchor-like spicules embedded in the skin (Fig 5.12d'). At Nacala Bay it swims under water with graceful, sinuous movements and entwines itself around the long stems of sea grasses, licking their leaves with its pinnate tentacles to remove the detritus and small organisms which coat the surfaces, or holding the tentacles aloft to intercept plankton. When stranded out of water these sea cucumbers collapse since they have only longitudinal muscles and no partitions in the coelom. A royal blue and red palaemonid shrimp, *Periclimenes rex* shelters in the cloaca of this cucumber, but it sometimes emerges and perches on top of it when out of water. The smaller synaptid sea cucumber *Opheodesoma mauritiana* (20 cm), which is grey with orange markings, may also be found in the sea meadows or on the reef (Fig. 5.12e).

The cushion star, *Culcita schmideliana* (diameter 150 mm) (Fig. 5.12f and f') may be stranded, away from the coral reef where it feeds on coral polyps. It has an orange colour, and resembles a partly deflated football for it has no arms, and the body is only roughly pentagonal.

The number of large brittle stars and the variety of their species in coral reefs exceed those of other habitats at Inhaca Island. Most of them are large having discs up to 20 mm in diameter and arms up to 150 mm long. Many species may be found at low tide in dead coral bases and some are seen on the surfaces of submerged colonies of living coral just under water. In the Ponta Torres coral reef, which is completely sheltered from wave action yet exposed to a strong tidal current, members of the family, Ophiocomidae, predominate, but these are less common on the west shore coral reef, opposite Barreira Vermelha, where the family Ophiotrichidae predominates (see Section 8.5.4) (Balinsky 1957). The two families differ in the details of oral papillae around the jaws of the mouth (Fig. 5.12h) which probably has a significance in food selection. Ophiocomidae have small, regular, pointed oral papillae on the jaws whereas Ophiotrichidae have more robust oral papillae.

Species of Ophiocomidae may be distinguised in the field by their colour; detailed differences in disc granulation and arms, spines and scales are discernible through a lens. *Ophiocoma erinaceus*, the largest and most spiny, is jet black in colour and the podia under the arms are bright red; the arms have stout spines on each segment. *O. pica* has a dark brown disc with fine golden lines on the disc and light bands on the arms, which bear long, slender spines. Both these species are seen most often among branches of living coral. *O. valencia* (also seen among sea grasses) in the warmer coves has a chocolate brown, velvet-like disk and green arms. *O. pusilla* is smaller than the other species, its disc is brown on the upper surface and lighter underneath. Dark spots occur on the disc and the arms are banded in light and dark shades of brown. Another brittle star, *Amphiura inhacensis* (Fig. 5.12g and h) of the family of small burrowers, Amphiuridae, has been described from sand at the base of the reef.

Many colourful species of large sea slugs have been described from this area (Macnae 1962). Apart from *Dolabella*, the sea hare, none of the large sea slugs is herbivorous. There *were* two species of this large, ungainly, green sea slug, *D. scapula* and *D. gigas* which differ only in the shape of their internal shells (Fig. 5.13a-c) (see Section 9.4.3). Recently, they have been considered to be one species, since there are intermediate shapes. The other opisthobranchs found here feed on soft-bodied animals and have evolved a variety of sucking mouthparts with or without a radula.

The only species of opisthobranch with a visible shell is the notaspid, *Umbraculum umbraculum*, (Fig. 5.13d) which has a bright orange, almost globular body capped by a thin, flat, circular white shell. Large individuals are 80-90 mm in diameter and over 30 mm high. This species browses on sponges. *Umbraculum* possesses a very broad radula which is said to have 150 000 teeth, making an ideal tool for rasping siliceous sponges.

Another notaspid, *Pleurobranchus inhacae* (150 mm) (Fig. 5.13e) is a broad flat sea slug, reddish purple in colour. It represents a transition stage in the evolution from shelled to non-shelled sea slugs. The shell in this genus is either very small and concealed or absent. There is no mantle cavity and the large pinnate gill lies on one side of the body in the fold between the body and the foot.

Dorid slugs (Doridacea) are nudibranchs without a shell and mantle cavity. They have evolved a number of external secondary finger-like gills surrounding the dorsal anus. *Hexabranchus marginatus* (Fig. 5.13f) is the largest dorid slug (120 mm) in the southern bay. The body is broad and flat and coloured scarlet with white markings. It is usually accompanied by

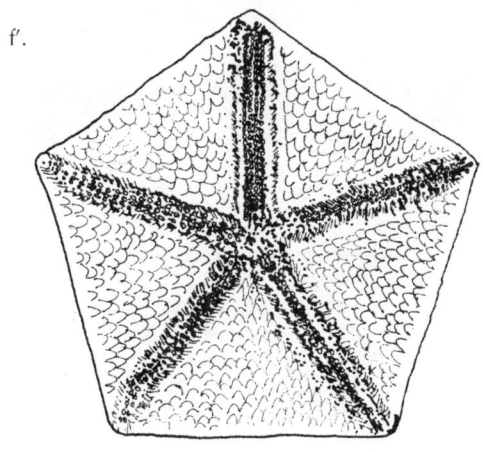
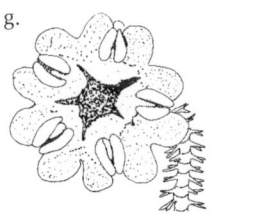
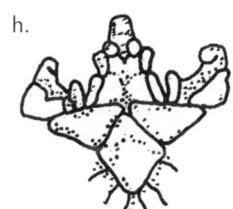

Fig. 5.12
f. starfish, *Culcita schmideliana* aboral view (x 0,3);
f'. oral surface (after Walenkamp 1990).

Brittle stars:
g. *Amphiura inhacensis* (x 0,2) (after J.B. Balinsky 1957);

jaw patterns:
h. ophiotrichid*;

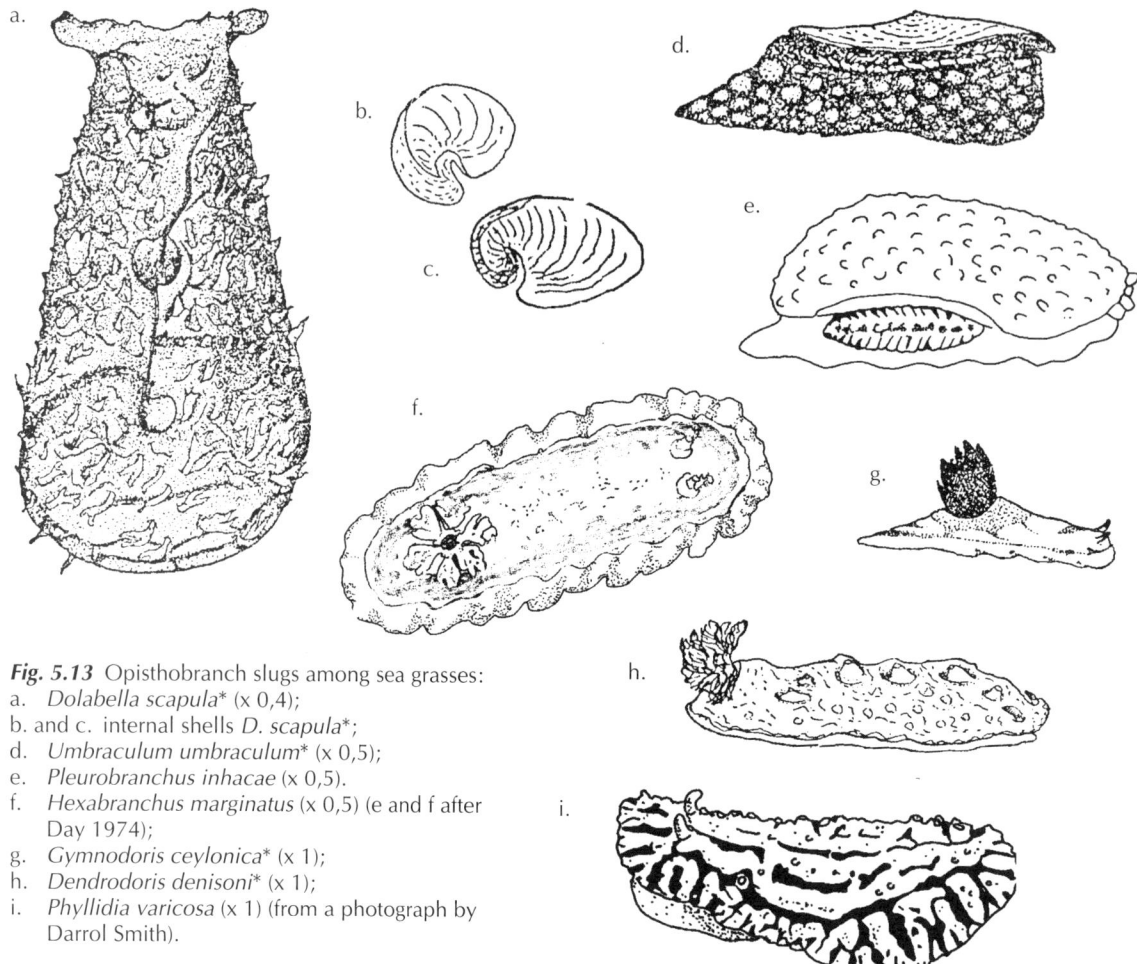

Fig. 5.13 Opisthobranch slugs among sea grasses:
a. *Dolabella scapula** (x 0,4);
b. and c. internal shells *D. scapula**;
d. *Umbraculum umbraculum** (x 0,5);
e. *Pleurobranchus inhacae* (x 0,5).
f. *Hexabranchus marginatus* (x 0,5) (e and f after Day 1974);
g. *Gymnodoris ceylonica** (x 1);
h. *Dendrodoris denisoni** (x 1);
i. *Phyllidia varicosa* (x 1) (from a photograph by Darrol Smith).

a commensal shrimp, *Periclimenes imperator*, which exactly matches the colours of the host. *Hexabranchus* retains jaws and radula and feeds on sponges. It has evolved a method of dealing with the indigestible siliceous spicules, keeping them segregated in a special sac leading out of the stomach, where they are compacted before being ejected in the faeces.

The most common dorid here, *Gymnodoris ceylonica* (50-70 mm) (Fig. 5.13g) has a pinkish-white, soft and supple body with bright crimson spots on the dorsal surface. The rhinophores are tipped with orange and a thin scarlet line runs down the rachis of each of the bipinnate gills. The bluish-brown digestive gland and the orange-red gonad shine through the translucent skin. *Dendrodoris denisonii* (Fig. 5.13h); about 60 mm in length, brownish in colour with iridescent blue spots is occasionally abundant.

Phyllidea varicosa (Fig. 5.13i) is a very poisonous sea slug which is about 6 cm long. It has a very tough, dark blue skin with pale blue longitudinal ridges and lateral patches, each with orange tubercles. The tentacles are orange and there are no gills. It feeds on coral polyps, and has strayed from the reef if found among sea grasses.

Many species of carnivorous swimming crabs of the genera *Charybdis* and *Portunus* (Fig. 3.21a) hunt among the sea grasses for fishes and other crabs. Female swimming crabs produce pheromones to attract males before mating (Hartnoll 1969). The chemical is released from the excretory pores at the bases of the antennae as a component of the urine. Males detect it by the chemoreceptors on the flagellae of the antennules. *Portunus sanguilentus* has a dark 'ox-blood' spot on the inner surface of each chela. When the male senses the pheromone released into the sea water, it exhibits searching behaviour in which these spots are displayed on extended chelae whilst it walks on tiptoe with body elevated on straightened legs. When it finds a female it carries her around for several days and then watches over her while she moults, then they copulate for several hours.

Hermit crabs in these sea meadows are larger species than those on the western sandy shore and they probably come up from the subtidal zone where they find appropriately large uninhabited snail shells. *Dardanus* sp. occupies a *Tonna* shell on which one or more anemones, *Calliactis polypus*, lodge (Fig. 5.14a). This genus of crab induces the anemone to settle on its shell by gently stroking it with its chelae. The anemone bends over, attaches its oral disc and tentacles and then releases its foot from the former perch. It somersaults into position. An order of dominance exists amongst these hermit crabs, the more successful acquiring one or even two anemones whilst the weaker ones have none.

A small flatworm *Stylochoplana iniquilina* (Fig. 5.14c) inhabits the umbilicus of the *Tonna* shell (Fig. 5.14b) in which the hermit crab lives. It is about 25 x 7 mm when fully grown and pinkish white in colour with two broad, median, longitudinal, yellow bands, speckled brown. There are no tentacles; two clusters of about 50 cerebral eyes occur above the brain and also lateral clusters of about 40 eyes, called tentacular eyes, since tentacles occur there in other genera of the family. It gains shelter from the shell and possibly protection by the sea anemone; the part played by the crab in this association is difficult to conjecture.

The hermit crab, *Aniculus strigatus* (Fig. 5.14d), has a flattened carapace so that it can slide easily in and out of the narrow slit of a cone shell in which it lives. The chelae are equal in size and deeply spooned. Both chelae and legs have scaly rings around the bases.

A hermit crab leaves its shell to moult; it swells to a larger size and then has to find a larger shell. In the selection of shells hermit crabs exhibit stereotyped behaviour. The crab turns a likely shell around so that the aperture faces it. The interior of the shell is investigated by means of the minor chela (or by both chelae) and by the first pair of walking legs, several times before accepting or rejecting it. Then it 'tries it on for size', it inserts its naked abdomen into the shell and then retracts inside it completely. It stretches out of the shell and contracts again, repeatedly, then it takes trial walks. Hermit crabs seem to have the ability to discriminate between species of shell, selecting for weight, volume and shape. A hermit crab which finds a suitable shell inhabited by another may fight over the possession of the shell using a ritualised combat procedure. The aggressor seizes the shell of the other crab with its chelae and turns the aperture to face that of its own shell. It rocks the victim's shell and bangs it with its own – a sound audible to the human ear. The non-aggressor leaves the shell 'voluntarily' and searches for another. Thus through bloodless, token aggression the shell resources are shared.

Courtship patterns are well developed among most crabs (Hartnoll 1969). Hermit crabs telegraph their motives and

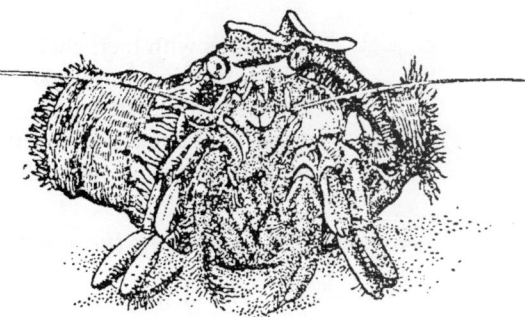

Fig. 5.14 Hermit crabs among sea grasses:
a. *Dardanus* sp. and commensal anemone, *Calliactis polypus** (x 0,5);

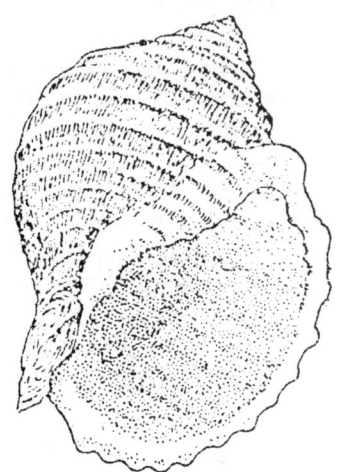

b. *Tonna costata** (x 1)

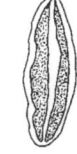

c. commensal flatworm *Stylochoplana iniquilina* (x 2) (after S. Prudhoe 1989).

d. *Aniculus strigatus* from a cone shell (x 1);

e. left chela, *Calcinus laevimanus* (x 1) (d and e after Barnard 1950).

intentions with their chelae. For example, *Calcinus laevimanus*, with a larger, smooth left chela which mimics a molluscan operculum, has striking brownish red marks on its heavily calcified, white chelae (Fig. 5.14e). It approaches a female with chelae raised displaying the stigmata. Females are aroused by gently rocking their shells and courtship proceeds with mutual tapping of the shells.

5.4.4 Fishes in the southern bay: mullet; pomfret; sand smelt; grunter; electric ray

The very sheltered southern bay has a special attraction for certain fishes, some to spawn and others to forage in quiet waters; they enter the bay from the Bay of Maputo or from the Indian Ocean (Smith 1958). There is also a population of coral reef fishes (although apparently fewer than in the west shore coral reef) and some of those fishes that frequent the sea meadows on the west coast (see Sections 8.5.6 and 9.4.7). The fishes described here are not confined to the southern bay.

Mullet are silvery-grey in colour, and their elongated bodies have two dorsal fins, the first one spiny and the second with soft rays; the scales are relatively large. These fishes do not have a visible lateral line and are mostly benthic feeders, consuming detritus, gastropods, foraminifera, diatoms and algae; they nibble on the sea grasses to scrape off epiphytic organisms. The mouth is small and has only fine, villiform teeth in the lips. The food is sorted by fine gill rakers and ground in the muscular, gizzard-like stomach. Mullet swim near the surface of the shallow water and on occasion many species repeatedly leap into the air – hence they are called 'springers'. About eight species of mullet occur in Inhaca Island waters. During September and October thousands of adult mullet, such as *Mugil cephalus* and *Valamugil buchanani* (Fig. 5.15a), swim into the southern bay from the Bay of Maputo seeking shallow water in which to spawn. Each fish weighs about 2 kg at this time and fishermen take advantage of the invasion by trawling from boats or catching the fish in nets spread across the permanent stakes that span the channels in the bay.

A similar migration of black pomfret, *Parastromateus niger* (Fig. 5.15b), is witnessed occasionally. This fish appears to be almost circular in outline since it has a deep body, flattened laterally with dorsal and anal fins reaching from mid-body to the tail; it swims on its side about a metre below the water surface. A shoal of large, glistening, silvery discs moving quickly under water is a spectacular sight. These fishes too provide a large haul for the fishermen.

The sand smelt (*Sillago sihama*) (Fig. 5.15c), a delicious fish, grows to about 25 cm long and has a stream-lined, tapering body, darker above and silvery below, and is covered with very small scales. This fish feeds on small benthic crustaceans and polychaete worms. The small mouth has fine teeth in bands on the jaws and palate. This fish is strongly gregarious, forming small shoals that are scarcely visible over sandy bottoms.

At the mouth of the bay feeding on the sand banks, there may be several species of grunter such as *Pomadasys furcatum* (Fig. 5.15d). These fairly large, silvery grey fishes feed on crustaceans and polychaete worms which burrow in the sand. Grunters often blow their prey out of the loose sand by squirting a jet of water from the small mouth and then they crush the prey with toothed plates in the pharynx. The grunting sound, which can be heard at night at Ponta Torres, is produced when these pharyngeal teeth are rubbed together. Grunters are caught on line by anglers, and are an important source of food to net fishermen.

Cartilaginous fishes (Chondrichthyes) such as sharks, dogfish and rays, are not as numerous as bony fishes, but the bay provides an excellent habitat for numerous species of ray. A less common but fascinating fish is the electric ray *Torpedo sinuspersici* (Torpedinidae) (Fig. 5.15e) which one may come across while wading in the southern bay or in the sea meadows on the west coast. It lives in fairly shallow water on a sandy bottom and it may enter estuaries or sheltered bays to give birth to its young in summer. It may lie immobile on sand or mud, stranded after the tide has gone out and partly covered with sand. The fish is almost disc-shaped with a small, muscular tail and a triangular tail fin. Dorsal fins lie far back above the root of the tail. The fish may grow to 50 cm in diameter but at Inhaca Island they are usually smaller.

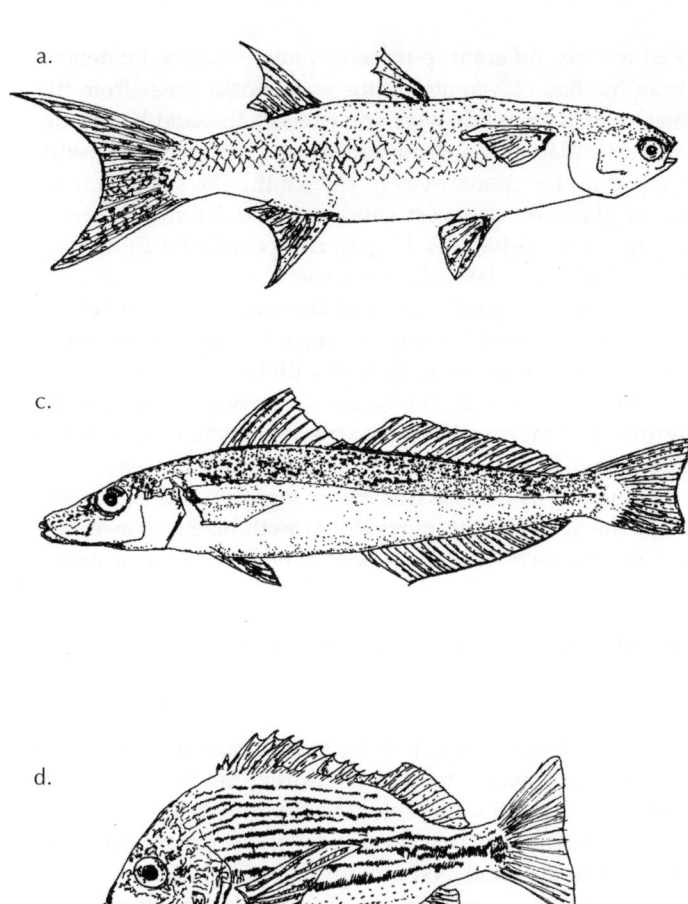
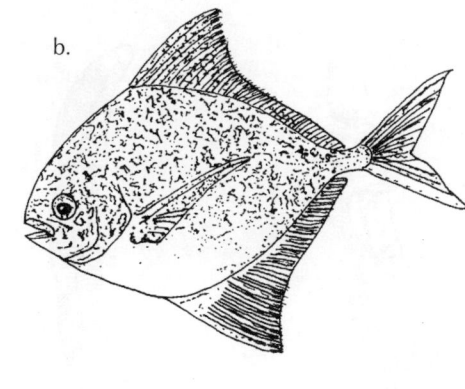
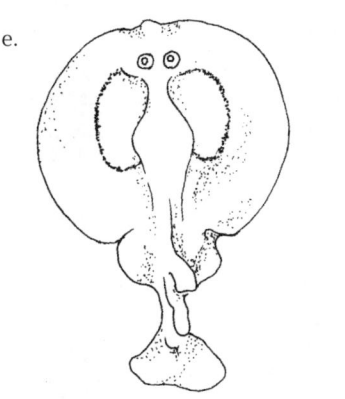

Fig. 5.15 Fishes in the southern bay:
a. Mullet, *Valamugil buchanani* (x 0,1);
b. Black Pomfret, *Parastromateus niger* (x 0,1);
c. Sand Smelt, *Sillago sihama* (x 0,5)
d. Grunter *Pomadasys furcatum* (x 0,2)
e. Electric ray, *Torpedo sinuspersica* (x 0,3) (a-e after M.M. Smith and P.C. Heemstra 1986).

The electric ray can generate an electric shock up to 100 volts when disturbed by a careless foot. The normal use of the electric shock is to stun fish prey which have been sighted or smelled. There are two large kidney-shaped electric organs placed dorsally on each side of the midline, under the skin. The electricity-generating cells are modified muscle cells, arranged in parallel so that the normal electric impulses from nerve endings are tremendously amplified.

In summer, game fishes such as the kingfish (*Caranx* spp.) and queenfish (*Scomberoides* spp.) may be caught in the entrance to the bay by trolling from boats, but numbers are few. Much better fishing is found in the open ocean off the east coast. Game fishes are described in Section 7.6 (Figs. 7.31a-h).

5.5 THE HABITATS OF THE NORTHERN BAY

The boundaries of the northern bay are the peninsula leading to the northern point on the east, the mangrove associations on the south, a subtidal sandy shoal and Portuguese Island on the west. The sand-bar leading from the northern point of Inhaca Island to Portuguese Island forms the northern boundary. Sea grasses and a coral reef have developed there in recent years. The bay measures about 10 km from east to west and 5 km from north to south.

The major habitats in the northern bay are similar to those in the southern bay, namely mangrove swamps, drier sand supporting *Dotilla*, the bubble crab, muddy sand with *Macrophthalmus*, the sentinel crab, and associations of sea grasses. The sea meadows are exposed to three different sources of water: estuarine influenced sea from the Bay of Maputo on the west, ocean water from the Indian Ocean to the north, which may breach the sand-bar in one, two or three places in different years, and also brackish water draining from the mangroves in the south. The sea meadows adjoining those on the west shore, formed around a former breach in the sand-bar, are largely composed of a *Cymodocea serrulata/Thalassodendron ciliatum* association with *Syringodium isoetifolia* as on the subtidal fringe of the west shore (see Section 9.1). The major part of the bay is covered by an association of *Halodule wrightii/Thalassia hemprichii* with some *Zostera capensis* and *Cymodocea rotundata*, the ocean water being tempered by mangrove drainage and being extremely sheltered, as is the southern bay (Fig. 5.1). The sandy mud in which the sea grasses grow is enriched by detritus from the mangroves. The lagoon just south of the sand-bar supports the growth of the pioneer sea grass, *Cymodocea serrulata*, at low tidal level, as on the west shore.

5.5.1 Bare flats: acorn worms; crab and polychaete associations

In the south-western part of the bay a sandy-mud belt runs parallel to the shoreline. In this area during the 1930s Professor van der Horst and his students found populations of a small enteropneust so dense that there was 'one (or more) in every spadeful of sandy mud'. Where are they now? He named the species *Saccoglossus inhacensis* (Fig. 5.16a) (Van der Horst 1940). This acorn worm is 50-70 mm long and has a yellow proboscis capable of extension to about 10 mm. It has an orange collar which is much shorter than the proboscis and almost as broad as long. The rest of the body is cream coloured. Unlike *Balanoglossus* and *Ptychodera* there are no genital wings and the gonads are unusually placed in one ventral row. The middle region tends to twist into a coil when the worm is extracted from its burrow but the abdomen is less muscular. *Saccoglossus* is easily overlooked since its wormcast is only about 2 mm high and soon spreads over an area of about 10 mm^2; and even if detected, the burrow is not easily traced in the soft surface sand. The habitat is described as having 'sand about 1 cm thick above the black mud'. The burrow can, however, easily be seen in the black mud below, since it is lined with shiny, smooth mucus to which clean yellow sand adheres. The upper part of the burrow is in the form of a loose spiral and the lower part is more tightly coiled at the base. There is one exit to the burrow and a shaft may lead to an entrance. *Saccoglossus* rarely protrudes its proboscis from the burrow at low tide but this is sometimes seen lying flat on the sandy surface, apparently immobile, but actually still feeding when the tide has just receded. Particles of detritus are trapped in the mucus on the proboscis and ciliary currents carry them towards the ciliary organ at its base. From time to time other small enteropneusts have been recorded from these flats such as the white *Willeyia delagoensis* and *Glossobalanus alatus* (Fig. 5.16b and c) and an unidentified pink species smelling strongly of iodoform.

Dotilla populations occupy the peripheral sandy area where sand is deposited from both north and south currents. The sand is stabilised here by large aggregations of small polychaete worms, *Mesochaetopterus minutus* (15 mm) which construct very fragile, upright sandy tubes about 30 mm long and 1-2 mm wide The masses of tubes form a dense mat in which the lower ends are stuck together with mucus in the damp sand. This worm is similar to other

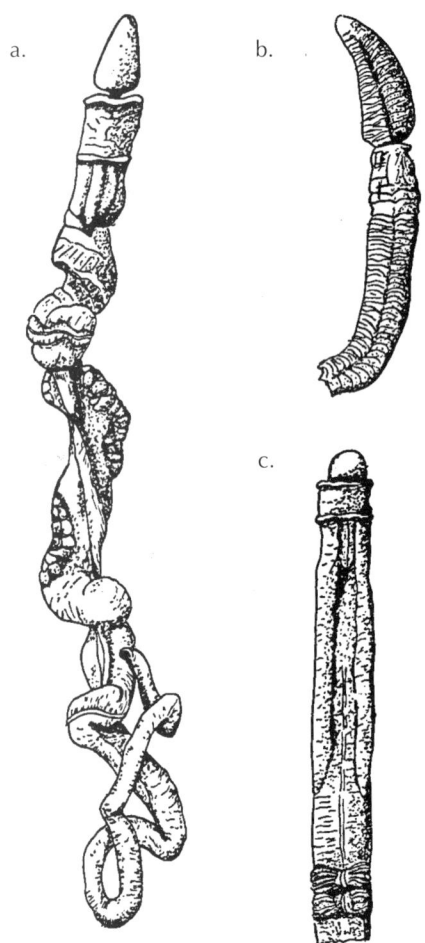

Fig. 5.16 Acorn worms in muddy sand in the northern bay:
a. *Saccoglossus inhacensis** (x 2);
b. anterior region *Willeyia delagoensis** (x 2)
c. *Glossobalanus alatus** (x 2) (a-c after van der Horst 1940).

chaetopterids (see Section 3.3.2). As a suspension feeder it is not in competition with the deposit feeding bubble crabs (see Fig 3.5e).

5.5.2 Sea meadows: pencil urchins and commensals; solitary corals; bivalves; in warmer pools: brittle stars; sea slugs

The pencil urchin, *Prionocidaris baculosa* (Fig. 5.17a) is common. It is parasitised by a degenerate snail, *Capula intortus* (70 mm) which attaches itself temporarily to the test and sucks out fluids. It resembles a tiny conical limpet in shape. The smaller pencil urchin *Eucidaris metularia*, with a test 2-3 cm in diameter, is sometimes accompanied by a commensal brittle star, *Ophiothela nuda*, entwined among the spines. The disc (5-9 mm) is brownish with white blotches and there are six pairs of spines on each segment of the arms. Is it a suspesion feeder? Oral and aboral plates as well as under-arm plates are covered with a thick skin, which is spiny on the oral surface. Another commensal, a little crab *Eumedomus granulosus* is sometimes present on this sea urchin.

The flow of the tidal current past the coral reef on the edge of the Inhaca Channel bring larvae from both corals and associates, some of which develop in the western sea meadows of the northern bay. A thick, flat, brownish coral, *Fungia*, has a skeleton 5-7 cm in diameter, the upper surface resembling the gills on the under side of the head of a mushroom, and it usually lies on sand. Many ridges radiate from the centre where the mouth of the single large polyp lies (Fig. 5.17b). The under surface of the coral is as smooth as a pebble and it is quite unattached. *Fungia* first grows in the coral reef as an attached colony, but buds in the form of tiny discs, made by one polyp, break off and are carried by currents to places of still water in the sea meadows where they sink to the bottom and grow. When feeding, cilia on the polyp surface beat towards the mouth so that detritus may supplement the zooplankton diet. During storms these corals may be buried under sand or on occasion can be turned completely upside down. They can clean away the sand and even right themselves in spite of the relatively heavy skeleton. The cilia on the disc beat away from the mouth when not feeding so that the disc is cleared of debris. The polyp turns right over by first swelling up (presumably to raise the centre of gravity) through taking in excessive water via the mouth by the reversal of the ciliary beat. It is lifted off the ground and can then by ciliary action alone turn itself right over. Another larger fungid coral *Herpetolitha limax* (Fig. 5.17c) is not so common. It has an elongated broad flat skeleton (20 x 100 mm) and more than one mouth in the furrows between the skeletal ridges. These have also been formed by budding of the single polyp.

Although these meadows are about 2 km from the coral reef some of the thousands of planula larvae of corals are carried in

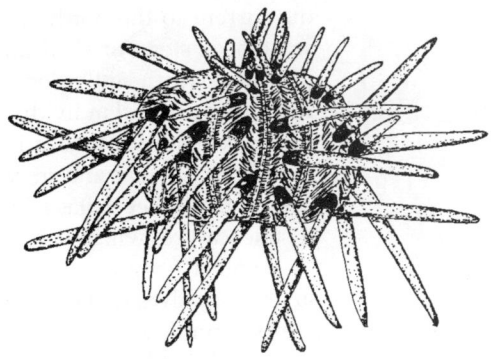

Fig. 5.17 Among sea grasses in the northern bay:
a. Pencil urchin, *Prionocidaris baculosa** (x 0,5);

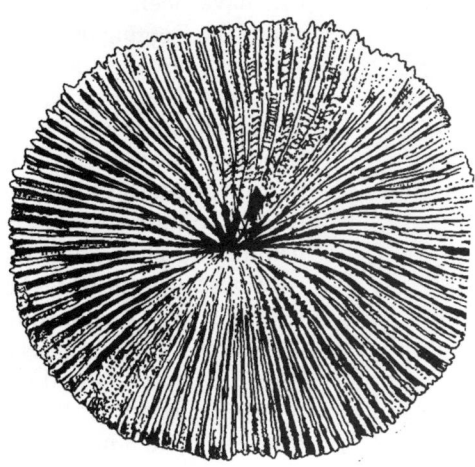

b. fungid corals: *Fungia actiniformes* (x 1) and

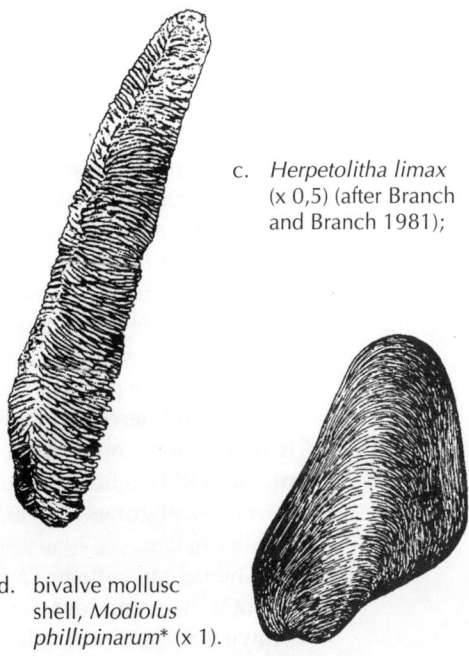

c. *Herpetolitha limax* (x 0,5) (after Branch and Branch 1981);

d. bivalve mollusc shell, *Modiolus phillipinarum** (x 1).

the current to the northern bay. Some of them, belonging to *Pocillopora* spp. (Fig. 8.4c), can settle on a stone or shell on bare patches between sea grasses and initiate a colony. Several planulae, which settle close together, fuse into one organism so that each colony is originally founded by *several* individual larvae. The colony grows very quickly by budding and secretion of the skeleton. At Inhaca Island *Pocillopora* (Fig. 8.4c) has been observed to grow 20 cm high in less than a year and in warmer water the rate is even higher. *P. damicornis* is rosy pink in colour, the skeleton has a bushy form with short stout branches growing in tufts and bearing distinct calyces which hold minute polyps, each with six tentacles.

Another feature related to the proximity of the coral reef and reef flat is the attachment of juvenile bivalves to the sea grass stems by means of slender byssus threads. These shells are less than 10 mm in size and very fragile. They belong to species such as pearl oysters and scallops which, when adult, inhabit crevices in the reef or rocks of the reef flats (see Section 6.4.4).

Fan shells (*Pinna* and *Atrina* species) occur in the sea meadows as in the southern bay (Fig. 5.8b and c). *A. bicolor* with the smaller, darker shell is more common.

Many live specimens of burrowing bivalves of the genera *Tellina, Codakia, Mactra*, and the like have been recorded here. They are more common subtidally and dead shells are abundant. Digging in the *Cymodocea* beds is, however so difficult that many species there may still be undiscovered. Two species lie on the surface sand. Single individuals of the brown mussel *Modiolus phillipinarum* (Fig. 5.17d) often contain commensal pea crabs, *Pinnotheres* sp. The hammer oyster, *Malleus anatinus* is strewn sparsely over the sea meadows. Its shell, 80-100 mm long, has extensions of the hinge line on both sides and the rest of the shell is long and narrow giving it the shape of a hammer. This species, when young, is attached by byssus threads to buried stones but like *Modiolus* it becomes free as it grows. The 'handle' of the hammer which encloses the body of the animal is very thick and incompletely lined with a nacreous layer, the edges being free of it.

The sea grasses in some large patches of the eastern half of the northern bay resemble those in the warmer coves of the southern bay (see Section 5.3) and host the same brittle stars. New species of notaspid sea slugs have been described: *Pleurobranchus inhacae* is orange with a red sole (Fig. 5.13e), *P. peroni* is dark purple, and *P. gemini* is white with chocolate patches (Macnae 1962). It has been suggested that these are colour varieties of the same species.

5.5.3 Fishes of the northern bay: wolf herring; eel-tail catfish; goat fish; pursemouth; fusiliers

The easiest way to gain insight into the large variety of fishes in the sheltered northern bay is to examine the catch in a trawl-net, landed by fishermen at the harbour from a sailing boat, which does not go as far afield into deeper water as a motorised fishing boat. The area fished will have included sea meadows and muddy areas and might include fishes that have strayed from the coral reef which is a protected area. The larger edible wrasses (labrids), emperors (lethrinids) and snappers (lutianids) described in Section 8.5.6, the rabbit fishes, half-beaks, milkfish and flat heads described in Section 9.4.7 and a few inedible strays from the coral reef such as box-fishes, tobies, scorpion fishes and file fishes may have been caught, but the smaller damsel fishes and coral fishes will have escaped the net (see Figs. 8.18 and 9.12). In addition there will be some smaller carnivores, more especially those that hunt for food in mud, and some planktivorous fishes, described below.

The wolf herring is carnivorous, but is related to the true, planktivorous herring; its body is much more compressed and almost ribbon-like. *Chirocentrus dorab* (Fig. 5.18a) has a large mouth and two fang-like teeth in the upper jaw and smaller canines behind them and in the lower jaw. It grows to over a metre in length and swims swiftly through the water to catch the smaller fishes.

The eel-tail catfish, *Plotosus lineatus* (Fig. 5.18b) is also a predatory fish, hunting near the sea bottom, using four pairs of feelers around the mouth to detect its prey. It has a wide dark brown stripe on each side of the body; the dorsal, caudal and anal fins are continuous around

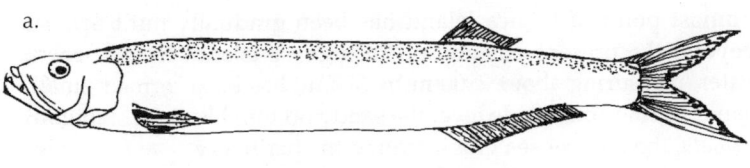
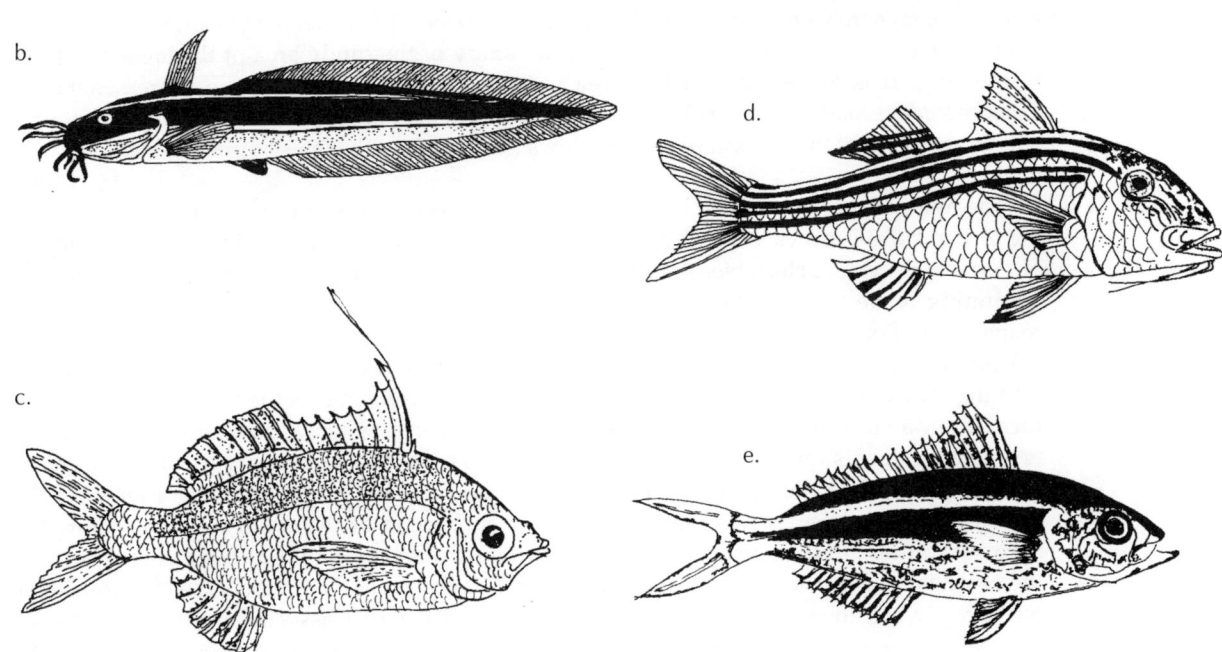

Fig. 5.18 Fishes in the northern bay, caught in fishing nets:
a. Wolf herring, *Chirocentrus dorab* (x 0,1);
b. Eeltail catfish, *Plotosus lineatus* (x 0,3);
c. Goatfish, *Parupeneus cinnabarinus* (x 0,3);
d. Pursemouth, *Gerres filamentosus* (x 0,5)
e. Fusilier, *Caesio caerulaureus* (x 0,3) (after M.M. Smith and P.C. Heemstra 1986).

the tail. Remarkably, the juveniles congregate in tight shoals that mill about in shallow depressions. Beware! Avoid the sharp, toxic spines!

A zoobenthos-feeder, the goat fish or red mullet, may also be caught in the nets as it occurs on the sandy mud around the sea meadows. This fish actively probes its barbels, located on the lower jaw, through the mud in search of small crustacean prey. *Parupeneus cinnabarinus* (Fig. 5.18c) which attains 28 cm in length, has one row of conical teeth in its jaws. They swim about near the bottom in pairs or groups, disturbing the bottom, often followed by wrasses and kingfish which feed on any released invertebrates. Many goat fish are red in colour with darker red longitudinal stripes on the sides.

There are three species of pursemouth around Inhaca Island. Most distinctive is the blue-spotted pursemouth, *Gerres filamentosus* (Fig. 5.18d). It has a downwardly protrusible mouth and fine teeth in the jaws, a forked tail and a single dorsal fin, the second spine of which is elongated. They swim in small shoals, pouncing on small crustaceans and tubeworms when these project above the sandy bottom. If a net should intercept them, a large haul can be expected.

Caesio caerulaureus, the 'beautiful fusilier' (Fig. 5.18e), feeds on plankton and swims around in small shoals. It is one of the most beautiful of the streamlined fishes, having a wide, bright blue lateral stripe between a yellow stripe dorsally and a red belly. Its yellow dorsal and red anal fins are long; the eye is large and the lower jaw projects a little beyond the upper jaw. It grows to 15 cm in length. Fusiliers are associated with the coral reef.

5.5.4 The lagoon at Portuguese Island: a new coral reef

During the last fifty years the western third of Portuguese Island has been gradually eroded and submerged. Sand to the north of it has been removed by currents and the sand-bar extending from the northernmost point of Inhaca Island has been gradually built up and extended westwards well beyond the present position of Portuguese Island. In this way a deep-blue lagoon of quiet water, measuring about 2000 m by 500 m, has been formed on the north coast of Portuguese Island. At low spring tide level the sandy bottom has been stabilised by a growth of *Cymodocea serrulata*, the pioneer sea grass, around the periphery.

The substratum in 1990 on the periphery of the lagoon was clean sand, deposited in the last ten years, and elements of the typical fauna of bare subtidal sand banks (see Section 9.5) were visible. There were large orange turret shells, *Terebra* sp., the leaping snail, *Strombus* spp., sea pens and burrowing sand crabs. The future study of the sandy area of the lagoon will be interesting. It is expected that the area will later be colonised by the large sea grass *Thalassodendron ciliatum* of the subtidal sea meadows when detritus accumulates and the fauna will be replaced by that typical of sea meadows (see Chapter 9). If this happens, how long will it take?

The more central area of the lagoon is about 1 m deep at low spring tides and the floor is the former coral bedrock which has a smooth, almost horizontal surface in which the pattern of *Porites* calyces is detectable. A new coral reef of perhaps five to seven years' standing is beginning to be formed. Precisely the same genera as were described in the study of rejuvenating Ponta Torres coral reef in 1976 are present (see Sections 5.4.2 and 8.6). The coral colonies are about 50-100 cm distant from one another and clusters of colonies are already forming. The only difference from the Ponta Torres rejuvenating reef is that this new reef is formed on a substantial, hard substratum of old *Porites*, whereas the new Ponta Torres regenerating reef is formed on a sandy substratum. There it was necessary to have a primary succession based on the intial colonisation of the sand by *Porites* spp. and only after a partial destruction of the *Porites* was there a secondary succession of the species which required a hard substratum. In the Portuguese Island reef the hard substratum is already present and so the secondary succession of *Pocillopora, Acropora, Echinopora,* and *Stylophora* (see Section 5.4.2) has already commenced. Coral fishes, starfish and other reef associates are as yet few in number but are present. Conditions for studying the growth of a coral reef could not be better, since the water is clear, there is no wave action and it is not much more than 100 m from the steep, sandy shore.

5.5.5 The sand-bar boundary

After the wealth of animal life seen in the northern sea meadows which includes many animals described in Chapter 3 on the western shore as well as from the sheltered northern bay, the sand-bar in the north of the bay is a complete contrast. Maps of the last fifty years have shown many changes in the height and extent of the sand-bar and in the number of places where the ocean water breaks through. The satellite photograph of May 1984 shows that the sand-bar had accumulated to a level above high tide along most of its length except where it is breached in two places. It extends from the north-east point across the original mouth of the bay, beyond Portuguese Island, and encloses the lagoon north of that island where the coral reef is developing. In the 1950s a large area of the sand-bar was washed away and a deep lagoon was formed, whilst in the 1940s the bar was almost complete. Thus the sand-bar is composed of shifting sand and, not unexpectedly, may be completely barren. Some rocks appear above tidal level in the eastern section.

6

LIFE ON SHELTERED AND VERY SHELTERED ROCKS

Communities of animals and plants on rocky shores are composed of very different species from those on sandy shores, described in the last chapters. A large proportion of animals are permanently attached to the rock surfaces. Thus they are exposed at low tides to the hazards of high temperatures, water loss and low content of dissolved oxygen. Their counterparts on sandy shores avoid most of these stresses by burrowing on both the upper and lower shores. On rocky shores in the middle and lower zones on sheltered rocks at Inhaca Island shelter is available under loose stones and slabs of limestone rock, whereas on the upper shelterd shores animals make their own shelters in shells or tubes in which they maintain more constant conditions, and they have evolved many adaptive devices. On shelterd shores, flat rocks are covered with silt and detritus which severely limits the settling of algal spores. Herbivorous feeders probably use detritus as well as microscopic algae, whereas on exposed shores some may feed on macroscopic algae (see Chapter 7).

Rock profiles are shown in Fig. 6.1a, b and c, which also indicate the distribution of the dominant animals. There are almost no macroscopic seaweeds. The gradients on the west shore, on the the well shaded rocks of Ponta Punduine and on the sunny Ponta Torres cliff differ, but the animals present are similar.

6.1 THE SUPRATIDAL FRINGE AND UPPER SHORE

Isolated groups of calcareous sandstone form outcrops along the upper shore at intervals of 100-200 m both north and south of the Marine Biological Station. At some places the rocks stand about 1,5 m high at the top of the shore just below the dune vegetation and slope gradually for 30 m to mid-tidal level where they disappear under the sand (Fig. 6.1a). Their surfaces are unevenly shelved with coarse cross-cut bedding, creviced and pitted by gentle, sand-laden waves. At equinoctial spring tides the whole outcrop is submerged while at ordinary high tides the gentle waves produce little splash and spray thus confining the supratidal fringe to a height of less than 50 cm. The barnacle zones and the oyster belt below have similar ranges, the latter spreading down the sloping rocks where the zone is truncated as the sand meets the rock.

6.1.1 Periwinkles; amphipods; displaced crabs; roving grapsid crabs; barnacles; predatory snails

Two species of periwinkles (Fig. 6.2a and b) span high tide on the western shore. *Littoraria (Littorina) kraussi* is the larger one (15 mm) that has a smooth, light pink and pointed shell; darker zig-zag markings are present on younger individuals. The grey nodulated *Nodilittorina natalensis* tends to select crevices on the less sunny parts of the rock whilst *L. kraussi* tolerates exposure to

Fig. 6.1 Rock profiles on sheltered shores and dominant animals (after Kalk 1958) (note different scales):

a. west shore, 600 m;

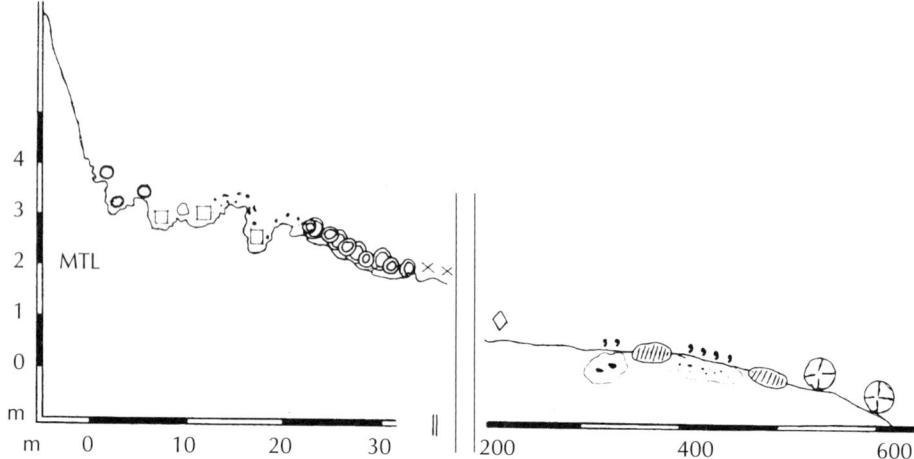

b. Ponta Punduine, south-west point, 60 m;

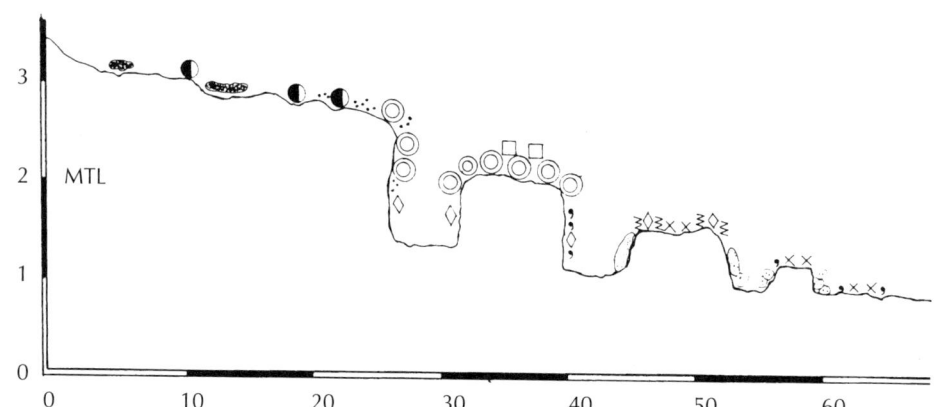

c. Ponta Torres, south-east point, 40 m.

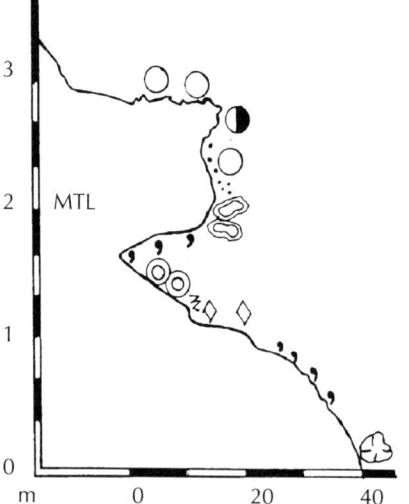

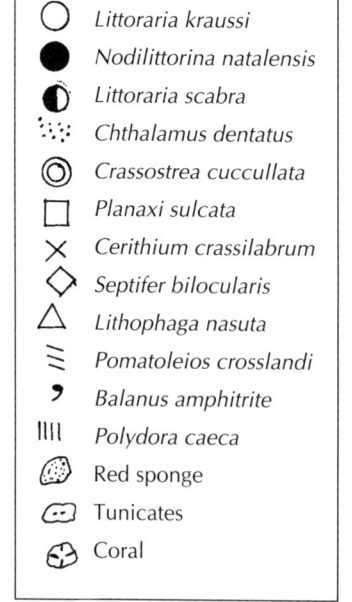

○ *Littoraria kraussi*
● *Nodilittorina natalensis*
◐ *Littoraria scabra*
∴ *Chthalamus dentatus*
◎ *Crassostrea cuccullata*
□ *Planaxi sulcata*
× *Cerithium crassilabrum*
◇ *Septifer bilocularis*
△ *Lithophaga nasuta*
≡ *Pomatoleios crosslandi*
, *Balanus amphitrite*
|||| *Polydora caeca*
 Red sponge
 Tunicates
 Coral

the sun. The density of each is about 50/m², much lower than on rocks exposed to strong wave action (see Section 7.2.1).

Between tides both species are inactive, closing their shells with horny opercula and attaching themselves to the rock by means of a mucous secretion which becomes dry and sticky around the edge of the shell. An air pocket left between the rock and the operculum provides some insulation from the heat of the rock. Periwinkles are cooled mainly by slow evaporation of water enclosed in their shells and mantle cavities, but the internal temperature (measured with a thermistor) may rise to over 40°C, with no ill effects. This is cooler than rock temperature on a summer's day but is often more than 5°C above air temperature. As water evaporates during the day from the spaces inside the shell, the operculum sinks inwards and the air pocket grows. The animals are usually immersed in the sea twice daily when the tide rises.

During neap tides the periwinkles migrate down the rock face for a few centimetres at night. This ensures that the encrusting algae on which they feed will have absorbed water. Although some animals may not become wetted, they survive since their store of water may last 14 days. In the week of spring tides they crawl up the rocks in small stages and thus avoid too lengthy an immersion. The distance covered here is only a few centimetres but on rocks exposed to spray they may travel 1-2 metres (see Section 7.2.1). They graze on the encrusting plants consisting of a mixture of microscopic filamentous green algae, diatoms and blue-greens, which together give the rock at that level its characteristic grey appearance.

The feeding mechanism of herbivorous snails is unique: it involves the radula, the odontophore and the lips (Fig. 6.2c). The radula is a flexible ribbon-like rasp bearing hundreds of rows of jagged, chitinous teeth which lie flat when the radula is at rest but are raised vertically on the portion of the radula which protrudes through the lips when in use (Fig. 6.2d). The radula is up to twice the length of the body and grows in the upper (supralingual) sac in the mouth. The lower end is secured anteriorly in a sublingual sac. Worn out teeth are resorbed in the sublingual sac as new ones move into position – a replacement system reminiscent of sharks' teeth. The odontophore is a cartilaginous support beneath the radula to which are attached a complex set of muscles from the body wall which move the radula and the ondontophore together, to and fro. The upper lip is reinforced by a pair of tough, chitinous 'jaws' which press firmly against the radula, clamping it to the odontophore when in action. Pressure of the radula against the rock surface is also aided by a flow of blood to the sinus in the head. The sides of the radula are curled up to form a groove so that the powdery particles are trapped by marginal teeth. The radula is then withdrawn into the mouth and the scrapings are swallowed.

In different species of herbivorous molluscs, this mechanism is modified for different feeding habits such as browsing lightly over surfaces, grazing on loosely attached organisms, rasping encrusting algae, fronds of seaweeds or seagrasses or even for sweeping detritus into the mouth. Small morphological differences in the three components of the feeding apparatus determine the ability to deal with specific food (Hawkins and Hartnoll 1983). Periwinkles have seven radula teeth in each row and they graze close to the rock.

Fig. 6.2 Periwinkles:

a. *Littoraria kraussi* (x 2);

b. *Nodilittorina natalensis** (x 2);

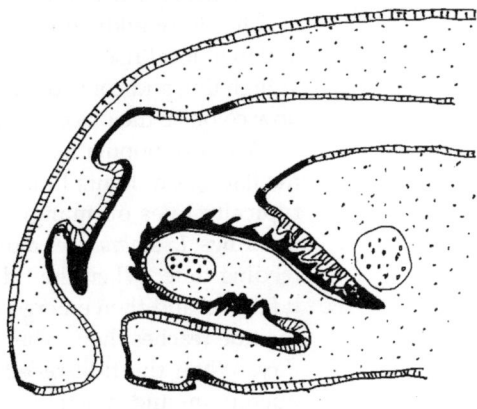

c. section through head showing grazing mechanism: odontophore, radula, lips (after Newell 1979);

d. one row of teeth in radula of littorinid;

e. veliger larva (x 40) (a, d and e after Kilburn and Rippey 1982).

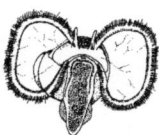

Both species adapt to the intermittently wet and dry habitat in several other ways. They move along on their trail of mucus, exhibiting a curious shuffling movement. The muscular foot is ciliated and split longitudinally so that each half moves separately. This mechanism can be used on both wet and dry rock and enables it to hold on securely while moving. Periwinkles supplement gill respiration by the dense vascularisation of the mantle cavity wall which allows for gas exchange in air. These species do not use water for excretion nor for reproduction. They excrete solid uric acid, which thus removes both ammonia and carbon dioxide in combination, as do land snails. They also have internal fertilisation like land snails and lay many eggs in lens-shaped, transparent capsules which are carried out to sea by tides. Early stages of development take place within the capsule and hatching is delayed until veliger larvae are formed. The larvae are planktonic, keeping afloat for several weeks by means of a large ciliated band, the velum, whilst they feed on phytoplankton (Fig. 6.2e). One wonders where they settle initially since shells smaller than 5 mm have not yet been seen on high level rocks.

The periwinkle zone is closely defined by high spring tide levels. In recent years the apparent local rise in sea level on the west shore ensures that higher rocks are now submerged at high tide and coincidentally periwinkles are found at higher levels than formerly. Barnacles now colonise the places where periwinkles congregated a few years ago.

A sparse population of a third species of periwinkle, *Littoraria (Littorina) scabra*, occurs on the flat rock at Ponta Punduine. It is the largest of the periwinkles (20-25 mm) (Fig. 4.13a) and typically lives on mangrove trees where it is a lighter brown colour, often with zig-zag markings. *L. scabra* uses a different pattern of migration from other periwinkles to optimise its feeding time. When the tide is out these periwinkles attach themselves to rocks above high tide level and then emerge with every incoming tide in response to increased humidity *before* the tide reaches them. They move down to graze on algae wetted by the high tide level of the day. When the tide recedes they travel back up again and seal their shells with horny operculum and mucus as do other periwinkles. Different species of periwinkles thus have genetically determined different *thresholds of humidity* to which they respond with migratory behaviour. *L. scabra* is also adapted to its semi-terrestrial existence by excreting uric acid as do other tropical species. In addition, they protect their young by incubating eggs in a temporary brood pouch, formed inside the mantle cavity, until the larvae have developed into the late veliger stage; only then are they released into the sea.

On buttress rocks in the southern bay the periwinkles are partly segregated. *L. krausii* is more common on the sunny north-facing sides and *L. scabra* on the more shaded south-facing sides. At Ponta Torres the supratidal fringe is horizontal and sandy (from sand blown from the dunes) and the population of periwinkles is very small, *L. kraussi* being about twice as common as *L. scabra*. The latter extends down the vertical face into the upper barnacle zone below. *Nodilittorina* seems to be completely absent from all very sheltered rocks.

At Ponta Punduine crowds of the nocturnal, whitish, rock-hopping amphipod, *Parhyale inyackae* (12 mm) are concealed during the day underneath flat slabs of rock (Fig. 6.3e) and emerge at night. This rock-hopper is much bigger than the sand-hopper, *Talorchestia anomala* of the sandy shores and has a more robust second pair of gnathopods, which are probably used for scraping algae and detritus off the rocks. *P. inyackae* is circum-tropical in distribution despite its local specific name, given when it was first described from Inhaca Island. It has a peculiar extra jointed spine on its third uropod which distinguishes it from similar amphipods. Horizontal migration and astronomical orientation mechanisms in sandy shore amphipods are well known (Pardi and Ercoli 1986). What clues do the rock-hopping amphipods use in their migrations?

Two species of crabs from the lower shore, quite surprisingly, shelter beneath these flat slabs of rock during ebb tides; these are the brown xanthid crab *Epixanthus frontalis* and the porcellanid *Petrolisthes lamarcki*. The crabs are quite inactive at this high level until the tide rises. These crabs will be considered further in Section 6.5.

The upper shore rocks are also inhabited by two species of grapsid crab which are agile climbers but spend the day between tides hidden in crevices among rocks. Grapsid crabs differ from ocypodids of the upper levels of the sandy shore in having an almost square, flatter carapace with a broad front between the short, stout eye-stalks. The more common *Grapsus fourmanoiri* (25 mm) is greenish brown in colour and there is one tooth on each side of the carapace behind the orbital tooth (Fig. 6.3f). *Metopograpsus messor* is smaller and brownish in colour, and has no lateral teeth on the carapace (cf. Fig. 4.12b). The crabs can also be distinguished by their chelae, those of *G. fourmanoiri* being spooned whilst the other has pointed chelae. Both species feed on the algal complex on the rocks using their chelae, and the differently shaped chelae may allow the two species to select different types of plant, but this has not been investigated.

The same species of small grey barnacle, *Chthamalus dentatus* occurs on the sheltered as well as on exposed rocks at Inhaca Island, but its density is reduced with increasing shelter. The majority of these barnacles lie above the oyster belt. The animal secretes a tough, calcareous, shallow cup with its base firmly cemented to the rock surface (Fig. 6.3a). The sloping walls are composed of six fused plates with crinkled edges. The roof is made of two pairs of movable and interdigitating plates. When submerged by the rising tide the plates are tilted upwards in the middle and six pairs of jointed legs are protruded. The animal inside appears to resemble a shrimp, but without head or tail (Fig. 6.3b).

Feeding may be easily studied at high tide on the western shore for the waves are gentle, the water is clear and the rocks below serve well as an observation platform. An intermittent flow of water is filtered through the numerous setae on the legs in the following way. When the roof plates are opened by muscles, a slow current is automatically sucked in through the very fine setae on the first two shorter pairs of legs which are bent over the inhalent opening. After a few seconds an exhalent current is created by the abrupt closing of the roof plates following rapid withdrawal of the legs. During these movements fine particles are washed off the fine setae of the short legs into the mouth, and the larger particles are flushed out from the coarse setae of the long legs and washed away. Finer particles which may have adhered to the coarse setae are brushed off into the mouth by the setae of the short legs. This species of barnacle collects the nanoplankton (down to 1 µm in size) and the smaller phytoplankton.

In the adjacent sandy slope at this tidal level the suspension-feeding wedge mussel, *Donax faba*, selects a range of *larger* diatoms from the gentle waves (see Section 3.2.2). Thus there is no overlap in the food requirements of the two dominant species on the western shore at this level. This is a remarkably clear example of the sharing of the phytoplankton food resources. A similar partitioning of phytoplankton between barnacles and oysters is mentioned in Section 7.2.2.

The true crustacean identity of barnacles, indicated by the jointed legs, is confirmed by the structure of the larvae which are typical crustacean nauplius larvae. The larva (Fig. 6.3c) has a triangular chitinous covering bearing curved spines and three pairs of jointed legs which keep it afloat whilst it feeds on phytoplankton near the surface of the sea. The characteristic

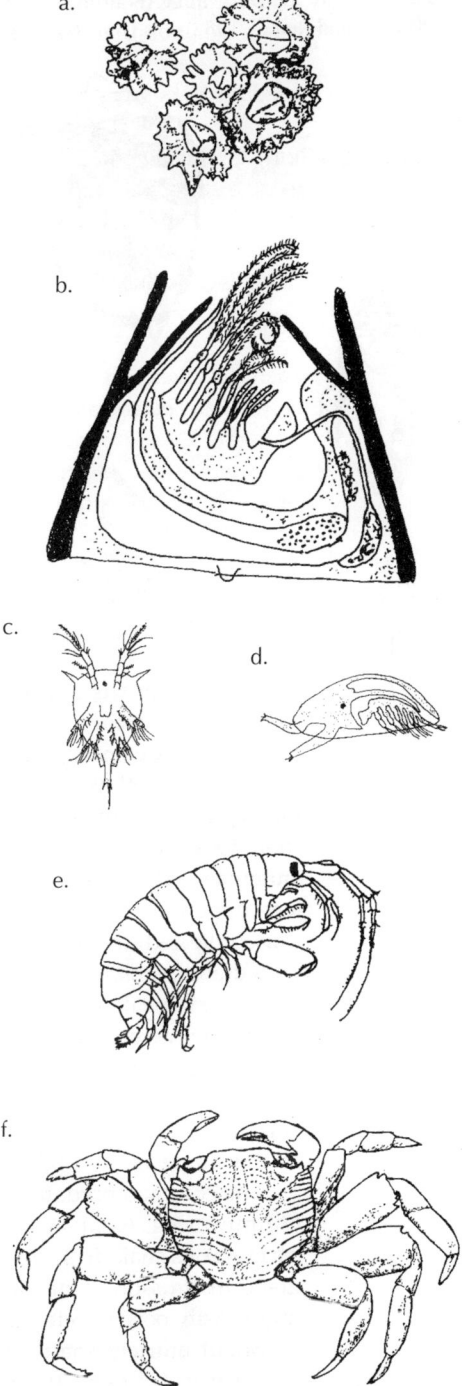

Fig. 6.3 Crustaceans on the upper shore:
a. Barnacle shell, *Chthamalus dentatus* (x 1);
b. shrimp-like stage inside shell (x 5);
c. nauplius larva;
d. cypris larva (x 40) (a-d after Branch and Branch 1981);
e. amphipod, *Parhyale inyachae** (x 3);
f. Rock crab, *Grapsus fourmanoiri* (x 0,5) (after Crosnier 1965).

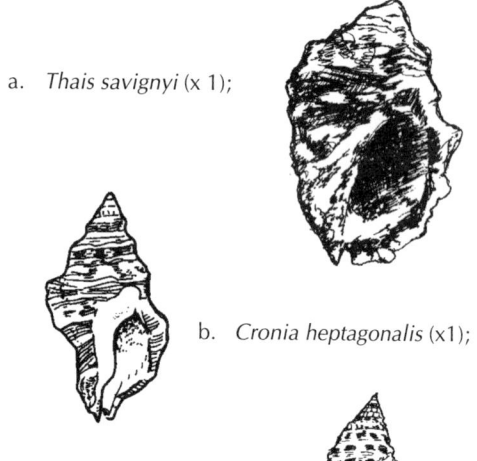

Fig. 6.4 Predatory snails on upper shore rocks (after Kilburn and Rippey 1982):

a. *Thais savignyi* (x 1);

b. *Cronia heptagonalis* (x1);

c. *Morula granulata* (x1).

large median eye prompted early naturalists to call the larva 'nauplius' after a one-eyed giant of that name in Greek mythology.

These pelagic larvae undergo a series of changes that are genetically programmed for settlement on rocks on the upper shore in the barnacle zone and nowhere else. The larva moults several times, each time adding one more pair of legs. A thin mantle grows around it in two sections between which the legs protrude ventrally. This stage is known as the 'cypris' larva, so-called because it resembles a small bivalved 'ostracod' crustacean of that name (Fig. 6.3d). Paired eyes are formed in addition to the nauplius eye. The response to light (which initially kept the larva at the surface) changes and becomes negative so that the cypris sinks and is carried inshore by the tide. The first pair of appendages (antennules) have, in the meantime, developed discs at their ends which are sensitive to touch and to a particular chemical secreted by adult barnacles so that when a larva makes contact with a rock in the 'right' zone, by chance, it clings to the rock. The odour arises from a specific protein known as 'arthropodin' produced in the barnacle exoskeleton (by a mantle secretion) and spreads to the rocks around, where the chemical is adsorbed. The larva now no longer feeds. Its gut degenerates and it lives on reserves of fat until it finally settles. The cypris larva creeps around the rocks testing the surface for texture with its sensory appendages, detecting shadows of other barnacles with its paired eyes and locating them with its sense of smell. When a vacant spot in close proximity to other barnacles has been selected, the cypris larva turns around several times before finally settling as though to ensure there is room for growth. The larva first stands on its head and applies the glands of the second pair of antennae to the rock which secrete the calcareous base of the shell. It then lies on its back with legs uppermost whilst the mantle secretes the calcareous wall and paired roof plates. When thus protected, the animal metamorphoses into the adult form of sessile barnacle. It loses its paired eyes, sense organs, head appendages and cement glands, all of which are resorbed. The alimentary canal is formed again and the six pairs of limbs differentiate into two pairs of short legs with fine setae and four pairs of longer legs with coarse setae.

Barnacles settle close to each other; they are hermaphrodite and have cross-fertilisation which necessitates the transfer of sperm through a long extensible penis tube. The fertilised eggs are incubated until the nauplius larvae hatch. At settling time the cypris larva is said to align its long axis with the direction of waves or currents.

The common predator on barnacles on the sheltered shores is the small snail *Thais savignyi* (30 mm) (Fig. 6.4a). A few may be seen 'nestling' in crevices when the tide is out. This carnivorous snail drills holes in the closed barnacle shell by means of a secretion from the foot gland that dissolves the matrix, loosening the calcium carbonate crystals so that the radula can more easily bore through the shell. The prey is then sucked out by the proboscis.

Small muricid snails always inhabit rocks and are not seen on the sandy shore. *Cronia heptagonalis* (25 mm) (Fig. 6.4b) for example, retreats under stones when the tide is out on the west shore. The shell has a sharp spire and a distinct siphonal canal so that it tapers above and below the 'shoulder'. *Morula granulata* (20 mm) (Fig. 6.4c), another muricid, hides in crevices. It has a thick squat shell with a worn conical spire and a narrow white aperture with two prominent and a few lesser teeth on the lip. It preys on barnacles.

6.1.2 Oysters; predatory snails; grapsid crabs; barnacles; and grazing snails

Large numbers of the oyster *Crassostrea* (*Saccostrea*) *cuccullata* cover the rocks below the barnacles down to mid-tidal level, forming a solid growth which excludes any other sedentary animal or alga (Fig. 6.5a). This species of oyster has a wide distribution from Natal to Tanzania and penetrates estuaries, being able to withstand variable salinity and a high silt load. It may live on horizontal or vertical faces of rock on both exposed and sheltered shores and grows best on gently sloping surfaces. The tough shells withstand erosion whilst the rock face itself above and below may be worn away by strong currents, resulting in the mushroom-shaped rocks of the southern bay. The oysters are very successfully specialised and at the same time the internal structure is degenerate.

All bivalve molluscs lose their 'heads' at metamorphosis from the larval stage, through the resorption of sense organs and cerebral ganglia. In addition an oyster loses its foot, including the pedal ganglion and the byssus gland which first attached the settling larva to the rock. It does not retain larval mantle eyes as swimming bivalves do. The hinge lacks teeth, there is only one movable shell and only one adductor muscle for closing it. There is no blood pigment, neither haemocyanin nor haemoglobin, which other molluscs usually have.

The animal is, however, efficiently protected. The lower valve of the shell (usually the left) is cup-shaped and very thick with a deeply scalloped margin. The shell is closed by the movable right valve which has peripheral teeth that closely fit the grooves (Fig. 6.5a). The remarkably tough shell is constructed on the same principle as bone i.e. crystals of calcite are embedded perpendicularly in a protein matrix. The upper shell of the young oyster is, however, fragile at the edges, and this is the weak spot where predators attack.

The feeding mechanism is highly efficient (Yonge 1960). Over 30 litres of water are filtered by large individuals in an hour at 24°C. This compensates for the short time of immersion on the upper shore, which is less than six hours in 24 hours. The ciliated gills are very large in proportion to the body and form a semicircle around it. There are no siphons but the mantle edges are connected at a point in front of the anus, dividing the mantle cavity neatly into inhalent and exhalent chambers. The exhalent current is augmented by an extra channel through the premyal space in front of the adductor muscle. A number of mobile interlocking tentacles, fringing the mantle edge, control the flow of water as is required for feeding, even more efficiently than do the siphons of other bivalves such as *Donax*. The horizontal gills rather than the labial palps, sort the food from sand and silt according to density. Should they become overloaded they can twitch and reject all the material deposited on them. Sensory cells on the labial palps have an additional function. They are sensitive to certain chemicals, rejecting bacteria, dinoflagellates and detritus and accepting diatoms from the phytoplankton in the range of 6-10 µm. Thus oysters sort the food for quality as well as for size. The range of size does not overlap that of barnacle food, so that the food resources are shared. In order to cleanse the mantle cavity of debris the oyster closes its shell abruptly about once a minute, ejecting faeces and silt. This is facilitated by a large proportion of the adductor muscle of the shell being composed of 'quick' muscle fibres; in this, oysters resemble swimming bivalves. This ability stands them in good stead when the surrounding water is laden with silt.

The ability to close the shell tightly is no less important for survival than efficient feeding, since the animal is exposed to air on the upper shore in considerable heat on sun-baked rocks. The closing is achieved by the 'slow-acting' fibres of the adductor muscle, which can be locked in the contracted state with little expenditure of energy, resembling giant clams in this ability. These muscle fibres are five times stronger than in the muscle of a frog's leg but they contract 20 times more slowly. These features are partly due to the greater thickness and length of the oyster's myosin filaments.

Crassostrea cuccullata is not viviparous as are some oysters on temperate shores, but it is said that a proportion of the population changes sex in winter. The synchronisation of the emission of eggs and sperm in the whole population (a feature of many marine animals with external fertilisation), has been well studied in oysters for commercial reasons (Yonge 1960). The mature animals are 'ripe' for some time before a rise in temperature initiates spawning which spreads rapidly through the population by chemical excitation. A pheromone is released in the seminal fluid which is carried into oysters by the feeding current and stimulates the release of eggs. Eggs also contain this substance which, when released, will stimulate the extrusion of sperm in neighbouring oysters, completing the positive feed-back mechanism. Oyster semen also releases another hormone 'diantlin' which relaxes the muscular framework of the gills and enlarges the cavity, so allowing the expulsion of eggs into the water current. The adductor muscle is also relaxed by this hormone and the shells gape wider, facilitating spawning. After a complete larval history in the plankton, including the trochophore stage (Fig. 6.5b) the veliger larvae (Fig. 6.5c) return to rocks and sink at the correct intertidal level to settle. Settling behaviour is similar to that of barnacles except that the first attachment is by means of the secretion of sticky byssus threads from the larval foot glands.

Predatory snails have not been observed in action on oysters, although empty oyster shells may be seen.

Herbivorous and detritus-scraping snails on the beach rock below the oyster belt require shelter from the sun. During the day aggregations of many species are found in large depresseions, under overhangs, on the shaded sides of rocks (at Ponta Punduine), or among loose stones at this level (on the western rocks). Groups of grazing snail, *Planaxis sulcata*, a tropical species of sheltered shores, occur in crevices and larger depressions on the rocks.Its shell somewhat resembles that of a large periwinkle in shape but the base is broader. It is spirally ridged and has a narrow siphonal groove in the aperture (Fig. 6.5d). The shell is speckled green and brown and has a horny operculum. This snail feeds on the microscopic algae on the rocks. It carries protection of the young a step further towards viviparity than have other sea snails on the upper shore. Eggs are brooded in a pouch of the mantle cavity until the larva reaches a creeping stage and only then is it released through a special pore behind the right tentacle.

The most numerous snails are those of the horn shells, *Cerithium crassilabrum* and *C. caeruleum* (Fig. 6.5e), although many are dead and merely shelter hermit crabs. *C. crassilabrum* has a wide thick flange on the lip of the aperture.

Other herbivorous snails are less numerous. *Nerita albicilla* is the sheltered shore species of *Nerita* (Fig. 6.5f). It has a thick, elongated dome-shaped shell with a dark patterned top and a flat, white columella shield bearing small tubercles underneath, giving the appearance of a slipper. Many are found among damp stones, feeding only at night. Very small egg capsules with a crystalline pattern are attached to the stones. *Turbo coronatus* (40 mm) is a turban shell with a crown of well-marked spiral, tubercled shelves and a large round aperture which can be closed by a thick, knobbly, calcareous operculum (Fig. 6.5g). *Monilia obscura* (25 mm) is another thick-shelled, drab species with a more tapering spire, but a squat shell, pale in ground colour with spiral markings and faint, spiral ridges (Fig. 6.5h). Most of these species are tropical species which have spread into Natal.

A few 'false limpets' *Siphonaria oculus* occupy small damp depressions on beach rock where they feed on filamentous green algae (cf. Section 7.3.1). This sheltered shore species has 50-60 white, radial lines on the dark shell and a white 'eye' in the centre of the brown interior of the shell (Fig. 6.5i). Although

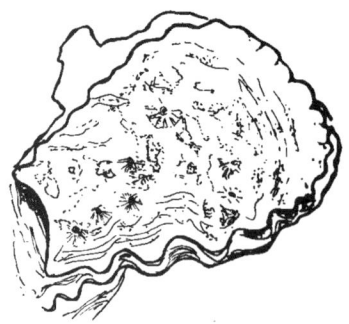

Fig. 6.5 Molluscs in the oyster zone:
a. oyster, *Crassostrea cuccullata* (x 1);

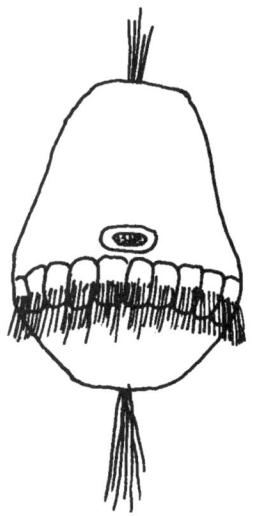

b. trochophore larva (x 1) (a and b after Kilburn and Rippey 1982);

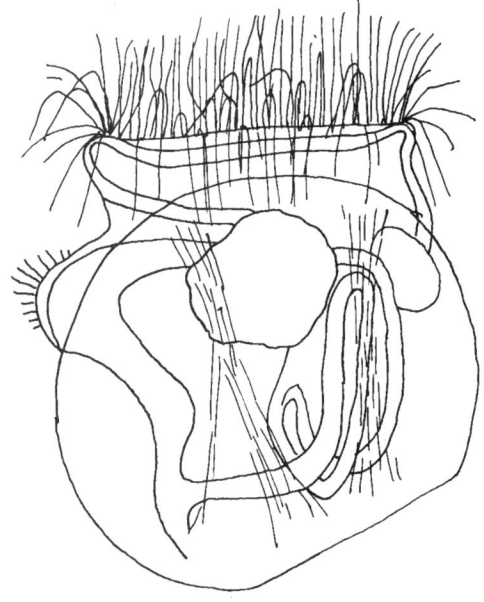

c. veliger larvae (x 100).

Siphonaria is limpet shaped, it is not a limpet but an air-breathing pulmonate, related to land snails, which has a vascrularised mantle cavity. There are no true limpets on sheltered shores. *Siphonaria* can be distinguished easily from a true limpet by the slight asymmetry of the shell on the right side where the siphon (air tube) from the mantle cavity exits.

Clibanarius virescens, a small hermit crab (20 mm), seeks shelter mainly in empty *Cerithium* shells. It has characteristic yellow-banded black legs protruding slightly from the aperture of the shell (cf. Fig. 3.12). This crab has a greenish carapace and the chelae are horny and spoon-shaped. It feeds on algae and detritus collected by scraping rocks and stones.

6.2 MIDDLE AND LOWER SHORE ROCKS

6.2.1 Bivalve moluscs; lower shore barnacles; sea cucumbers; ocypdid crabs.

Several species of bivalve molluscs, which are ciliary suspension feeders, are attached to rocks by byssus threads which persist into adulthood. The brown shell of the sheltered-shore rock mussel *Septifer bilocularis* (40 mm) has many fine ridges radiating from the narrow hinge at the pointed end (Fig. 6.6a). A very small internal ledge at the hinge that supports the single strong adductor muscle typifies the genus. The valves of the shell fit closely since the edges are finely crenellated.

On the western shore in limestone rock, a number of specialised bivalves occur. The ark shell *Barbatia foliata* (25 mm) (Fig. 6.6b) contrasts in shape to *Septifer*. Its hinge line is broad and straight with several teeth. It has the added protection of a thick hairy periostracum. The most secure of the bivalves is the smaller date-stone mussel *Lithophaga nasuta* of which there are one or two in every limestone rock (Fig. 6.6c). At first sight the animal is hardly visible under the rock since it burrows right into the limestone, but its presence can be detected by the figure-of-eight opening of its vertical burrow through which its siphons are protruded while feeding. The shell is almost cylindrical in shape with tapering rounded ends, the hinge being at the lower end. *Lithophaga* is covered by a dark periostracum and this may have an adhering chalky layer derived from the rock as it is abraded by the shells. It rotates from right to left and back again repeatedly and succeeds in making a hollow nest for itself, softening rock by means of an *acid secretion* of mucus from the mantle edges. This species does not live in the sandstone rocks of Ponta Punduine, in which it could not burrow.

Numerous small pink barnacles occur on the sides of rocks and on some of the old coral slabs. These are *Balanus amphitrite var. communis* which is characteristic of quiet water (Fig. 6.6d). It is usually even smaller here than *Chthamalus* of the upper shore and its calcareous wall is white or pink with dark red lines running from apex to base of each of its six plates. Its distribution seems to be discontinuous.

The light brown sea cucumber *Holothuria (Lessonothuria) insignis* is found under rocks, attached by tube-feet or even partly buried in the sand below the rocks (Fig. 6.6e). It occurs higher on the shore than the sea cucumbers, such as *H. scabra* and *H. atra* (see Section 3.6.4) which inhabit sand. This sea cucumber is never exposed to light and has a relatively thin, light

Fig. 6.5 grazing snails:
d. *Planaxis sulcata** (x 1).

e. *Cerithium caeruleum* (x 2);

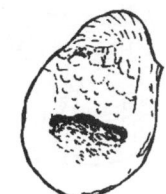

f. *Nerita albicilla* (x 1);

g. *Turbo coronatus* (x 1);

h. *Priotrochus (Monilea) obscura* (x 1)

i. *Siphonaria oculus* (x 1) (e-i after Kilburn and Rippey 1982).

Fig. 6.6 Fauna on lower shore rocks:

a. *Septifer bilocularis* (x 1);

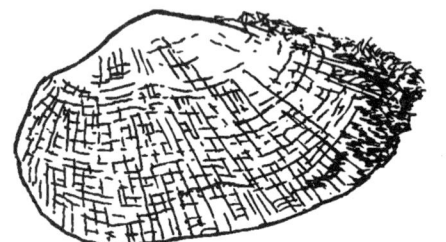

b. *Barbatia foliata* (x 1);

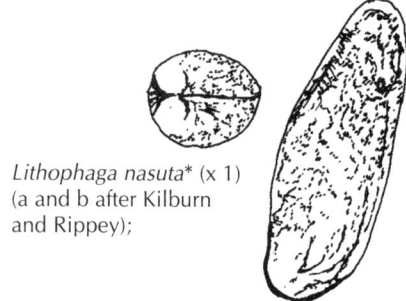

c. *Lithophaga nasuta** (x 1) (a and b after Kilburn and Rippey);

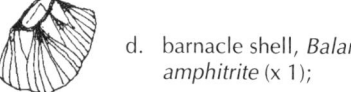

d. barnacle shell, *Balanus amphitrite* (x 1);

e. sea cucumber under rocks, *Holothuria insignis* (contracted) (x 0,7);

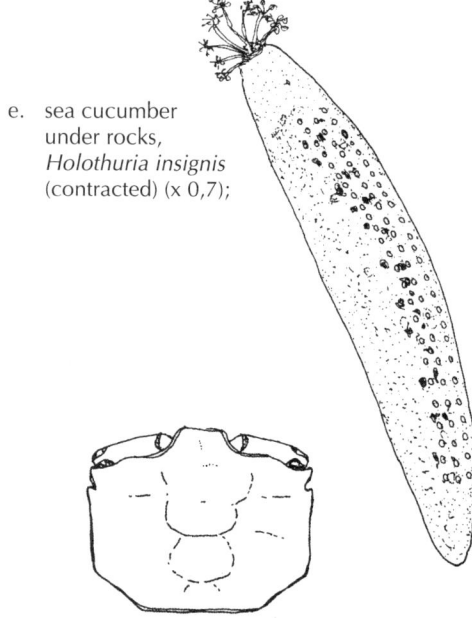

f. ocypodid crab, *Macrophthalmus bosci* (x 2) (after Crosnier 1965).

brown, speckled skin and small peltate (club-shaped) tentacles. It feeds on detritus as do other species of *Holothuria*. This species should not be confused with suspension feeding sea-cucumbers which have arborescent tentacles and occur lower on the shore under rocks (see Section 6.5.1).

A small ocypodid crab, *Macrophthalmus bosci* (10 mm) (Fig. 6.6f) haunts the rocks at this level which is similar to that on sand where its sibling species, the sentinel crab *M. grandidieri* burrows. *M. bosci* does not burrow but keeps moist by sheltering under rocks when the tide is out. It has the familial character of long eye stalks. Its short legs are fringed with long hairs and the carapace has a low pile of microscopic hairs, probably to retain water. Each chela has fine teeth on the finger that is used to scrape detritus from rocks.

6.3 THE SUBTIDAL FRINGE ON THE REEF FLATS

The lowest zone on the sheltered vertical rocks at Ponta Punduine is covered by a wide band of encrusting colonial animals, red sponge above and pink compound tunicates below. These also occur on the vertical faces of the coral debris on the west shore.

6.3.1 Red sponge, compound and solitary tunicates; burrowing polychaete worms

Brilliant red sponge is about 10-20 mm thick (Fig. 6.7a) and is permeated through and through with black streaks. These mark the burrows of a species of the polychaete *Polydora normalis*. In this small worm (10 mm) the burrowing setae on the fifth segment are fairly blunt, smooth and slightly hooked. These setae appear to be adequate for boring into the relatively soft tissue of the sponge. In this environment the worm has the advantage of the supply of food collected by the irrigation of the sponge, described below. Another common commensal is the small green brittle star *Ophiactis savignyi* (see Section 6.5.1), which is a filter feeder.

The sponge itself is a highly organised mass of a very few types of cells and it does not have the systems of organs typical of invertebrates. The surfaces are covered by flat cells which conceal a solid matrix penetrated by innumerable branching canals (Fig. 6.7b). The matrix contains a large number of amoeboid cells with many functions. Some secrete the protein matrix (spongin); some create spicules of silica of specific shapes which are secreted around a protein axis and lie embedded in the matrix, adding to the animal's support. The amoebocytes synthesise the red pigment that gives this sponge its colour and they may also give rise to the reproductive cells. They are probably responsible for the toxic effect of a sponge on animals confined with it in a bucket of sea water. The sponge decays extremely quickly and fouls the water.

The smooth free surface of the sponge is raised in uneven hillocks and valleys which play a large part in the efficiency of the internal irrigation system (Fig 6.7a). The surface is perforated by thousands of microscopic pores (ostia) each of which is a hole right through a single cell. These lead into narrow channels which ramify through the sponge and converge into wider canals ending in the large visible holes (oscula) on the tops of the surface hillocks. At intervals along the narrow channels there are enlarged spherical chambers lined by flagellated collar cells (Fig. 6.7b). Water is drawn into the sponge through the ostia by the beating of the flagella, and is expelled through the oscula. The oscula are closed, when the tide is out, by a ring of cells containing muscle fibrils.

Sponges feed on bacteria filtered from the current of water that flows through the animal. The flagellated cells lining the chambers, called 'collar' cells, not only create the current of water but also filter it (Fig. 6.7c). They engulf the captured bacteria and digest them in the following way. Each collar cell (choanocyte) rests on the internal matrix and from its free surface hundreds of raised microvilli form a 'collar' around the flagellum. The microvilli are 0,2 µm apart and are joined loosely by cross bridges, thus forming a cylindrical sieve. The single flagellum protrudes from the cell body through the centre of the collar. Spiral waves, generated by the cell, travel along the flagellum from the base to the tip creating a gentle movement of water towards and into the collar. The water escapes through the spaces between the microvilli, whilst the bacteria are selected from those particles small enough to have passed through the ostia, which constituted the first selective barrier. The microvilli in the collar ingest the bacteria by enclosing indentations of the cell membrane to which they adhere, that is by phagocytosis. The vacuoles so formed pass down the microvilli to the cell body where lysosomes containing digestive enzymes fuse with them. The digested food diffuses to other cells.

The flow of water through the sponge due to the flagella alone would be slow but it is enhanced by water movements such as waves, currents or tidal flow over the uneven sponge surface. Two physical phenomena enhance the flow, the Bernouilli effect and the Venturi effect (Alexander 1979). The Bernouilli effect (Fig. 6.7d) states 'Fluid flows faster over mounds than in the depressions between them and the pressure is lower when the velocity is higher'. It follows that the pressure between the mounds is higher. For the sponge the effect of this property of the water current outside is that the entry of water through the ostia receives a boost. The exhalent current from each much larger osculum on the top of a hillock is also augmented by the Venturi effect of 'viscous entrainment', by which the fluid flowing fast over the raised surface will join the exhalent current and pull it along with it by reason of the viscosity of water molecules (Fig. 6.7e). Thus water tends to be drawn out of the osculum at a faster rate than would be produced by the flagella alone. The flow of water through the sponge also serves as a respiratory supply of oxygen and removes carbon dioxide and the excretory product, ammonia.

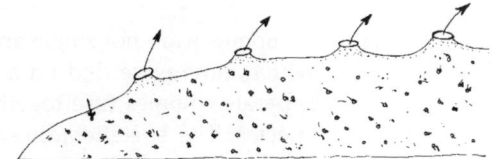

Fig. 6.7 Animals on rocks of reef flats:
a. red sponge, fragment of colony with oscula and ostia;

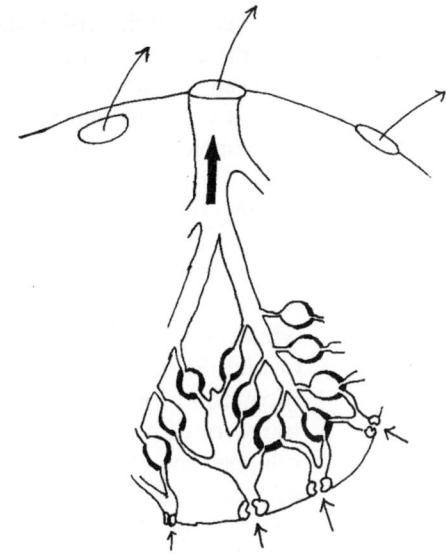

b. section of sponge showing canal system (× 100) (after M. Wells 1967);

c. flagellated collar cell (× 1 000);

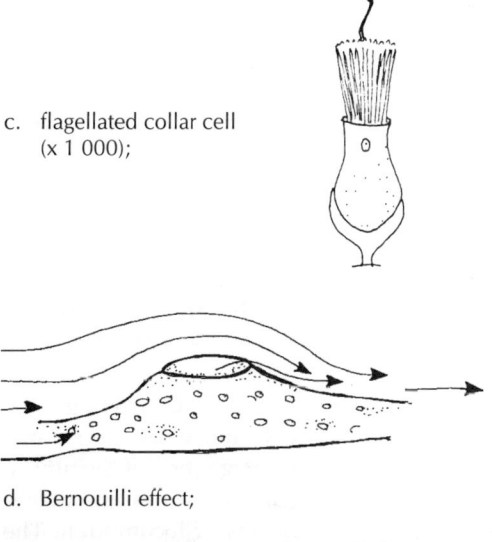

d. Bernouilli effect;

e. Venturi effect; (c-e after Alexander 1979);

Sponges are not single animals but are formed by the fusion of several developing larvae which, having settled on a rock, increase in size by cell division and organised growth. Separate colonies fuse together by virtue of the adherent properties of the cells of the same species. The specific adhesion properties of cells was first demonstrated in the 1920s by experiments on cells of sponges. A sponge was pressed through the fine pores of bolting silk and washed into a dish of sea water. Microscopic examination showed that the cells had been individually separated. Later, cells were separated by immersing the sponge in sea water from which calcium and magnesium ions had been removed, and more recently by ultrasonic vibrations. In all cases the sponge reconstituted itself with the flat cells on the outside, the collar cells on the inside and amoebocytes between the layers. If cells from two species were separated and then mixed, each species reconstituted itself separately. Specific adhesion between cells through surface forces is now recognised as a universal phenomenon for tissues as well as species and depends on the glyco-proteins of the plasma membranes.

Gametes are formed from amoebocytes or archeocytes (undifferentiated cells). Sperm is ejected in the current through the osculum in a strong stream and enters another sponge through the ostia. A sperm entering a choanocyte is transferred to an amoebocyte and carried into the middle layer of tissue to an egg with which the amoebocyte fuses. The sperm then enters the egg and fertilises it. Rapid cell division leads to the formation of a larva with an external layer of flagellated cells and a mass of cells inside. It may be incubated by the parent while it grows. It is called a 'parenchymula' larva because of the central mass of 'parenchyma' or undifferentiated tissue. It increases in size, ruptures the matrix and escapes in the current through an osculum. It swims around briefly and then settles on a rock. The larva is transformed into the adult type of structure by apparently turning inside out. The flagellated cells become internal as the 'parenchyma' cells migrate to the surface layer. Differentiation into canals and flagellated chambers proceeds with growth as the cells multiply. Small colonies just visible on the rocks then fuse to form an adult with highly organised canals and chambers.

The pink, or sometimes white, compound tunicate (ascidian) *Eudistoma rhodopyge* is often attached to the same rock face as the red sponge and takes the form of thick lumpy sheets resembling sponges (Fig. 6.8a). In contrast to sponges however, tunicates are characterised by many *pairs* of minute holes which can be distinguished with a lens even when they are closed. These mark the inhalent and exhalent siphons of the individual animals (zooids) which comprise the lobes of the sheet. The internal organs of the zooids are similar to those of higher invertebrates, although the colonies superficially resemble sponges. The zooid has an alimentary canal, sex organs and a heart. The nervous system is merely a single group of nerve cells in a dorsal ganglion which supplies nerves only to sense organs and muscles of the siphons. Significantly, some organs resemble those of vertebrates, but this is more pronounced in the larva. The internal organs are better seen in the solitary tunicates described below, some of which contain larvae that are visible under low magnification.

The largest solitary species of tunicate on these rocks are *Ascidia incrassata* (60 mm) and *A. arenosa* (30 mm) (Fig. 6.8b). The tunic around each zooid is a tough, white, translucent bag, which, surprisingly, is made of cellulose fibres embedded in protein. The tunic is extended at one end into two finger-like tubes which cover the siphons. The two species are distinguished externally, as their names suggest, by the larger size of *A. incrassata* and the sandier impregnation of the tunic of *A. arenosa*. In the latter the siphon tubes are closer together. These adult tunicates are attached to the rock along the whole of one surface and they lack any means of locomotion. They can, however, open and close the siphons. Before closing the siphons the mantle muscles contract a little and squirt out a stream of water, which has earned them the name of 'sea squirts'.

The animal inside the tunic is covered by a mantle which secretes the tunic and encloses a space (atrium) around the organs. The largest organ is the pharynx, a perforated 'branchial basket' lined with cilia, which leads to the alimentary canal (Fig. 6.8c). The branchial basket cilia create a current of water that passes into the inhalent siphon and through pores to the atrium from which it is removed through the exhalent siphon. Suspended food particles are filtered through the branchial basket, collected in a special groove (endostyle) and passed down the alimentary canal. The 'basket' also absorbs oxygen from solution in the sea into the blood spaces in its strands.

Tunicates have a unique type of heart, which is tubular and enclosed in a rigid wider tube, the pericardium, each of which is only one cell thick (Fig. 6.8d) (Kalk 1970). Each heart cell has muscle fibrils next to the lumen only. The heart has the extraordinary ability to change the direction in which the waves of contraction travel. This reversal occurs every 2-3 minutes, after the heart has come to a stop for a few seconds, resulting in the direction of blood flow being reversed intermittently. The blood travels into the blood spaces in the wall of the branchial basket for oxygenation and then, on reversal of flow, to the viscera where oxygen is given up and dissolved food is collected for distribution. This reversal is all the more remarkable because the heart has no nerve supply but is under the control of one hormone which is secreted into the lumen by small papillae on the heart cells themselves (Fig. 6.8e). The mechanism for the reversal of the heartbeat involves two pacemakers, one at each end of the tube. A supply of hormone is built up in these cells and is secreted in turn into the lumen; each alternately initiates the direction of the muscular contractions. When the supply of hormone at one end is exhausted the heart stops beating until the secretion at the opposite end has accumulated enough to be released. Then the waves of contraction start at that end.

The synthesis of cellulose is another peculiar feature which is linked with the extraordinary capacity of *Ascidia* species (and a few other ascidians) to collect vanadium from the sea where its concentration is only about 2×10^{-5} p.p.m (Kalk 1963a). Vanadate ions are adsorbed by the mucus on the branchial basket, then sequestered in vacuoles in wandering amoebocytes where it is concentrated and reduced (Kalk 1963b). V^{+++} ions are chelated with sulphate and a protein, forming the green pigment haemovanidin, and the cell becomes a green vanadocyte. These cells travel through the blood vessels around the intestine absorbing sugars from which cellulose fibres are synthesised. The precise biochemical role of haemovanidin has not yet been elucidated. It does not carry oxygen as do similar iron-containing compounds such as haemoglobin. It is probably involved in the synthesis of cellulose. The vanadocytes are conveyed to blood spaces in the tunic where they migrate into the soft protein material of the inner layer of the tunic, apparently laying down cellulose fibres as they move along.

One may wonder how newly settled larvae are able to form tunics since there is scarcely sufficient time to acquire enough vanadium *de novo* to synthesise haemovanidin and cellulose. Electron microscopic studies (Kalk 1963c) have shown that provision is made to build up a reserve in the ovary since the eggs of these tunicates are surrounded by 'nurse' cells derived from the parent, containing a very high concentration of haemovanidin from which the developing larvae acquire the ready-made pigment.

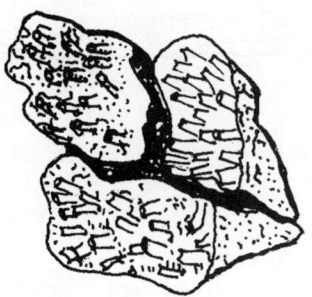

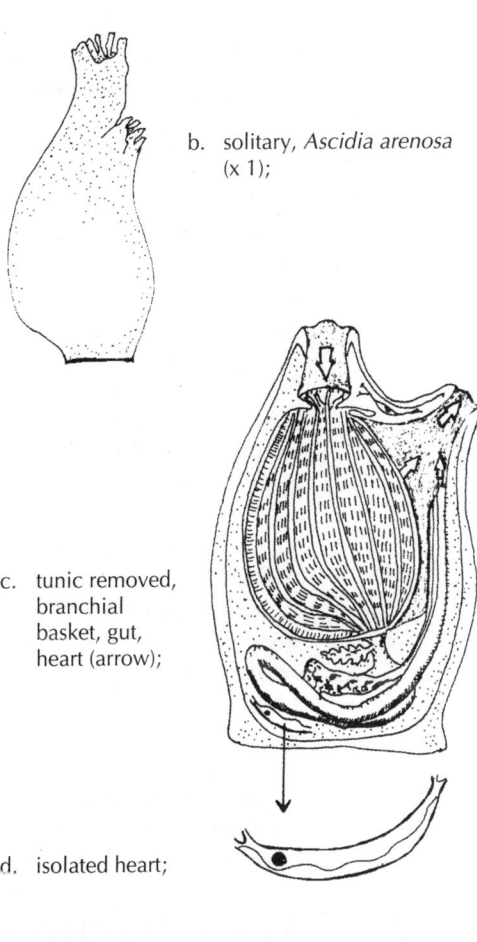

Fig. 6.8 Tunicates on the reef flats:
a. *Eudistoma rhodopyge** (x 2) fragment of colony;

b. solitary, *Ascidia arenosa* (x 1);

c. tunic removed, branchial basket, gut, heart (arrow);

d. isolated heart;

e. ultrastructure: section of heart showing muscle cells (b-e after Kalk 1970);

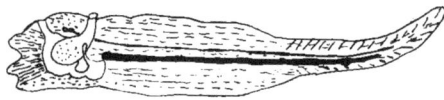

Fig. 6.8

f. tadpole larva (x 10) (after Branch and Branch 1981).

g. burrowing polychaete, *Polydora caeca**, anterior end (x 5)

g'. shoulder seta.

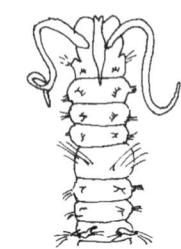

Another solitary tunicate species on the rocks, *Styela marquesana* (2 cm) has a brown leathery tunic and red-tipped siphons. *Clavelina enormis* (1,5 cm) the most beautiful of tunicates, has a royal-blue body that shines through the translucent tunic. This species occurs in clusters of finger-like zooids, joined together by a stolon attached to the rock. The larvae developing inside the zooid are especially large (600 μm) (Fig. 6.8f). They may easily be found by coarse dissection and examined in sea water under low magnification, when distinct chordate features can be seen. The larvae are tadpole-like with muscular tails supported by a notochord. Rhey have sense organs: a statoycyst for balance and an ocellus sensitive to light. Sensations and locomotory activity are co-ordinated by a dorsal nervous system. After a short period of free swimming the larvae settle and metamorphose into adults by resorption of larval features and asymmetrical growth.

Undisturbed old coral rocks, likely to display a variety of established organisms for study, are covered with furry felt buried in fine silt. The 'fur' is composed of hundreds of tiny soft tubes, a few millimetres long, protruding from slightly wider tubes which penetrate vertically into the rock for some 6-10 mm. These tubes are made by the slender polychaete worm *Polydora caeca* (20 mm) (Fig. 6.8g and g'). The two grooved and ciliated tentacles on its head are characteristic of the spionid family, one member of which, *Scolelepis*, is common in muddy sand and has already been described (Section 3.5.2). The ability of this species of *Polydora* to burrow in limestone rocks seems all the more remarkable because, as far as is known, it does so mechanically without the aid of acid or enzyme. Excavation is carried out using only the characteristic polychaete structures, i.e. setae on parapodia. The parapodia are reduced to mere muscular swellings on each segment but the setae on the fifth segment are strong and highly modified (cf. *Polydora normalis* in Section 6.3.1).

The process of burrowing begins when the worm settles in the post-larval stage in the vicinity of already established worms of the same species and secretes its short mucoid tube. It is said to turn upside down to apply the stout 'shoulder' setae on the fifth segment to the task of excavation. It uses these as though on a circular file lubricated by sea water and the filings are continuously swept out with the battery of complex brushes on other segments. The setae on the 'shoulder' are unusually robust with slightly curved tips bearing rows of minute hooks and even smaller hairs inside the curves. The swollen circular area around the anus, the pygidium, is glandular and one wonders whether the secretion might aid in softening the limestone to help in burrowing.

When they become adult, *Polydora* worms lay eggs in tough capsules from which larvae are liberated into the sea only when they have developed to a fairly advanced stage. The time spent swimming freely in the sea before selecting a site for settling is therefore very short.

6.4 THE 'REEF FLATS' AND CORAL REEF ASSOCIATES

The area called the 'reef flats' on the western shore and on the south-western shore beyond Ponta Raza at Inhaca Island resembles the flats adjacent to a *mature* coral reef in origin and position. The rocks are all derived from dead coral and lie shoreward of the coral reef that flourished near Ponta Raza in the 1930s and along the western shore, parallel to the edge of the coral reef, that spread along the banks of the Inhaca Channel in the 1940s. It is richest near the existing coral reef opposite Barreira Vermelha, midway between the Marine Biological Station and the fishing village.

In *mature, tropical* coral reefs the reef flats form a very large intertidal pool with a horizontal floor. When the spring tides are low, water remains standing about 10-30 cm deep and the pool never dries out. Rocks of dead coral litter the floor and may protrude above the surface of the water, most of them supporting a rich growth of macroscopic algae. Small colonies of living coral grow on a small proportion of the coral debris that remains submerged. Towards the seaward edge the living coral increases and the algae become correspondingly reduced.

At Inhaca Island there are several differences from the reef flats of a mature reef. The sea remains between tides only in pools around rocks but they support a similar fauna. Here surface water is less than 10 cm deep between tides. Very few macroscopic algae are present and sparsely distributed on the silt-covered coral debris. They are more common in the area near the central part of the reef, which has been continuously growing for at least fifty years. Living coral does not colonise the dead coral slabs in this area since their surfaces are not continuously submerged. In places where the long-stemmed sea grass *Thalassodendron ciliata* (see Section 9.1) has colonised sandy places between the rocks the pink coral *Pocillopora* is frequently found on very small rocks.

The variety of animal species on these limestone rocks increases towards extreme low spring tidal level. No two rocks shelter the same members of the community and repeated studies often reveal new records of species that have ventured up from the coral reef. The differences in microhabitats on the reef flats depend partly on the contours of the skeletons of the coral species which formed the rock and partly on the length of time that the rock has been available for colonisation on the shore. Some slabs of rock have been so recently deposited that there is not yet a coating of silt on the upper surface and they are discoloured orange and yellow by microscopic growths of algae. The under surfaces will not have attached forms, but mobile animals and those burrowing in the sand beneath may be found. Very old or 'weathered' rocks with flat surfaces are suitable for encrusting growths of compound species whereas rocks derived from branching species such as *Acropora* will offer crevices for small animals to hide from predators. Smaller rocks are unlikely to have remained undisturbed, since the local people have sought supplementary food for centuries, until the declaration of the Nature Reserve in 1965. The ban on collecting food is even now not strictly enforced. Sometimes students or tourists have overturned rocks and not replaced them so that the newly exposed animals attached to them die. The richest rocks are half buried in sand and can be overturned with difficulty since they are permanently lodged in sand. The pools around them yield many mobile species.

6.4.1 Echinoderms: starfish; sea urchin; sea cucumbers; brittle stars

The small green starfish, *Linckia multifora* (disc 10 mm in diameter) has slender, tapering arms extending from a small disc, all lightly speckled with white (Fig. 6.9a). Strangely, this species usually has six arms, and one arm may be longer than the others. This departure from the 5-membered pattern of radial symmetry in echinoderms arises from the habit of autotomy when an arm breaks off at the disc. The arm regenerates the rest of the disc and five new arms, so that the new individual has six arms (Walenkamp 1991). A disc is first formed completely at the top of the regenerating arm, from which small bulges then grow into arms until they are almost equal in length to the original arm. Many stages of regeneration may be found (Fig. 6.9b).

The ambulacral grooves along the under surface of the arms are narrow, the tiny spines are hidden, the tube feet are small and movements are slow. Two species of minute ectoparasitic snails often occur in the groove. *Thyca ectoconcha* (Fig. 6.9c) has a cap-shaped shell with a curled spire and a very small foot. It has neither jaws nor radula but the proboscis pierces the starfish skin which is thin and vulnerable in the groove. It remains permanently fixed to it whilst it feeds on internal tissues. *Stilifer linckiae* (Fig. 6.9d) has a coiled shell, periwinkle-shape but much smaller, and it is embedded below the surface of the starfish tissue. The parasite develops a flask-shaped sac around itself, which opens at the surface by a pore through which water is pumped for respiration. Its proboscis sucks food from the host.

Another not so common starfish on the reef flats that also occurs on the coral reef is *Asterina burtoni* of various shades of brown, which has a thick disc (c. 30 mm in diameter) tapering to thinner margins, and short, thick arms (Fig. 6.9e). The upper surface is covered with very short spinelets and patches of greyish crystals may occur. Sometimes it may have

164 NATURAL HISTORY OF INHACA ISLAND

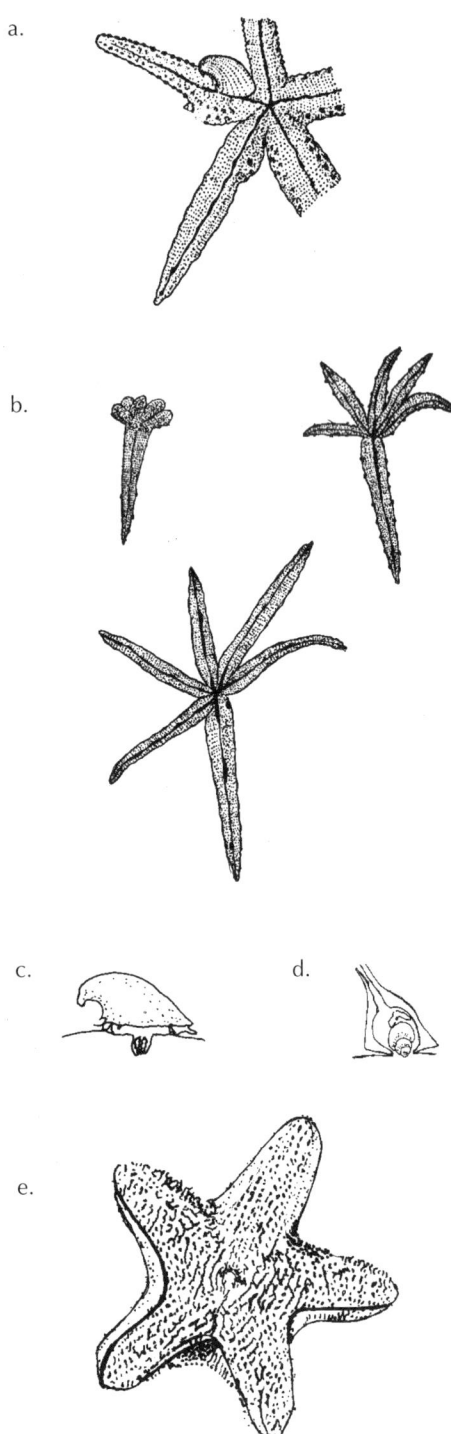

Fig. 6.9 Starfish on rocks of reef flats:
a. *Linckia multifora** and parasites (x 1);
b. stages in regeneration from one arm;

parasites:
c. *Thyca ectoconcha* (x 3);
d. *Stilifer linckiae* (x 3);
e. *Asterina burtoni* (x 1) (after Walenkamp 1990).

six arms. Details of starfish and sea urchin structure are described in Section 9.4.

The common sea urchin here is *Echinometra mathaei*, which has an oval test (80 by 120 mm) bearing stout, solid, slightly tapering spines. The spines and test are all of one colour, which may be grey, purple, brown or reddish (Fig. 6.10a). This species may be found on the upper surface of rocks or sheltering on the sides of a rock as well as underneath it, very firmly wedged into crevices and held by spines and tube feet. It removes the microscopic algae on the rock with its 'Aristotle's lantern' (five large pyramidal teeth slung on the sides of a skeletal pentagon, decribed in Section 9.4.1). A slender commensal alpheid shrimp, *Arete indica*, often perches between the spines and adopts precisely the same shade of colour as its host. It is well camouflaged since it mimics the host's spines in length as well as in colour. It leaves the host to feed but returns to the same sea urchin for protection.

The sea cucumber *Stolus buccalis* (about 150 mm when extended) has a pinkish skin and tube feet irregularly distributed over the whole body surface instead of being concentrated in wide strips or in a 'sole' as in other rock-inhabiting sea cucumbers (6.10b). This appears to be an adaptation for its suspension-feeding habit; firm attachment is made by peculiar tube feet which have a curious concertina-like, conical shape when contracted (Fig. 6.10b). This cucumber has ten long, black finely branching tentacles around the mouth which are extended under water and form an arborescent cage for trapping suspended particles on mucus. Food is conveyed to the mouth by pushing the crown of tentacles right into the mouth without any sorting mechanism. When exposed to air between tides the tentacles are completely retracted and the whole body is somewhat contracted to about 100 mm by means of the longitudinal muscular bands on the internal surface of the skin. The skin is toughened by numerous 'knobbed button' spicules in its inner layers (Fig. 6.10c). There are 5 small white triangular teeth around the anus. This Indo-Pacific species should not be confused with the rarer Natal species *Pseudochnella (Cucumaria) sykion* which is black dorsally and light brown ventrally. Its fairly long, retractile tube feet are concentrated into bundles in five longitudinal rows, two on the dorsal surface and three on the ventral surface. The spicules in the skin are densely packed, round, knobbed plates (Fig. 6.10d). This subtropical species prefers quiet water as does *Stolus* and is also a suspension feeder, but it is capable of more movement and tends to be gregarious.

Ophiactis savignyi (22 mm) (Fig. 6.10e) is a small species of brittle star, said to be the commonest in the world, occurring on all tropical and sub-tropical shores of both Atlantic and Pacific Oceans. It is abundant under coral debris, hiding in crevices and hollows among fine rubble and boring into red sponges. The disc diameter is about 3 mm and the arms are three times as long. Like other members of its family, Amphiuridae, which burrow in sand (see Section 3.4.3) the disc is covered with tiny scales and the jaws have no dental papillae. The colour is green

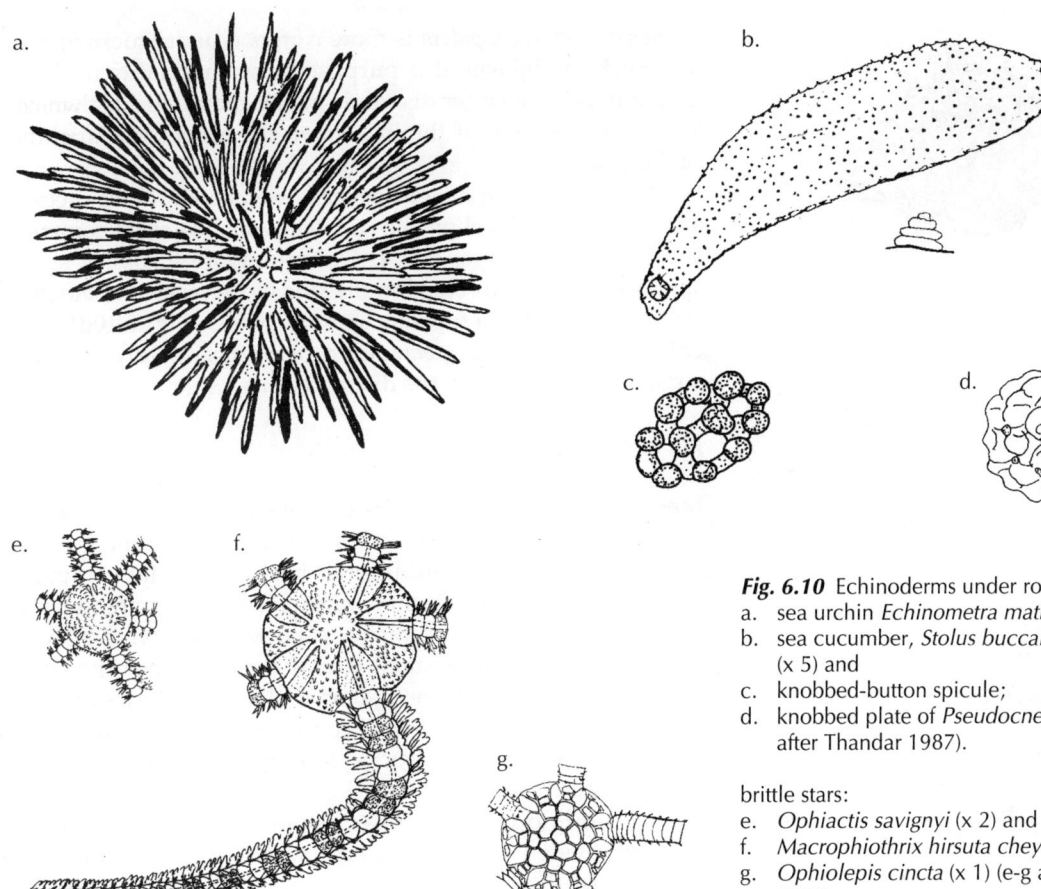

Fig. 6.10 Echinoderms under rocks on reef flats:
a. sea urchin *Echinometra mathaei* (x 0,5);
b. sea cucumber, *Stolus buccalis* (x 1), tube foot (x 5) and
c. knobbed-button spicule;
d. knobbed plate of *Pseudocnella sykion* (b-d after Thandar 1987).

brittle stars:
e. *Ophiactis savignyi* (x 2) and
f. *Macrophiothrix hirsuta cheyeni* juvenile (x 2);
g. *Ophiolepis cincta* (x 1) (e-g after Balinsky 1957).

with white bands or patches. Usually there are six arms, rarely five, with six spines on each segment, but regenerative stages have not yet been found at Inhaca. This species also inhabits the dead bases of living coral where young stages might be found.

Three or more large species of brittle star, having a 150 mm span of the arms, may be found in the pools under various rocks. They drop off the rocks as one lifts them. Most of them are also common in the coral reef and are members of the ophiotrichid family in which the spines, projecting at various angles along the arms, are sharp and conspicuous. The disc is covered by very tiny spines. The mouth has seven oral papillae but no dental papillae. *Ophiothrix foveolata* has distinguishing dark red radial lines on the disc including the bare radial shields (Fig. 8.13a). Each segment of the arm has several longish spines, a dark dorsal spot and a thin transverse line. The general effect of the pattern may be purplish red or bright red in younger individuals and variations in colour from violet to shades of orange may occur in one individual. It is very common in the deeper parts of the coral reef.

Macrophiothrix hirsuta cheyeni is very dark purple in colour except for a narrow, yellow midline along the dorsal surface of the arms, which is more noticeable towards the tips (Fig. 6.10f). The disc including the radial shields are covered with stumpy thorns that are seen only with the aid of a lens, and it seems the most robust of the species encountered here. A commensal scale worm *Polyeunoa (Hololepidella) nigropunctata* (15 mm) occurs on the undersurface of the disc, almost equal in length to the disc. It has 55 segments with 25 pairs of elytra (transparent scales) on the dorsal surface. The worm has a pair of dark stripes on the dorsal surface which matches the brittle star in colour, concealing it unless it moves. Occasionally a molluscan parasite, *Stilifer* sp., has also been found embedded in *M. hirsuta*, the most common brittle star in the reef itself although there are fewer under rocks on shore. The latter are sometimes lighter in colour.

Macrophiothrix aspidota is more common under rocks on the shore than in the reef. It is purplish brown and fairly uniform in colour. It has a larger disc and smaller radial shields than *M. hirsuta* and no sign of the line down the arms present in the sibling species.

Occasionally one may discover a very smooth brittle star at a very low tidal level on the reef flat, which is more common subtidally. *Ophiolepis cincta* (Ophiolepidae) has very short spines smoothly adpressed to the side of the arms and the disc (15-20 mm in diameter) has smooth plates (Fig. 6.10g). The colouration is quite spectacular, being orange-brown with large, white splashes on the disc and arms.

6.4.2 Crustaceans

Most of the numerous species of crabs living under rocks on the lower reef flat and in the coral reef belong to the xanthid family. Typically, a xanthid crab has a fairly flat carapace, which is not vaulted to include space for aerial respiration; it is completely aquatic. The carapace is fan-shaped, having a broad front and bearing four small lateral teeth between the short eye-stalks. The legs are very short with sharp dactyls and the crab is slow moving. This shape seems ideal for living in confined spaces beneath rocks, hiding in small crevices and clinging firmly.

The xanthid crabs of Inhaca Island (±40 species) can be roughly divided into three groups according to size, carapace pattern, feeding habits and position on the shore. Many intertidal species are about 10-35 mm in length; larger crabs over 80 mm in length occur in the subtidal fringe and subtidally in the coral reef, whilst the very small species (c. 10 mm) are associated with living coral.

The carapace of the first group is characteristically divided by grooves into small distinctly granular areas (areoles), although the details are sometimes concealed by hairs. *Pilumnus vespertilio*, the woolly crab is shrouded in long hairs matted with silt, the only bare areas being the spoon-shaped chelae (Fig. 6.11a). When the carapace has been denuded of hairs the typical features are seen (Fig 6.11a'). Two other *Pilumnus* species have fewer hairs. *Xantho* spp. have a slightly wider carapace with more pronounced areoles and the chelae are pointed (Fig. 6.11b). In *Actaea* spp. granulation is exaggerated on the carapace and the chelae are spooned. *Etisus* spp. have a more oval, strongly dentate front, with shaggy hairs on the legs and spoon-shaped chelae (Fig. 6.11c). *Epixanthus frontalis* (40 mm) is the only xanthid crab that migrates with the incoming tide to high tide level where at Ponta Punduine it rests between tides under flat stones (see Section 6.1.1). The brown carapace is smoother and the chelae are pointed (Fig. 6.11d). All the species of xanthids are listed in Appendix A. Their close similarities may be baffling to the general biologist in the field, but the crab species are distinguishable by the shape of the male pleopods (under the bent abdomen) which are used in copulation. Some of these are illustrated in Fig. 6.11e. This group of small xanthid

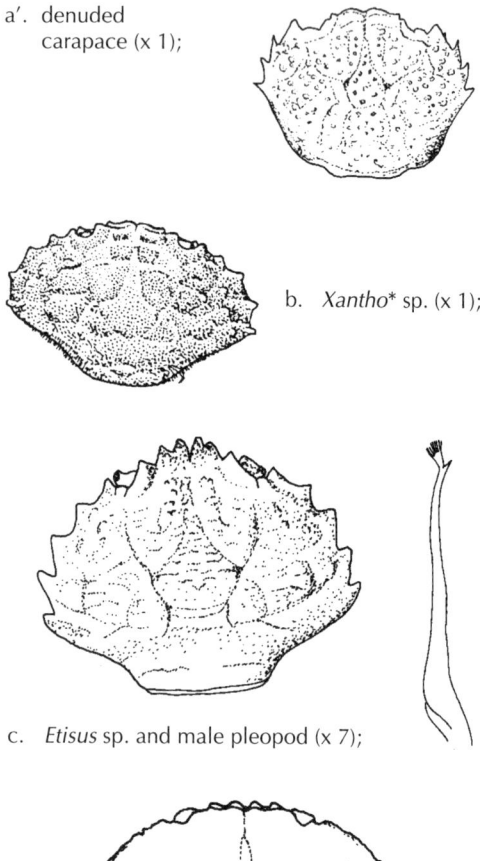
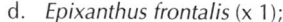

Fig. 6.11 Xanthid crabs under rocks on reef flats:
a. woolly crab, *Pilumnus vespitilio** and

a'. denuded carapace (x 1);

b. *Xantho** sp. (x 1);

c. *Etisus* sp. and male pleopod (x 7);

d. *Epixanthus frontalis* (x 1);

e. male pleopods of xanthid crabs.

crabs feeds by scraping the rocks under which they hide when the tide is out. It is suspected that the diet of those with spoon-shaped chelae differs from that of crabs with pointed chelae. The stomach contents might be investigated.

Curiously, some individuals of *Xantho voeltzkowii* have round external swellings between the abdomen and the ventral carapace (Fig. 6.11f). These crabs have been infected by the crustacean parasite *Sacculina* which is a cirripede, classed with barnacles. The life-history of *Sacculina* commences with the hatching of a nauplius larva from the fertilised egg. This grows into a cypris larva, thus resembling barnacles (see Section 6.1.1). The cypris larva has no cement glands, but the first antennae are modified into sharp, hooked blades each resembling a tin-opener. A female cypris larva uses these to pierce a hole in the chitin of the abdomen of the host, *Xantho*, when she makes contact with a hair and holds fast with her appendages. The internal organs of the parasite degenerate into a mass of simple amoeboid cells which migrate slowly into the crab's body and multiply rapidly, spreading branches between the tissues on which it feeds. After a time the parasite becomes mature and one of the branches pierces the soft skin between segments of the abdomen and grows into the hard sac which is seen from the outside. The cells in the sac differentiate into an ovary and a brood chamber with a very small pore. At this stage a male cypris larva is attracted to the female parasite and attaches itself at the pore. Its cells dedifferentiate and a small mass of cells enters the brood chamber. These cells then differentiate into testes so that the female *Sacculina* becomes hermaphrodite. Sperm fertilise the eggs that then develop into larvae, which are released from the brood chamber and complete the life-cycle by infecting xanthid crabs. The parasite

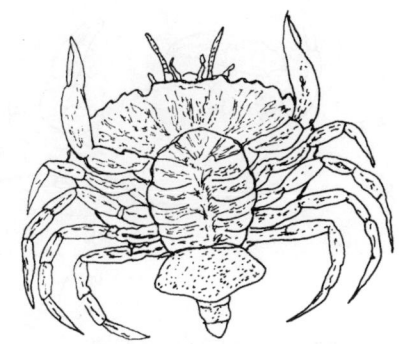

Fig. 6.11
f. *Xantho* sp. (x 2) parasitised by cirripede, *Sacculina* sp.;

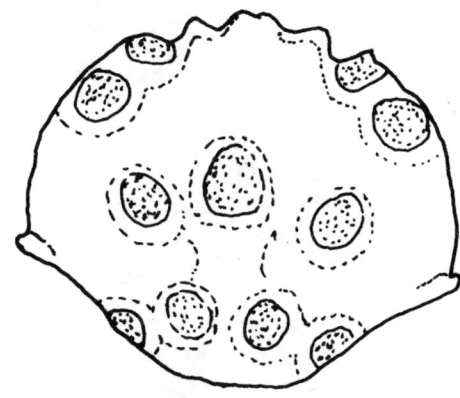

g. *Carpilius maculatus* juvenile (x 0,5) (a', c-g after Barnard 1950).

not only feeds on the crab's tissues, but has a pronounced effect on the crab's metabolism. It interferes with the normal hormones which control growth and sex in the host so that the crab does not moult and becomes sterile. The narrow abdomen of an infected male crab becomes wide like that of the female. The host crab remains alive, despite the drain of nourishing the parasite and the hormonal changes.

The second group, the large xanthid crabs, is characterised by highly coloured, smooth-polished carapaces and more robust chelae. For example, *Carpilius maculatus* (85 x 115 mm) (Fig. 6.11g) is orange coloured with eleven or more large blood-red spots on the dorsal surface. The chelae are massive and have a molar-like tooth on the thumb, suggesting it feeds on molluscs. Another species, *Atergatis roseus* (61 x 102 mm), is rose-red with a whitish border and has black chelae with white tips; it is also carnivorous.

The third group of very small, smooth xanthid crabs inhabits the branches of living coral only. Five species occur on the reef at Inhaca Island and are described in Chapter 8 (see Section 8.5.2).

The other carnivorous crabs found on the reef flats are several species of the swimming crab *Thalamita* which hunt small xanthid crabs in this area and take refuge under stones when the tide recedes instead of burrowing in sand as do the other genera of swimming crabs (see Section 3.5.2). *T. crenata* (47 x 72 mm) (Fig. 6.12a) is most common but several other species occur which differ mainly in the pattern of the lateral teeth (see Appendix A). These crabs have roughly hexagonal carapaces and four lateral teeth behind the orbital tooth.

The tiny jet black spider crab, *Elamena mathaei*, is common here, and has a triangular carapace with tiny eyes peeping on each side of the narrow rostrum (Fig. 6.12b). Long slender legs enable it to escape rapidly into hiding when a rock is overturned.

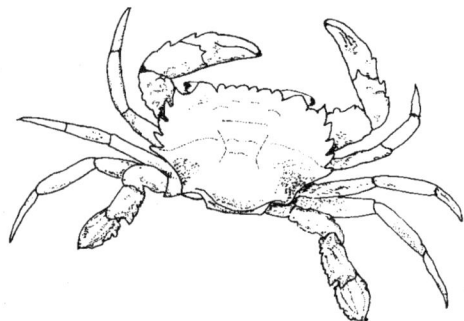

Fig. 6.12 More crabs on the reef flats:
a. swimming crab, *Thalamita crenata* (x 0,5) (after Crosnier 1962) note 5th swimming leg and stout chelae;

b. spider crab, *Elamena mathaei** (x 1);

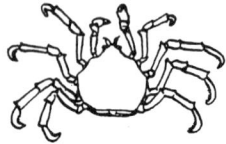

porcellanids:
c. *Petrolisthes lamarcki** (x 1,5)

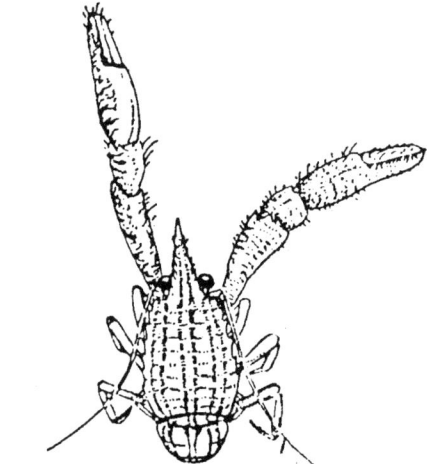

d. *Galathea picta** (x 3).

The very strange crab *Petrolisthes lamarcki* (Fig. 6.12c), which may be seen at Ponta Punduine displaced to high tidal level under rocks (Section 6.2) usually lives under rocks on the reef flats. It is a porcellanid crab related to the hermit crabs; i.e. it is an 'anomuran' with an 'odd tail' and not a true brachyuran crab with 'short tail'. The abdomen has a tail-fan with uropods like a shrimp or prawn and it is loosely bent under the thorax. It appears to have four pairs of legs instead of five pairs, since the fifth is very small and upturned. Its flattened chelipeds are twice as large as the body and are held stretched out on the rock in front of the animal. The crab movements are fairly slow and clumsy, handicapped by the large chelae, and one cannot imagine them seeking food. In fact porcellanids do not search for food but remain stationary for suspension feeding. The family is sometimes referred to as 'porcelain' crabs and indeed they are very fragile and lose limbs easily. The name 'porcellanid' actually refers to the resemblance to the 'porcine' shape of the carapace, resembling 'a little pig'. The characteristic long whip-like flagella of the second antennae are waved alternately to augment the respiratory current, as in some hermit crabs. In so doing the chemo-sensitive first antennae may detect edible suspended matter. These movements then cease and are followed by alternate sweeping movements of the third maxillipeds, which are fringed with setae. Particles are collected on the setae of all the joints, which are spread fan-like, as they move towards the mouth, gradually closing to retain food whilst the current of water escapes. The brush at the tip of each second maxilliped is thrust against the setae and removes the food, which is then pushed into the mouth without further sorting. The feeding can be watched if the crab is placed under water in which there is some detritus scraped from rocks. *Galathea picta* (Fig. 6.12d), a tropical crab allied to the porcellanids, but not a suspension feeder, is also found in the coral reef and has large chelipeds. A pair has been seen feeding on the large starfish *Linckia multifora*, taking two days to dig out the soft tissue inside the tough skin (Fig 8.14a).

There are a few shrimps under rocks. The only alpheid shrimp here is *Alpheus edwardsii* (25 mm) which has the enlarged snapping chela of the alpheid family (see Section 3.5.3). (Many other alpheids occur in the coral reef.) This species is reddish brown in colour and its large chela differs from that of sand burrowers in having thin margins to the palm. It is also more robust in body. A new species of alpheid shrimp, *Athanopsis rubricinctuta* has recently been discovered in the sand on the reef flat (Berggren 1991). This species is commensal with an echiurid worm and shares its burrow. This confirms the suspicion (Section 5.4.3) that echiurid worms do burrow in the sand flats although they had so far eluded capture. The alpheid shrimp is about 40 mm long and has an opaque whitish body with red and yellow transverse stripes on the carapace and abdomen. The echiuroid belongs to the subfamily Ochaetostomatinae like the echiurid at the Ponta Torres coral reef. The animals were extracted with a small 'Yabbi' suction pump (used in Australia for extracting freshwater crayfish). Using a similar method, a new species of thalassinid shrimp, *Naushonia lactoalbida* was found in the gravel between coral debris just below lower water of spring tides. It is similar in general structure to *Upogebia* in the mangrove swamp (Fig. 4.12e) but is more robust and has one large hairy and spiny third maxilliped and is white in colour (Berggren 1990).

A large shrimp, *Saron marmoratus* (60 mm) with a brownish variegated colour (Fig. 6.12e) occurs among these rocks as well as on the east shore. It has a cryptic colour pattern. The zebra shrimp *Gnathophyllum americanum* (20 mm) has a broad body and horizontal black stripes. The name *Gnathophyllum* describes its peculiar leaf-like jaws, in this case the third maxillipeds.

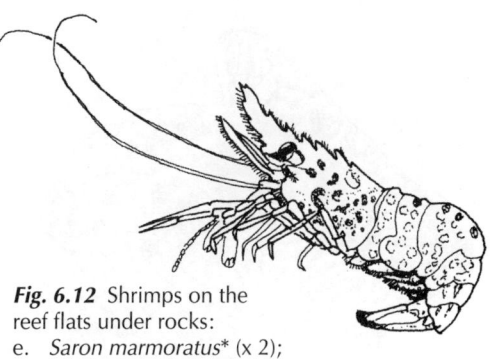

Fig. 6.12 Shrimps on the reef flats under rocks:
e. *Saron marmoratus** (x 2);

The dark green mantid shrimp or stomatopod, *Gonadactylus glabrous* is the most voracious carnivore of the reef flat although it is only about 7 cm long (6.12f). It is at first recognised by the speed at which it disappears into the shade under the disturbed rock. This escape behaviour was analysed experimentally at Inhaca Island (Bolwig 1954). It swims towards dark objects preferentially even if it has to pass through a well-lit field. This type of behaviour is known as 'skototaxis' or negative photo-taxis. The mantid-shrimp seeks shelter among the rocks and comes to rest when many points of contact can be made by the legs and body, both ventrally and laterally. This is known as 'thigmokinesis'. Thus light and touch are two stimuli which trap the mantis-shrimp under rocks during the day. When the light intensity falls, it swims between the rocks and is most active in darkness.

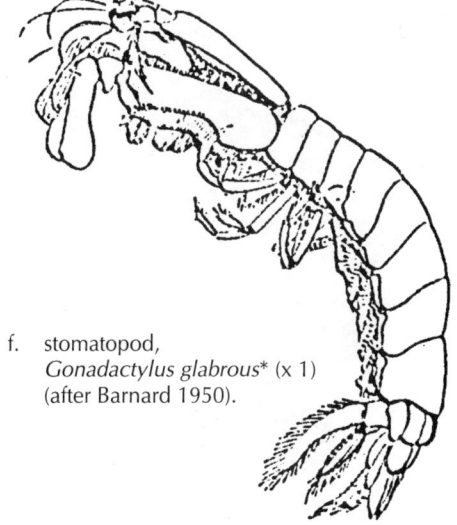

f. stomatopod, *Gonadactylus glabrous** (x 1) (after Barnard 1950).

The jack-knife chelipeds of this mantid shrimp lack the barbed teeth of its counterparts that burrow in sand and spear their prey on the move (see Section 5.3.2). It may appear to be less harmful, but this is misleading. The hunting technique of this rock mantid-shrimp is that of stalking, pouncing on and smashing almost-immobile prey. It hammers a crab or a mollusc with the 'heel' of its knife-shaped chela. The first strike stuns the crab, the second breaks open the carapace or shell. Subsequent hammering breaks off the legs and chelae. The prey is dragged to the special lair, a cavity in the rock, and the soft tissues are devoured. The skeletal remains are removed and placed outside the retreat. Shells are similarly destroyed and the mollusc or hermit crab within is eaten.

Gonadactylus exhibits aggressive behaviour towards species of sand mantid shrimp and drives them away. Serious fighting for territory may also occur between individuals of the same species. This is preceded by an aggressive display of a conspicuous coloured spot on the merus, the fourth segment, of the cheliped. This may cause one mantid shrimp to retire, defeated. When they fight no harm is done since attempts made to strike are parried by the heavily armoured telson which is used as a shield and is presented by a rapid flexure of the body musculature. A return blow is received similarly on the telson of the other without doing any damage. Then one appears to be victorious and the other retreats. If one holds a mantid shrimp in the hand a spiking stroke from the pointed finger of the chela is provoked, which can inflict a deep cut.

6.4.4 Molluscs: bivalves attached to rocks; grazing and carnivorous snails

Eight genera of bivalve molluscs representing as many families live on the rocks of the lower reef flat (Kilburn and Rippey 1982). These differ in details of the shape of the shell and its hinge, which reflect the degree of attachment to rock surfaces. They are all filter feeders and have the typical bivalve gills, palps and mantle but the development of the tubular siphons for the inhalent and exhalent currents varies with the positions of different genera on the rocks.

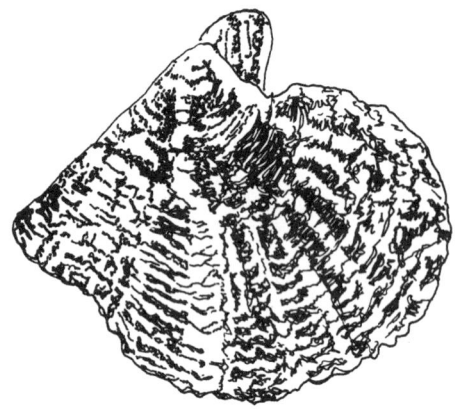

Fig. 6.13 Bivalves on the reef flats under rocks:
a. *Pinctada nigra* (x 1);

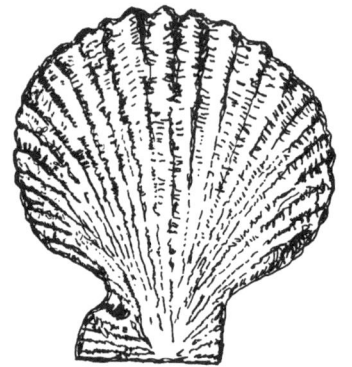

b. *Chlamys senatoria** (x 1);

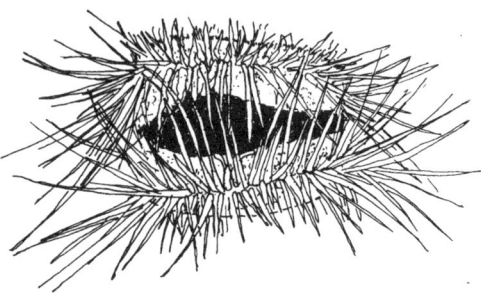

c. *Lima lima* (x 1);

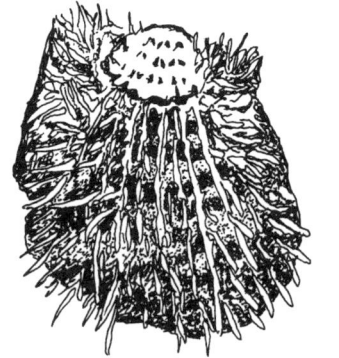

d. *Spondylus nicobaricus* (x 1);

Some shells are permanently attached, others nestle in crevices, some bore into rock whilst others, when adult, surprisingly swim away to escape predators.

'Pearl oysters' such as *Pinctada nigra* (40 mm) are attached to rock by means of short byssus threads which emerge near the hinge through a deep notch in the right shell, on which the animal rests. The foot is very small and the mantle gapes widely when feeding. The thin shell is black with cream blotches and the hinge is toothless (Fig. 6.13a). Characteristically, the shells have an iridescent nacreous lining of 'mother of pearl'. This is the genus which makes precious pearls by covering minute foreign bodies with similar layers of nacre, secreted by the mantle. In some tropical regions where they occur in large numbers they were exploited for jewelry by divers but now they are cultivated for the production of pearls. There are only a few individuals at Inhaca Island located here and there on the reef flats and subtidal fringe.

'Scallop' shells have a narrow toothless hinge with extensions on each side called 'ears' and the interior is not nacreous. They are attached to rock by byssus threads while young but as adults they lie free and can jump and swim clumsily to avoid predators. The motive force in swimming is the contraction of the 'quick' muscle fibres in the single adductor muscle. The mantle edge is fringed by little tentacles sensitive to chemicals. At their bases are highly coloured simple 'eyes' which detect shadows since they are sensitive only to light and dark. *Chlamys senatoria* has one 'ear' longer than the other and a reddish, ribbed and slightly prickly shell (Fig. 6.13b).

The most spectacular of the bivalves is the file shell *Lima lima* which jumps into the pool beneath a rock when it is lifted, revealing a bright orange mantle with a fringe of long mobile tentacles (Fig. 6.13c). It swims away in jerks by clapping its valves firmly together and expelling jets of water on each side of the hinge so that it moves forward. The long orange tentacles at the edge of the mantle contract and bend inwards to augment the movements of the shell and mantle. These tentacles secrete an acrid, sticky mucus in defence. When attacked by a predator the tentacles are easily broken off and they wriggle away, distracting the predator while the prey animal escapes. The shell is white and ribbed, and has two small asymmetrical 'ears'. Its lower half is slightly thorny. The strong ligament is interior, but it is somewhat larger than that of *Chlamys*.

The thorny oyster, *Spondylus (hystrix) nicobaricus* (40 mm) resembles scallops in shape but is completely sedentary, with its right shell firmly cemented to the rock surface. The upper shell is elaborately sculptured with serrated spines and sharp hooks which daunt carnivorous molluscs (Fig. 6.13d). It closes its shell securely with a ball-and-socket kind of hinge. It too can 'see' predators by means of a row of small simple eyes on each mantle edge. When closed, the shell resembles the rock on which it lies but when it gapes for feeding the orange mantle is conspicuous. The 'thorny oyster' occurs singly on rocks and not in clusters as do true oysters.

The 'false cockle' *Cardita variegata* is more oval than a true cockle and is slightly asymmetrical with the hinge near one end instead of being central. The thick shell is similar as it is coarsely ribbed radially and is covered with tiny scaly outgrowths (Fig. 6.13e). It is distinctive in colour, having a buff background with reddish brown blotches. It does not burrow as a cockle does, and when young it is attached to rock by byssus threads. Older ones nestle in the crevices in rocks and on sand beneath them. *Cardita* is exceptional among bivalves in that it has yolky eggs and broods its young until they hatch complete with shells.

Another less common 'nestler' is *Eastonia solanderi* (20 mm) which does not belong to a family of its own as all the others here do, but is a modified burrowing clam (Mactridae) although it does not burrow. It nestles on the surface, its shell slightly gaping at both ends. The shell is also radially ribbed, one valve forming a keel at one end (Fig. 6.13f). The prevalence of radial ribbing amongst the rock bivalves is said to be an adaptation for water run-off.

Two genera of bivalves burrow into the rock in different ways. *Petricola divergens*, the smaller one (Fig. 6.13g), uses a rocking movement produced by the anterior and posterior pedal muscles, the foot between them acting as a fulcrum. It files the rock slowly (and for short times) by means of its radial ridges, more especially those on its anterior surfaces. The older ridges become worn and only the current year's growth is markedly ridged. The burrow is fairly shallow and is made when the ribs press against the rock with the edges gaping, being pulled open by the elastic ligament. The force is all the stronger for its being applied external to the hinge, which is secured by peg-like teeth inside the chalky white shell. There are short siphons which reach the rock surface for feeding.

The other burrowing genus, *Gastrochaena* spp. are the only bivalves to have concentric ridges on their round shells (Fig. 6.13h). Its burrow, deeper than that of *Petricola*, is worn perfectly smooth by the ridges on the shells. Two siphons about 30-40 mm long extend to the surface of the rock and their openings are lined by a smooth ivory-like calcareous secretion. *G. cuneiformes* (20 mm) is the common species.

There are few species and individuals of herbivorous snails on the rocks of the reef flat, possibly because of competition for microscopic algae with large and successful populations of xanthid crabs which have strong chelae. A flattened brown 'topshell', *Stomatella sulcifera* (15 mm) has a very large aperture and spiral ridges on the shell (Fig. 6.14a). The 'horn shell' *Cerithium dialeucum* (30 mm) has a speckled brownish shell with complex spiral granular ridges and 'suture' lines across them; its siphonal canal is unusually bent abruptly sideways (Fig. 6.14b).

Omnivorous and carnivorous snails are also few in number of individuals, but several species are represented (as one might expect since they are selective in the choice of prey). A small whelk *Engina mendicaria* (15 mm), which has a smooth black shell marked by a yellow line around the middle, is a scavenger. Its shell tapers towards both the spire and the base (Fig. 6.14c). A snail with a shell of similar shape and size, *Peristernia forskali*, buff-coloured and much roughened by ribs and spiral grooves, has a purple tinge (Fig. 6.14d). There are 2-3 weak plates on the columella next to the aperture through which a red foot is protruded. It is related to the tulip shell *Fasciolaria trapezium* (Fig. 9.16g) found among seagrasses, but it feeds on tubicolous polychaetes by inserting its long proboscis into the tubes. *Mitra litterata* (30 mm), a mitre shell with 4-5 pleats on the columella, is whitish with irregular, darker blotches and very fine spiral lines (Fig. 6.14e). It feeds on the small sipunculid worms which hide in rock crevices, which it swallows whole.

Fig. 6.13
e. *Cardita variegata* (x 1);

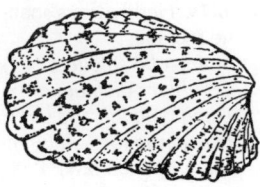

f. *Eastonia solanderi* (x 1);

g. *Petricola divergens* (x 1);

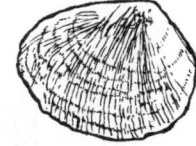

h. *Gastrochaena* sp.

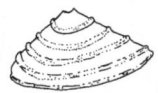

Fig. 6.14 Herbivorous snails on the reef flats:
a. *Stomatella sulcifera* (x 1);
b. *Cerithium dialeucum**(x 1);
c. *Engina mendicaria* (x 1);
d. *Peristernia forskali* (x 1);
e. *Mitra litterata* (x 1);

Carnivorous snails:
f. *Conus textile* (x 0,5);
g. *Cypraea erosa* (x 1) (a, c-g after Kilburn and Rippey 1982);
h. *Cymatium pileare* (x 1).

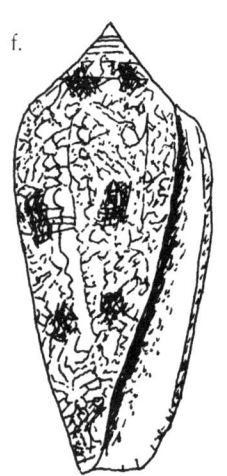
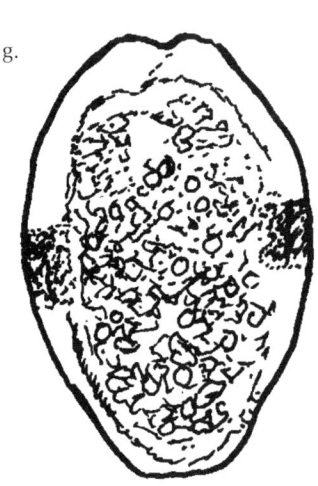

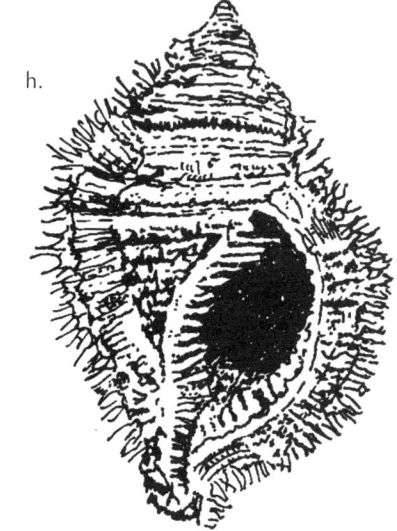

A few species of the poisonous cone shells occur under rocks, notably *C. textile* which feeds on molluscs (Fig. 6.14f). This species should be handled with special care since the toxin from the radula teeth is poisonous to man and has proved fatal in some cases. The whitish shell (40 mm) has a fairly high spire (for a cone) and is covered with an irregular network of fine red lines and scattered reddish-brown blotches. Another cone shell, *C. miliaris* feeds on polychaetes. The shell (30 mm) is greyish with tiny red dots and dashes arranged spirally and has a marked shoulder crown.

Cowries are always seen, but apart from *Cypraea annulus* and *C. moneta*, which also live among sea grasses (see Section 3.5.3), no species is very common. *C. erosa* (35 mm) is the easiest to recognise by the pair of large brown rectangular blotches spreading from the centre of the sides to the white area below (Fig. 6.14g). *C. vitellus* is said to feed on the compound pink tunicate, *Eudistoma rhodopyge*. Shells of adults faintly retain the light- and dark-brown banding which is conspicuous in young shells and it also has small irregular white spots dorsally. *C. lamarcki* (30 mm) is brownish dorsally with pretty pale blue spots surrounded by white rings. *C. isabella* is smaller (25 mm), almost cylindrical in shape with orange-red tips at each end of the drab polished shell. Most cowries are unspecialised browsers but they should be observed carefully to find out whether they have food preferences. Sometimes a cowrie is found sitting on white eggs in a flat mass about equal in size to the area of the foot. The eggs are enclosed in soft capsules attached to rock by short stalks. The cowrie incubates the eggs under its large foot for 2-3 weeks until they hatch (Fig. 3.19b).

The largest carnivorous molluscs are the 'triton' shells said to resemble in shape the trumpets of minor legendary Greek gods with fishes' tails. *Cymatium pileare*, the hairy triton, has a papery periostracum with many loose shreds. The shell is fairly narrow (70 x 35 mm) with a prominent, smooth and wide, longitudinal ridge along one side, a long tapering siphonal canal and a red interior (Fig. 6.14h). It probably feeds on starfishes or sea urchins by dissolving a hole in the test with sulphuric acid in its saliva. There are several tropical species of *Cymatium* (see Appendix A) in the shells of which live large hermit crabs such as *Pagurus euopsis*.

6.4.4 Many kinds of worms: flatworms; sipunculids; suspension-feeding polychaetes

The groups of worms associated with rocks are Turbellaria (flatworms), Nemertea (bootlace worms), Sipunculida (peanut worms) and Polychaeta (segmented worms).

When a boulder of coral rubble is overturned as the low tide is rising to cover the rocks, leaf-thin flatworms emerge from crevices and swim away slowly to escape the light, with undulating, muscular motion of the whole body or they crawl into crevices slowly, propelled by the action of cilia on the lower surface. They are sometimes brilliantly coloured purple, yellow and green but may be dull grey or brown, each species with a distinct pattern of lines or dots. They are called Turbellaria because of the turbulence set up in the anterior olfactory ciliated pits just behind the simple eyes on the head. Flatworms are carnivores or scavengers, detecting their food by the chemo-sensory cells in these pits. Food is engulfed by a mobile, protrusible pharynx shaped like a bell, situated on the ventral surface. Flatworms are structurally simple worms without a body cavity, the space between the skin and the branched gut being packed solid with cells, including muscles. A network of fine nerves branching from the cerebral ganglia co-ordinates their movements.

Thirty-six species of flatworms have recently been described from Inhaca Island (see Appendix A), most of which are known from coral reefs around the Indian Ocean and others that were first found at Inhaca Island (Prudhoe 1989). The most common genus is *Pseudoceros* of which 25 species live on the coral reef or reef flats (see Fig. 2.2a and b, Section 2.4). It belongs to the Order Cotylea in which the ventral sucker is posterior to the genital pore. The mouth is in the anterior third of the body and the pharynx has a ruffled border capable of much expansion when swallowing animals larger than itself. Clusters of eyes occur on the tentacles on the anterior edge of the body and above the cerebral ganglia. The species of *Pseudoceros* are usually fairly broad and between 20 mm and 100 mm in length. *P. mossambicus* is black dorsally with many greenish spots of many shapes and sizes on the dorsal surface and a marginal orange band, whilst the under surface is green (Fig. 6.15a). *Pseudoceros inhacensis* is greyish with a tinge of yellow, a narrow black marginal band and a deep brown median band, tapering at the ends. In the order Acotylea in which there is no ventral sucker, a new genus has been described: *Gabiella inhacensis*, which is only 12 mm long and 3 mm wide, light brown in colour with a darker median band. It was named in honour of Dr Vivian Gabie whose 1958-1963 collection of flatworms and vivid paintings were used in taxonomic studies (Prudhoe, 1989).

Nemertines are represented by at least one slender, flattened, string-like species. The common name for this type of nemertine is 'bootlace' worm. *Baseodiscus hemprichi* (200 mm) is pink in colour with a mid-dorsal, red pin-stripe (Fig. 6.15b). Its proboscis is shot out from the sheath in its head to catch small motile prey (amphipods and isopods). Its activity is restrained compared with that of the giant sand nemertine *Cerebratulus* (see Section 3.4.3).

Sipunculids, the non-segmented, cylindrical worms without appendages, are sometimes known as peanut worms. This epithet fits the rock-inhabiting species well since they contract into fat cylinders on exposure to air between tides, with crumpled skin like the shells of peanuts. The worms have an internal structure and habit of feeding on detritus similar to those described for the much larger, active *Siphonosoma cumanensis* which burrows in sand (see Section 3.3.2). The latter certainly does not resemble a peanut shell. The common species of sipunculid on the rocks of the reef flats is *Physcosoma scolops* (Fig. 6.15c), which is wedged into crevices and cannot be extracted without breaking the rock. When the tide returns the worm relaxes and stretches out its proboscis to over twice its length for feeding. This sipunculid probably does not bore into rock as do some coral reef species. There are usually 2-3 individuals of *Physcosoma* sp. on each rock on the reef flats.

Fig. 6.15 Worms on the reef flats:
a. flatworm, *Pseudoceros mossambicus* (x 1) (after Prudhoe 1989);
b. Nemertine, *Baseodiscus hemprichi* (x 1);
c. sipunculid, *Physcosoma scolops* (x 1);

Carnivorous polychaetes on the reef flats:
d. polynoid *Ipnione muricata** (x 1) covered by elytra;
d'. three pairs of elytra removed showing head and parapodia*;
e. fireworm, *Eurythoe complanata*, anterior end (x 1);
e'. a parapodium with poisonous bristles;
f. iridescent worm, *Leocrates claparedi* (x 1) (e and f after Day 1974).

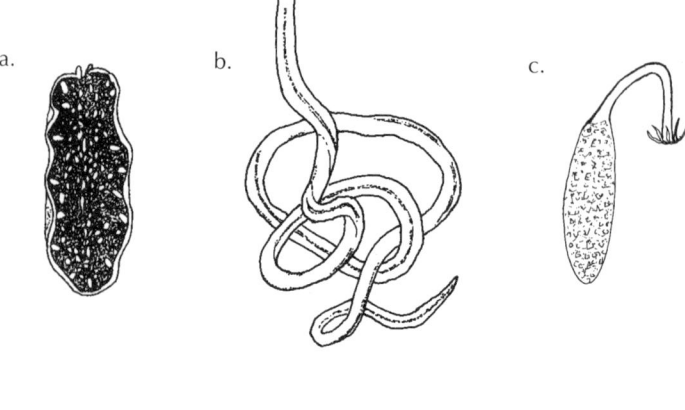
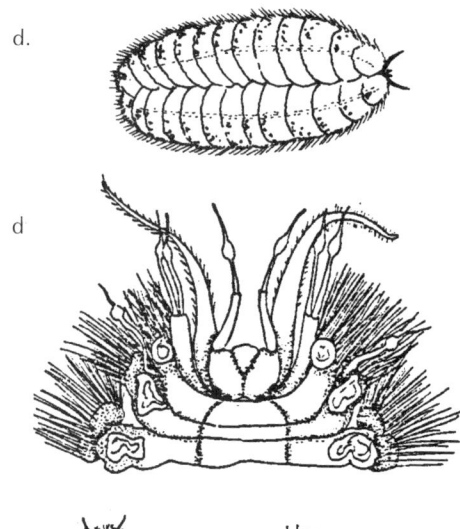
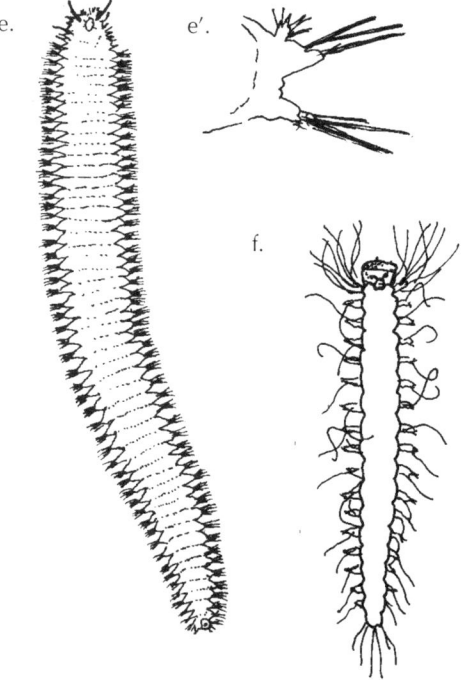

There are many species of polychaetes on rocky shores. Some are small and inconspicuous, often creeping among algae, where there are larger numbers than on the reef flats. Among them are robust and active predators, burrowers and suspension feeders, which differ in species from those in sand.

On the reef flats about eight families of polychaete worms are represented by several species each (see Appendix A). Many have been described from type species found at Inhaca Island (Day 1967).

One family, the polynoid worms, known as scale worms, have relatively large pairs of overlapping chitinous scales (elytra) growing on every other segment of the body on the dorsal surface. They are active predators on smaller animals which they seize with four jaws on the proboscis as they creep over the rocks with slightly sinuous motion, using their parapodia. *Iphione muricata* (20 mm) is a common rock scale worm with 13 pairs of elytra on a broad, oval, somewhat flattened body (Fig. 6.15d). Its parapodia are concealed by the elytra (Fig 6.15d'). Each parapodium has a single whip-like gill and is equipped with a tuft of fine setae and another of strong bristles with microscopic serrations providing a good grip on the rock as the worm creeps. Elytra of different species of polynoids have different patterns, for example, *Iphione* has large tough plates decorated with a polygonal pattern, bearing two rows of thorns and fringed with setae posteriorly. The fringe is said to prevent the silting up of the gills beneath the elytra.

One representative of the amphinomid family of worms, the bristly fireworm, is common under rocks. This is *Eurythoe complanata* (80 mm); it is injurious to man and should not be handled (Fig. 6.15e and e'). Its whitish, calcareous bristles break off easily and pierce the skin, releasing a toxin which causes a painful irritation. The pain persists even after the bristles have been removed. This worm feeds on soft sedentary animals, such as tunicates, which are sucked dry. They have neither jaws nor teeth. Every parapodium has a bunch of short filamentous gills and tufts of setae (Fig. 6.15e'). The setae are of many kinds and those that penetrate the skin of possible prey, the 'harpoon' setae, have recurved serrations down one side which ensure that they cannot easily be removed from the victim.

The hesionid worms are short and robust with beautiful iridescent colours. *Leocrates claparedi* (50 mm) at first glance resembles a hairy centipede since it has only 16 segments from which relatively long parapodia protrude, each bearing a long, whip-like jointed dorsal cirrus (Fig. 6.15f). It is much more active than the other two worm types described above and is carnivorous. Its head bears eight jointed, whip-like tentacles with sense cells. The strong protrusible proboscis has one central dorsal, and one central, ventral tooth.

Several species of long thin worms, the nereids and syllids, live under rocks but they are more common where the rocks are coated with seaweeds and are discussed in Section 7.5.3 which describes invertebrates on rocks exposed to wave action.

Although they are not common, the robust eunicid worms that are typical of the dead coral bases of the coral reef may be mentioned here because they are unusually large, and may be found at very low tides. The worms have the powerful jaws of the eunicid family especially well developed (see Fig. 3.10d). The 'mandibles' are gouge-like and partly calcified and the 'maxillae' are strongly toothed. *Lysidice collaris* (70 mm) burrows right through the limestone of the coral debris and is said to feed on it. This worm has three short tentacles, two eyes on its bilobed head, and strong setous parapodia that do not project much beyond the body segments. Gills are absent.

The tube-dwelling families of suspension-feeding worms on rocks are the sabellid and serpulid polychaetes that both protrude feathery crowns from the openings of their tubes (Fig. 6.16a). These are used for both suspension feeding and respiration. A sabellid worm constructs a fairly wide, straight, horny tube which is coated with sand or mud and lodged in a rock crevice at its lower, closed end. A serpulid worm secretes calcium carbonate to make a much smaller, narrow tube which in different genera is more or less coiled. It is wholly attached to rock or weed and may be closed at both ends, the anterior opening being temporarily guarded by an operculum when exposed to air (Fig. 6.16c and d). Sabellids do not have an operculum and are confined to the lowest levels of the shore and most occur subtidally. Three distinct regions of the body are recognisable in both families: the head with crown and collar (a fold of skin at the base of the crown), a short thorax and a longer, tapering abdomen, both bearing setae of extraordinarily diverse shapes.

One of the common sabellids, the beautiful *Sabellastarte longa* (150 x 15 mm), has a brownish body shading to purple and a branchial crown, banded orange and brown (Fig. 6.16a). The branchial crown is formed by the radial and lateral branching of a pair of spirally twisted tentacles. Each axis bears 4-6 whorls of radioles, each of which is pinnately branched and coated with a regular pattern of cilia. The plumes are held open rigidly in the form of a fluted funnel by means of tiny internal skeletal rods holding the pinnules at 90° to the axis of the radiole. The crown can be withdrawn instantly at the threat of a shadow, detected by sensitive ocelli placed at intervals along the radioles. A current of water is drawn into the funnel from *below* by the beating of small cilia lying along the under-surfaces of the pinnules (Fig. 6.16b). Large lateral cilia force the current *upwards* between the pinnules and particles fall from the eddy on to their upper surfaces where they are trapped in mucus. Small cilia on these upper surfaces beat towards the bases of the pinnules and convey the particles towards the central groove of the radioles (Fig. 6.16b). Each central groove (Fig. 6.16b') is divided

Fig. 6.16 Plankton-feeding polychaetes:
a. sabellid in tube, *Sabellastarte longa**;

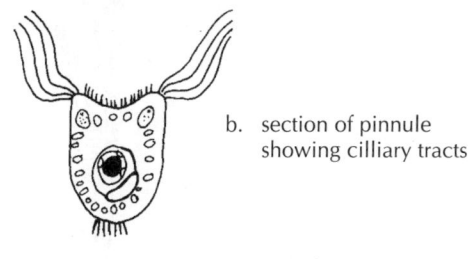

b. section of pinnule showing cilliary tracts

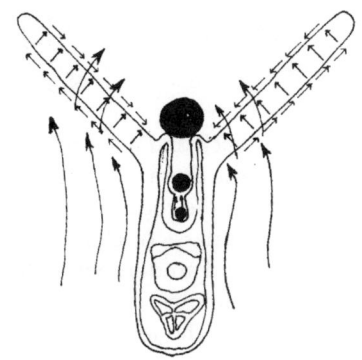

b'. T.S. tentacle midrib showing directions of currents for collecting and sorting food and sand (after Newell 1979);

Fig. 6.16 Plankton-feeding polychaetes:

c. chitinous thorns on operculum of *Hydroides* sp. (x 3);

d. tube of *Serpula vermicularis*

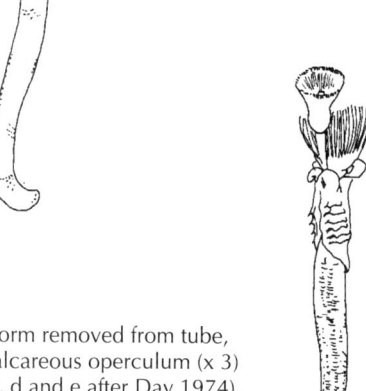

e. worm removed from tube, calcareous operculum (x 3) (c, d and e after Day 1974).

vertically into three compartments: the narrowest at the bottom where food particles of 1-2 μm congregate; a middle region where larger particles suitable for tube building are collected; the largest diameter at the top of the groove carries the larger particles destined for rejection. Cilia in the three compartments of the groove beat towards the collar. On the collar three corresponding tracts of cilia convey fine particles to the mouth, medium particles to an area of the collar concerned with tube secretion and plastering and the third leads to the small palps on the top of the head, which rejects the largest particles. Such an exquisite functional design is perhaps unparalleled among invertebrates. The parapodia are concerned with holding the worm firmly in its tube. Eight thoracic parapodia have flat setae on the dorsal lobe and a plate of hooked setae (uncini) on the lower lobe. On the abdomen this arrangement is reversed, possibly to prevent twisting the body.

The little serpulid worms (cf Fig. 6.16c and d) have one radiole modified into a stalk which may develop a thorny calcareous or chitinous operculum on the end. This exactly fits the opening of the tube when the animal is withdrawn. The calcareous tube in which the worm lives lies flat on a rock and is securely cemented to it. The mechanism of filter feeding is similar to that in sabellids, and the crown is often highly coloured. The collar is partly glandular and secretes the calcareous cement that forms the tube. The manner in which the worms settle on rock is reminiscent of that of barnacle larvae (see Section 6.1.1). After a short time as a free-swimming larva, it settles on a rock and crawls around. Then it lies on its back and a special larval attachment gland on the mid-dorsal surface of the posterior part of the thorax secretes a milky fluid in which the larva rolls while the fluid hardens and adheres to the rock, forming the initial part of the tube. It prefers rocks with a film of bacteria on the surface, or chooses the fronds of specific seaweeds or sea grasses. The larva lies on its back and undergoes a dramatic metamorphosis into its adult form within the space of 5-10 minutes.

The form of the tubes and opercula of the serpulid worms are highly specific, as is the substratum chosen. The most common on coral rock are several species of *Hydroides* in which the operculum bears chitinous thorns (Fig. 6.16c), and *Spirobranchus* in which the thorns are calcareous. The tubes in these two genera are straight and ridged. *Serpula* is represented by the cosmopolitan *S. vermicularis* (70 mm) in which the tube is coiled at its closed end and its operculum is chitinous and almost flat (Fig. 6.16d and e).

6.4.5 Sparse seaweeds on the coral debris: 'blue-green', green, brown and red algae

Most rocks on the reef flat are coated with a slimy growth of blue-green algae which is more pronounced in summer. *Chrococcus turgidus*, *Aphanotheca stagnina* and *Merismopedia glauca* (Fig. 6.17a-c) are common.

Very few species of macro-algae occur on sheltered rocks. The crevices of some rocks, still retaining the shapes of the coral skeleton of which they are made, may be occupied by the small, lumpy, glassy cushions of *Valonia macrophysa* and the more compact *Dictyosphaeria versluysii* (Fig. 6.17d and e). Some sand-covered, smooth rocks have a conspicuous growth of the fluted, fan-shaped thalli of the green alga *Udotea orientalis* (Fig. 3.16a), which are much larger than those in deep sandy pools on the exposed north-east rocks. Similarly, the brown alga *Padina boryana* (Fig. 3.16c), which has a fan-shaped, greenish thallus, occurs on flat sandy rocks and is larger than individuals on the exposed rocky shore.

Very few bushy tufts of the red alga *Acanthophora muscoides* (Fig. 7.17f) with stiff, tiny branches, are attached to rocks, but many more are seen cast ashore in the drift. In addition the pale, slimy tufts of the bushy *Liagora ceranoides* (Fig. 7.17g) the delicate *Champia compressa* and the very small *Gelidiopsis rigida* represent the red algae (cf Section 7.5.1).

This association of algae is impoverished, since the rich growth of small red algae, the long fronds of *Sargassum* spp. and the jointed cactus-like green alga *Halimeda*, which are typical of more tropical reefs, have not been found. These algae flourish on the north-east rocks at Inhaca Island where wave action is brisk and silt does not accumulate (see Sections 7.4 and 5).

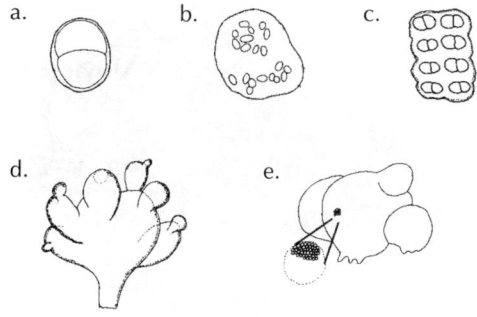

Fig. 6.17 Algae on coral debris on reef flats:
Blue-greens
a. *Chrococcus turgidus* (x 1 000);
b. *Aphanotheca stagnina* (x 1 000);
c. *Merismopedia glauca* (x 1 000) (a-c after S.F.M. Silva 1991);
Green algae:
d. *Valonia macrophysa* (x 1);
e. *Dictyosphaeria versluysii* (x 1) (after Seagrief 1980);

6.4.6 Small stray fishes from the coral reef: stonefish; firefish; coral reef fish

Small fishes are quite frequently seen in pools under large slabs of coral debris on the reef flat; some are resident and others come in with the tide from the coral reef and are stranded temporarily in the pools when the tide recedes. Some are juveniles using this sheltered habitat as a nursery area.

Everyone who walks along the muddy shore of the reef flat should be warned to wear shoes, because the most poisonous fish in the coral reef, the stonefish, *Synanceia verrucosa* (150 mm) (Fig. 6.18a) may lie half-buried in muddy sand between rocks, exposing the poisonous, grooved spines of its dorsal fin. This fish is almost completely sedentary and certainly so when the tide is out; it is exceptionally well camouflaged, being much the same colour as a small, encrusted rock, greenish brown with an uneven, lumpy surface. The skin is toughened with collagen fibres but the scales are not visible. Resemblance to a small rock or stone is enhanced by the covering of detritus, algae and even very small hydroids, which it is able to shed intermittently when, quite unusually for a fish, it moults the skin. The stonefish belongs to the family Scorpaenidae, noted for their venom, which is present in the skin as well as in the spines. The venom may have an advantage for the fish of being an antibiotic that destroys bacteria which might infect the bare skin. The venom is an effective protection from predators that would find these slow-moving fishes easy prey. When touched, even accidentally, the stonefish exposes its dorsal fin spines by pressing back the skin and discharging venom into their grooves from pairs of glands. A stream of venom is injected under pressure into a would-be predator such as a squid or a shark. The venom contains a neurotoxin which may be *lethal to man* within a few hours! The poison in the afflicted limb may be denatured by immersion in very hot water, and this gives some relief from the extraordinarily intense pain. Clearly, it must be re-emphasised that footwear and gloves should always be worn when studying the rocks on the reef flats at low tidal level, as well as on the coral reef itself, although an antidote has now become available.

The juveniles of a related scorpionfish, the 'devil's firefish' *Pterois miles* (Fig. 6.18b), one of the many beautiful coral reef fishes, may by chance be found trapped in a rock pool. It has a vivid red body with intersecting dark, oblique lines, dark spots on the fins and tentacles on the head. The pectoral fins are exceptionally large and supported by strong spines so that they can fan out at rest in an apparently threatening pose. The devil's firefish has venom in the spines of all its fins that would sting predatory fishes which might attempt to capture it despite the warning colouration. It is carnivorous and, unusually for a scorpion fish, it hunts actively for small crabs and shoal fish over a limited territory from which intruders may be warned off by its colour and sting.

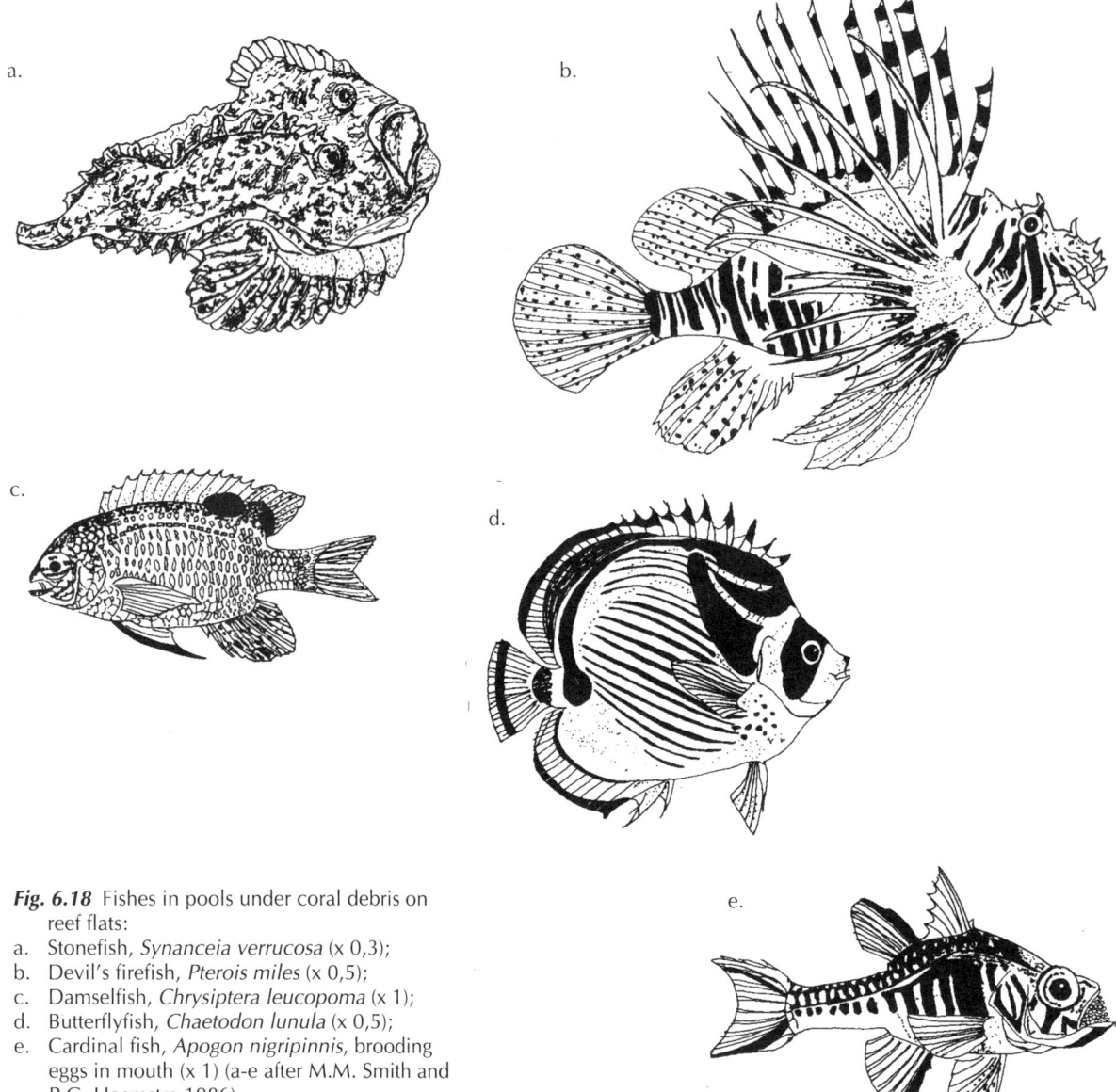

Fig. 6.18 Fishes in pools under coral debris on reef flats:
a. Stonefish, *Synanceia verrucosa* (x 0,3);
b. Devil's firefish, *Pterois miles* (x 0,5);
c. Damselfish, *Chrysiptera leucopoma* (x 1);
d. Butterflyfish, *Chaetodon lunula* (x 0,5);
e. Cardinal fish, *Apogon nigripinnis*, brooding eggs in mouth (x 1) (a-e after M.M. Smith and P.G. Heemstra 1986).

This fish should *not be handled*, since the venom injected by the spines is very painful to man for some time, though seldom dangerous.

Small fishes that normally rest at night under rocks in the coral reef may be disturbed on the reef flat when a large rock at very low tidal level is overturned and the pool is exposed. These may be damselfishes (Pomacentridae) (Fig. 6.18c), butterfly fishes (Chaetodontidae) (Fig. 6.18d), or striped cardinal fishes (Apogonidae) (Fig. 6.18e). These are described in Section 8.5.6 on fishes of the coral reef.

These small fishes with different feeding habits and activity times give a slight foretaste of some of the kinds of fishes which are present in the coral reef, except for the large carnivores and coral grazers. They do not by any means indicate the variety or the large numbers of individuals one may study while snorkelling over the coral reef itself when it is covered by the tide.

7

LIFE ON ROCKS EXPOSED TO STRONG WAVE ACTION

7.1 THE ENVIRONMENTAL EFFECTS OF STRONG WAVE ACTION

Strong buffeting by waves from the ocean has several effects other than the sheer force of impact. Spray generated by breaking waves reaches higher up the shore than the tide on a sheltered shore. The effects of high spring tides are thereby extended by several metres and the time of dryness between tides is reduced. Strong ocean winds bring cooler and more humid air, reducing evaporation on the shore to some degree. Ocean water carried in by every tide is also cooler than the sea in sheltered bays and more quickly reduces the temperature of the environment after hours of exposure to the sun. The churning of the waves stirs up sand on the shore, removes and carries away finer grains, leaving coarse sand to abrade the rocks and the organisms on them. The agitation of the seething waters enriches their oxygen content and prevents silt and detritus from settling, so that algae can flourish. In the tropics most seaweeds (except green filamentous species sheltered by a sand deposit when dry) are confined to the lower shore or to pools on the middle shore, owing to their intolerance of heat and high levels of radiation from the sun. Seaweeds are broken up by wave action and reduced by bacterial decay to very small particles of detritus. Bacteria and phytoplankton in the aerated waters are plentiful.

7.1.1 Topography of the exposed rocks at the north-east point

A complete range of rocks from above high tide to low tide is exposed only in the shallow cove between Cabo Inhaca and Ponta Mazondue. Attention is concentrated mainly on this area in this chapter. In this small bay below the lighthouse the coastal dune is 80 m high; the sloping sandstone base of the dune is visible to a height of 30 m above the middle shore and it levels out to form a high platform at 5 m, covered by shifting sand, which is splashed by spray from wave action at high tides. A bare rock cliff descends to mid-tidal level. Neap tides reach about halfway up the rock face where a shallow pool is formed on a ledge under an overhang. From the base of the cliff an almost horizontal platform extends seawards for about 40 m, forming the middle shore. In some places a vertical drop of about 1 m leads to the lower shore whilst in others there is a gentle slope of broken rocks. The lower platform slopes very gently to extreme low spring tide level for about 40 m or so. The three platforms are wave-cut terraces exposed at intervals during the retreat of the sea in the late Pleistocene period (see Section 1.3.1). The profile of the shore is shown in Fig. 7.1, in which the dominant organisms in each tidal zone are indicated.

Wave action on these exposed shores is classified as being 'moderate'; there is a coral reef 5 km offshore which reduces its force. The surfaces of the rocks on the three levels of the shore differ considerably. The rock cliff of the upper shore is deeply pitted and has very small, sharp

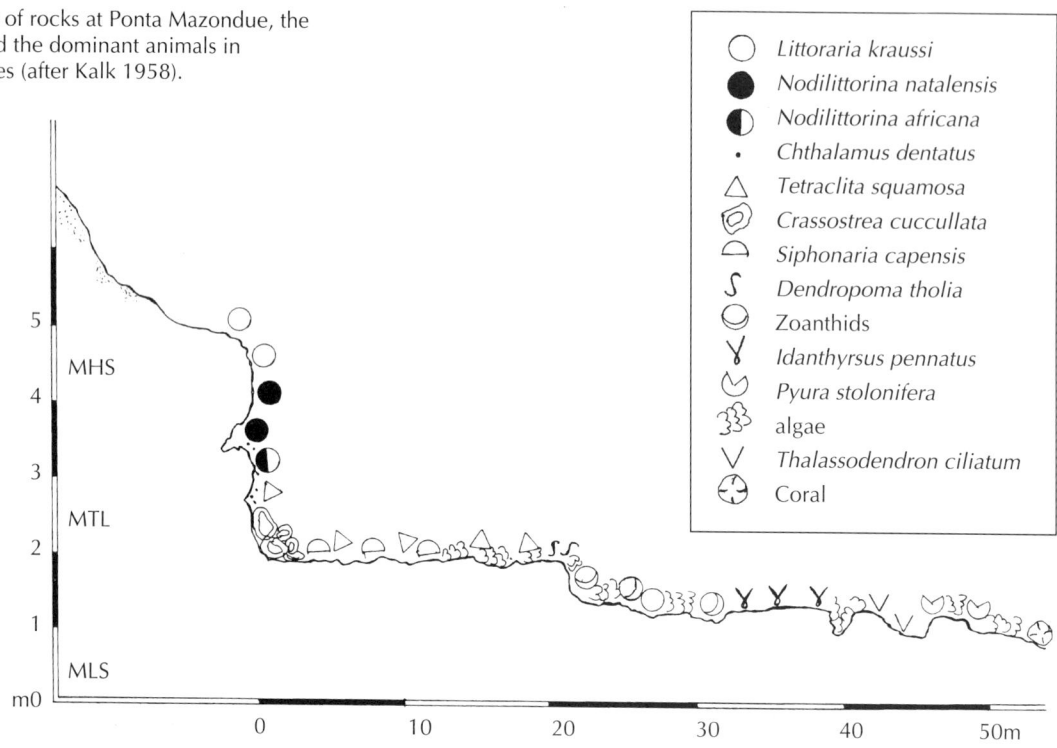

Fig. 7.1 Profile of rocks at Ponta Mazondue, the north point, and the dominant animals in successive zones (after Kalk 1958).

pinnacles; the middle platform is smooth and has large and small shallow pools bounded by raised rims; the lower platform is almost completely covered by seaweeds and is eroded into deep pools, wide gulleys and small caves.

7.2 THE UPPER SHORE ON THE CLIFF

Biological zones on the cliff are clearly demarcated. The darkened rock of the supratidal fringe ends just below the level of the ledge about halfway down the face. Below this is the upper barnacle zone where the rock appears greyish brown. The white oyster belt lines the foot of the cliff and may continue for about 50 cm on to the initial slope of the middle platform.

7.2.1 The supratidal fringe: microflora; periwinkes; nerites

The rock surface of the upper part of the cliff is darkened by encrustations of microscopic Cyanophyta, commonly called 'blue-green algae'. The 'blue-green' cells occur in colonies embedded in mucilaginous sheaths that stick to the rock and even penetrate its pores. They lose water between tides and readily absorb moisture again from spray and humid air. Their quick recovery from desiccation is an adaptation to this high tidal level because the time available for photosynthesis is prolonged while they are wet. The pigments phycocyanin and phyco-erythrin mask the chlorophyll so that photosynthesis is possible even in intense sunlight as in red algae, which have a similar complement of pigments. The products of assimilation by blue-green algae are glycogen and glyco-protein which may constitute a food source for grazing molluscs. Cyanophyta reproduce rapidly by fragmentation or by forming spores and so are not decimated by grazing.

In association with blue-green algae there are assemblages of several species of epilithic diatoms (attached to rock surfaces), as well as unicellular green algae such as *Chlorella* spp. which penetrate rock pores even at this high tidal level. It is not yet known what proportion of diatoms, blue-green or green algae are digested by the different molluscs that graze on them.

Some littorinid species on temperate shores are known to select diatoms as food preferentially.

Three species of periwinkle (littorinids) and three species of *Nerita* snails feed on the microscopic flora at night or when the rocks are wet or damp. The majority of individuals of each species of periwinkle is found at a different tidal level from the others as shown for one day of spring tide in Fig. 7.2 (Kalk 1958). Individuals of the smooth, pink *Littoraria (Littorina) kraussi* (Fig. 6.2a) are all attached above the level of Mean High Spring Tide. The grey periwinkle, *Nodilittorina natalensis* (Fig. 6.2b) is about ten times more numerous than the pink species, spreading over a belt about 1,5 m wide both above and below the highest tide mark. Very few shells of the little, smooth, blue periwinkle, *Nodilittorina (Littorina) africana africana* (Fig. 7.3a) occur in the supratidal fringe but they extend downwards to the level of 1 m above mean tidal level into the barnacle zone. They are well known from the shores of Natal and the Cape in South Africa, but are not seen in the warmer sheltered bay of Inhaca Island.

The distribution of periwinkles depends largely on their relative endurance of exposure to sun, evaporation and strong wave action. The species with the smallest store of water, *N. africana* must be wetted by every tide in order to replenish its water supply and, in addition, to remove its excretion of ammonia. Every day it migrates down the rock to meet the high tide of the day and then back again, unlike the other two species that can withstand drying for several days since they excrete solid uric acid on access to water (see Section 6.2.1). The relative resting positions of the three species on the cliff correspond with their latitudinal distribution. *L. kraussi*, at the highest level is more successful on most tropical shores whereas *N. africana*, at the lowest level, extends furthest south.

The *Nerita* species occur on the cliff face, scattered and almost concealed in crevices. On more tropical exposed rocks they may equal the littorinids in number and even predominate on tropical shores in Somaliland (Chelazzi and Vannini 1980). They are better equipped to withstand excessive heat and water loss than are periwinkles since their shells are thicker and larger, almost globose in shape, and they store relatively more water behind the operculum. The shells of two species are also deeply ridged and thus contribute to cooling by reflecting heat from the enlarged surface area. The aperture of the shell is overlapped by a thick rim and an enlarged ventral lip so that water is lost slowly. The operculum is

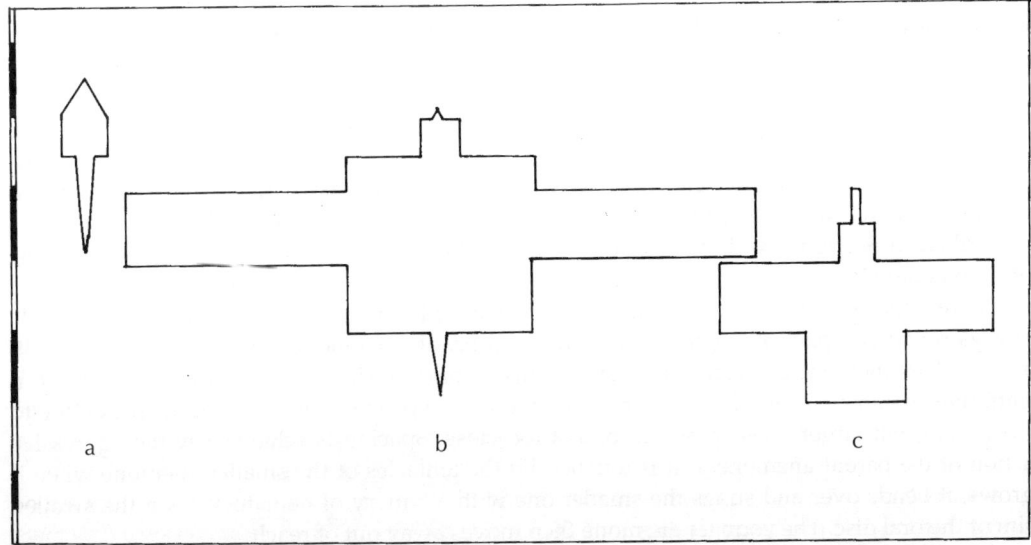

Fig. 7.2 Histogram: distribution of 3 species of periwinkle at spring tide on rocks at Ponta Mazondue (after Kalk 1958):
a. *Littoraria (Littorina) kraussi*;
b. *Nodilittorina natalensis* (maximum density 1000/m^2);
c. *N. africana*.

Fig. 7.3 Snails on the cliff at high tides:

a. *Nodilittorina africana* (x 2);

b. *Nerita plicata** (x 1);

c. *N. textilis* (x 1);

d. *N. polita* (x 1) (a, c and d after Kilburn and Rippey 1982).

calcareous but fairly thin and closely fitted against the toothed aperture. The snail withdraws further into the shell when water is lost but, as in periwinkles, the living tissues do not become desiccated.

Nerita plicata (20 mm) has a white shell which is deeply sculptured by spiral grooves and ridges (Fig. 7.3b). *N. textilis* (35 mm) (Fig. 7.3c), which occurs slightly lower on the cliff than *N. plicata*, is similar in shape and has rows of black spots along the grooves. *N. polita* (35 mm) has a smooth shell, rounder in shape with dark markings (Fig. 7.3d). It feeds still lower on the rock face at about mid-tidal level above the sand into which it descends and hides by day, returning to the rock to graze at night.

Two littorinid and two neritid species migrate down the rock at night in stages to meet the high tide as the levels change from springs to neaps; they retreat up the rock as tides change from neaps to springs (see Section 6.2.1). Species find their way by means of a variety of cues to which they are sensitive.

Both littorinids and neritids have adaptations to enable them to survive on land, viz. supplementary air-breathing, solid uric acid excretion, control of temperature by 'transpiration' and the ability to tolerate heat. All of them are, however, tied to the sea for the completion of their life-histories since they possess free-swimming larvae. In the neritid family some species have adapted to fresh water, but there are none in the littorinid family.

7.2.2 Barnacle and oyster zones: tropical barnacles; oysters; tropical mussels; air-breathing Siphonaria; anemone

Below the zone of the periwinkles two species of barnacles cover the rock face of the cliff. They feed on phytoplankton. The small grey *Chthalamus dentatus* (Fig. 6.3a) which also occurs on the sheltered rocks (see Section 6.3) forms a dense mass of shells interrupted here and there by scattered individuals of the large pink barnacle which displaces it from below. *Tetraclita squamosa rufotincta* (Fig. 7.4a) is typical of tropical East African shores exposed to strong wave action (Kalk 1959). The method of feeding is similar to that of *Chthalamus* (described in Section 6.3). Is it able to select a different range of particle size amongst the phytoplankton?

In the middle of this zone a ledge about 50 cm wide, lying under the overhanging rock, always contains a pool of water in which live red beadlet anemones, *Actinia equina* (Fig. 7.4b and b'). This anemone is viviparous, retaining the embryos inside the body cavity (coelenteron) until they are about 2 mm in diameter. When they are expelled, they settle on the rock close to the parent, but larger ones appear to be not too close. Spacing is achieved by the aggressive action of the parent anemone. If it is touched by the tentacles of the smaller anemone when it grows, it bends over and stings the smaller one with a battery of nematocysts on the swollen rim of the oral disc. The younger anemone then moves away out of reach.

The oyster seen on sheltered shores, *Saccostrea cuccullata* (Fig. 6.5a) (see Section 6.3.2) appears among the barnacles about 50-75 cm from the base of the cliff. At first the oysters are scattered and then form a dense white band which continues on the slope of the middle platform for about 50 cm on the eastern, more exposed side of the bay. Is it perhaps due to the warmer temperature that the oyster belt occurs *below* the upper barnacle zone at Inhaca Island instead of above it, as in Natal? In some of the empty oyster shells cluster a few individuals of the very small tropical mussel, *Parviperna rupella*, which is common on shores in Tanzania and in northern Mozambique. The shells of *P. rupella* are thick, squarish and black externally with a purple interior (Fig. 7.4c).

Predatory snails are similar to those on the western shore, with the addition of *Purpura panama* (Fig. 7.5a) that feeds on barnacles and limpets.

The grazing snails, *Cerithium, Planaxis* and others seen on the sheltered rocks of the west coast are *absent* from the exposed rocks. On wave-beaten tropical shores small air-breathing *Siphonaria* species replace them. The pulmonary mantle cavity under the shell exits through a very short siphon anteriorly on the right side. As in limpets, which they partly replace on the middle shore, resistance by the animals to vigorous water movements is reduced to a minimum by the flat domes and ridges of their shells. At Inhaca Island there are four species of *Siphonaria* known as 'false' limpets (and only two species of true limpet) that spread from the lower cliff to the middle platform. *Siphonaria capensis* (Fig 7.5b) and the smaller *S. dayi* are found on the cliff as well as the platform below. The oval greyish shells resemble each other superficially but differ in the number and distinctness of the ridges and the colours of the internal surfaces.

Fig. 7.4 On the cliff:

a. pink barnacle, *Tetraclita squamosa rufotincta** (x 1);

b. anemone on a ledge in a pool, *Actinia equina* conracted; offspring released (x 1);

b'. tentacles extended;

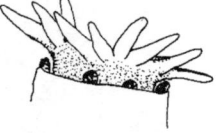

c. *Parviperna rupella*, inside shell (x 1) (after Kilburn and Rippey 1982);

7.3 MIDDLE PLATFORM – BETWEEN MID-TIDE AND LOW NEAP TIDE

During weeks of neap tides the middle platform is wet even at low tides, but a veneer of coarse sand covers it.

7.3.1 Smooth sand-scoured rock: green algae; false limpets; a limpet; roving grapsid crabs

The smooth rock platform at the foot of the cliff is coated with small, filamentous green algae (Chlorophyta) including *Enteromorpha* species, mainly *E. prolifera* and *E. compressa*. The adult form of *Enteromorpha* is composed of several thin, rounded, hollow fronds which grow from a tuft of rhizoids (Fig. 7.6b). The thallus itself has little resistance to wave action; however, the alga has considerable powers of regeneration since the meristematic region is just above the holdfast and even the segments that may by broken off may regenerate rhizoids to form new attached individuals. *Enteromorpha* species are also able to survive changes in salinity both above and below that of the sea, due either to evaporation or to rainfall during the period of low water.

In this region of the shore, epilithic diatoms are outnumbered by those growing epiphytically on the green filamentous algae, thus making a rich feeding ground for grazing molluscs.

The thin sandy covering of the smoothly eroded rock is frequently streaked by greenish tracks. These are made by species of the air-breathing *Siphonaria* (false limpets) when they slowly plough their way through the sand feeding on the exposed algae at night before the tide rises. The tracks are reinforced when they return to the 'home scar' after feeding under water in the early morning. The 'home scar' is a slight depression in the rock, abraded by the

Fig. 7.5

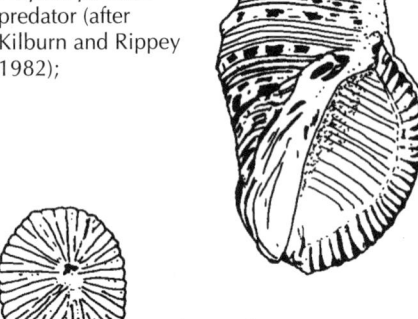

a. *Purpura panama* — predator (after Kilburn and Rippey 1982);

b. *Siphonaria capensis* grazing snail (x 1) (after B. Allanson 1958).

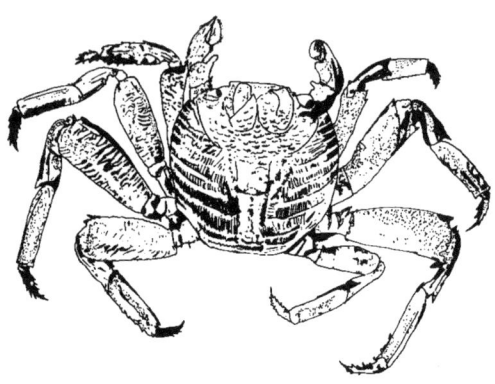

c. Rock crab, *Grapsus tenuicrustatus* (x 0,5) (after Crosnier 1965).

edge of the shell, into which it fits securely when the tide is out. Each false limpet makes random excursions to feed on the algal crop over a radius of 15-20 cm and returns over the *same* path. The sense organs used to return to the 'home scar' appear to be sensitive to chemicals that a limpet can detect in its own initial mucous trail and can distinguish it from that of others.

Siphonaria carbo prefers rocks partly covered by sand. Its shell has a slightly toothed edge, 40-55 weak white ridges and a mottled appearance. Internally the shell is slate blue in young individuals and darker brown in older ones, except for a lighter, peripheral band which has brown radial streaks. The shells of *S. anneae*, the smallest of all the species is dark brown on the outer surface instead of the usual grey. There are only 29 radial ribs, which bear weak spines, and the apex of the shell is curved to the left.

The only true limpet found at this level is *Cellana capensis* (Fig. 7.6a). It is distinguished from all other limpets, including the false limpets by the intense lustre of the nacre lining the shell and its smooth but sharp edge. The upper surface of the shell has very thin, slightly granular lines radiating from a distinct apex. Respiration in true limpets is aquatic and carried out by rows of very small secondary gills on the edge of the mantle. These extend into long filaments when the animal is under water. The animal retracts under its shell between tides and, as in all true limpets, the ability to adhere strongly to the rock surface is proverbial. The force of the contraction of vertically aligned muscles in the foot is multiplied many times by that of the surface tension in the very thin layer of mucus between the foot and the rock. False limpets are less firmly attached since their mucus is thicker. When threatened by a carnivorous snail, *C. capensis* raises its mantle to cover its shell and exudes repellent mucus.

Roving grapsid crabs lurk in the crevices in the cliff when the tide is high. When the tide recedes from the cliff they emerge and descend to the water's edge during the day as well as at night. They are similar to other grapsids, both in the mangrove and on the sheltered shore, in shunning submergence in water. *Grapsus tenuicrustatus* (60 mm) (Fig. 7.5c) is light green in colour, mottled with yellow and red. Dorsally the carapace has many ridges that run transversely and facilitate water runoff. The dactyls are highly chitinised and bear several short spines. The chelipeds are short and the chelae are stout; they are used for scraping algae from rocks. On the exposed rocks the xanthid crab has stronger, spiny and granular chelae whilst the sheltered rock crab has smooth chelae. *G. tenuicrustatus* prefers the lower shore and has a flatter carapace, which has thinner, tougher chitin. These grapsid species are remarkably agile and can flatten themselves against the rocks in the crevices, holding on with their sharp dactyls as the waves wash over them.

7.3.2 Shallow pools: barnacles and coralline algae; brown algae

A number of shallow and deeper small circular pools occur on the middle platform.

The pink barnacle *Tetraclita* which is present on the vertical cliff, occurs on the raised rims of these circular pools. An advantage for microphagous feeding may accrue from this peculiar habit since being raised about 20-30 mm above the surface of the rock enables the barnacles to reduce the amount of coarse sand particles collected whilst filter-feeding under water.

Red algae (Rhodophyta), of the articulated coralline family, occur in small tufts on the rims of the pools and also on the sides of the shells of the larger barnacles. Normally they are found lower down with algal turf. Articulated corallines have slender, repeatedly branching, jointed thalli. Internally, each branch is composed of many long cellular filaments, aggregated together and covered by a pseudo-epidermis. The cell walls of the filaments have thick deposits of calcite except at the joints where one segment connects with the next. The joints or 'genicula' make the thallus very flexible so that it is not damaged by the force of the waves (see Section 7.3.3).

The main species here is *Jania intermedia* (Isaac 1958) which has a number of thin jointed axes attached to the rock by rhizoids. Each axis repeatedly bifurcates to form a tuft of branches at the end. The joints are very close together and cross striations mark the segments (intergenicula) (Fig. 7.6c). The thick growth of algae breaks the impact of the waves and protects the rims of the pools from erosion.

In some of the shallow sandy pools monospecific stands of different species of brown algae (Phaeophyta) normally occuring on the lower shore flourish. Other pools are worn deeper by abrasion and have a floor covering of sand several centimetres thick; they appear to be barren. Still other pools are colonised by a single species of blue-black, compound tunicate, *Eudistoma caeruleum*, in the form of stiff stalks (Fig 7.6d). Tunicates also belong on the lower shore.

In the shallowest pools the floor is covered by many flat thalli of the tropical, brown alga *Padina boryana* (Dictyotales) which looks yellowish green (Fig. 3.16c) and is called the 'Peacock Fan'. A thin deposit of aragonite crystals of calcium carbonate in the cell walls reinforces the thallus, the distal margin of which is curled over, protecting the meristematic cells on the periphery. It is rare on the sheltered shore (Section 6.3.3).

In deeper pools, species of brown seaweeds belonging to the order Fucales occur; these are the most specialised of the brown seaweeds. Tissues are differentiated so that the thallus has a distinguishable stipe (stalk) and blades or fronds. The cell walls have an inner cellulose layer and an outer 'gummy layer' of alginic acid which may lose or gain water in response to the alternate exposure to air and water with the ebb and flow of the tides.

The tropical fucoid *Turbinaria ornata* (Fig. 7.7a) may be found in some pools of this lower barnacle zone, although it is more common and larger in size in the pools of the lower platform. The thallus is light brown in colour and takes the form of thick stipes from which grow erect branches of about 20 cm in height, that bear stiff saucer-like (peltate) fronds, mounted in turn on short 'pedicels'. Another morphologically differentiated species, *Cystoseira myrica*, has a bushy habit reaching only 3-4 cm high (Fig. 7.7b) (Isaac and Chamberlain 1958). The thallus has a tough stipe bearing many branches; the lower portion of the thallus has lateral branches which are flattened and have narrow, flat lobes projecting on each side; the upper branches are cylindrical and terminate in small vesicles. The dark brown thallus of *Hormophysa triquetra* (Fig. 7.7c) is simpler in form yet it may grow to over 20 cm in length

Fig. 7.6 On the middle platform:

a. Limpet, *Cellana capensis* (x 1) (after Kilburn and Rippey 1982);

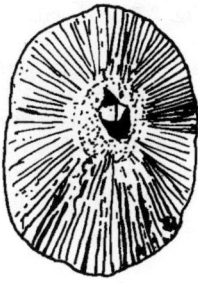

b. green alga, *Enteromorpha compressa* and detail (x 2) (after Simons 1976);

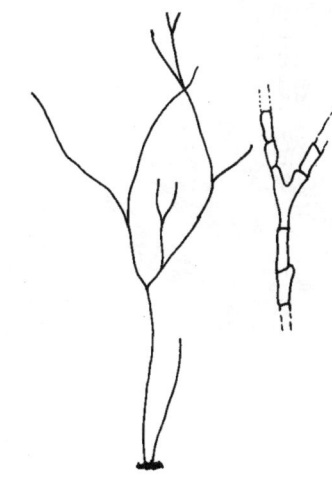

c. red alga, *Jania intermedia*, habit (x 1) and detail (after W.E Isaac 1957).

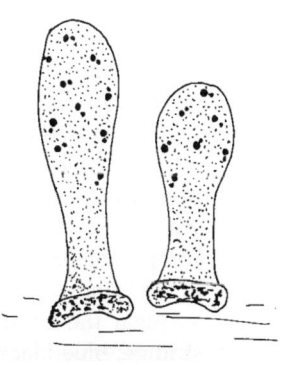

d. *Eudistoma caeculeum*, a blue-black compound tunicate.

Fig. 7.7 Seaweeds in mid-tidal pools:

a. brown algae: *Turbinaria ornata* (x 0,5) and detail;
b. *Cystoseira myrica* (x 0,5) and detail (after Isaac and Chamberlain 1957);
c. *Hormophysa triquetra* (x 0,5) (a and c after Isaac W.E 1956).

if the pool is deep enough. The thallus is supported by a wiry midrib from which flat, narrow triangular blades project alternately and irregularly on each side.

Brown algae have chlorophyll *a* and *c* and other pigments including carotene and the brown fucoxanthins. The latter absorb light in the blue, green and yellow wavelengths and pass on the energy to chlorophyll *a*. The mixture of brown and green pigments, which are present in different proportions in different species is responsible for the many colour variations observed in this order. The Fucales are poorly represented intertidally on lower tropical shores by only a few species in pools that remain partially full of water when the tide is low. This is in marked contrast to temperate shores in Europe and South Africa where there are several zones dominated by fucoids on the middle and lower levels. The only genus comparable in density is *Sargassum* which is confined to the subtidal fringe and subtidal zones in the tropics.

Near the edge of the middle platform a few pools sometimes contain twenty or more strange, blue-black club-shaped, thick fleshy stalks 50-60 mm long. They grow upright from the rock beneath the sand and bear pairs of minute openings on the higher, thicker part of the stalk. These are the typical inhalent and exhalent apertures of tunicates (see Section 6.4.3). *Eudistoma caeruleum* is the only tunicate on this shore to have an erect habit in contrast to species on the lower shore which form flat sheets on sloping rocks. By growing erect the feeding apparatus of *E. caeruleum* is raised above the sand and well-placed to take advantage of the turbulence above the pool to extract its planktonic food.

Inhaca Island: The west shore faces the bay of Maputo and east shore faces the Indian Ocean (Satellite photo courtesy of CSIR).

LEFT: Sea anemone and a commensal Two-bar anemone fish, *Amphiprion allardi* (Robin Harris).

BELOW: Ahermatypic coral *Dendrophyllia sp.;* the red polyps expand during both the day and night.

ABOVE: A plume worm, *Sabellastarte longa,* emerging from tube in sand between corals (Robin Harris).

BELOW: A hard coral, *Acropora,* with the polyps expanded at night (Robin Harris).

RIGHT: Cardinal fish, *Apogon flagelliferus* (Robin Harris).

BELOW: Spotted boxfish, *Ostracion meleagris* (Robin Harris).

ABOVE: Blue surgeon, *Acanthurus leucosternum* (Darrol Smith).

LEFT: Devils firefish, *Pterois miles;* the spines will sting if touched (Robin Harris).

ABOVE: A fiddler crab, *Uca inversa,* on the outer edge of the swamp; the tooth is inverted on the upper part of the chela (Haring Swart).

LEFT: Crowded life on a coral fragment: striped shrimps, white hydroids, green sponge.

BELOW: A leaf eating crab, *Sesarma smithi,* carrying an *Avicennia* leaf to its burrow (Pat Berjak).

ABOVE: A Boomslang among the tree branches, *Dispholidus typus* (Geoffrey Nicholls).

RIGHT: Bushveld Rain Frog, *Breviceps adspersus*. (Microhylidae) This terrestrial frog lives in a subterranean burrow, appearing above the ground only when it rains.

BELOW: Gecko, *Hemidactylus mabouia* (Geoffrey Nicholls).

ABOVE: Curlew sandpiper, a predator on burrowing worms (Geoff Lockwood).

LEFT: Spotted Eagle Owl, *Bubo africanus* (Geoff Lockwood).

ABOVE: Little Stint, *Calidris minuta,* on a sandy shore (Geoff Lockwood).

RIGHT: Black-bellied Khoran, *Eupodotis melanogaster* (Geoff Lockwood).

Avicennia marina tree on the edge of the swamp; note the aerial roots (Dick Pienaar).

Natural deposition of wind blown sand covering and killing the dune vegetation (E.M.C. Groenendijk).

Exposed east coast rocks: Middle platform with shallow pools and seaweed; algae turf on distant rocks (Robin Harris).

7.3.3 On flat rocks: articulated coralline algae

Towards the lower edge of the platform the number of articulated coralline algae increases forming a rough carpet on the rocks. These algae of the middle shore are subjected to high evaporation for two or three hours in the middle of the day during the weeks of spring tides. The plants do not dry out since their growth form of slender, overlapping branches retains moisture between them and the calcification of the cell walls may have a protective function.

There is a mixture of several species of the genera, *Jania* and *Amphiroa*, *Cheilosporum* and *Arthrocardia*. These genera are distinguished by the shapes of the intergenicula (calcified segments) of their slender, jointed thalli, as shown in the diagrams. *Amphiroa* (Fig. 7.8a) has a bushy form and the intergenicula are somewhat flattened; the thallus is often of a greyish red colour. The flattened axes of *Cheilosporum* (Fig. 7.8b) are prickly since the intergenicula are winged, projecting laterally like little arrow heads. *Arthrocardia* (Fig. 7.8c) also has flattened intergenicula but the whole plant is much more robust than the others.

The calcification process in the cell walls of the coralline algae is reminiscent of the formation of the skeleton of reef-building corals. Calcification is faster in light than in darkness since photosynthesis is involved, as in corals through their symbiotic algae (see Section 8.2). In coralline algae photosynthesis provides the energy necessary for the active transport of calcium ions from the cytoplasm to the cell walls and also creates an alkaline medium in which precipitation of calcium carbonate can occur. The outer surface of the coralline alga is weakly acidic and it is postulated that hydrogen ions, generated from the photosynthetic process, diffuse out of the cell leaving an excess of OH- ions inside the cell. The occurrence of the enzyme carbonic anhydrase, that accelerates calcification in corals, has been postulated but not verified in coralline algae. Its *absence* would be compatible with the very slow rate of growth and calcification in algae in comparison with that of corals.

The genera described above may also occur in the algal turf of the lower platform, in gullies and subtidally. Red algae possess chlorophyll *a* and *d*, carotenes, and auxiliary pigments, red (phyco-erythrin) and blue (phycocyanin). These pigments, after absorbing energy from the blue and green wavelengths of light, respectively, transfer it to chlorophyll *a* for photosynthesis. This mechanism compensates for the relatively low penetration of light into calcified thalli. Similarly these accessory pigments promote photosynthesis under water; light rays of blue and green wavelengths of the white light spectrum penetrate water further than do the red and violet wavelengths. The proportions of red and blue pigments within the cells of red algae produce many shades of colour, from lilac-grey to bright red through shades of pink, purple and brown that mask the green chlorophyll; however, there are exceptions, such as the brilliant green of *Hypnea viridis* on broken rocks. Members of a single species may also exhibit different shades of colour

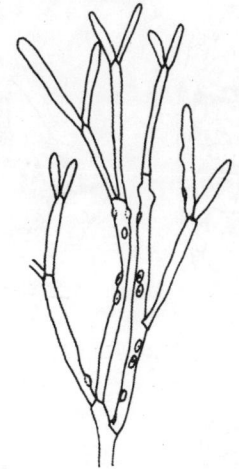

Fig. 7.8 Coralline algae, middle platform:
a. *Amphiroa* sp. (x 2);

b. *Cheilosporum* sp. (x 1) and detail (after Isaac 1957);

c. *Arthrocardia* sp. (x 1) and detail (a and c after Simons 1976).

188 NATURAL HISTORY OF INHACA ISLAND

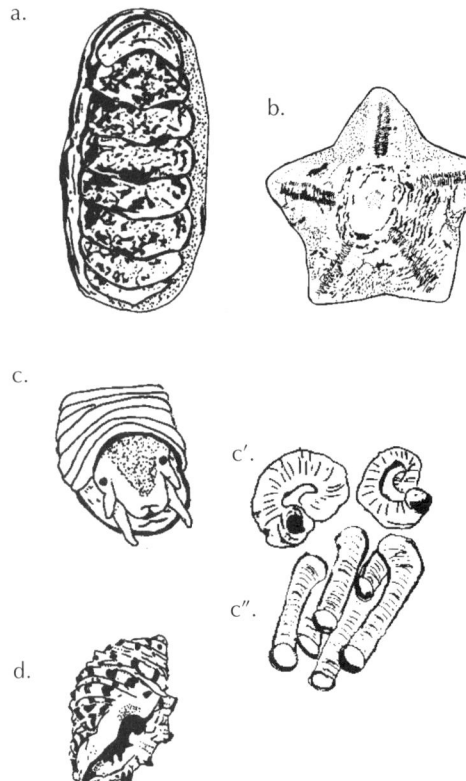

Fig. 7.9 On the seaward edge of the middle platform:
a. chiton, *Ornithochiton literatus* (x 1);
b. starfish, *Patiriella exigua* (x 1) (after Walenkamp 1990);
c. snail in encrusting tubes, *Dendropoma tholia*, head (x 5) and
c'. mass of tubes (young coiled tubes enlarged);
c". older tubes;
d. predatory snail, *Morula uva* (x 1) (a, c, and d after Kilburn and Rippey).

according to the level of light to which they are exposed. In a gulley, where the amount of shade varies from zero in open water to perhaps 70% in a small cave, the amounts of accessory pigments are adapted to the amount of light they receive.

7.3.4 Encrusting coralline algae; chitons and starfish

Encrusting coralline algae occur only in small patches on rocks towards the edge of the middle platform and at lower levels, forming a thin coating of a lilac-grey or a pinkish calcareous layer.

Starfish and chitons are said to prefer grazing on encrusting coralline algae. One small species of chiton, *Ornithochiton literatus* (25 mm) (Fig. 7.9a) may be found in small hollows at the edge of the middle platform. This is the only chiton to occur on the shores of Inhaca Island since the large tropical, hairy chiton of high tidal level and the chitons of temperate shores are absent. Chitons are symmetrical molluscs with a large ventral foot; they are covered dorsally by eight arched, calcareous plates overlapping like tiles on a roof, held in position by a peripheral muscular band, the 'girdle' which joins the muscular foot below. This support of the shells is flexible so that the chiton can curl into a ball or fit itself into irregular spaces. The shells of young chitons are perforated above the minute sense organs which are sensitive to light. The young animal avoids direct sunlight and emerges to feed on encrusting algae only at night. Older animals, in which the shells are eroded and the eyes are no longer functional, may be found fixed to open rock surfaces during the day.

Another coralline algal grazer is the small cushion-starfish, *Patiriella exigua* (diameter 4-5 cm) (Fig. 7.9b) which occurs in groups matching the colour of the rock and the encrusting coralline red algae in tints of grey, speckled with green or brown. It feeds by everting its stomach on to the rock surface to remove algae and then withdraws it and continues digesting the algae or diatoms. The starfish adheres to the rock by means of the short tube feet in the ambulacral grooves and by the spiny under surface of the skin. (Starfish are described in detail in Section 9.4.1.)

7.3.5 On the edges of the platform: tube-building snails

The edge of the middle platform towards the more sheltered western part of the small bay is plastered with hundreds of bluish-grey calcareous tubes that form a band about 5-10 cm thick and about 30 cm broad made by the strange worm-like snail, *Dendropoma tholia*, a member of the vermetid (worm-like) family (Fig. 7.9c, c' and c"). The tubes are 20-30 mm long, a few millimetres in diameter, intertwined and partially cemented together. The head, foot and mantle cavity of the snail within, are near to the opening of the tube which can be closed by a horny operculum. The visceral hump of the snail is uncoiled and the body is fixed to the tube by a columellar muscle, just as in a true snail. The closed end of the tube, made by the very young snail, is very slightly coiled.

The position occupied by *Dendropoma* is ideal for feeding on plankton which is netted in a unique way. The foot secretes a very sticky mucus from a gland between the two pedal tentacles, which are peculiar to vermetid snails. Mucus is carried along the grooves of the tentacles by ciliary action; at intervals of a few seconds it is flicked away from the tips of the tentacles by their twitching and quivering movements but the mucus still remains attached to the tentacles. The mucous threads so formed spread out under water and form a net that is held afloat for a few minutes, thereby intercepting microscopic food. Suddenly the tentacles retract, the proboscis moves forward and the jaws nip off the net at its source. The net is quickly hauled into the mouth by the backward movement of the radula in which the teeth are hook-like. The rhythm of secretion, netting and hauling is repeated many times between short, resting intervals during the hours when the animals are submerged.

Dendropoma eggs are fertilised by sperm swept into the mantle cavity by the respiratory current created by the gill. In the early summer about ten fertilised eggs may be found in a single row inside the mantle cavity. They are retained until the larvae have grown their tiny spiral shells and have a reserve of fat to support them through metamorphosis. The free-swimming stage lasts only a day or so and the larvae settle on a rock near the colony or upon other tubes of their own species. The larvae may select an area overgrown by a thin covering of lilac-grey, encrusting, coralline algae, in which each larva makes a shallow groove with its radula and adheres to the substratum by sticky mucus secreted from a gland in the foot. The larva completes its metamorphosis in two days: the head tentacles shrink, the foot is reduced to a pad bearing the mucus gland and the two pedal tentacles grow. Three days later a white rim around the aperture of a new shell (visible under a microscope) indicates the beginning of the almost straight tubular shell of the adult which is marked by irregular transverse ridges.

Many calcareous tubes of the snail may appear to be empty since no operculum is visible. The snails may have died of desiccation but more likely they have fallen prey to the muricid snail, *Morula uva* which specialises in a diet of *Dendropoma*. This carnivorous snail (20 mm) is white with rows of blunt, black tubercles on ridges with spiral threads between them (Fig. 7.9d).

The empty *Dendropoma* tubes provide a refuge for a large number of juvenile and small invertebrates. About 100 individuals comprising 17 species were found in one block of worm tubes within an area of 25 cm^2 and a thickness of 10 cm. Most of them were juvenile polychaete worms of the nereid, syllid and eunicid families which normally live among the algae on the lower rocks. Some animals which lie in crevices on the lower shore were also found; these were a sipunculid, two species of the brittle star, *Ophiactis* sp., isopods and two very small species of xanthid crab.

7.4 THE LOWER PLATFORM – BELOW LOW NEAP TIDES

A gentle rocky slope leads to the subtidal fringe where it is incised by gullies up to 1 m deep, some of which lead to small cave-like overhangs. Deep pools with sandy floors are numerous.

7.4.1 Mixed algal turf: green, brown and red algae

An almost continuous cover of small seaweeds, mostly red and green with but few brown and some blue-green (Cyanophyta) species, form a thick sward on the rock surface of the lower platform. Fifteen genera have been listed in the algal turf on Inhaca rocks, some represented by more than one species (A. Critchley pers. comm.), although within each common genus there are fewer species than on rocks within the tropics (Isaac and Isaac 1967, 1968).

Algal turf is dominated by aggregates of both small filamentous and fleshy algae and includes developmental stages of large algae. Many of the green fleshy algae have a creeping stolon which supports erect axes of the thallus about 20 mm high. The majority of the red species are very small and are members of the Order Ceramiales. The most conspicuous green, brown and red algae with erect thalli and a spreading habit are described below.

Two orders of green algae (Chlorophyta) are represented. They differ primarily in their mode of cell division, resulting in a totally different macroscopic appearance. Species of Caulerpales are conspicuous whilst those of Siphonocladales are smaller and some are cryptic in habit; all reproduce sexually.

Species of *Caulerpa* are essentially single-celled since the nuclei divide without the formation of cross walls, resulting in multinucleate tubes or coenocytes, consolidated to form

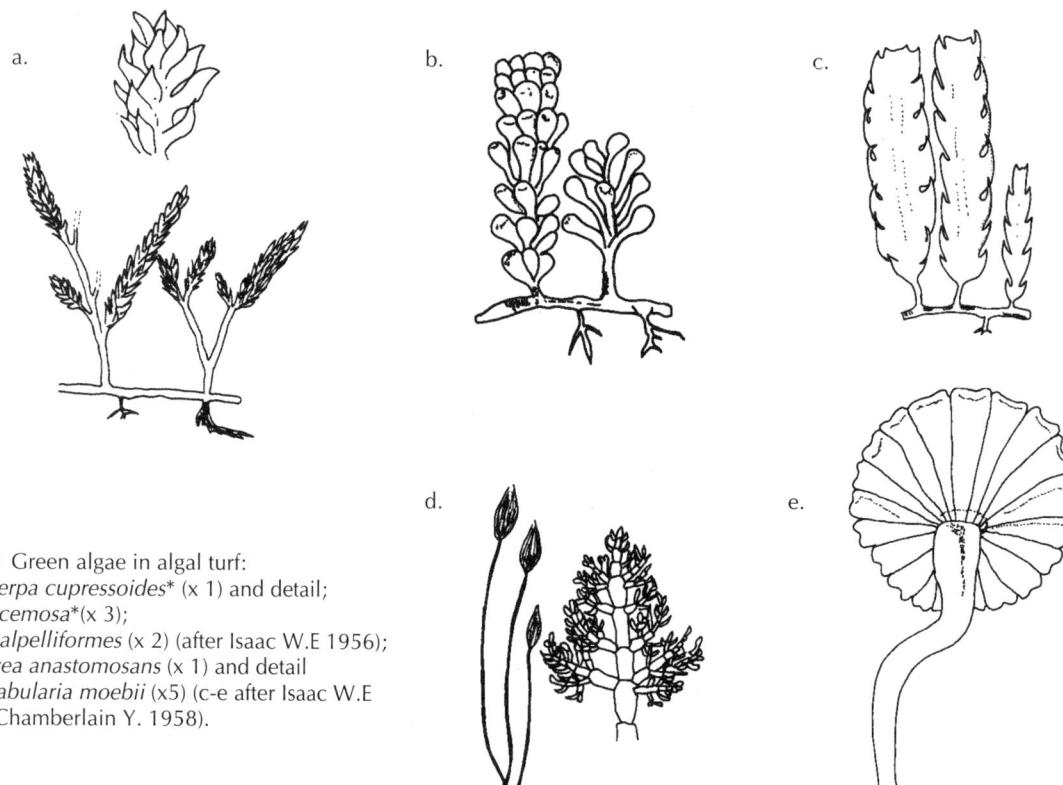

Fig. 7.10 Green algae in algal turf:
a. *Caulerpa cupressoides** (x 1) and detail;
b. *C. racemosa**(x 3);
c. *C. scalpelliformes* (x 2) (after Isaac W.E 1956);
d. *Struvea anastomosans* (x 1) and detail
e. *Acetabularia moebii* (x5) (c-e after Isaac W.E and Chamberlain Y. 1958).

the thallus. A distinct toughness is imparted to the coenocytic structure by irregular crisscrossing ingrowths (trabeculae) on the cellulose walls. The macroscopic result is a thallus with a firm prostrate stolon, rhizoidal attachments at intervals and erect photosynthetic blades which may be a few centimeteres high. The latter are modified in a great variety of shapes and arrangements in the various species.

The moss-like *Caulerpa cupressoides* (Fig. 7.10a) has very small leaf-like 'pinnules' in four rows, crowded towards the ends of the blades (Isaac 1956). Small clumps of club-shaped pinnae on erect axes, each about 20 mm high, grow upright from a long stolon in *C. racemosa* (Fig. 7.10b). The largest, *C. scalpelliformis*, has flat fern-like fronds with scalloped margins that grow 10-30 mm high from the stolon (Fig. 7.10c). The delicate, dark green, feathery species *Bryopsis pusilla* is also a member of the Order Caulerpales.

The coenocytic structure of these algae allows free flow of nutrients through the plant and, if pierced by the jaws of a worm or the radula of a nudibranch, wounds are healed by the formation of a plug synthesised from pre-existing protein granules, which migrate to the site of injury. Many species of *Caulerpa* produce a poison, 'caulerpcin', which is known to enter the food chain although it is said to have an anti-herbivore function.

The other Order Siphonocladales is distinguished by 'segregated cell division' a process whereby the protoplasm cleaves with each nuclear division into portions of unequal size but without complete cell walls being formed. In the case of *Struvea anastomosans* (Fig. 7.10d) this results in a long string-like thallus the end of which branches into triangular cells (Isaac and Chamberlain 1958). The ends of these cells anastomose with other similar branches by means of rhizoid-like projections from the cells. The terminal branches thus resemble intricate lacework. The whole plant is mixed intimately with other algae and, owing to its small size, it is seen only when carefully dissected out from other species.

The peculiar single-celled green alga, *Acetabularia* (Dasycladales) (Fig 7.10e) finds support in the shelter of other algae on rocks. It has a 'stalk' about 10 mm long with a little scalloped 'cap'. This unique giant cell has proved useful in experimental cell biology in the study of the interactions between nucleus and cytoplasm; the nucleus is in the basal end of the cell near its attachement by rhizoids and it can easily be removed by cutting the 'stalk'. The enucleated cell, i.e. the upper part, carries out photosynthesis for many weeks, but does not reproduce. By grafting a part containing the nucleus to an enucleated part of another species it has been demonstrated that the specific shape of the cap' is determined by the nucleus. Stored cytoplasmic molecules of mRNA, derived from the nucleus, control its development in light.

The only brown algae (Phaeophyta) common in the algal turf are small species of *Dictyota* and *Dictyopteris delicatula* which have flat, strap-shaped, dichotomously branching, multicellular thalli characteristic of the Dictyotales (Fig. 7.11a and b). These two genera differ in the possession of a midrib in the blades of *Dictyopteris*, and its absence in *Dictyota*. When under water *Dictyota* may take on a faint blue iridescence.

Among red algae (Rhodophyta) the coralline species mentioned in Section 7.3.3 are present and, in addition, several non-coralline species of the Orders Ceramiales and Gelidiales. The most conspicuous of the former is *Laurencia natalensis* (Fig. 7.12a) which has a compressed axis and pinnate branches that are bright green in colour, bearing red-tipped globular 'pinnae' (Isaac 1956). *L. complanata* resembles the frond of a fern and is bright red, whilst *L. pumila* has short paler branches and stumpy lobes.

The thallus of *Gelidium* is not well differentiated and is much stiffer. *G. reptans*, the major species, has flat, blunt axes bearing scattered, final branchlets reduced to blunt lobes (Fig. 7.12b). The consistency of the thalli is distinctly cartilaginous owing to the quantity of polysaccharides *surrounding* the cells. Species of this genus are used worldwide as a source of agar jelly. Larger brown and red algae occur in deep pools, permanently underwater.

Most of the species of red algae are capable of regeneration after being grazed by herbivorous molluscs.

7.4.2 Animal competitors with algae: herbivores, colonial animals; filter feeders

Competition between animals and algae may take three forms: direct herbivory which may limit the spread of algae but often stimulates growth; rival occupation of space on the rock and lastly interference with the reproduction of algae.

Amongst the algal turf there are small patches of rock which are apparently bare. The seemingly bare rock is covered with sporelings of red, green and brown algae on which the limpets graze and thus it is kept clear of macroscopic forms of algae. The limpets are fairly similar dark-shelled species: *Patella pica* (formerly recorded as *P. barbara*) (Fig. 7.13a) and another unidentified species that differs mainly in the presence of a dark brown peripheral band on the light interior of the

Fig. 7.11 Brown algae in algal turf:

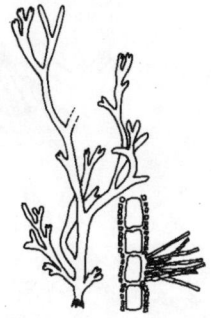

a. *Dictyota* sp. (x 2) and detail;

b. *Dictyopteris delicatula* (x 1) (a and b after Seagrief 1980).

Fig. 7.12 Red algae in algal turf:

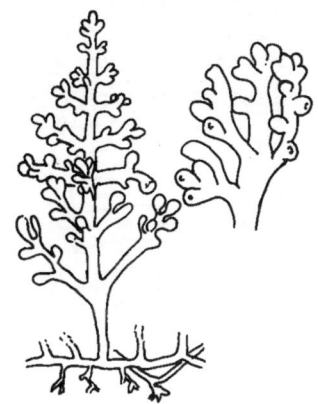

a. *Laurencia natalensis* (x 2) (after Seagrief 1980);

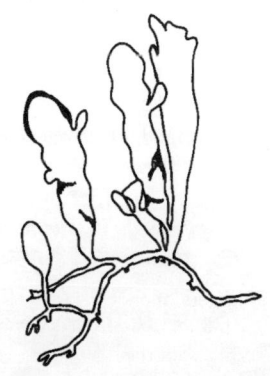

b. *Gelidium reptans* (x 5) (after Isaac W.E 1956).

Fig. 7.13 Animals among algal turf: limpets

a. *Patella pica** (x 1);

b. *P. granularis* (?) (x 1);

polychaetes:

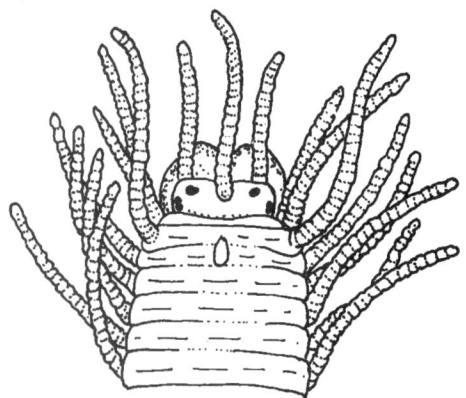

c. *Syllis variegata*, head enlarged;

d. *Platynereis* sp., jaws with grazing teeth (c and d after Day 1951 and 1934);

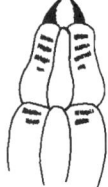

e. isopod, *Dynoides serratisinus* (x 2)

e'. telson enlarged (after Kensley 1978).

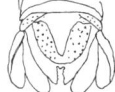

shells of the latter similar to that in *P. granularis* (Fig. 7.13b). Scattered tufts of *Gelidium reptans* grow on the shells of the two species of limpets, which are not very numerous.

The invertebrates that feed on mature algae are small polychaete worms of the syllid family (Fig. 7.13c) and small sphaeromid isopods, *Dynoides serratisinus*, which shelter in the algal turf or in the worm-like *Dendropoma* tubes. Some species of nereid worms are omnivorous, consuming algae, worms and crustaceans.

The cosmopolitan polychaete worms *Syllis variegata* (Fig. 7.13c) and *S. armillaris* are most common. These worms feed by engulfing a small pinna of a coenocytic alga with the pharynx and piercing it with a single dorsal tooth, then sucking the alga dry by means of its muscular gizzard which acts as a suction pump. The specialised *Platynereis* (Fig. 7.13d) has file-like chitinous teeth on the proboscis which are used to rasp its algal food.

Dynoides serratisinus (Fig. 7.13e) is the common sphaeromid isopod in the algal turf, which belongs to the family of 'sea lice' that can roll themselves into balls, like woodlice. It is distinguished from other sphaeromids by the peculiarities of the telson which, in the male, has a deep slit in the centre and seven teeth on each side of it, and it bears numerous sharp granules on the upper surface. The crustacean breaks off part of the algal fronds with its mandibles but leaves the basal parts of the plants, which are able to regenerate.

Large areas of rock around the pools and on the edges of gullies towards the lower limit of the platform are entirely covered by sheets of compound animals: zoanthids, tunicates and sponges that have ousted the algal turf. Zoanthids (compound anemones) are typical of Natal rocks. The polyps are intermediate in size between those of anemones and corals and they are embedded in a fleshy base, very firmly attached to the rock surface.

Palythoa nelliae is the most common species (Fig. 7.14a), each polyp is 10-20 mm wide and stands about 20-30 mm high when fully expanded with tentacles extended. The tentacles are brown and the disc around the mouth is pale green. Both the polyps and the base are encrusted with grains of coarse sand. *Zoanthus sansibaricus* (Fig. 7.14b) has a thin greyish-green basal sheet and small flat polyps with violet tentacles and a green disc. Sandy tubes of the polychaete *Mesochaetopterus minutus* stand upright between the polyps of *Zoanthus sansibarcus* on the sandier rocks. This tropical filter-feeding worm, seen on the sheltered sandy shore (see Section 3.5.1) uses a mucous funnel to catch phytoplankton. Between the sheets of zoanthids rows of solitary red sea anemones live in shallow crevices on the rock beneath. These anemones are carnivorous and extend their tentacles under water to catch small wandering crustaceans.

Another zoanthid *Isaurus spongiosum* tends to occur separately in shallow depressions or in pools. The large upright polyps are permanently extended 30-40 mm above the base and are brown in colour. Each polyp is curved asymmetrically at the top where little nodules surround the disc.

These coelenterates occupy rock surfaces which are also suitable for colonisation by algal turf species. The animals are able to adhere to the substratum and to grow more quickly, leaving no room for algae.

Filter-feeding tunicates, sponges and mussels compete with algae by preventing algal spores and gametes from settling, through the activity of their feeding currents. Although the filter feeders may not select the spores and gametes as food, since they prefer finer particles, the reproductive cells become entangled in mucus and are rejected.

Unlike the rocks in Natal, among the zoanthids, there may be patches of rock about 15 x 25 cm in area, covered by compound tunicates. *Diplosoma modestum* has a bright yellow basal sheet in which the zooids are marked by rose-pink spots. The brilliant green sheets are *Symplegme viridis* (Fig. 7.15a) and the black sheets are *Aplidium lubricum*. In addition, three species of encrusting sponges, not yet identified, form striking patches of purple, yellow and red among the algae near the deep pools. Very occasionally groups of purplish-brown mussels, *Perna perna*, take the place of algal turf where the sea is especially turbulent (Fig. 7.15b), but the areas covered are small.

7.5 THE SUB-TIDAL FRINGE: POOLS, GULLIES AND BOULDERS

The mixed sward of smaller turf-forming algae that was so well developed on the horizontal surface of the rocks gives way to luxuriant growths of macro-algae in the deep pools and gullies at the low water level of spring tides. Siphonous green algae are prominent and large brown algae occur, with the spaces between them filled by a larger number of red algae, which are smaller in size and varied in species. Most of the animals in the pools are concealed during low tides. The flat rocks beyond the pools are encrusted with coarse, sandy worm-tubes from which the heads of the worms protrude only when under water. Other rocks are covered with crowds of a large solitary tunicate among which few algae are found.

7.5.1 Seaweeds in pools: green, brown, and red algae; sea grasses; dinoflagellates

Several species of the dark green *Codium* of the Order Caulerpales are attached to the walls of the pools and are conspicuous. The genus *Codium* is unmistakable since the surface feels velvet-like, soft and spongy, due to the ends of the intertwining coenocytic filaments being swollen into tiny, closely packed, club-like projections. These are known as 'utricles' and contain chloroplasts (Fig. 7.16a). The chloroplasts migrate towards the surface of the thallus under low light intensity when submerged and inwards when the tide is out, thus avoiding strong light. *C. duthiae* (Fig. 7.16a), a common species, has cylindrical axes that branch repeatedly as it attains a height of 10-

Fig. 7.14 Colonial coelenterates on rocks:

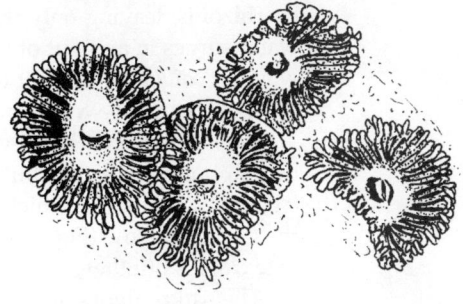

a. *Palythoa nelliae* (x 2);

b. *Zoanthus sansibaricus* and tubes of polychaete, *Mesochaetopterus minutus* (x 2) (a-b after Branch and Branch 1981).

Fig. 7.15 Filter feeders on rocks:

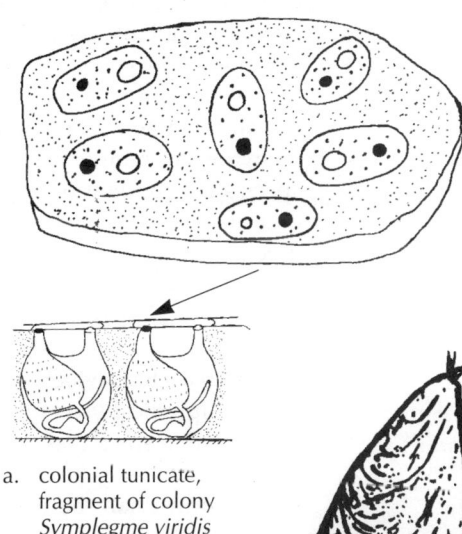

a. colonial tunicate, fragment of colony *Symplegme viridis* surface view and arrow to section (x 3);

b. black mussel, *Perna perna* (x 1) (b after Kilburn and Rippey 1982).

12 cm. *Codium lucasii* has a thick, lobed, cushion-like form and its texture is somewhat rubbery; it occurs in pools but more frequently on the vertical sides of gullies. The alga is securely attached by rhizoids, leaving only the margins of the lobes free, so that it cannot be torn when the sea water surges in and out of the gullies. The thallus of *Pseudocodium devriesii* branches in a similar way to *Codium duthiae*, but the axes are slender and a paler green colour (Fig. 7.16b). It lacks the spongy texture of a true *Codium* since the utricles are laterally fused at the surface.

Halimeda cuneata has a very distinctive, erect and robust thallus, constructed internally on the 'siphonous' pattern but impregnated with calcite. Each axis of the thallus is made of flat, triangular segments, which are inverted so that the sharp apex of each is joined to the base of the triangle below (Fig. 7.16c). In Inhaca pools these plants are smaller in size than those in the true tropics but they may reach about 13 cm in height.

The sandy floors of some of the deep pools are sometimes covered by several individuals of *Udotea orientalis*, a siphonous, calcified, green alga with a fan-like thallus (Fig. 3.16a). This species also occurs on the sand of the reef flat on the west coast of the island where, in these sheltered waters, it grows to twice the size (see Section 3.6.3).

Chamaedoris delphinii, a member of the Order Siphonocladales which occurs on the rocks in clumps, has been called the 'paint-brush alga' because it has a circular tuft or brush of filaments, tipped with white deposits of calcite at the end of each slender axis (Fig. 7.16d).

Another alga of the order Cladophorales, *Microdictyon kraussi* grows epiphytically on *Codium*, the rhizoids obtaining a hold between the easily separated utricles (Isaac 1956). *Microdictyon* resembles a bright green transparent leaf superficially, but on closer examination it appears opaque and beautiful, composed of almost microscopic, prolific, blunt branches, overlapping one another, resembling an intricate lace pattern (Fig. 7.16e).

Two cushion-like species of the Cladophorales may be found in rock pools or on the flat rocks between them and they may also be a constituent of algal turf. More often they are found on the vertical faces of gullies. *Valonia macrophysa* (20 mm) (Fig. 6.17d) has a metallic grey-green colour and a number of distinct, tiny vesicles at the ends of a few stumpy branches. The other, *Dictyosphaeria versluyii* (Fig. 6.17e) has two or three blunt, swollen lobes composed of large cells, about the size of pin-heads, under a smooth and glassy surface. These two species also occur on rocks near the coral reef.

Two genera of the third order of green seaweeds, Dasycladales, are abundant. These have radially symmetrical whorls of lateral branchlets. *Neomeris vanbossiae* has a long axis on which the whorls of minute branchlets appear granular and whitish with green tips (Fig. 7.16f). This species is abundant on the sides of rock pools and even more so on the floor of the broad gullies where they open to the sea at extreme low tide level. *Dasycladus ramosus* has very dark green, sparingly branched, flaccid axes, 20-40 mm long and 4-8 mm wide, which appear to writhe like worms in the moving water (Fig. 7.16g) (Chamberlain 1958). It grows in clumps at the bottom of pools and when grasped in the hand its chlorophyll is released so that the water in the vicinity becomes streaked with bright green.

The red algae which coat the vertical walls of the rock pools and gullies in profusion or grow epiphytically on other algae are classified into orders according to their microscopic reproductive structures. In the three major orders many genera have similar growth forms. Some species may be recognised in the field. Gigartinales generally have somewhat more robust forms than the filamentous or feathery Ceramiales. *Rhodymenia natalensis* (Fig. 7.17a) (Order Rhodymeniales) is often dominant in pools and is one of the few algae to extend its distribution on to the rocks covered by the solitary tunicate *Pyura stolonifera* (redbait) at extreme low level on the shore. Several species of *Plocamium* have relatively broad, flat and feathery, bright red thalli which differ from each other in the details of the very short marginal branchlets (Fig. 7.17b). These plants exhibit beautiful streaks of iridescent blue and purple when viewed in direct sunlight. Several genera with thread-like, but firm axes and branches, are epiphytic on other algae. One of the most elegant is *Hypnea rosea* in which each little branchlet on its sparse branches curls at its end (Fig. 7.17c).

Many genera of the Ceramiales such as *Polysiphonia* and *Platysiphonia*, have delicate filamentous forms resembling a tuft of fine hairs, whilst *Ceramium* spp. have small delicate,

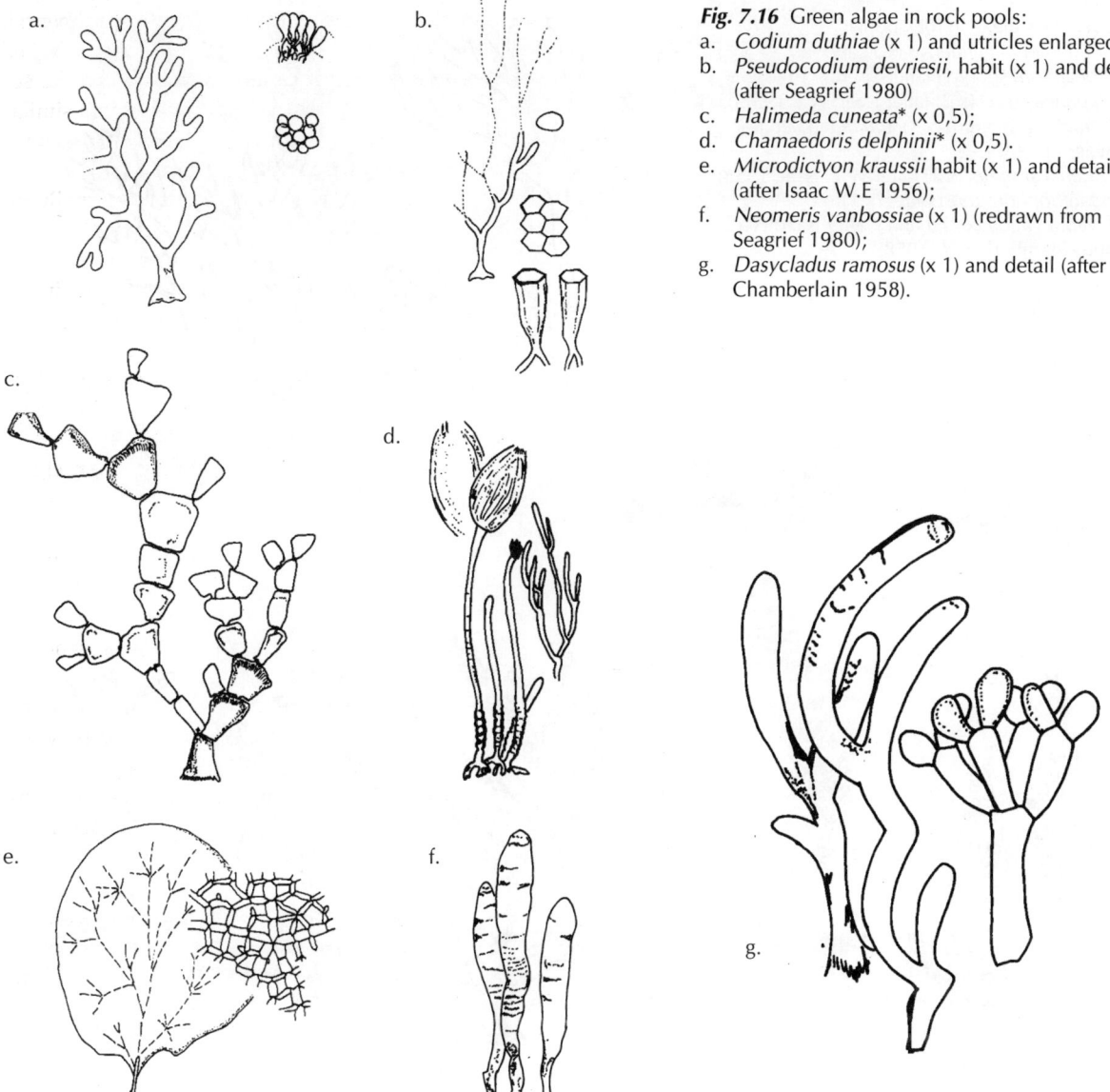

Fig. 7.16 Green algae in rock pools:
a. *Codium duthiae* (x 1) and utricles enlarged;
b. *Pseudocodium devriesii*, habit (x 1) and detail (after Seagrief 1980)
c. *Halimeda cuneata** (x 0,5);
d. *Chamaedoris delphinii** (x 0,5).
e. *Microdictyon kraussii* habit (x 1) and detail (after Isaac W.E 1956);
f. *Neomeris vanbossiae* (x 1) (redrawn from Seagrief 1980);
g. *Dasycladus ramosus* (x 1) and detail (after Chamberlain 1958).

feathery forms (Fig. 7.17d). Most are epiphytic in habit. A few members of the Ceramiales are more robust: for example, *Digenia simplex* which grows up to 10 cm high, has branches resembling a bottle brush. Very narrow, short and closely applied stiff blades surround both the main axis and each branch (Fig. 7.17e). *Acanthophora muscoides*, of similar height, has a bushy form (Isaac and Chamberlain 1958). The branches are coarse, stiff and prickly, often seen with spore-containing cystocarps forming small swellings less than 1 mm long at the bases of the reddish spiny branches (Fig. 7.17f).

A complete contrast to *Acanthophora* is shown by *Liagora valida* in the Order Nemalionales. It also has a bushy form but it is soft and pliable, pale in colour with pink-tipped branches (Fig. 7.17g). It is internally impregnated with calcite, yet its surface is slimy to the touch.

The brown algae are larger and more robust than the green and red algae, but there are fewer species in these warm pools. In the Order Dictyotales, *Stypopodium zonale* is the largest species and grows up to 30 cm long (Fig. 7.18a). *Zonaria subarticulata*, a smaller, similar species is also common, recumbent amongst other algae (Isaac 1958). Its many-branched flat axes are narrow near the base and then broaden into flat spatulate ends (Fig. 7.18b). Smaller species in

Fig. 7.17 Red algae in rock pools:
a. *Rhodymenia natalensis* (x 1) and
b. *Plocamium* sp. (x 1) (after Simons 1976);
c. *Hypnaea rosea* (x 2) (after Isaac W.E 1957);
d. *Ceramium planum* (x 1) and detail (after Seagrief 1980);
e. *Digenia simplex* (x 1) (after W.E. Isaac 1956);
f. *Acanthophora muscoides* (x 1);
g. *Liagora valida* (x 2) (f and g after Isaac W.E and Chamberlain Y. 1958).

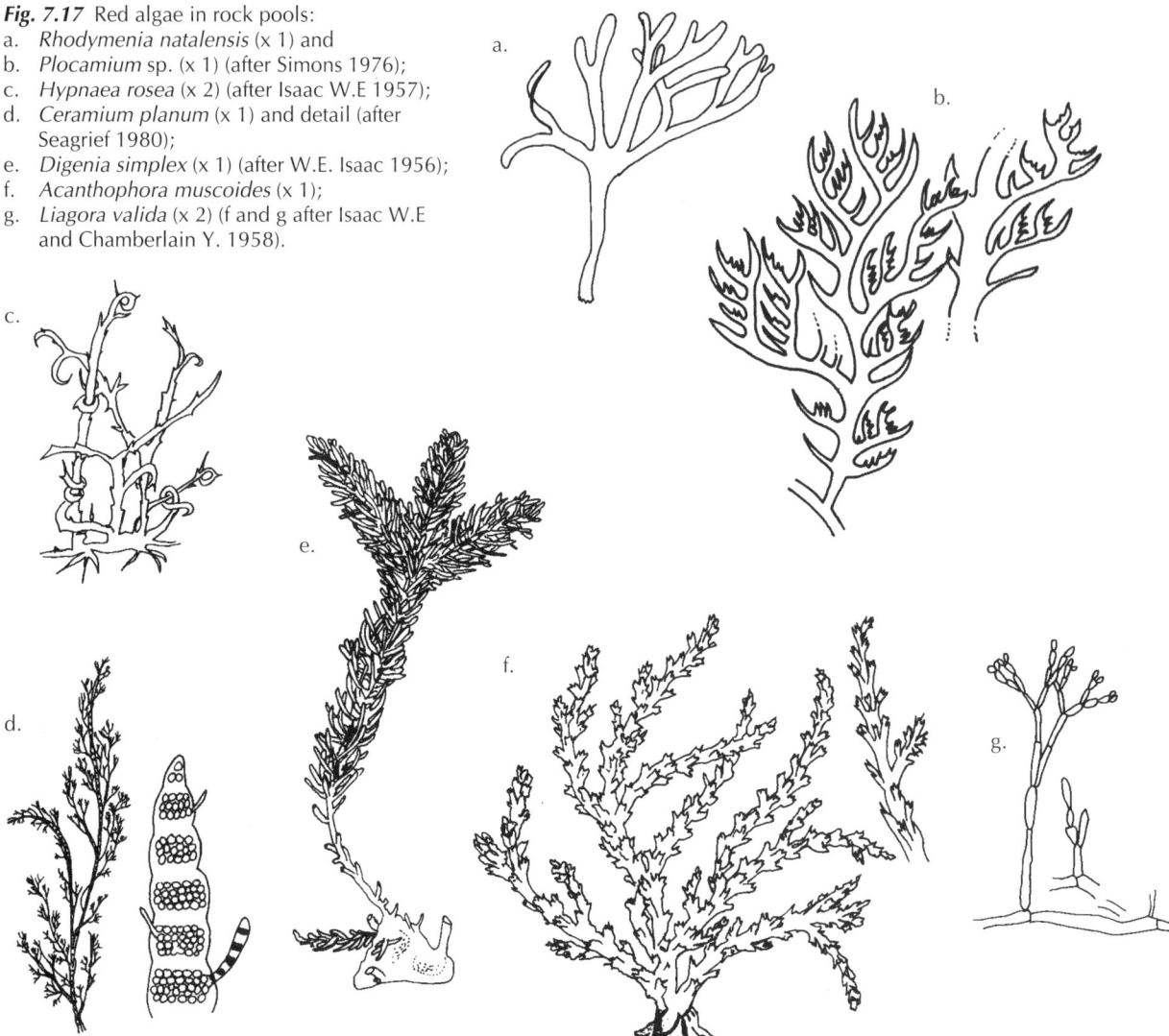

the same order occur in the pools and gullies of which the dark brown *Dictyopteris longifolia* is perhaps the most common. The thallus has a short stipe which supports narrow strap-like blades each with a tough midrib and divided into two at the tips (Fig. 7.18c).

The largest of the brown seaweeds belong to the Fucales. *Hormophysa triquetra* and *Turbinaria ornata*, which are also displaced to the shallow pools of the middle platform (Fig. 7.7c and a), grow more luxuriantly in the deeper pools of the lower platform. In the openings of the gullies at extreme low tidal level the rocks are covered by the trailing thalli of several species of *Sargassum*. One species (Fig. 7.18d) has several cylindrical axes bearing flattened, tapering, leaf-like laterals. Another has much longer trailing, straight blades and shorter blades bear bladders at the bases. These are the tropical counterpart of the temperate species of *Fucus* spp., the wracks, on cooler South African and northern hemisphere shores. A large number of species of seaweeds at Inhaca island also occur on Kenyan shores (Isaac and Isaac 1967, 1968), although others are found in Natal.

In some pools at the lowest tidal level, the seaweeds are replaced by rich growths of the long-stemmed sea grass, *Thalassodendron ciliatum* (Fig. 9.1). These tropical angiosperms are not known to flower on the exposed coast of Inhaca Island, although they do so on its warmer sheltered shores (See Section 9.2). This is the only species of sea grass on the north-eastern

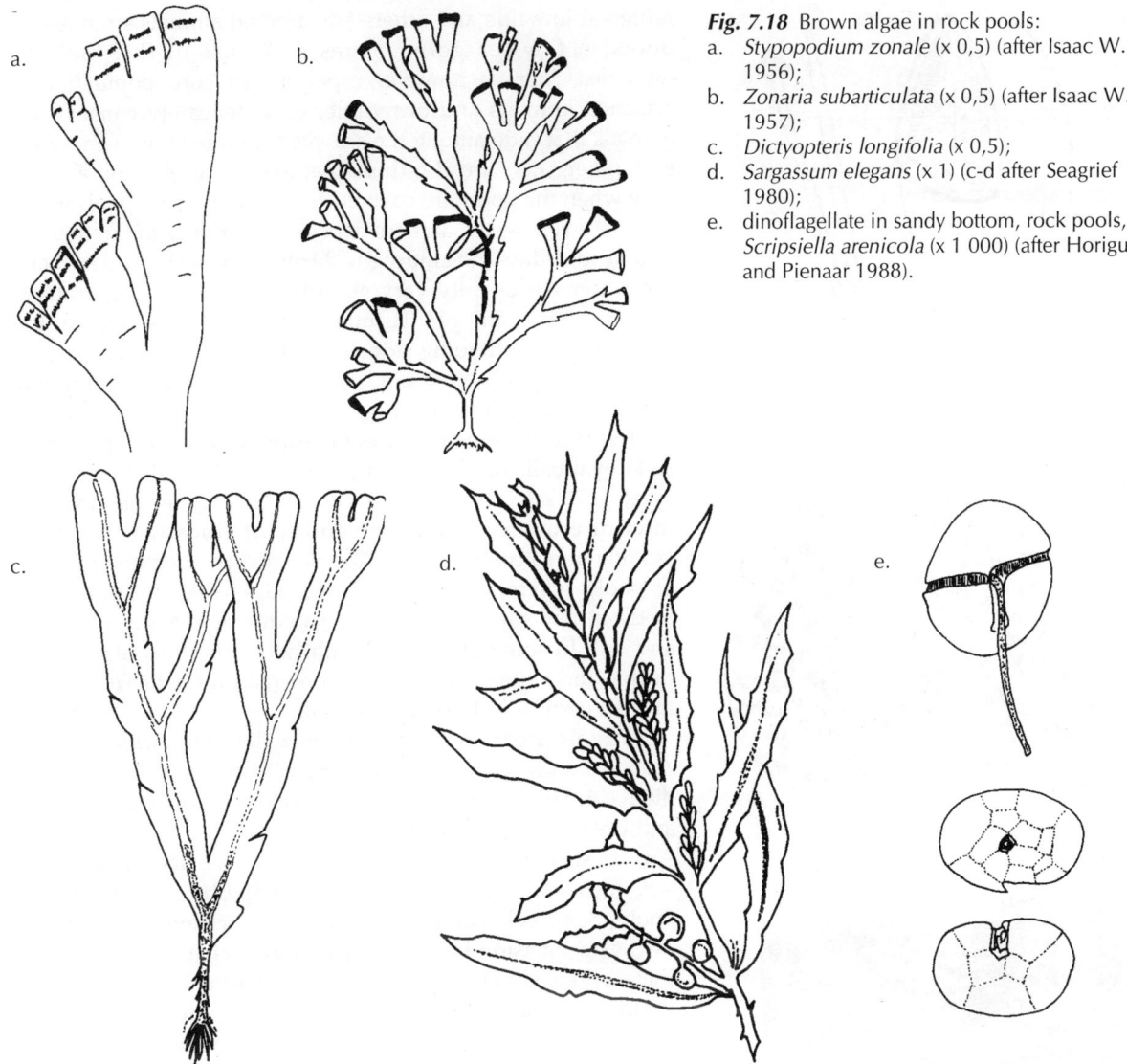

Fig. 7.18 Brown algae in rock pools:
a. *Stypopodium zonale* (x 0,5) (after Isaac W.E 1956);
b. *Zonaria subarticulata* (x 0,5) (after Isaac W.E 1957);
c. *Dictyopteris longifolia* (x 0,5);
d. *Sargassum elegans* (x 1) (c-d after Seagrief 1980);
e. dinoflagellate in sandy bottom, rock pools, *Scripsiella arenicola* (x 1 000) (after Horiguchi and Pienaar 1988).

coast at Inhaca Island, although on Mozambique Island in northern Mozambique, which has semi-exposed shores in the shelter of a large coral reef, other species of sea grasses flourish in sand pockets on the lower rocky shore as they do on the sand of the west coast of Inhaca Island (Kalk 1959).

In tidal pools with a sandy substratum that are flooded daily during high tide, microscopic dinoflagellates abound in the sand (Horiguchi and Pienaar 1991). Very large numbers of *Scripsiella arenicola* (Fig. 7.18e) discolour the surface of the sand at low tide, making large brown patches on the yellowish sand that disappear as the tide rises. It exhibits a diurnal migration to the surface and down again, that is mediated by its flagella.

7.5.2 Sedentary invertebrates: sponges; corals; hydroids; calcareous tube-building worms

The rock-pool communities of animals around the low water levels of spring tides include representatives of almost all marine animal groups, occupying all trophic levels. This variety is partly a consequence of their mixed origin. Some are robust intertidal animals seeking

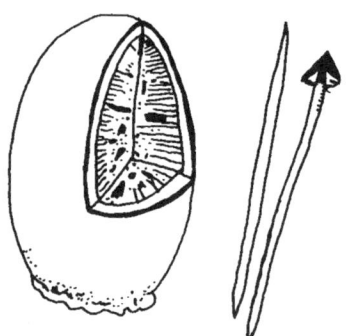

Fig. 7.19 Animals in rock pools:
a. black sponge, *Geodia* sp. (x 1) and spicules enlarged (after Day 1974);

b. coral, *Anomastrea irregularis* (x 1)

b'. calyces enlarged.

refuge at low tide and others are subtidal animals stranded in pools. A few sea-going fishes make use of the pools as nurseries. There is, however, a permanent 'core' population of attached animals and of mobile, very delicately constructed animals that hide among the seaweeds and feed on them or on each other; there are also those that live in crevices and emerge only when the pools are covered by the tide. These pools never dry, not even partly, and are exposed to air only for an hour or two at mid-day and midnight during low spring tides. Thus the water varies only very slightly in salinity, temperature, oxygen content and pH. Turbulence is confined to times of the incoming and outgoing tides and almost subtidal conditions prevail. The pools have sandy floors available to the very few detritus feeding animals.

The rims of some of the pools support a row of apparently solid oval balls of black sponge about 30 mm in height, *Geodia* sp. (Fig. 7.19a), often hidden by algae. This sponge has a yellow interior crowded with large spines of silica which project radially into a central cavity. The elevated position on the edges of pools, where sponge larvae settled, enables the sponge to feed on new supplies of micro-plankton as the sea washes in and out. In some places the brightly coloured encrusting red, purple and yellow sponges, encountered on the flat rocks, also line the vertical walls of the gulleys.

A typical coral of the pools is a flat encrusting species, *Anomastrea irregularis* (Fig. 7.19b and b'), often about 10-15 cm in diameter. This 'anomolous' astraeid coral is peculiar to rock pools and does not occur in the coral reef; it is also endemic to East African shores. The polyps are usually brown and the tentacles, which cluster closely around the oral apertures, have greyish knobs. Some colonies are green in colour or have green patches owing to invasion by green filamentous algae. The tissues of these corals show signs of erosion, a pathological condition. Budding of the polyps in this species is intratentacular, which accounts for the irregular sizes of the calyces to which attention is drawn by the specific name. The septa of each calyx are joined by a circle of small bridges, the synapticulae (see Fig. 8.3f).

There are also bushy growths of some coral genera that occur in the coral reef, such as *Pocillopora, Stylophora* and *Acropora*, (see Fig. 8.4) but the species (see Appendix A) are those tolerant of cooler water and are also found in the rock pools of Natal coasts. The main reef-building coral species are not present (see Section 8.3), although sometimes a brain coral can be seen under an overhang in a small cave.

Unlike the western sheltered rocks on the island, small delicate colonies of 18 species of hydroids have been reported here of which a few colonies are large enough to be visible to the naked eye (Millard 1975). Hydroids are coelenterates with minute polyps growing on branching stems supported by slender horny skeletons and attached to a substratum by a fibrous holdfast. In thecate hydroids, each polyp is surrounded by a skeletal cup, the hydrotheca, whereas the species with athecate hydroid polyps lack this protection. Most of the rock pool hydroids are thecate and attached to seaweeds, to other hydroids or to sponges and to the redbait, *Pyura*, as well as to sea grasses.

Hydroids feed in the characteristic, carnivorous manner of coelenterates, stinging very small planktonic animals with nematocysts whilst catching them with a circle of tentacles (see

Section 8.1). They differ from other coelenterates in life history (Fig. 7.20a-d). The polyps in the attached colonies are asexual and grow by budding. Some buds (gonathecae) (Fig. 7.20a) do not grow tentacles but give rise to many sexual individuals which become free swimming. These are minute, umbrella-shaped 'medusae' with peripheral tentacles and sense organs, resembling jellyfish (Fig. 7.20c). Each medusa develops male or female gonads, and sheds eggs or sperms into the sea where fertilisation occurs. The fertilised egg gives rise to a microscopic ciliated larva, the planula (as in corals and anemones) (Fig. 7.20d), which soon settles and founds a new colony by budding.

Thecate hydroids are more common in pools subtidally and most genera are less than 10 mm in height, whilst a few grow to over 100 mm and are fairly conspicuous. Nine families are represented of which three are common: Campanulariidae, Sertulariidae and Plumulariidae. They differ from one another in the toughness of the horny exoskeleton and the degree of suppression of the sexual phase of the life history, as well as in the shape and positions of the hydrothecae, polyps and nematocysts, as typified by the larger common examples described below.

The cosmopolitan species *Obelia geniculata*, the familiar textbook hydroid, is a typical member of the Campanulariidae in which the hydrothecae are an inverted-bell-shape, i.e. campanulate. The polyps grow on short ringed stalks branching alternately from a slender upright stem (Fig. 7.20a). Each polyp has a single row of tentacles above from which protrudes a trumpet-shaped tube, the hypostome, ending in the mouth (Fig. 7.20b). The theca around the bud which gives rise to medusae, the gonotheca, is pear-shaped. The colony as a whole has a delicate structure and remains upright only under water.

In contrast, genera of the family Sertulariidae are rigid even when taken out of water. The skeleton is more robust being thickened between hydrothecae, which have no stalks and lie in pairs closely adpressed to the axis of each twig or hydrocladium. Each hydrotheca has a toothed margin and can be closed by a hinged operculum. *Thyroscyphus aequalis* (Fig. 7.21a and b) is one of the larger species which grows to 70-100 mm in length. The stem is stiff and woody and the fascicled holdfast is attached to the leaves of the sea grass *Thalassodendron ciliatum* or to large brown seaweeds. The main stems and hydrocladia are all in one plane and lie flat above the substratum. The hydrothecae are on short pedicels growing alternately along the hydrocladia; they are cylindrical in shape and have opercula. The gonatheca is a little larger than the hydrotheca and contains one degenerate medusa with gonads, that is not set free. After fertilisation, planula larvae develop and escape from the gonotheca, thus the free medusoid generation is omitted.

In the Plumulariidae, the colonies have a feathery appearance; there is a stout main axis of many strands along which closely spaced branches grow (Fig. 7.22a-c). These bear very small, parallel hydrocladia, along which flattened hydrothecae are closely spaced. The hydrotheca is not closed by an operculum and the polyp has a conical hypostome. Special

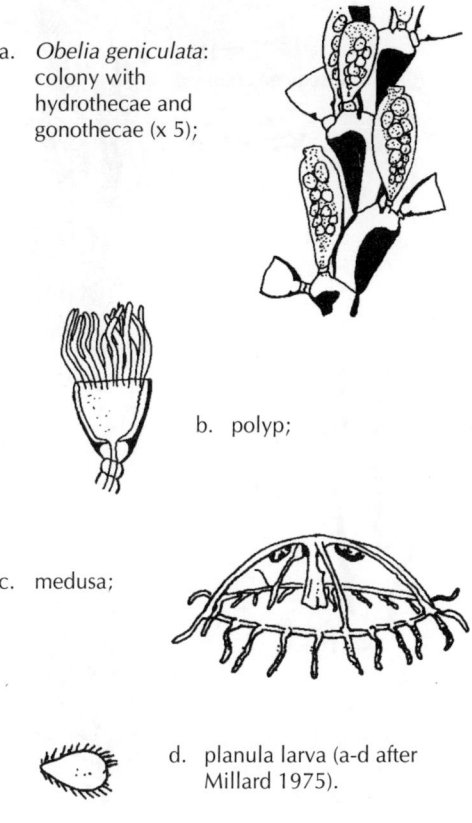

Fig. 7.20 Hydroids in rock pools: Campanulariidae

a. *Obelia geniculata*: colony with hydrothecae and gonothecae (x 5);

b. polyp;

c. medusa;

d. planula larva (a-d after Millard 1975).

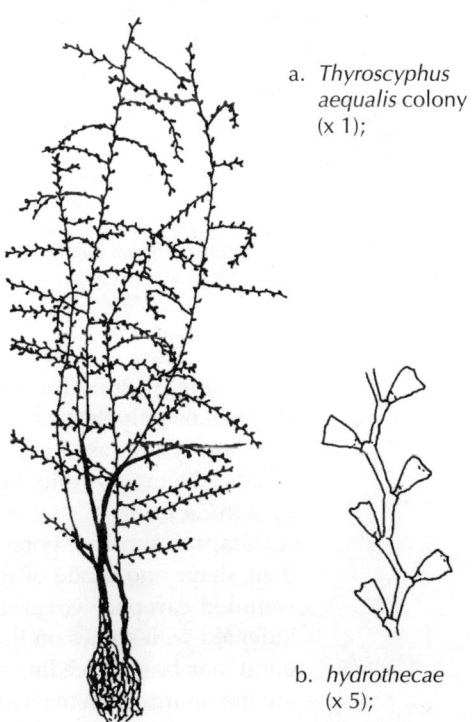

Fig. 7.21 Hydroids in rock pools: Sertulariidae

a. *Thyroscyphus aequalis* colony (x 1);

b. *hydrothecae* (x 5);

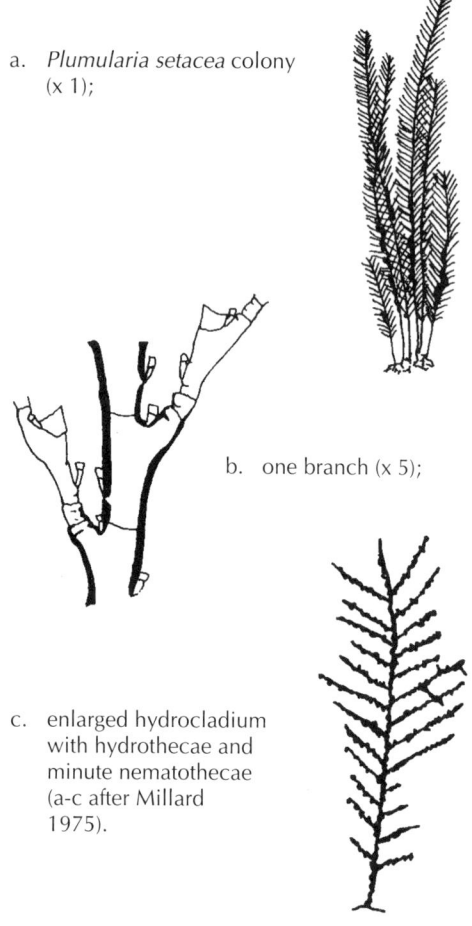

Fig. 7.22 Hydroids in rock pools: Plumulariidae

a. *Plumularia setacea* colony (x 1);

b. one branch (x 5);

c. enlarged hydrocladium with hydrothecae and minute nematothecae (a-c after Millard 1975).

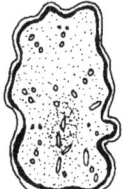

Fig. 7.23
a. flatworm, *Pseudoceros decorus* (x 1), purple with white spots, yellow and purple marginal bands (after Prudhoe 1989).

small movable, stinging polyps (nematothecae) without mouth or tentacles, which specialise in producing nematocysts, occur near the feeding polyps. One of the common species, *Plumularia setacea*, attaches itself to a sponge by means of rhizoids. If grasped in the hand, sharp stings may be felt from the activity of the numerous nematophores. The gonathecae are elongated in this species and contain only gonadal tissue. About eight eggs are fertilised within the gonatheca and develop into planula larvae *in situ*, the medusa stage being completely suppressed. Is this an economy in the turbulent sea, a safeguard for reproduction?

The most conspicuous sedentary polychaetes are the serpulids (see Section 6.5.4), which form minute white calcareous tubes, scattered singly on large blades of submerged seaweeds. The tubes of *Spirorbis foraminosus* are curiously coiled in an anticlockwise direction and are ornamented with five longitudinal ridges and transverse slits on the grooves between them. When under water the feathery crown of tentacles protrudes for ciliary feeding. After fertilisation of the eggs, one tentacle, modified into an operculum to close the tube, carries the developing embryos in a membranous sac. On the other hand, larvae of *Serpula* and *Hydroides* settle on rocks and shells and then form smooth white tubes coiled in a clockwise direction (see Section 6.4.5). The only colonial species *Pomatoleios kraussi* which forms a tangle of bluish tubes on rocks in Natal, is not common on Inhaca Island shores; where it does occur, it is at the level of low neap tide.

7.5.3 Mobile invertebrates: flatworms; polychaetes; crustaceans; pycnogonids; sea slugs; bubble shells; cones; cowries; echinoderms

Although flatworms are not as numerous here as on the coral reef, some of the beautiful and graceful, tropical species are found among the seaweeds. *Pseudoceros decorus*, a recently described species is oval in outline and 20 mm in length (Prudhoe 1989). It is purplish dorsally with a pattern of white spots of various sizes and shapes (Fig. 7.23a). The margin of the flatworm is outlined by a yellow band enclosing a broad band of deep purple. Flatworms are described in Section 6.5.4.

Hidden amongst the seaweeds, about 20 species of actively swimming or crawling segmented worms abound. These are mainly members of three families of polychaetes and differ in sensory appendages, masticatory apparatus and in locomotory organs, all of which are related to diet. They belong to the mainly algivorous syllids, the omnivorous nereids and the carnivorous eunicids and have been described in Section 7.4.2.

Grapsid, xanthid, porcellanid and spider crabs are common and easily distinguished by their shape and mode of feeding. A grapsid crab, *Plagusia depressa tuberculata* (45 cm) has a rounded carapace covered with tubercles (Fig. 7.24a). The antero-lateral edge is deeply indented with spines on the second joint and the dactyls have many spines. When the tide is out it may be seen feeding on rock surfaces. It has small chelae and is probably herbivorous as are the sesarmids in the mangrove swamps.

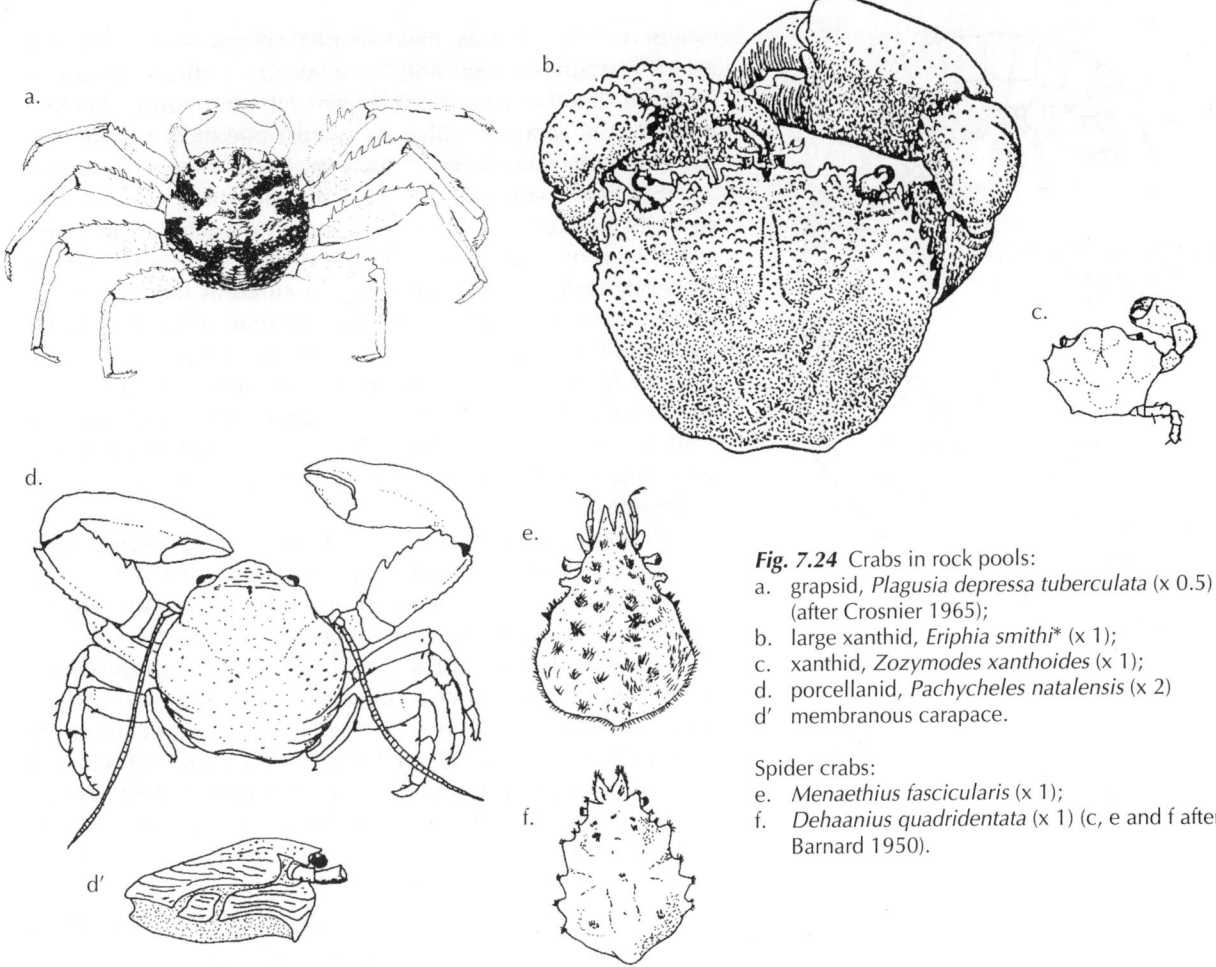

Fig. 7.24 Crabs in rock pools:
a. grapsid, *Plagusia depressa tuberculata* (x 0.5) (after Crosnier 1965);
b. large xanthid, *Eriphia smithi** (x 1);
c. xanthid, *Zozymodes xanthoides* (x 1);
d. porcellanid, *Pachycheles natalensis* (x 2)
d' membranous carapace.

Spider crabs:
e. *Menaethius fascicularis* (x 1);
f. *Dehaanius quadridentata* (x 1) (c, e and f after Barnard 1950).

There are two groups of xanthid species, large and very small. The large ones are three species of *Eriphia* in which the carapace is deep with convex lateral margins and the chelae are powerful. These crabs are usually first detected by recognising a bright blue or red eye protruding from a crevice. The reddish brown crab *E. smithi* (50 mm) with blue eyes is the most common (Fig. 7.24b). It is not unexpected that these crabs return to their home holes after nocturnal feeding excursions over the rocks, using only visual cues. It feeds on molluscs by cracking the shells with its heavy, toothed chelae.

An assortment of much smaller xanthid crabs such as *Zozymodes xanthoides* (Fig. 7.24c) have spoon-shaped chelae and make up the other group (see Appendix A). They scrape the rocks for algae and any small animals entangled in them. They are best found in quantity when the tide is out, by opening up the sandy worm tubes of *Idanthyrsus pennatus* on exposed rock where they shelter. The tubes are described in the next section.

Porcellanid crabs are represented by four species (see Appendix A). They are anomurans, related to hermit crabs. They are more common than on the sheltered shores for they are filter feeders and so gain in efficiency from the turbulence of the water (see Section 6.5.2). The most common porcellanid on the rocky shore is *Pachycheles natalensis* (Fig. 7.24d and d'). Its carapace (6 mm) is mottled blue and red on a greenish background and it differs from all other species by the presence of peculiar membranous patches on the calcareous sides of the carapace. It would be interesting to find out their function. Do they have a respiratory function as in *Dotilla fenestrata* (see Section 3.5)?

Six species of small spider crabs of the genera *Menaethiops*, *Menaethius* and *Dehaanius* (see Appendix A) have been found sheltering among algae or in empty *Idanthyrsus* worm tubes

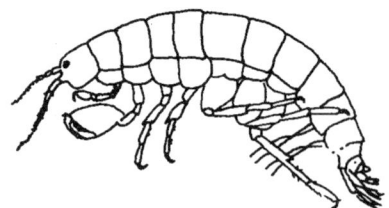

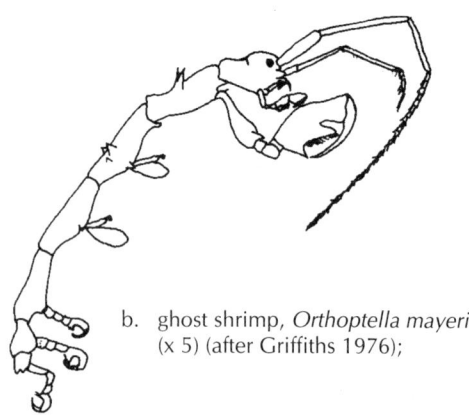

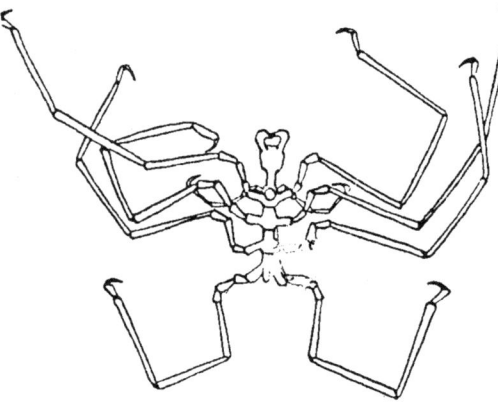

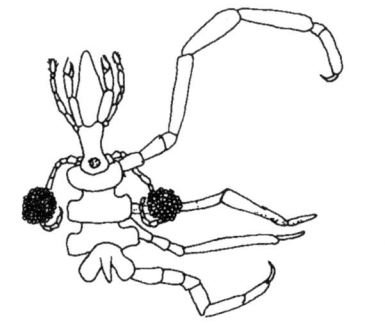

Fig. 7.25 Amphipods in rock pools:
a. *Hyale grandicornis* (x 2) (after Branch and Branch 1981);
b. ghost shrimp, *Orthoptella mayeri* (x 5) (after Griffiths 1976);
pycnogonids:
c. *Parapallene hodgsoni** (x 2);
d. *Pycnogonum* sp. (x 3).

(see Section 7.5.5). The shield-shaped carapaces are produced into a rostrum in front and have specific patterns of lateral teeth on the carapace. Most species have numerous hooked hairs on the carapace either as a short pile or in patches. To these are attached pieces of algae or, in the case of *Dehaanius quadridentata*, small hydroids. The dactyls of the legs are suitably hooked for clinging to algae or rock surfaces and resisting wave turbulence. The beak-like chelae are used for gathering algae and hidden minute animals as food. Common species *Menaethius fascicularis* and *Dehanius quadridentata* are shown in Fig. 7.24e and f. The life history of the spider crab is peculiar in that it moults few times. There are only two zoea larval stages (instead of five as in other crabs), and after the adult form has been attained the crab does not moult again. Very small xanthid crabs such as *Zozymodes xanthoides* (Fig 7.24c) also occur here.

The common amphipod is the circumtropical species, *Hyale grandicornis*, which is smaller than *H. inhacae* of the sheltered rocks. It has enlarged second gnathopods that are almost square in outline and have few spines (Fig. 7.25a).

Highly modified caprellid amphipods or ghost shrimps such as *Orthoptella mayeri*, have a long slender thorax armed with an enlarged second pair of robust gnathopods, and a minute abdomen (Fig. 7.25b). They hide between the branches of hydroids or algae from which they are barely distinguishable without using a lens. They cling securely with the hooks on the last three pairs of legs while the upper part of the body sways freely in the water and catches small motile prey such as small worms and crustaceans. Caprellids are also found on hydroids among sea grasses (see Section 9.4.2).

The pycnogonids, sometimes erroneously called sea spiders, are a group of marine arthropods which in detail, are quite unlike any other class of arthropods although they resemble them in having a chitinous jointed exoskeleton and many jointed legs (King 1973). They may have evolved from the earliest ancestors of spiders long before the latter left the sea. The long body consists of a narrow head and thorax fused together, the abdomen being merely a tiny posterior protuberance, as in ghost shrimps. There are four pairs of very slender nine-jointed walking legs. In the male, a much shorter pair of ovigerous legs, which pick up masses of eggs when they have been externally fertilised, is usually present. It carries them until they hatch into three-legged larvae, but unlike a crustacean nauplius, the larval legs have only three segments and are the precursors of chelicerae, palps and the first pair of legs, instead of antennae and mandibles. There is no nauplius eye. Larvae may still cling to the egg mass while they moult several times, adding another pair of legs each time.

The pycnogonids usually live in deep water in temperate or cold seas. A few species may be found on tropical shores hidden among branches of the hydroids on which some species feed. Others feed on sponges, tunicates or anemones. In spite of having many-jointed legs, locomotion is very, very slow. The legs move only in an up and down direction in one plane, except for a central joint which has a little movement forward and backward. Individuals are usually quite stationary, the body hanging from the long legs and swaying in the water. They can swim slowly by slow consecutive beating of pairs of legs with the dorsal surface of the body held foremost.

Two species have been recorded from Inhaca pools: *Parapallene hodgsoni* is a slender form (Fig 7.25c) and a *Pycnogonum* sp. (Fig. 7.25d) has a stouter body and shorter legs. *Parapallene* has no palps on its head and the appendages normally bearing chelae, the cheliforae, are very small, jointed but blunt. In *Pycnogonum* sp., the stouter form, cheliforae are absent. Both species feed on hydroids or sponges by making holes in their prey with horny lips on the end of a proboscis and sucking out fluids. So narrow is the body of a pycnogonid that the branches of the gut and the reproductive organs are encased in the legs. In late larval stages both the legs and the internal organs can be regenerated.

The molluscs in rock pools include the herbivorous slugs and carnivorous bubble shells (opisthobranchs) and both herbivorous and carnivorous snails (prosobranchs) but no bivalves.

Two species of delicate, highly coloured sea slugs belong to the order Sacoglossa, a name that indicates that the used teeth on the radula are not discarded but are retained in a sac. They have no gills, but the delicate skin is permeable to gases. Each species of Sacoglossa is restricted to feeding on a single genus of alga. *Elysiella (Halimeda) pusilla* (Fig. 7.26a) is found on the siphonous green alga *Halimeda*, on which it feeds exclusively. Tiny white tracks are left on the seaweed when the animal scrapes away the green surface tissue so that the underlying calcite is exposed. The curved teeth in a single row on the radula are adapted for scraping by means of the serrations on their inner edges.

A similar, even smaller, sacoglossan *E. punctata* feeds on the green alga *Codium* whose dark green colour it matches. Its method of feeding fits the anatomy of this genus of seaweed. Single utricles on the surface are grasped by the lips and foot of this tiny animal and pierced by a smooth, dagger-like tooth. The contents are squeezed and sucked into the gut and the collapsed utricle remains when the slug moves on to drain another.

Sea slugs in another Order, Nudibranchia, also feed on algae, but in a less specialised manner; they graze by means of a many-toothed radula. Nudibranchs have neither shells nor mantle cavity and have evolved exposed, feathery gills around the mid-dorsal anus instead of having gills inside a mantle cavity as in snails (see Fig. 5.13h). They are more robust and larger than sacoglossans and are often of one colour, such as the red *Dendrodoris rubra* and the black *D. nigra*.

Among snails in the pools the herbivores are the cowries which graze on young stages of algae and, inadvertently perhaps, also on small hidden animals. In places where wave action is brisk, *Cypraea caputserpentis* (35 mm) is common; the shell resembles the colouring of a snake's head (Fig. 7.26b). In the more sheltered places the smaller *C. helvola* (20 mm) occurs. The shell is greenish with white spots dorsally, the sides are brown and it is lighter underneath. The reddish-brown mantle is flecked with light green and the mantle papillae are bright red.

Fig. 7.26 Molluscs in rock pools:

a. sacloglossa (slug) *Elysiella (Halimeda) pusilla** (x 4);

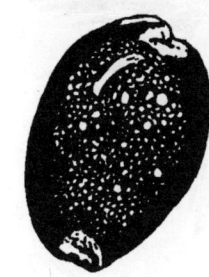 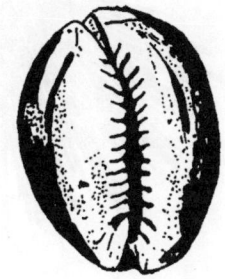

b. cowrie, *Cypraea caputserpentis* (x 1);

c. opisthobranch, *Hydatina physis* shell (x 1) (b and c after Kilburn and Rippey 1982);

carnivorous gastropods:

d. *Conus ebraeus** (x 1);

e. *Bursa granularis** (x 1).

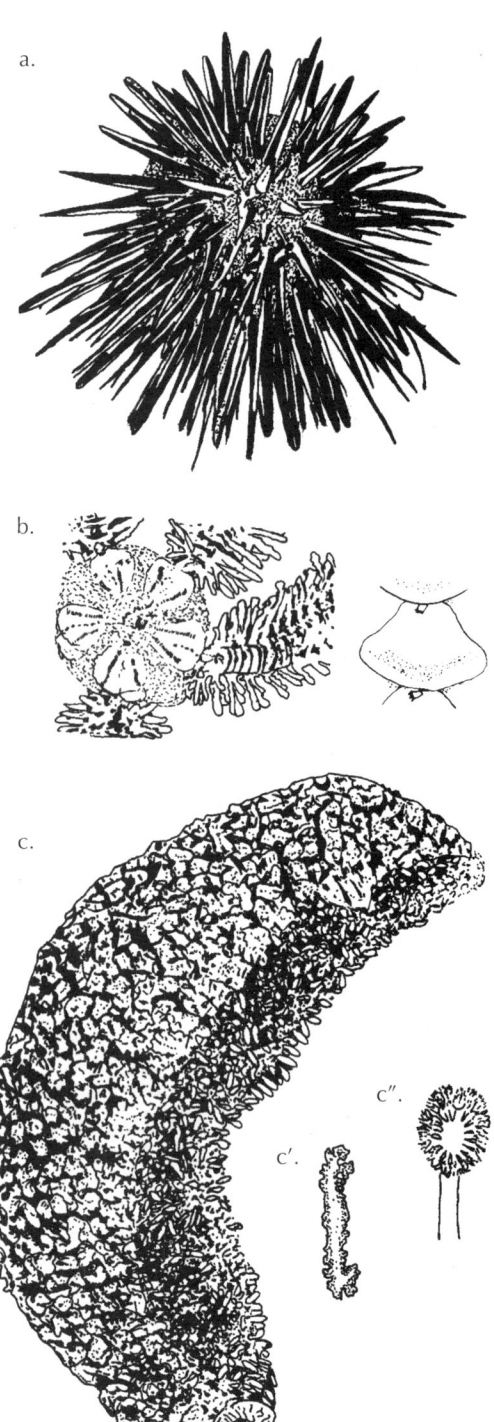

Fig. 7.27 Echinoderms in rock pools:
a. *Stomopneustes variolis* (x 0,5) (after Branch and Branch 1981);
b. brittle star, *Ophiothrix echinotecta* (x 4), upper arm plate enlarged (after J.B. Balinsky 1957);
c. sea cucumber, *Actinopyga mauritiana* (x 0,5),
c'. one tentacle and
c". a spicule enlarged (c-c" after Thandar 1987).

The carnivorous bubble shells, which are primitive opisthobranchs (Cephalaspidea), have gills inside a posterior mantle cavity. Their shells are very delicate, smooth and globose with a wide aperture into which the animal cannot completely withdraw its whole body. *Hydatina physis* (Fig. 7.26c) occurs on the exposed shores. It has a deep rose-coloured foot edged with blue and the shell has several longitudinal rust-red wavy streaks, sometimes over a whitish background and sometimes over faint, concentric brown pin-stripes. *Hydatina* feeds on small polychaete worms which it swallows whole, the foregut being protected from laceration by the bristles of the prey by a chitinous lining.

Among the carnivorous snails, noted for their well-developed olfactory osphradium in the mantle cavity, *Conus ebraeus* (25 mm) is common. This shell is covered with a bright yellow periostracum through which rows of small rectangular black spots are conspicuous (Fig. 7.26d). This cone-snail feeds on the ubiquitous nereid and eunicid worms concealed among the algae. *Bursa granularis*, the tropical frog-shell (40 mm) (Fig. 7.26e) rests in pools between tides; it feeds on worms, molluscs and even sea urchins, using its extremely acidic saliva to dissolve calcareous shells.

At first sight echinoderms cannot be seen in the rock pools and gullies, but by searching near the upper parts of the walls, horizontal crevices may be found that are just large enough to conceal sea urchins and sea cucumbers. Unlike the rocky pools of cooler shores these pools do not contain starfish. The sea urchins are the dark, oval-shaped *Echinometra mathaei* which have the stronger, flatter spines and the round, dark *Stomopneustes variolis*. The former species also occurs on the rocks of the reef flat on the sheltered shore (Fig. 6.10a) whereas the more robust *Stomopneustes* is confined to rocks exposed to strong wave action (Fig. 7.27a). Underneath them, lives nearly always, a small greenish brittle star, *Ophiothrix echinotecta*, appropriately named since the sea urchin provides a welcome roof in the turbulent water of the incoming tide (Fig. 7.27b). There are almost no suitable cracks or loose stones where brittle stars may rest securely on the exposed shore, but *Ophiactis carnea*, *O. plana* and *O. savignyi* can be found in empty worm tubes. This is in complete contrast to the sheltered coral limestone rocks of the reef flat and coral reefs, where both large and small brittle stars are very common (see Sections 6.5.1 and 8.5.4).

Two species of sea cucumber, which do not occur on the sheltered rock or sand, usually live in separate groups in rock crevices. Both species are brownish with no tube feet dorsally but the ventral 'soles' have a crowded mass of very adhesive tube feet which hold them securely in the rocky crevice when the tide rushes in and out. The tube feet are also used to creep around at night to feed on the *rocks* above at low tide, possibly on algae or detritus. The gut contents should be examined since most sea-cucumbers feed on sand. In *Actinopyga mauritiana* (Fig. 7.27c, c' and c") the sole is pale purple; *A. plebeja* has a rose red sole. Both species have five triangular calcareous plates embedded in the external skin around the anus.

7.5.4 Fishes in the rock pools of the exposed shore (Smith and Heemstra 1986): klipfish; blennies; gobies; flagtail

A number of species of small bony fishes, most of them not more than 10 cm in length, inhabit the rock pools. Some are permanent inhabitants, some are young stages of fishes that frequent the rocky reefs of shallow water as adults and make use of the pools as nurseries. Others are small species usually found in coral reefs (see Section 8.5.6). Larger fishes that live in deeper water may visit the pools in search of prey (see Section 7.6).

Three families of fishes found at Inhaca Island pools are especially well adapted to life in rock pools, having the ability to leave the water temporarily and skip or crawl to adjoining pools. Their pelvic fins are adapted in three different ways to assist with crawling or supporting the body on the rock. Most are exceedingly well camouflaged with colours and patterns that match their habitat – be it seaweed, sea grass or encrusted reef.

Klipfishes (Clinidae) are largely endemic to southern Africa and most inhabit cooler shores among the giant strap-like brown seaweeds, known as kelp. Although this alga does not occur in the warmer rock pools at Inhaca Island, some of the klip fishes have extended their range northwards, possibly assisted by the erratic cooler counter-current which flows northwards close inshore during the winter. The clinids, such as *Pavoclinus graminis* (10 cm) (Fig. 7.28a) which is common on the east coast of South Africa, have smooth elongated bodies, covered with small scales. The pelvic fins of the clinids have evolved into narrow, *jointed 'legs'* that support the body when at rest and with which the fish is able to crawl over the rocks. The camouflage of adults is achieved by a pattern of darker and lighter brown patches and stripes and by the ability to change colour, becoming lighter or darker according to the background of seaweeds in which they conceal themselves. Fishes change colour very rapidly since the chromatophores, like those in reptiles, are under nervous, rather than hormonal control, as in invertebrates.

The rock-hoppers or blennies (Blennidae) are a common and widely distributed family on tropical shores – including Inhaca Island. They are small and have elongate bodies but lack scales, and the spines of the median fin are feeble. The anal fin is short and the pelvic fins are *short curved rods*. The most conspicuous species of rock-hopper are *Antennablennius bifilium* (Fig. 7.28b) and *Omobranchus fasciolatus* (Fig. 7.28c). *O. fasciolatus* is a common carnivore of warmer rocky shores.

The gobies (Gobiidae) are also elongate fish but normally have a more rounded head and have scales on the body. They do not swim far when disturbed but quickly 'perch' again using the *joined, cup-shaped pelvic fins* which form an adhesive disc. The small *Priolepis inhaca* (4 cm) (Fig. 7.28d) may be recognised by its overall reddish colour and the reticulate colour pattern of its large scales, which are edged with red lines. There are small canine teeth in the jaws and it feeds on small invertebrates.

Many other rock-pool fishes are beautifully streamlined and dart gracefully about the pools, hovering at the rock face. For example, the flagtail (Kuhliidae), *Kuhlia mugil* (12 cm) (Fig. 7.28e) has a brilliant silvery body and five black radial bands on the tail fin which clearly shows the 'flag'.

The bluestreak cleaner wrasse, *Labroides dimidiatus* (100 mm) (Fig. 7.28f) may occur in rock pools. It is common over the coral reef and in deeper water. Several have been observed nibbling the teeth of the ragged toothed sharks (Fig. 7.30h) that collect around the wreck at the north east end of the island. The cleaner weasses remove mucus and parasites from their host's teeth and gills. The conspicuous wide longitudinal royal blue stipe advertises its services and predatory fishes do not attack it or drive it away. Like many other species of wrasses (Labridae) the cleaner wrass can change sex; small shoals are comprised of a harem of several females with one slightly larger male.

7.5.5 Reef-building polychaetes; green algae; tunicates

Most of the horizontal rock surface at the edge of the lower platform is covered by an encrustation of very coarse sand, 150-200 mm thick, which is composed of horizontal,

Fig. 7.28 Fishes in rock pools:
a. clinid, *Pavoclinus graminis* (x 0,5);
blennies:
b. *Antennablennius bifilium* (x 1);
c. *Omobranchus fasciolatus* (x 1);
d. goby, *Priolepis inhaca* (x 1);
e. flagfish, *Kuhlia mugil* (x 0,7);
f. cleaner wrasse, *Labroides dimidiata* (x 1) (after M.M. Smith and P.C. Heemstra 1986).

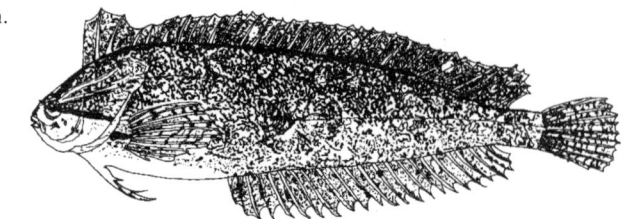
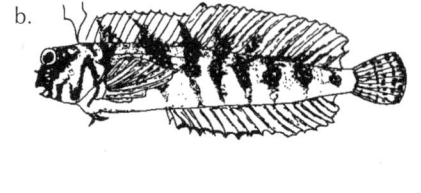
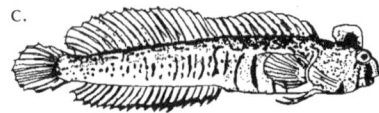

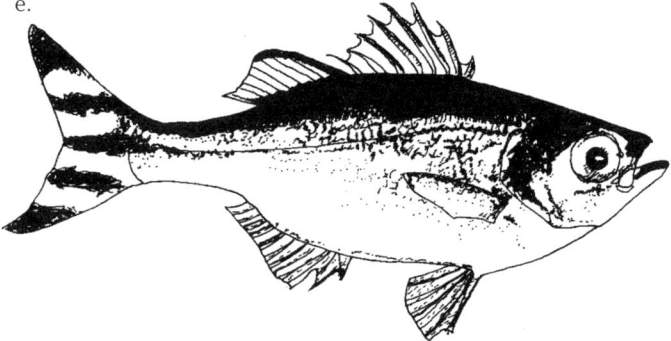
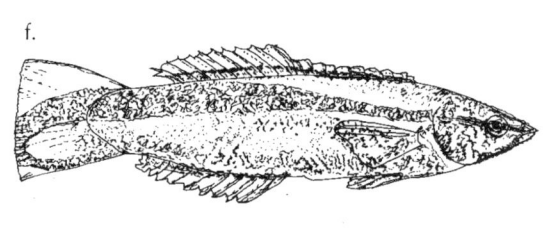

overlapping worm tubes, strongly cemented together. Each tube is about 70 mm long and closed at one end; the other end has a flanged aperture, 10-15 mm in diameter, that always faces landward. These tubes are constructed by the sabellariid polychaete worm *Idanthyrsus pennatus*, the Natal Reef worm, which grows to a length of about 100 mm (Fig. 7.29a). Its range extends to the latitude of Mozambique Island (15°S) When the rock is exposed to air for an hour or so at low spring tides the worm withdraws into its tube. When the worm feeds under water the head is protruded and the cleft is opened, tilting the two halves apart and revealing bunches of slender grooved and ciliated tentacles (cirri) alongside the mouth. These are extended to catch plankton on their mucoid surfaces. Two shorter, stouter tentacles on the head select very coarse sand grains from the backwash and cement them to the edge of the tube as the worm tilts its head into the retreating, sand-laden surf. It occupies the same niche as does *Gunnarea capensis* on the rocks around the Cape coast of South Africa.

Empty worm tubes are inhabited by a large number of a variety of small invertebrates since the tubes provide the only available firm shelter from the surf on these shores. On one occasion it was found that a chunk of worm tubes about 30 x 20 x 10 cm^3, from which 20 living *Idanthyrsus* worms were extracted, sheltered about 200 other invertebrates. There were 38 polychaete worms of 18 species belonging to the syllid, nereid and eunicid families, which usually feed among algae. In addition there were two species of scale worms, i.e. polychaete worms that are covered by pairs of large, raised, transparent scales (elytra). These are the carnivorous, brown *Lepidonotus durbanensis* and *L. cristatus* both of which have 12 pairs of elytra, which in *L. cristatus* carry a crest. This genus has a complex set of antennae on the head (cf. Fig. 6.15d).

There were also about 80 small crabs, mainly spider crabs, porcellanids of several species and a few very small xanthid species. The small, stout pycnogonid, *Pycnogonum microps*, was also found here (Fig. 7.25d). Eleven species of small molluscs, including cones and cowries, key-hole limpets and sea-slugs were present. There were small brittle stars, the majority being the tropical greenish, six-armed, *Ophiactis savignyi* (Fig. 6.10e), the reddish-brown, five-armed *Ophiactis carnea* and the slightly larger, velvet-brown *Ophiocoma valencia*, all of which are common on the reef flat and coral reef. The rare *Ophionereis vivipara* may sometimes be found; it gives birth to fully formed young that have been incubated internally and emerge through radial slits on the oral surface of the body. These are all small species; the larger species of the coral reef do not appear to be present.

One green seaweed species, *Ulva rigida* (Ulvales) (Fig. 7.29b), is able to attach itself to the very coarse cemented sand of the worm tubes, despite the removal of algal spores in the feeding currents of the worms (Isaac 1956). This alga has a brilliant green thallus in the form of a flat sheet supported by a very short fleshy stipe. The sheet grows to about 15 x 25 mm and remains completely rigid when out of water. The thallus is not fluted as in most temperate species of *Ulva* (known as the sea lettuce); it is composed of only two layers of large cells, embedded in thick mucilage. The chloroplasts in these cells have been shown to rise to the surface of the cells for photosynthesis under water and to sink to the bottom of the cells at low tides when they are exposed to strong sunlight. This adaptation to exposure may well be of critical importance in its exposed habitat.

Some of the rocks at the very low intertidal level, including the single large vertical boulder, small caves and overhangs, are almost completely covered by hundreds of the large, almost spherical, black tunicate, *Pyura stolonifera* (Fig. 7.29c), known as 'redbait' in Natal. Internally the body structure is similar to that of *Ascidia* described in Section 6.4.2. The red alga *Rhodymenia natalensis*, which is dominant in some low level pools, occupies the rocks nearby.

7.6 FISHES OF THE OPEN SEA (Smith and Heemstra 1986)

The food chain of the open sea culminates in the large, carnivorous fishes which are tempting game for anglers, since they are fast and powerful. Some of them can be caught on line from the shore. They come fairly close inshore on the continental shelf in search of smaller fish prey such as bream, snappers and the smaller rock cod, which haunt the rocky reefs, or shoaling planktivorous fishes, such as herring-like fishes and needle fishes. The food chain around the island is enriched by the fishes of the offshore coral reefs where the fish *genera* are much the same as in the sheltered reefs, but the *species* may be different as is the case in Kenya coral reefs (Talbot 1965). Many of the carnivorous fishes are also present in the Bay of Maputo and here attention is focused on the larger game fish species which delight anglers.

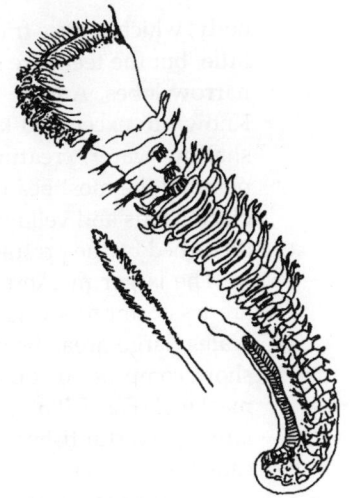

Fig. 7.29
a. A reef-building polychaete, *Idanthyrsus pennatus** (x 1); and single palea from crown of operculum;

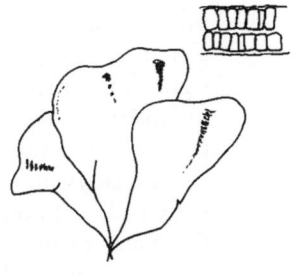

b. green alga, *Ulva rigida* (x 1) and cell detail (after Seagrief 1980);

c. solitary tunicate, *Pyura stolonifera* (x 0,5).

7.6.1 Carnivores: Kingfishes; Mackeral and bonitas; Barracuda; Ladyfish; Marlin; Rock cod and Sea bass; Ragged-toothed shark

The kingfishes (*Caranx* spp.) (Carangidae) grow to a large size – up to a metre long – and have deep bodies, flattened laterally. They have thin, flattened or needle-shaped scales covering the body, which reduce friction. The mouth is moderately large and the lower jaw may protrude a little, but the teeth are small. The dorsal and anal fins extend to the tail, which is forked with narrow lobes. Among the most powerful of gamefish is the giant kingfish *Caranx ignobilis*. Known to exceed 50 kg, this aggresssive predator will vigorously attack its prey, often in shallow water, creating a visible and exciting display. *Gnathodon speciosus* (Fig. 7.30a) is possibly the most beautiful of the family. It has a bright yellow or silvery colour with narrow black bands and yellow fins. It feeds on benthic crustaceans, molluscs and small fishes that are disturbed by the protractile mouth when it prods the sand.

The larger mackerel (Scombridae) have more elongated bodies than the kingfishes; they are less compressed laterally and thus more cigar-shaped. The body has even fewer minute scales, large areas being naked. The mouth is large and, unlike kingfishes, mackerel have short, compressed cutting teeth and feed on fishes. Both *Scomberomorus plurilineatus*, the queen mackerel (Fig. 7.30b), and *S. commersoni*, the king mackerel, occur around Inhaca Island. These large, powerful fishes feed on shoaling fishes and squid. Sometimes mackerel are mistakenly called barracuda.

The tropical striped bonita (*Sarda orientalis*), like the tuna of cooler waters, is a member of the mackerel family. The mouth is large and, unlike the tuna, it is armed with formidable teeth; the body is mainly brownish in colour (Fig. 7.30c).

The true barracuda (Sphyraenidae) are species of the genus *Sphyraena*; these fishes have long, torpedo-shaped bodies, almost cylindrical in cross section, small scales, two small dorsal fins and a large mouth with strong 'canine' teeth in the jaws and on the palate; the lower jaw projects a little beyond the upper jaw. Barracuda may often be encountered, from the safety of a boat, in large groups and some may leap high out of the sea into the air – a most spectacular sight, which may continue for an hour before the 'school' swims away. *Sphyraena barracuda* of the Indian Ocean has dark blotches on the sides of its greyish body; the fins are dusky with paler tips. It may attain a length of 150 cm (Fig. 7.30d).

The ladyfish, *Elops machnata* (Elopidae) (Fig. 7.30e), is similarly torpedo-shaped, but has large scales and only one short dorsal fin; it has a large mouth with many minute teeth on all the bones of the mouth and on a bony plate below the lower jaw. Large breeding shoals congregate in shallow water in Mozambique during winter months. It has a more primitive ancestory, as shown by its unusual life history; the eggs hatch into transparent ribbon-like young (resembling eel larvae, but with a forked tail) which soon change into the normal adult form when they grow.

The black marlin, *Makaira nigricans* (Fig. 7.30f) belongs to the family that includes the sailfish and spearfish (Istiophoridae). It has a robust body, built for speed and strength, and it may grow to a length of 5 m. The vertebrae are clearly adapted for speed in having large dorsal and ventral crests to which large body muscles are attached. The pectoral fins are slender and the dorsal fin has a large anterior crest followed by a very low, long fin; the tail has extremely narrow dorsal and ventral lobes. Characteristically the marlin has an upper jaw produced into a short spear. It is rarely known to ram its prey with the spear; it slashes its prey rather than impaling it. The fish prey is usually swallowed whole, sucked into the large mouth without attempting to chew it; it has only very fine, villiform teeth.

The garrupa or rock cod family (Serranidae) is well represented at Inhaca Island and includes both large and smaller species. There are many species of *Epinephelus* which have deep, robust bodies, not built for speed, having a square-cut tail and a cryptic coloration. When threatened they take refuge under rocks from which the angler has great difficulty in dislodging them because they anchor themselves between rocks by expanding the large opercula covering the gills. The colourful and common *E. fasciatus* (35 cm) (Fig. 7.30g) is a reddish colour with the edge of its dorsal fin conspicuously black. *E. tauvina* is a large species and may grow to 70 cm in length. *Cephalopohlis*, a genus of small rock cod (15-25 cm), is represented at Inhaca Island by several species. *C. boenack*, a brownish rock cod frequents the old reefs offshore on the east coast of Inhaca Island.

The brindlebass *E. lanceolatus* is one of the largest rock cod, and may grow to over 2 m in length, becoming one of the largest fishes in the sea. It is reputed to attack divers. It feeds on

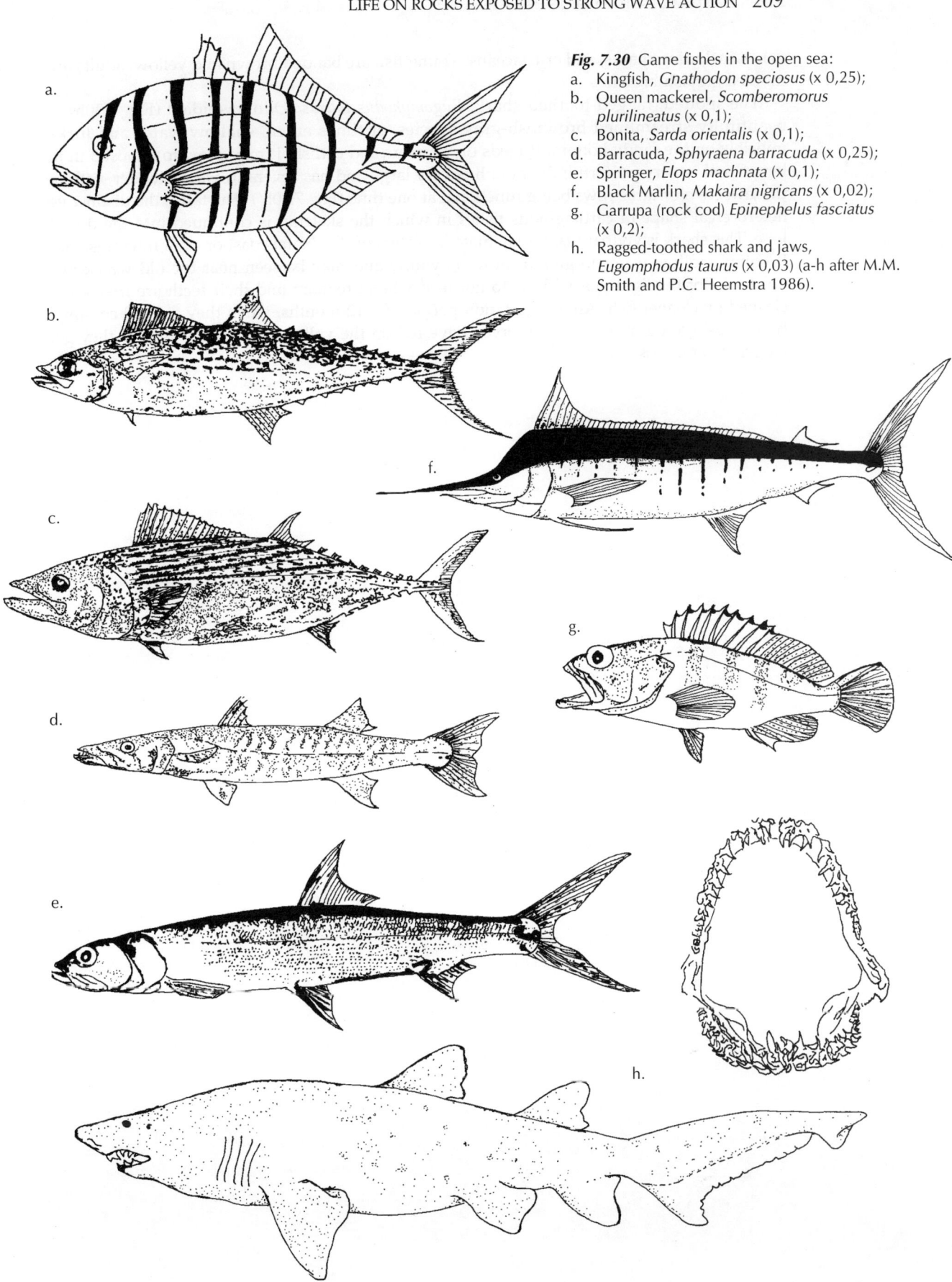

Fig. 7.30 Game fishes in the open sea:
a. Kingfish, *Gnathodon speciosus* (x 0,25);
b. Queen mackerel, *Scomberomorus plurilineatus* (x 0,1);
c. Bonita, *Sarda orientalis* (x 0,1);
d. Barracuda, *Sphyraena barracuda* (x 0,25);
e. Springer, *Elops machnata* (x 0,1);
f. Black Marlin, *Makaira nigricans* (x 0,02);
g. Garrupa (rock cod) *Epinephelus fasciatus* (x 0,2);
h. Ragged-toothed shark and jaws, *Eugomphodus taurus* (x 0,03) (a-h after M.M. Smith and P.C. Heemstra 1986).

fishes, including sharks, and crustaceans. Young fish are banded brown and yellow; adults are mottled greyish-brown.

The spotted, ragged-toothed shark, *Eugomphodus taurus* (Odontaspidae) (Fig. 7.30h), a beautifully streamlined brownish-grey species, at times enters shallow water and lurks invisibly on the sandy bottom. It feeds on grunters and catfish. It may grow to almost 3 m in length and is very powerful; the mouth is very large and has several rows of slender teeth in the jaws, two or three rows being functional at one time (Fig. 7.30h'). The tail is the distinctive heterocercal shape of cartilaginous fishes in which the skeleton is continued into the dorsal lobe. This shark is viviparous; it may mate in waters off the Natal coast or even Inhaca Island. It enters shallow water to give birth to its young and may be seen near the old wreck just north-east of the lighthouse. They do not feed when pregnant and their teeth are frequently cleaned by cleaner fish. After a gestation period of 9-12 months, when they are 100 cm long, two 'pups' are born, which as embryos have fed on the yolk of many eggs and, as they get bigger, on embryos.

8

CORAL REEFS

The main criteria for the building of a successful coral reef are present on the sheltered coasts of Inhaca Island, despite its geographical position outside the tropics. The temperature of the sea is rarely below 18°C; the required hard substratum occurs below low tide in the form of old coral rock at the edge of channels in which strong currents continuously sweep away threatening deposits of sand. However, these reefs are in a state of turnover through intermittent cyclonic destruction and rejuvenation, that is more rapid than in mature tropical reefs (see Section 8.6). Patches of coral also occur offshore, 5-15 m deep on the east coast on a substratum of old reef.

During the day, the coral reefs at Inhaca Island resemble compact shrubberies of low stony bushes of many pastel shades. Some 'shrubs' have short, stout branches, a few are more delicate; some are hard, round mounds, large or small, and others are like flat shelves or fluted plates. The branches of the corals may appear to overlap but not so as to exclude light from any one. Coral reefs are as dependent on light as green plants because the cells of reef-building coral animals contain thousands of unicellular algae, an imprisoned phytoplankton species, containing photosynthetic chloroplasts. These algae must be exposed to light so that the corals will grow. The multitude of large pores (calyces) in the coral rock, which contain the coral polyps, allow for penetration of light into the internal tissues of the coral in daylight. At night polyps extend their tentacles to feed.

Tropical reefs in deeper water differ in dimensions and in age from those at Inhaca Island. Viewed from above while one is swimming, the tropical reef near the Isle of Goa in northern Mozambique, resembles the canopy of a forest through which one may glimpse, in the clear water below, tall rust-coloured pillars, large round mounds of coral and delicate fronds of a much greater diversity of species. It will be seen in Section 8.7 that in structure and species the coral reefs at Inhaca Island resemble the edges of the inner fringing reefs on the sheltered shores of the Indo-Pacific Region. The species composition is similar but somewhat impoverished.

At Inhaca Island over a hundred species of animals are associated with living coral as commensals, or merely as lodgers, and as predators, whilst many more inhabit dead coral bases. This large number indicates the kind of variety of life, which is even greater in larger, more permanent and mature reefs. Communities in the coral reef ecosystem have been described as the most diverse in the world. It includes not only the reef itself but the reef flats, the sea meadows and the bare sand-banks of the subtidal fringe that surround them (Fig. 8.1). It is partly exposed at very low spring tides, more especially during the equinoctial tides in April and September (unless there is a strong onshore wind). The whole ecosystem can be viewed well by snorkelling over the area when it is submerged to a depth of about 1 m during neap tides as well as spring tides, or viewed when exposed to air for a width of about 10 m at extremely low spring tides. The presence of the different components of the ecosystem depends on the *slope* and *composition* of the substratum: rock, muddy sand or clean sand (Fig. 8.1a, b and c).

The biota of the reef flat on the west shore is described in Section 6.5 in some detail; this area of scattered coral rubble is accessible to study between tides and the animals live in pools

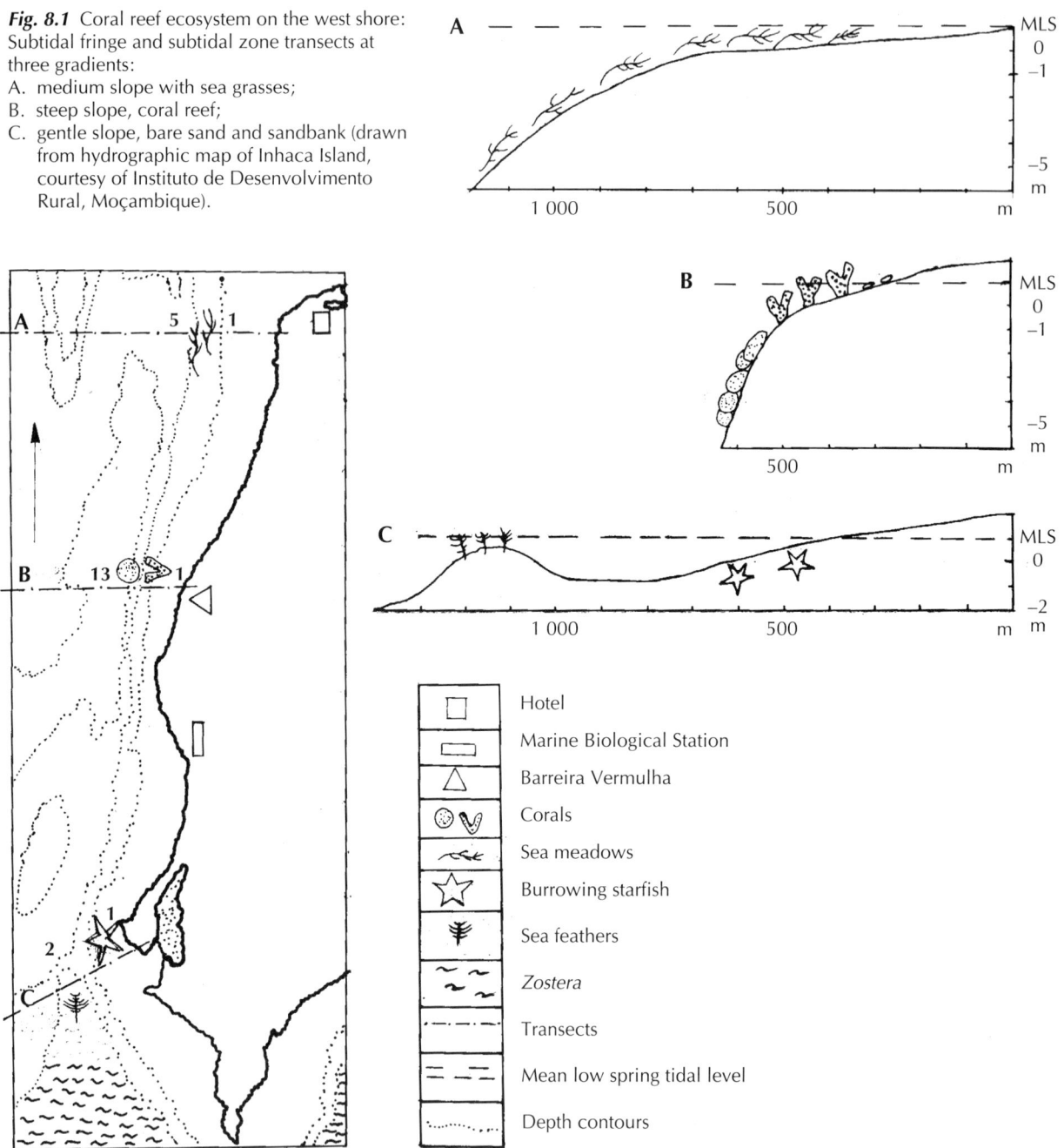

Fig. 8.1 Coral reef ecosystem on the west shore: Subtidal fringe and subtidal zone transects at three gradients:
A. medium slope with sea grasses;
B. steep slope, coral reef;
C. gentle slope, bare sand and sandbank (drawn from hydrographic map of Inhaca Island, courtesy of Instituto de Desenvolvimento Rural, Moçambique).

beneath the coral debris or buried in the old coral rock. The sea meadows and bare sandbanks of the subtidal fringe are described in Chapter 9.

8.1 THE DISTRIBUTION OF CORAL REEFS ON SHELTERED SHORES

Three small fringing reefs occur along the coasts of Inhaca Island in shallow water, their total length being about 3-4 km (Fig. 1.1). One reef is within direct access by foot from the Marine Biological Station on the west coast opposite Barreira Vermelha (Red Cliff). In the southern bay, the very shelterd Ponta Torres reef that is only 50 m from high tide mark, extends at intervals along the east bank of the channel from Ponta Torres Strait into the bay. This is described in

Section 5.5. A small new reef is being formed in a lagoon on the northern side of Portuguese Island and at present the coral heads appear to be a few years old, dispersed 50 cm to 1 m apart.

The species composition in these reefs has recently been described in a comprehensive publication (Boshoff 1981) based on forty years of records in which he was assisted by the Transvaal Underwater Research Group of divers. This list has been updated and revised (Schleyer 1995). Differences between Inhaca Island reefs can tentatively be attributed to the age since formation, the frequency of disturbance by storms and the degree of silting. Before these aspects of the reefs can be discussed adequately, attention must be given to the basic question of how coral animals are able to construct reefs. The descriptions of common coral genera at Inhaca Island in Section 8.3 will enable the observer to recognise them in the reefs.

8.2 THE BIOLOGY OF REEF-BUILDING CORALS

The unique reef-building properties of corals are all the more remarkable since coral animals resemble the more familiar sea anemones, but are much smaller and more delicate. They have a ring of mobile tentacles around a mouth, located at one end of a tubular body which is just a hollow sac used for digestion, absorption of food, disposal of waste and excretion. They have none of the organs present in other multicellular animals. The phylum to which they belong is called Coelenterata to indicate the simple body plan of each animal or 'polyp', in which a single cavity (coel-) is also the gut (enteron) (Fig. 8.2a). These animals are also known as Cnidaria (meaning nettle-like) because they all possess peculiar stinging cells, the cnidocytes (8.2b), which form explosive, poisonous nematocysts (Fig. 8.2c) or adhesive spirocysts (Fig. 8.2d).

The miniature coral polyps have six tentacles (or multiples of six) with which they capture zooplankton prey. The enteron is incompletely divided into as many compartments by means of vertical folds of the internal layer of cells (called mesenteries), each bearing a longitudinal strip of muscle and many long filaments bearing digestive cells. The coral polyps differ markedly from sea anemone polyps in having a very thin middle layer of the body wall (mesogloea), which in sea anemones is thick, stiff and elastic. Coral polyps lack the supportive and elastic properties of mesogloea, but instead of it they have a hard stiff calcareous skeleton with sheets (septa) between the mesenteries. This provides support and also raises the polyps on branches well above the surface of the rocks on which they grow and thus into the ambience of zooplankton, which swarm near the surface of the sea at night, and on which most species feed. Protection is provided by calyces or skeletal cups into which polyps retract during the day by contraction of the mesenteric muscle strips. Since they lack elasticity, emergence of the polyps at night for feeding depends on inflation of each polyp with seawater brought in by the ciliary activity of cells lining the enteron and filaments, whilst the muscles are relaxed.

The biggest difference between reef-building coelenterates and all other coelenterates (including non-reef building corals of deeper or colder seas) is their *rate of growth*. In reef corals growth is accelerated in many ways. Firstly, each colony is a complex aggregate of many polyps most of which develop a bud that, without separating from the parent, grows tentacles around the mouth. The coral colony has thus an enormous feeding area of tentacles which collect zooplankton. The feeding area is matched by an equally extensive area for digestion and absorption on the many long gastric filaments on the mesenteries in the enteron. Secondly, the coral polyps are able to absorb dissolved organic substances through the body wall to supplement their carnivorous diet. Lastly and most importantly, the corals rely on thousands of microscopic, single-celled algae (Fig. 8.2e) containing chlorophyll, which live *within* the cells of the inner layer and manufacture supplementary food for corals as by-products of photosynthesis. These are dinoflagellates known as zooxanthellae (Yonge and Nicholls 1931).

The capture of minute individual crustacean and worm-like animals in the plankton and the transfer to the mouth by the muscular tentacles is co-ordinated by the nerve net underneath the outer layer of cells. An extreme case of co-ordination is seen in some corals which engage in 'tentacle concerts' in which all the polyps in the colony alternately stretch and curl their tentacles simultaneously and repeatedly, thus circulating water. Each tentacle has batteries of 'cnidocytes' (Fig. 8.2b), derived by repeated division of 'mother cells' spaced out along their edges. Each

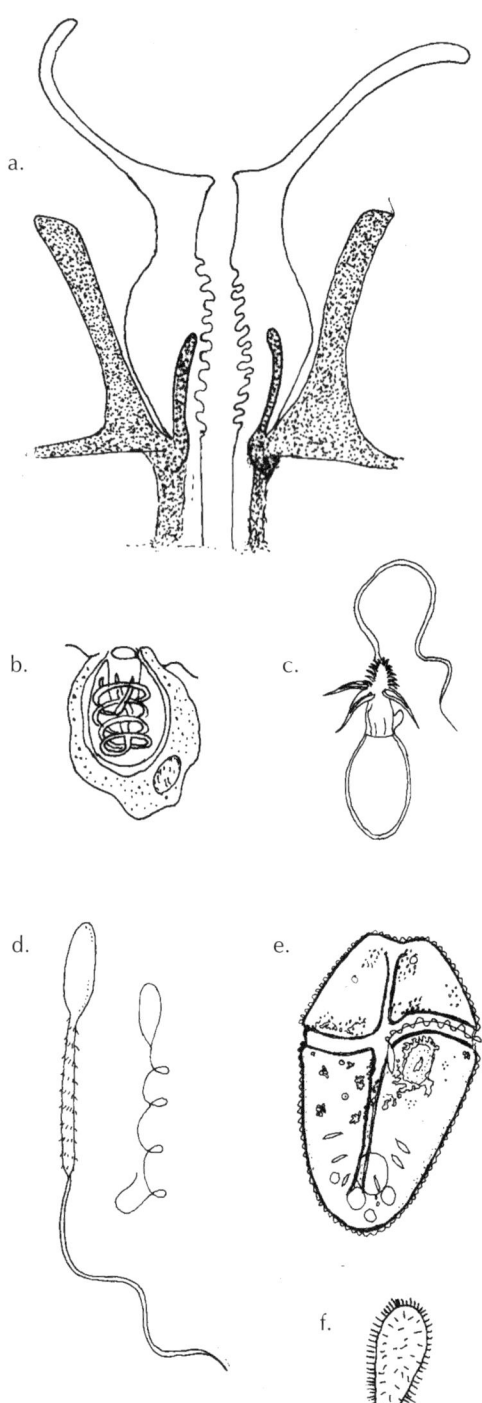

Fig. 8.2 Structure of coral:
a. L.S. polyp showing tentacles, skeleton and mesenteries (x 10);
b. cnidocyte;
c. nematocyst;
d. spirocysts (b-d after Branch G. and Branch M. 1981);
e. symbiont, *Symbodinium (Gymnodinium) adriaticum* (x 1000) (after Wickstead 1965);
f. planula larva of coral (x 100).

cnidocyte develops an intracellular cyst through the activity of the Golgi bodies, one membrane of which forms the wall of the stinging organelle, the nematocyst, or of an adhesive organelle, the spirocyst (Fig. 8.2 c and d). Inside each cyst a long tubular thread is formed and spirally coiled. The free edge of the cnidocyte has a static sensory flagellum projecting beyond a low wall of stationary cilia. This sensory organelle acts as a trigger for the expulsion of the nematocyst or spirocyst. It is sensitive to the tripeptide, glutathione, and to the amino acid, proline, which, to the coral, indicate contact with animal protein. In response to these stimuli the operculum or lid of the capsule, which in corals consists of three flaps, is opened. Then the nematocyst is shot explosively into the prey. The coiled thread turns inside out, exposing rows of backwardly projecting barbs, which prevent its removal after piercing the prey and injecting the toxic contents. The explosion may possibly be caused by osmotic absorption of water when the capsule is opened since the contents have a much higher concentration of salts than that in the sea. In most corals the poison is a mild neurotoxin which paralyses the prey. An exception occurs in the hydroid coral, *Millepora*, known as the 'fire coral' in which the nematocysts can penetrate human skin, causing extreme pain but the effect does not last long. *Millepora* is common in reefs of northern Mozambique but does not occur at Inhaca Island (Kalk 1959). In corals there are far more spirocysts than nematocysts; the former are smaller, the threads are longer and not venomous. When everted, the spirocyst thread releases folded adhesive microfibrils which envelop the prey and hold it fast while it is stung by the nematocysts.

Digestion is extremely rapid, no food remaining in the enteron by the morning after a night of feeding. Crustaceans of 10 mm in length can be digested in a few hours by the larger polyps and the empty skeletons are ejected. This efficiency depends on the activity of both an extracellular proteolytic enzyme, working at a neutral pH, and intracellular enzymes which dissolve fine particles ingested by phagocytosis. Both digestive and absorptive areas occur on the very numerous mesenteric filaments.

Carnivorous feeding is supplemented, to an extent of about 50% of the coral's needs in the actively growing parts by the supply of excess metabolites produced by the symbiotic algae. These are thousands of the dinoflagellate, *Symbodinium (Gymnodinium) adriaticum*, imprisoned in the non-motile stage of the normally motile phytoplankter, which rapidly reproduce asexually in the cells (Fig. 8.2e). Plant nutrients are obtained by the absorption of ammonia, sulphates and phosphates provided as waste products of coral metabolism. Excess glycerol, glucose and amino acids, derived from the metabolism of the algae are stored in 'accumulation bodies'; they gradually diffuse into the animal cells and are used as food. Thus the coral gains in two ways: by the supply of food and by the removal of its waste products so that no energy is expended in excretion (Yonge 1968). Feeding symbiosis is only *one* aspect of the relationship between corals and their algae. The growth of the calcareous coral skeleton is accelerated by the biochemical co-operation of the algae, so that hermatypic corals (containing algae) are able to grow 14 times faster than non-reef building corals (ahermatypic) which either have no algae or very few.

The way in which calcium combines with carbonate to form the skeleton may be described as a series of steps. The coral polyp adsorbs calcium ions from the sea on the mucoid coating of the internal layer of cells, which then actively absorb them. The calcium ions then diffuse to the inner boundary of the cells where they are actively transported to the external layer of the cells in which they accumulate and are bound to membranes, thereby avoiding an upset of the ionic balance of the cells. The entry of calcium is accelerated by this removal. Two thirds of the carbon dioxide required to make carbonate is derived from coral respiration and that of the algae at night; the rest is absorbed by diffusion from the sea. The solution of carbon dioxide in water is a slow process, but in corals it is accelerated by the special enzyme, carbonic anhydrase, present in the cells, which catalyses the combination of the gas with water to form carbonic acid. Carbonic acid spontaneously ionises to bicarbonate and hydrogen ions, followed by the formation of carbonate ions, carbon dioxide and water. Symbiotic algae accelerate the formation of carbonate in daylight because they remove the released carbon dioxide by their high rate of photosynthesis.

In the external layer of cells of the coral polyps the carbonate combines with calcium ions to form calcium carbonate in small vesicles that migrate to the external surface. Vesicles containing aragonite crystals of calcium carbonate have been viewed, by means of electron microscopy, in the process of emptying their contents outside the cells. The crystals are deposited on the very thin organic layer, which coats the polyps, and the skeleton is formed by their accumulation. It follows from the accelerative action of the algae that skeleton formation is most rapid in the presence of sunlight. Symbiosis is completed by yet another algal aid to the coral. The supply of oxygen produced during photosynthesis in the symbiont is more than sufficient for the coral's respiratory requirements. These biochemical pathways are not hypothetical, they have all been traced by means of ingenious experiments using radio-active tracers (Goreau and Goreau 1959).

The skeleton of the coral takes the shape of the budding coral polyps. The calyx of each polyp is a cup with several incomplete calcareous partitions (septa) that alternate with the mesenteries and thus completely support the delicate polyp. The calyces are connected by areas of skeleton that are formed on extensions of their walls. Beneath the polyps the stem tissues secrete calcium carbonate making the skeleton continuous and raising it higher and higher as it grows. The patterns of branching and the growth rate vary with the species and are genetically programmed for the manner of budding and the union of the septa. Budding may be extratentacular (Fig. 8.3a) or intratentacular (Fig. 8.3b); the junction of the calyces may be plocoid, cerioid or meandroid (Fig. 8.3c – e) and the septa may by joined by synapticulae (Fig. 8.3f).

Corals grow quickly in the first few months of their lives, reaching a diameter of 1 cm in the first month. Branching corals increase in length by a few centimetres a year. Compact corals

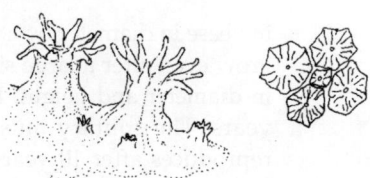

Fig. 8.3 Patterns of budding in coral polyps and resultant calyx patterns:
a. extratentacular budding and calyces*;

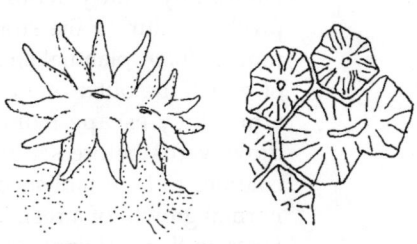

b. intratentacular budding and calyces*;

skeletal patterns:

c. plocoid*;

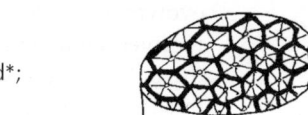

d. cerioid*;

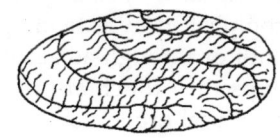

e. meandroid*;

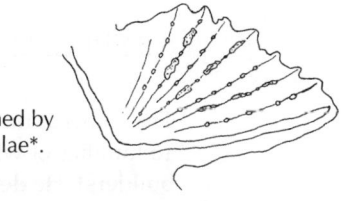

f. septa joined by synapticulae*.

increase in diameter by about 1 cm a year, becoming mature in about 8-10 years and continue to grow even after a large size has been achieved, some colonies at Inhaca Island becoming 3-4 m in diameter and height. It is likely that the branching species mature earlier, probably at 3-4 years. The solitary mushroom coral *Fungia* grows only 7 mm in diameter in a year and reproduces after 10 years when the diameter has reached about 10 cm, a size that is rare at Inhaca Island (see Section 5.6.2).

During reproduction, the colony as a whole acts as one individual. Most corals have either ovaries or testes in every polyp, although a few species are hermaphrodite. Fertilisation is usually internal, when sperm released from neighbouring colonies is drawn into the polyp's enteron by ciliary action during nocturnal emergence. Several thousand fertile eggs are produced during the summer months in most species and start developing internally into multicellular 'planula' larvae about 250 µm in diameter; they are differentiated into two layers of which the outer one is ciliated (Fig. 8.2f). The planulae swim upwards towards the light when they escape from the enteron; after two days they change direction and swim downwards before settling on old coral bases. This would be time enough to enable new colonies to be established along the edge of the same channel at Inhaca Island during the normal growth of a reef. Some larvae have a longer life and may be carried by ocean currents far from their original sites and repopulate Inhaca reefs. Mortality among larvae is high since tidal and other currents remove them from the vicinity of the reefs.

Detailed studies of recruitment of very small corals have not been made at Inhaca Island, but on the Great Barrier Reef of Australia marked metre quadrants have been photographed annually for many years (Connell 1973). The rate of establishment of new colonies, visible at 1 cm diameter, appeared to be on average about 5 per square metre per annum. High mortality may reduce the number of those established by unusually severe exposure to air, rain, sedimentation or wave action and sometimes by heavy predation. The recruitment rate on the new Portuguese Island reef is much slower.

It is difficult to imagine interspecific competition in sedentary corals, but they have evolved two ways of spacing themselves so that there is no overlapping and sufficient light is received by all survivors. When fast-growing branching species grow above massive species, the polyps of the latter, which are in the shadow of the branches above, die from lack of light but the colony as a whole survives and grows in another direction. When the massive or encrusting species find themselves within reach of others, they prevent any further encroachment by killing off polyps of their neighbours in a border about 2 cm wide. This is accomplished by extending the mesenterial filaments from the polyps either through the mouth or through adjacent tissue and laying them on the threatening coral growths. The polyps they touch are then destroyed and digested. Coral families can be arranged in hierarchical order, according to their powers of destruction of neighbours, such that those with larger polyps or massive skeletons dominate the others (Connell 1973).

8.3 SPECIAL FEATURES OF THE COMMON CORAL GENERA

The corals at Inhaca Island have been classified by Dr P H Boshoff as belonging to 45 genera in 16 families of which most are hermatypic (reef builders) and a few are ahermatypic (non-reef builders). He described about 15 genera in nine families as abundant or common or as having some special biological interest (Boshoff 1981). These have been selected for discussion in this chapter and in Section 5.5.2. The taxonomic classification of hermatypic corals depends on intricate details of the cleaned skeleton, such as the form of the skeletal calyces, the presence or absence of pores, the number and kinds of septa and the connection between them (trabeculae) and the pattern of budding. There is a key to the identification of coral families and some genera in the first and second editions of this book. An updated list of species is being completed (Schleyer 1995). It is, however, possible to recognise common genera in the field by macroscopic observation and to correlate the gross form with the feeding habits of the polyps. This is also sufficient to pinpoint certain species that have a special relationship with commensals or parasites. The positions of common species on the reefs or in rock pools on the east coast give some indication of their tolerance of certain environmental conditions.

The most abundant and widely distributed corals in the reefs of the whole world are species of *Acropora* of the family Acroporidae in which the branching skeleton has *terminal* calcyes (Fig. 8.4a). In mature reefs, the beautiful stag's horn coral, *Acropora cervicornis*, forms continuous forests of coral over many square kilometres in lagoons. Its topmost branches were visible in clear water at a metre or so below the surface (at low tide) from a boat skimming the sea between islands off the coast of Mauritius, for example, or in the Great Barrier Reef of Australia. They grow on submerged rocks and their branches may attain a metre in length, the main 'trunk' being 20 cm in diameter. In the miniature reef on the west shore of Inhaca Island smaller stag's-horn coral is also relatively plentiful, patches of it extending over several square metres; it is easily recognised by the pattern of branching. It is absent in the Ponta Torres reef possibly because of silt, abrasion by sand in the strong current or because of the variable temperature. In addition, about 30 species of *Acropora* have been identified in all the reefs of Inhaca Island (Riegl 1994). The occurrence of *Acropora* in reefs with marked speciation and the well-developed areas of *A. cervicornis* epitomise the tropical nature of the reefs.

The genus *Acropora* may be recognised by the distinctive, smooth cylindrical calyces, each about 2-5 mm long, which are closely arranged in irregular rows around the axis of the branch, leaving little space between them (Fig. 8.4a). Each twig of the branch has a larger *terminal* calyx and its polyp is the parent of all the other polyps borne on the twig. This is confirmed by the fact that the twig is hollow and the intact septa of the terminal calyx extend right down the tube and remain continuous with the septa of the lateral calyces. A broken twig will show this feature distinctly. The terminal polyp has six tentacles and the lateral polyps have six long and six short tentacles. Each tentacle has white spots which are batteries of spirocysts and nematocysts so that the tentacular surface, exposed at night, is a very efficient organ for trapping zooplankton. In these sheltered reefs during the day the branches of different species may appear mauve, green, pink, red, brown or deep purple when under water and the polyps are contracted. The colours can be seen down to a depth of 5 m at low tide at Inhaca Island, but when the sea is turbid a fine slimy deposit of silt discolours their surfaces and further reduces the penetration of light. The true colours in all their vividness reappear within 48 hours after the water has cleared and the cleaning operations of the corals have been completed by means of ciliary currents, which have swept the debris down the column. In the lee of the Isle of Goa in northern Mozambique, in exceptionally clear water, colours are visible down to a depth of 15 m or so.

Some species of *Acropora* such as *A. cervicornis* are killed by slightly warmer temperatures because of their high oxygen demand, and for this reason they are not found in the shallows of the reef or in rock pools but in water at least 1-2 m deep at low tide. Other species of *Acropora*, such as *A. ocellata*, the most abundant species of this genus, occur in rock pools at Cabo da Inhaca and also in pools as far south as Durban, in Natal.

Another common genus in this family, *Montipora*, is represented by many species at Inhaca Island. The calyces are only 1 mm in diameter, sunken below the surface and the skeleton is very porous (Fig. 8.4b) The specific forms of the skeleton are very variable, being branching, leaf-like or forming mounds with an undulating surface resembling a cumulus cloud.

The first corals to recolonise the edge of a potential reef are species of *Pocillopora* (Fig 8.4c) of the family Seriatoporidae. It is also one of the fastest growing corals, growing 20 cm in height in 7-8 months. *Pocillopora* spp. also occur among the sea grasses near reefs, attached to small stones. Some of the species occur in rock pools at Cabo da Inhaca and also on the coast of Natal, whilst others grow abundantly only in coral reefs. This genus has a low bushy form with many blunt branches and a rough surface. Its branches are covered by groups of many close-set, tiny calyces about 1 mm in diameter which, without a lens, appear to be merely holes in the skeleton. As usual the polyps are protruded only at night and are very small, having 12 very short tentacles. Short tentacles are a handicap in the capture of sufficient zooplankton and their action is supplemented by *ciliary feeding*. Ciliary currents carry detritus and stranded zooplankton embedded in surface mucus *up* the column of the polyp to the

Fig. 8.4 Coral genera:

a. *Acropora* (x 1) and enlarged terminal calyx;

b. *Montipora* (x 2);

c. *Pocillopora* (x 1);

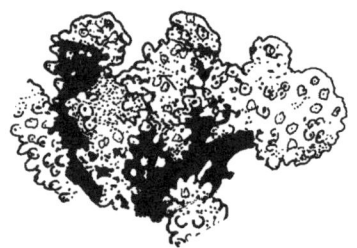

d. *Porites* (x 1)

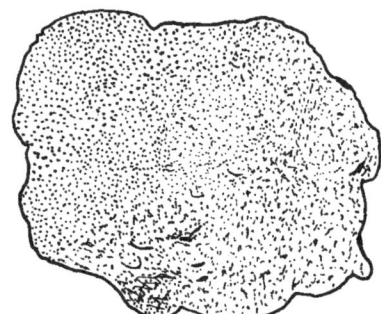

d'. calyces enlarged (c-d' after Branch and Branch 1981)

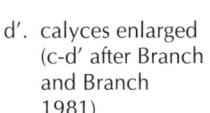

mouth. When not feeding the ciliary mucus currents are *reversed* and carry debris away from the mouth, becoming cleaning currents.

Reproduction in *Pocillopora* occurs during every month of the year instead of annually and, unusually, during different phases of the moon. This timing provides opportunities for the planulae larvae to settle over the subtidal fringe at many low tides. Planulae larvae often settle together in large numbers on the base of a dead colony at the edge of low spring tide. They then fuse together as they grow, each producing buds that continue to grow and bud again as one 'colonial' individual.

The polyps are said to be sensitive to dilution of sea from freshwater drainage. This may have excluded them from the upper reaches of the reef in the southern bay, into which some fresh water drains from the mangroves. They would die following exceptionally heavy rain. Species of *Pocillopora* which occur in rock pools are *less sensitive* to increased temperature and dilution than are most species of *Acropora*.

Stylophora is another genus in the same family with a similar growth form, except that its branches are somewhat flattened (Fig. 8.10d). In the centre of each calyx, visible in a cleaned skeleton, the top of a supporting central pillar (stylus) is visible as a central disc or stylet. *Stylophora* is also a pioneer in reef-building and resembles *Pocillopora* in distribution. One species, *S. pistillata*, gives shelter to very small female crabs of the species *Hapalocarcinus marsupialis*. The newly settled gall crab larva so irritates the surface of a crevice in the coral, when it clings between branches, that the coral responds by growing perforated walls around the intruder, imprisoning it for life in a gall (see Section 8.5.2).

Another genus of this family, *Seriatopora*, is the main fossil coral in the Ponta Raza fossil bed seen at mid-tide near the exit of the Ponta Raza mangrove channel (see Section 1.3.1).

The genus *Pavona*, in the family Agaricidae in which internal septae are reduced, forms a foliaceous colony in the form of flat, stony plate-like branches (see Fig. 5.11e). On the upper surface, or on both surfaces if the plates are upright, there are irregular rows of star-shaped openings about 1-2 mm in diameter. Each star has 12 points marking the ends of the septa, which continue as ridges linking up closely spaced 'stars'. These ridges give the surface a rough, striated texture but the calyces do not protrude above the surface (Fig. 5.11e). Buds of new polyps occur within the ring of 12 tentacles of some polyps so that their openings may be elongated if the bud has not yet been separated by a developing wall. Other holes are much smaller than those of the parent polyp indicating new buds that have been separated. The polyps are very small and the tentacles invisible to the naked eye. They therefore resemble *Pocillopora* in their method of feeding, depending largely on the secretion of mucus and the reversal of cleaning currents to carry trapped *detritus* to the mouth. The mucus seems to stretch across the little tentacles and acts as a net to catch small particles from the moving water above. Four of the five species of *Pavona*, including *P. cactus* with spiny plates, occur in the very shallow low tide water of the middle reaches of the Ponta Torres reef (see Section 5.5.2). This genus is one of the least sensitive to shade and silt. It is probably for this reason that isolated groups of *Pavona* spp. have survived in the more southerly parts of the Ponta Torres reef when other genera may have been killed by a period of especially high silt or sand load in the current.

The corals which form numerous solid, rounded, rock-like colonies are species of *Porites* of the family Poritidae in which the polyps are very small, the calyces being well defined but very close together (Fig. 8.4d and d'). *Porites* species are interspersed between branching species as mounds below low tide in water at least 1 m deep. The largest at Inhaca Island are 3-4 m in diameter and 4 m high, but in foreign mature reefs *Porites* may grow to about 10 m across. They are hardy corals, slow-growing but resistant to mechanical forces that destroy the branching genera. The surface of each 'boulder' colony is covered with very small hexagonal calyces, so closely adjacent that there are no spaces between them. The predominant colour may be purple or brown. Each polyp has 12 tentacles that, when withdrawn, are loosely folded into the mouth of the polyp (instead of merely contracting), thereby excluding sand and silt.

A closely related species in the same family, *Goniopora savignyi*, is the only coral in the Inhaca Island reefs which fully *expands* its large polyps in daylight (Fig. 5.11 b-d). It has a similar skeletal form to *Porites* although the mounds are smaller and the calyces are much larger. The polyps have 24 long, yellow, finger-like tentacles arranged in a single ring. It is best seen in the vicinity of the cactus coral, *Pavona*, in the Ponta Torres reef, but it occurs in deeper water and is rarely uncovered at extreme low tides.

In the family Faviidae, a transition is seen from the simple calyces of large polyps to the intricate skeletal pattern and fused calyces of 'brain' corals. *Favia* and *Favites* form small boulders 20-30 cm in diameter from which *large* close-set polyps and their calyces protrude from the surface. In *Favia* (Fig. 8.5a) the calyces are rounded and about 10 mm in diameter; they are united by the peritheca, i.e. the arrangement is plocoid (Fig. 8.3c). The calyces of *Favites* (Fig. 8.5b) are somewhat larger, polygonal and cerioid in arrangement, i.e. united by the calyx walls (Fig 8.3d). The septa are very conspicuous, 12 larger ones alternating with 12 smaller ones, which extend as small ridges from one calyx to the next. Budding is *intratentacular* (Fig. 8.3b) so that some calyces appear distorted as new buds form and new septa grow before the new polyp is separated by a vertical constriction. These species with large polyps are *zooplankton feeders*, using tentacles, spirocysts and nematocysts. The ciliary currents are always directed away from the mouth and thus free the surface from silt and debris; These genera always occur in deeper water of about 2-3 m and are never exposed to air.

The 'brain corals' form colonies in which the skeleton has a pattern of valleys and ridges, reminiscent of the convolutions of the cortex of the mammalian brain (Fig. 8.5c). The ridges are

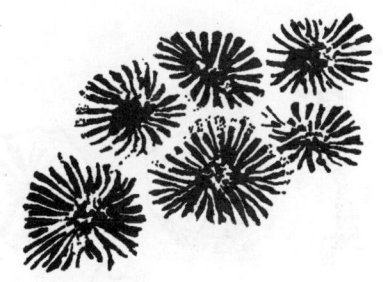

Fig. 8.5 Coral genera
a. *Favia*, calyces (x 1)*;

b. *Favites*, calyces (x 1)*;

c. *Meandrina* type (brain coral)* (x 1) and

c'. complex polyp everted at night with rows of tentacles on ridges and mouths in the valleys (x 0,3).

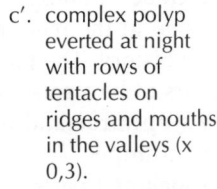

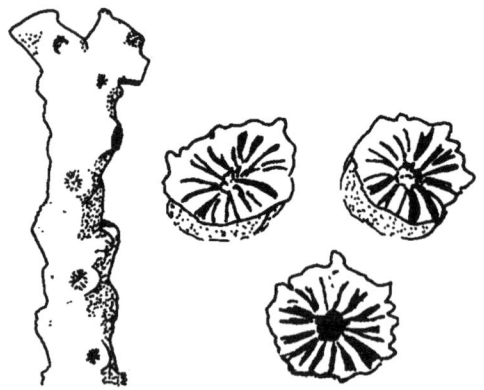

Fig. 8.5 Coral genera
d. *Galaxea* stem (x 0,5) and calyces enlarged;

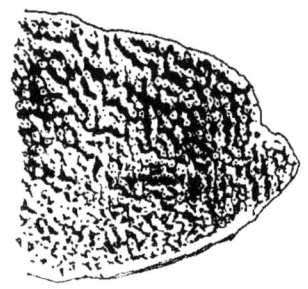

e. *Turbinaria* (x 0,3)*.

the fused lateral walls of rows of calyces in which adjacent walls, which would have separated the polyps, are not formed; when expanded the polyps also appear joined to each other; this arrangement is 'meandroid', and the result of intra-tentacular budding. The tentacles cover the ridges as a continous wall; in the sinuous valleys between them, at intervals of about 3 mm, lies a row of small mouths (Fig. 8.5c'). The long brown tentacles are extended at night and cover the whole colony. The gastric cavity of each polyp opens into that of adjacent polyps so that there is one continuous gastric cavity for the whole 'colony', which is unusual. The genus *Symphyllia* is represented at Inhaca Island by two species found only in water 5 m deep.

Some families of coral specialise in a skeletal form in which the distance between large polyps on the stem or boulder is long. *Galaxea fascicularis* is a local example of the family Oculinidae at Inhaca Island that has this form (Fig. 8.5d). The calyces, 10 mm high and 5 mm in diameter stand upright on the long stem. Each polyp has 24 tentacles and is supported by 24 septa which project above the calyx wall; the budding of the polyps is extratentacular. Although this species is not common in Inhaca reefs its brilliant emerald green colour is conspicuous. A specialised alpheid shrimp lives in association with this species (see Section 8.5.2). *Galaxea clava* in the warmer waters of northern Mozambique grows into robust, rust-coloured pillars several metres high, resembling church organ pipes in height and diameter (Kalk 1959).

In the fungid family, *Fungia*, the mushroom coral, resembles an inverted mushroom head in having conspicuous septa radiating from the centre on the convex upper surface, the lower surface being flat and smooth (Fig. 5.17b). The whole coral is made by one polyp only and the *adult form is free*, becoming detached from the parent stock when only a few millimetres in size. The free form grows to a diameter of 25-30 cm among sea grasses but at Inhaca Island it may reach only 10 cm, although most of them are smaller. The simple large polyp has a slit-shaped mouth surrounded by a very large number of long tentacles slightly swollen at the tips. These are arranged in concentric rings on the surface between the septa. The polyp tissue also covers the smooth under-surface of the coral skeleton. Internally the gastric cavity has as many mesenteries as septa, with which they alternate. The septa are peculiar in having many calcareous projections (synapticulae) forming bridges between them which penetrate through the mesenteries (Fig. 8.3f). The colour of the fungids is often brown, sometimes green, whilst the tentacles are white. There are nine species of *Fungia* at Inhaca, four of which occur amongst the sea grasses on sand, two species among coral rubble, two in pools and one in the reef among live corals. This strange distribution is explained by their peculiar life history. The planula larva of a fungid attaches to rock in the reef and grows into what appears to be a small solitary cup-like coral with 12 septa. The circular free edge of the coral grows upwards and more septa are formed, the shape of the coral becoming like the mouth of a trumpet. The polyp breaks off leaving a stalk behind, and is then carried by a current to sheltered parts of the subtidal fringe where it grows, becoming thicker and flatter and eventually convex on the upper surface. The living stalk that is left behind on a rock develops lateral buds, which in turn grow, break off and are distributed to the various habitats by different currents. One species, *F. fragilis*, is abundant in one area of the sea meadows in the southern bay in an area of about 50 x 100 m. The inverted mushroom shape and large gastric cavity, which can be inflated by centripetal ciliary action, enable the fungid coral to right itself when turned over. The reversed beat of the cilia, away from the mouth, then frees the surface of silt.

In the same family the brownish *Herpetolitha* (serpent stone), which is also a detached coral, grows into an elongated, slightly twisted colony (Fig. 5.17c). It has similar septa to *Fungia* but they radiate from a number of separate points along a central groove. In this genus many polyps with green mouths occur in a row in the groove. Colonies found in the shallows on sand among the sea grasses at Inhaca Island are 10-15 cm long.

The family Dendrophyllidae is represented by three quite dissimilar genera. *Tubastrea* (*Dendrophyllia*) and *Heteropsammia* are ahermatypic and *Turbinaria* resembles other reef corals in being hermatypic. *Tubastraea* (*Dendrophyllia*) is a large, branching coral with large polyps 10-15 mm in diameter, and fairly long tentacles, which grow separately from a common stem. Each stem is really the parent polyp from which the other polyps grow and has developed a thick skeleton (Fig. 5.11a). The numerous septa in the calyces are thin, calcification being much less than in the previously described hermatypic corals, and growth is much slower. Because of the lack of symbiotic algae, they must rely entirely on zooplankton for food and the polyps expand in sunlight. There are 3 species at Inhaca Island in two types of habitat. *Turbastraea micranthus* has a jet black skeleton and grows in the shade of a large overhanging rock on the Ponta Torres cliff; it is also frequently seen in the small coral reefs off the east coast of the island. It has large blood-red polyps and the tentacles appear to be spiked. The other two occur in rock pools. *Turbinaria* is hermatypic and needs light to speed up the calcification of its heavier skeleton. One species, *T. peltata*, spreads out its polyps on a slightly concave bowl on a broad base (Fig. 8.5e). The polyps on the rim of the bowl bud continuously so that the diameter of the colony increases. Another species builds a number of slightly fluted plates and has smaller, crowded polyps obliquely arranged on a short stalk. This genus always grows in water 2-3 m deep or more at low tide.

Heteropsammia, a small unbranched coral has usually one large calyx with many septa and is plentiful on the sand at the bottom of the Inhaca Channel in 10 m of water. It has a curious life history. The planula always settles on an empty sea snail shell which is occupied by a small sipunculid worm, *Aspidosiphon corallicola*. The coral polyp and its skeleton grow over the shell and completely enclose it except for a small hole through which the worm protrudes its head to consume detritus. The association with the sipunculid worm benefits the coral by the provision of its adopted shell as a hard substratum and also as a means of escape from being buried in shifting sand. The muscular worm drags the coral to the surface again after they have been buried and moves along to fresh sediment, leaving 'drag trails' behind it. This coral collects detritus as food.

8.4 PHYTOPLANKTON AND ZOOPLANKTON IN THE SEA OVER THE CORAL REEF

The microscopic organisms: plants, animals and bacteria, that float in the surface waters of the sea constitute the plankton, the main source of free-living food for corals. Plankton is also an essential, basic component of the food chains on the seashore, as well as in the open sea (see Fig. 2.1). It is the sole food of adult filter feeders such as sponges, coelenterates, serpulid and plume worms, barnacles, some hermit crabs and porcellanids, some amphipods, bivalve molluscs, brittle stars and crinoids, tunicates and some fishes. The pelagic larvae of many shore animals feed on the phytoplankton and themselves contribute to the zooplankton. In addition, a large population of permanent zookplankters thrives in surface waters; these are protozoans, mainly foraminiferans and radiolarians, crustaceans such as copepods and ostracods, some mating and young polychaetes, some tunicates and arrow worms.

The inshore phytoplankton is much denser than that of the open sea because the water is enriched by nutrients washed out from the shore by tides. The total plankton in the vicinity of a coral reef has an even greater density of organisms since the nitrogen components of the excreta of the large populations of coral reef associates contribute to the food of phytoplankton on which the zooplankton depends.

Several studies of plankton density have been carried out in the Bay of Maputo and more recently in waters just beyond the coral reef on the west coast of Inhaca Island, at monthly

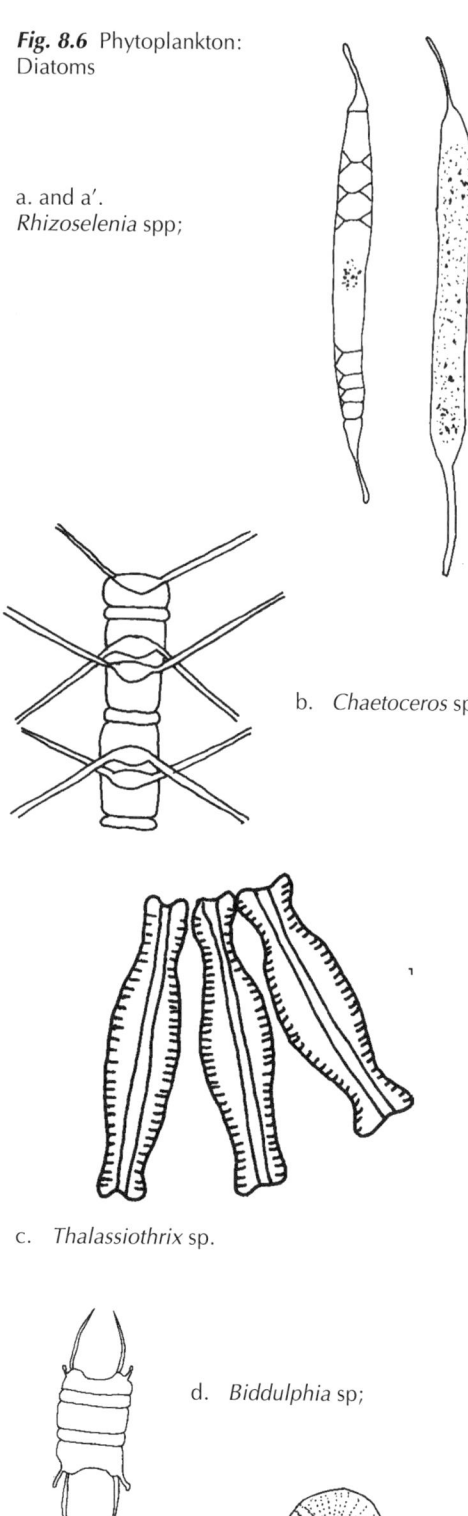

Fig. 8.6 Phytoplankton: Diatoms

a. and a'. *Rhizoselenia* spp;

b. *Chaetoceros* sp.

c. *Thalassiothrix* sp.

d. *Biddulphia* sp;

e. *Coscinodiscus* sp., terminal and side views (all approx. x 1 000) (a-e after Wickstead 1965).

intervals over a year, by biologists at the Universidade Eduardo Mondlane (Gove and Cuambo 1990). The depth of the water sampled was about 12 metres and the temperature of the sea ranged from 18,3°C in July to 28,2°C in January. The temperature was almost constant in the water column during the sampling times. A plankton net of 100 µm mesh size was used, which collected major phytoplankton and zooplankton (i.e. mesoplankton) but excluded smaller organisms such as foraminifora, microplankton, nanoplankton and bacteria.

The phytoplankton appeared to be dominated by diatoms (Bacillariophyceae) since only the larger dinoflagellates were trapped and the smaller species (known to be abundant in tropical waters from their nocturnal luminescence) escaped the sampling procedure.

The diatoms have rigid silica-impregnated cell walls of different shapes with intricate patterns (see Section 3.3). The genus *Rhizoselenia* was abundant throughout the year, *Chaetoceros* and *Thalassiothrix* were abundant from April to December, *Biddulphia* and *Coscinodiscus* were present throughout the year (Fig. 8.6 a-e). These genera have a variety of flotation adaptations: *Rhizoselenia* has an elongate tubular cell with four terminal points; such a needle-shaped cell would float when its axis is horizontal. *Chaetoceros* has an almost cuboid cell with four projecting hair-like spines at the corners; many cells are often inter-meshed and float as a mass. *Coscinodiscus* has the shape of a drum and its large vacuole is filled with a low-density fluid (Boney 1975).

The Dinoflagellate species, *Triceratium*, *Peridinium* and *Noctiluca* (Fig. 8.7a-c) occurred in greater numbers from August to December. The organisms are single-celled and swim freely, propelled by two flagella one of which lies in the 'girdle', a groove around the cell, and the other is directed posteriorly in the longitudinal groove or sulcus. These minute plants are covered by a specific armour of cellulose plates of various proportions and their chlorophyl in a chloroplast is masked by yellow xanthophyll, peculiar to the phylum, called peridinin. Each cell is able to shoot out darts (trichocysts) when disturbed. Some genera, e.g. *Noctiluca* (Fig. 8.7b) are luminescent so that on a calm, dark night the whole surface of the sea may shine brilliantly. The dinoflagellate species responsible for the poisonous 'red tide' was not reported here.

Another dinoflagellate in the plankton, *Peridinium balticum* (Fig 8.7d and e) exhibits a remarkable symbiosis in which the dinoflagellate itself is the host, although it is only 30 µm in length (Horiguchi and Pienaar 1991). The host has external calcareous plates, its own plasma membrane and organelles *except* for chloroplasts. A comparatively large chrysopyhte *endosymbiont* alga occupies the peripheral portion of its cytoplasm. This has well-developed chloroplasts that have taken over the function of photosynthesis for the host. The endosymbiont has a plasma membrane surrounding its own organelles. In addition, it contains virus-like particles in its cytoplasm. In the cytoplasm of the host dinoflagellate, symbiotic bacteria were also seen in electron microscopic studies (Fig. 8.7d and e).

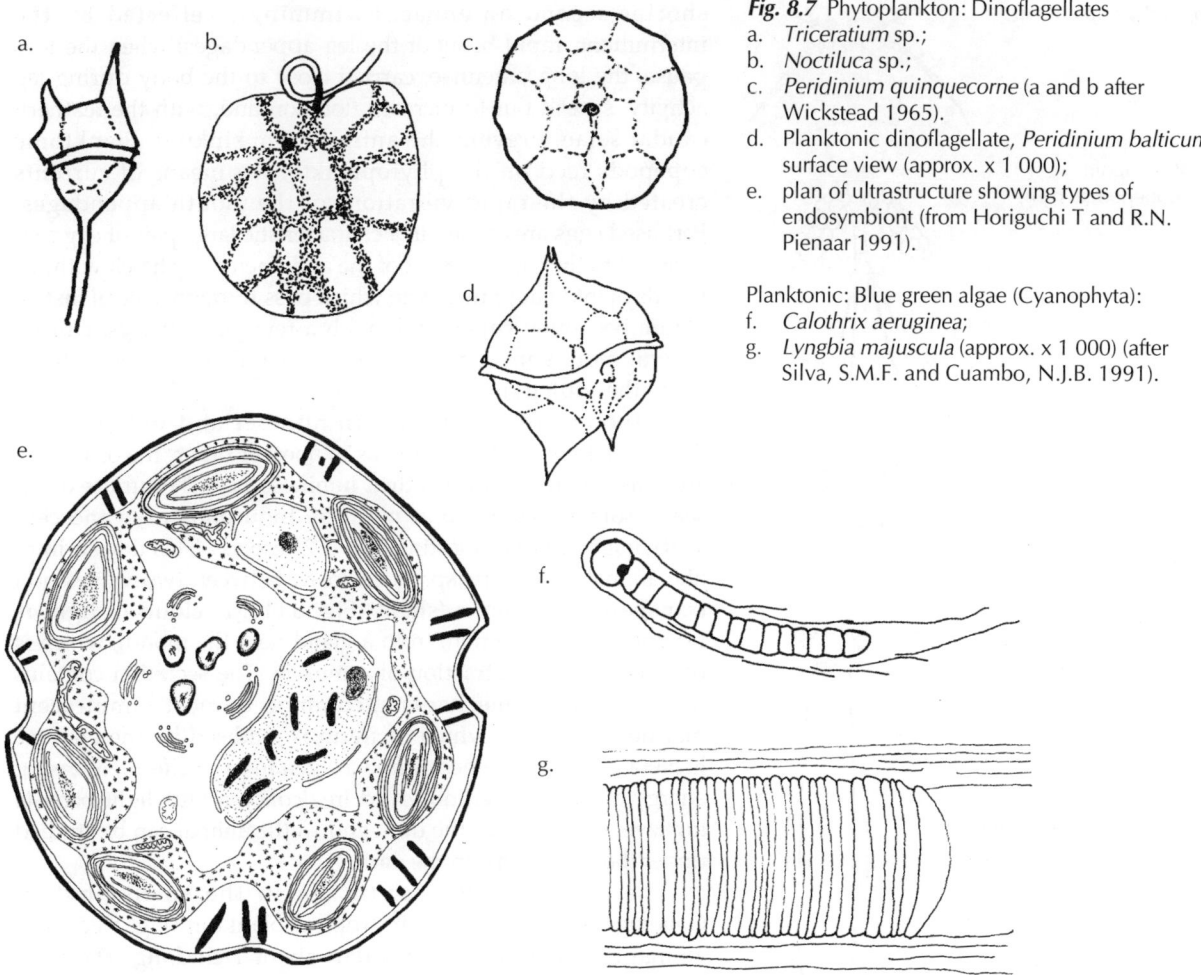

Fig. 8.7 Phytoplankton: Dinoflagellates
a. *Triceratium* sp.;
b. *Noctiluca* sp.;
c. *Peridinium quinquecorne* (a and b after Wickstead 1965).
d. Planktonic dinoflagellate, *Peridinium balticum* surface view (approx. x 1 000);
e. plan of ultrastructure showing types of endosymbiont (from Horiguchi T and R.N. Pienaar 1991).

Planktonic: Blue green algae (Cyanophyta):
f. *Calothrix aeruginea*;
g. *Lyngbia majuscula* (approx. x 1 000) (after Silva, S.M.F. and Cuambo, N.J.B. 1991).

The free-living species, *Symbodinium (Gymnodinium) adriatica* (Fig. 8.2d) becomes symbiotic in the cells of all reef-building corals (see Section 8.2), in some other coelenterates and in *Tridacna*, the giant clam (see Section 8.5.1). They provide food for their host by their metabolism and reproduce by fission inside the host cells as well as in the free state. In addition blue-green algae (Cyanophyta) such as *Lyngbia majuscula* and *Calothrix aeruginea* are always present in the sea (Fig. 8.7 f and g).

In the plankton assay the numbers of the phytoplankters varied little throughout the year, as might be expected from the tropical temperatures and the absence of seasonal changes in light, turbidity and nutrients. During every 24 hours, however, different species of phytoplankton peak in numbers at different times of the day or night, so that the numbers of each species found in the samples are variable; an exact assessment requires hourly sampling for 24 hours. The apparent increase in total numbers of plankton in July depended on the numbers of permanent zooplankters, such as *Sagitta* (see below), and the hatching of larvae from the eggs of shore animals at that time rather than on an increase in phytoplankton.

The permanent zooplankters are, in order of density, crustaceans, arrow worms, tunicates, polychaetes and protozoans.

Among the crustaceans the copepods are the most numerous zooplankters (Fig. 8.8a). They are about 1 mm in length, the segmented body being broad and cylindrical, bearing jointed appendages; the abdomen is narrow and devoid of appendages except for the terminal uropods, which have long feathery setae. A characteristic median eye, retained from the nauplius larval stage, occurs on the head, which bears a pair of long jointed, setous first antennae and

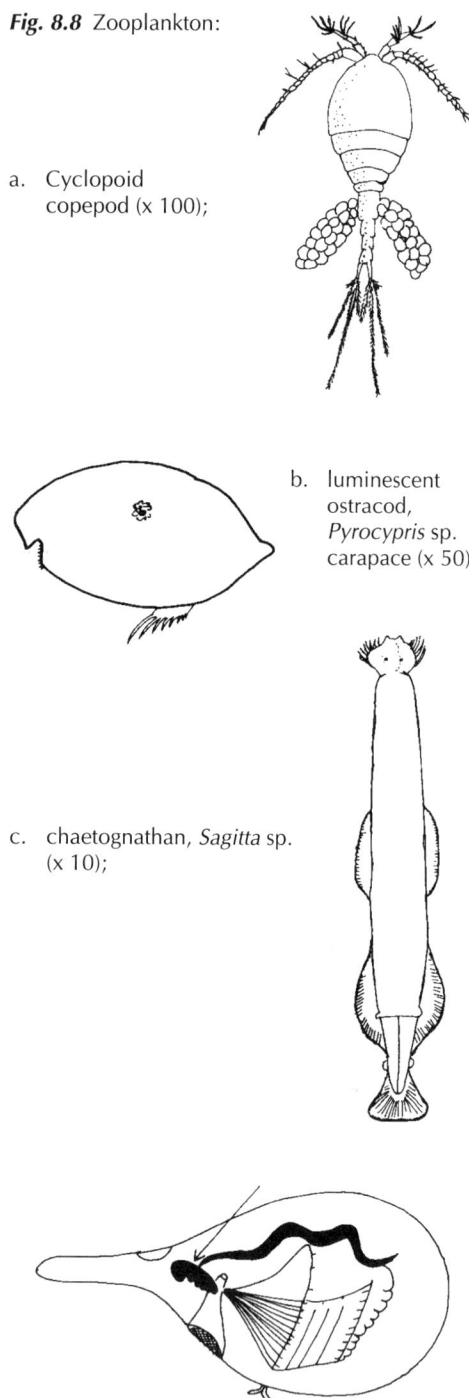

Fig. 8.8 Zooplankton:

a. Cyclopoid copepod (x 100);

b. luminescent ostracod, *Pyrocypris* sp. carapace (x 50);

c. chaetognathan, *Sagitta* sp. (x 10);

d. Larvacea: pelagic tunicate in mucoid 'house' with sieves (arrow to animal) (x 10) (a-d after Wickstead 1965).

shorter second antennae. Swimming is effected by the intermittent, rapid beats of the leg appendages; when the legs pause, the long antennae, carried close to the body during leg activity, spread out to increase flotation and, with the feathery caudal setae, prevent the animal from sinking. Planktonic copepods feed on the phytoplankton by means of currents created by the rapid vibrations of the mouth appendages. Fertilised eggs are carried in a characteristic large pair of egg sacs attached to the first segment of the abdomen; they hatch within a few days into nauplius larvae which pass through several instars before becoming copepodids with a few pairs of legs. During several moults more pairs of legs are added and they finally adopt the adult form.

Although relatively few in number and not constant throughout the year, crustaceans in another order, the ostracods, are conspicuous because of their luminescence. The surface of the sea is sometimes lit up at night by thousands of evanescent sparkling points, each lasting only 1-2 seconds. This tiny shrimp-like animal, *Pyrocypris* sp., is enclosed between two valves of a transparent chitinous 'shell' (Fig. 8.8b). A cloud of light is produced by a secretion, from a gland near the mouth, which is released by the contraction of a muscle. The secretion contains the biochemical mechanism, common to most luminescent marine animals, in which luciferin is oxidised by the enzyme luciferase in the presence of free oxygen to oxyluciferin, releasing energy as light. When observed in alcohol, a blue light persists for several minutes in the dark and so the animal can be studied under the microscope in the dark.

The arrow worm, *Sagitta* sp. in the small phylum Chaetognatha, is common in tropical plankton (Fig. 8.8c). The animal resembles a feathered dart about 1 cm long. The head has a pair of eyes and many larger and smaller curved chitinous spines around the mouth, which can be concealed when a hood is pulled over them and lessens resistance while swimming. Horizontal fins, with chitinous rods that stiffen them, border the trunk and a lobed horizontal fin ends the tail; these serve as flotation devices. The arrow worm darts forward by contraction of longitudinal muscles and captures prey, such as larvae and protozoans, with the oral spines. The taxonomic position of this small phylum is ambiguous; each animal has several pairs of coelomic sacs which act as a circulatory system and as a pseudoskeleton and there are no respiratory or excretory organs; it might by placed near Aschelminthes. During embryology it shows some affinity with vertebrates in the formation of the coelomic sacs. The numbers of *Sagitta* in the plankton hauls peaked in February and July.

Planktonic tunicates (Urochordata) of the class Larvacea (Appendicularia) are present throughout the year, as they are in plankton throughout the oceans. The adults retain the larval (tadpole) shape of a large tail attached to a small body with few gill pouches (Fig. 8.8d). Instead of a cellulose tunic, which characterises sedentary tunicates on the shore (see Section 6.4.2), the body secretes a gelatinous bag or 'house' much larger than the animal, in which it floats. The to and fro movements of the tail causes a current to enter the 'house' through an inhalent pore covered by a fine grid, and to leave through another pore. The current is strained a second time through a finer net inside the 'house' and finally filtered through the small pharynx as in shore tunicates. The gelatinous 'house' is used for feeding, then shed and replaced in a few hours.

During moonless nights in August and September the adult reproductive phases of polychaete worms on the shore, the Nereidae, Eunicidae and Syllidae, may invade the sea and swarm at the surface. In Nereidae and Eunicidae, the eyes enlarge, the posterior segments change considerably in shape to adapt to swimming; they become larger, the parapodia develop fans of setae for floating as the eggs and sperms become ripe (Fig. 8.8e). In *Trypanosyllis gemmipara* clusters of reproductive buds occur under the last segment. These mature and swim away, later producing gametes.

All the members of one species swim to the surface simultaneously and release their eggs when ripe. The females secrete a hormone that stimulates the males to release sperms, which in turn release a male hormone that stimulates the shedding of eggs. Fertilised eggs develop into trochopore larvae (Fig. 8.8f). Later larvae (Fig 8.8f') spend several months in the sea before returning to the shore to burrow among seaweeds.

The most common protozoans in the samples of plankton were the very small ciliates in the order Tintinnida, particularly the genus *Tintinnopsis* (Fig. 8.8g). This is a planktonic ciliate which covers itself with a basket (lorica) made from sand grains, cemented together by mucus. It has the shape of a vase from which protrudes a large disc fringed by strong tufts of cilia around the mouth, used for feeding and swimming.

Foraminifera (see Section 3.7.1), with calcareous shells (Fig. 3.22 a-e) and occasionally Radiolaria, with siliceous shells, are present in the plankton; they are both modified amoebae in the Order Rhizopoda.

Many eggs of the invertebrates that live on the shores were found in the plankton samples from January to September, with a peak in April and May. These develop into invertebrate larvae of many kinds, such as the nauplii of barnacles (Fig. 6.3d), zooea and megalopa of crabs and shrimps (Fig. 3.1e and f), veligers of bivalves (Fig. 3.2d) and gastropod molluscs (Fig. 6.2e), bipinnaria, pluteus or auricularia of echinoderms and tornaria of enteropneusts (acorn worms) (Fig. 3.17e). The planula larvae of corals would probably have been too small to be caught in the 100 μm mesh net. These larvae all feed on various sizes of phytoplankton according to their feeding apparatus.

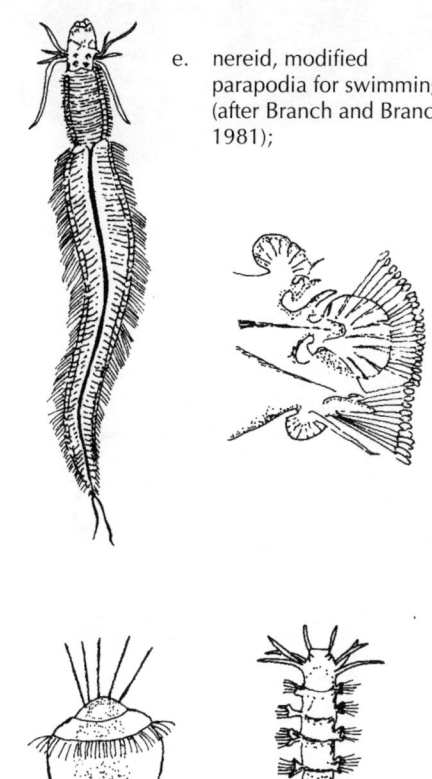

Fig. 8.8 Planktonic breeding phase in polychaetes:

e. nereid, modified parapodia for swimming (after Branch and Branch 1981);

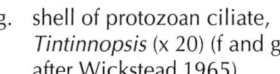

f. early and f' late polychaete larvae (x 100);

g. shell of protozoan ciliate, *Tintinnopsis* (x 20) (f and g after Wickstead 1965).

8.5 CORAL REEF ASSOCIATES

8.5.1 Large sedentary animals: Giant clams; coelentrates; sponges; worms

Giant clams, *Tridacna maxima* and *T. squamosa*, measuring up to 30 cm long, are scattered among the corals on the west coast. They have very thick, whitish shells with fluted edges and weigh several kilograms (Fig. 8.9a). These bivalves are related to 'cockles' which burrow into sand but the giant clams lie exposed on the surface, wedged between corals and attached to dead coral by a bunch of tough strings, the byssus. When covered with water the shells are wide open and upright (unusually for bivalves); a large area of the mantle is spread between the

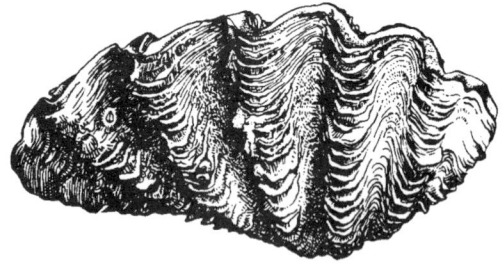

Fig. 8.9 Coral reef associates:
a. giant clam, *Tridacna maxima* one valve of shell* (x 0,2);

b. shell open, showing mantle, siphons (arrows).

shells and thus exposed to the sun (Fig. 8.9b). The mantle is highly coloured, green to brown and blue to purple, forming intricate, bright patterns. The pigment is contained in fixed cells called iridiophores which protect the tissues underneath from the lethal effects of the harmful wavelengths of light to which they are exposed. The mantle is continuous over the gap between the shells except for two large openings: the inhalent siphon at one end and the exhalent siphon in the centre. The inner of the three layers of the mantle near the siphons contain thousands of zooxanthellae, i.e. the microscopic dinoflagellates of reef corals. They live within the wandering amoebocytes which occupy the blood spaces in these parts of the mantle near the siphons. Light is focused on them by groups of superficial, hyaline cells which act as lenses; these are similar in structure to those in the 'eyes' of cockles and other bivalves (see Section 3.6.4). One advantage, gained by the giant clam from algal photosynthesis, is the use of the consequent excess metabolites to supplement the normal bivalve ciliary feeding on phytoplankton. The zooxanthellae multiply rapidly by asexual division, then some of the cells grow old, become incapable of photosynthesis and cannot multiply. These are not ejected as in corals, but they are used as *food* by the clam in a remarkable way, *without* the use of the digestive system. Amoebocytes leave the mantle blood spaces and transport the senile algae by means of the blood stream to the places in the body where metabolic requirement is high, such as the gills or the byssus glands of the foot and the active cells in the stomach that produce the digestive enzyme of the 'crystalline style'. These amoebocytes digest the algae and pass on amino acids and sugars to the surrounding tissues. In clams the symbionts themselves thus eventually become food for their hosts, a fate unlike that in corals, in which the dying symbiont cells are ejected.

The giant clams have a peculiar anatomy. No other bivalve mollusc rests on its hinge and opens its shell wide at the opposite end; nor is the foot, with its byssal gland for secreting attachment threads, so conveniently placed near the hinge; normally it protrudes through the gape of the shell. This topsy-turvy arrangement in clams, which enables the animal to display the tissues containing algae to the sun, has been brought about by the rotation of the shell with respect to the foot during the course of evolution over 600 million years. The mantle has also increased in size, the two halves have fused and grown thicker and the siphons have become enlarged. This rotation was accompanied by the suppression of the development of the anterior adductor muscle of the shell. One large posterior adductor muscle remains near the inhalent siphon. This muscle has many extremely strong 'slow-acting' muscle fibres as well as 'fast' fibres so that the shell closes quickly and remains watertight on exposure to air when the tide recedes. It is said that extraction of a person's trapped foot is impossible without first killing the animal in boiling water.

Tridacna maxima is more common on the reefs at Inhaca Island than *T. squamosa*. The rib scales on the outside of the shell, called 'squamae', are more scroll-like in *T. maxima* and they extend into the depressions between the ribs.

Patches of colour of soft corals (Octocorallina) (green) and ahermatypic coral polyps (red) occur among the hard corals on the reef of Inhaca Island. Two species of Alcyonacea are particularly conspicuous, forming leather-like sheets 50 cm in diameter or more, their bases being attached to old, dead coral. On both these species there are hundreds of small polyps, open during the day. In *Sarcophyton*, the thick sheet grows into a large number of scalloped folds, standing several centimetres high with crowds of small polyps on the crests of every fold

(Fig 8.9c), thus presenting an enormous feeding area. *Sinularia* may be as large but is more rubbery in texture and the surface area is increased by the projections of a large number of small finger-like lobes each with many small polyps.

In contrast, the colonies of the ahermatypic coral, *Tubastrea* (*Dendrophyllia*) have brilliant red polyps, each measuring 30 mm in diameter, growing on an erect black skeletal stem and resembling a prickly inflorescence when expanded fully. Each polyp contracts into a compact red ball when disturbed by a passing fish. The skeleton is depicted in Fig. 5.11a.

The gorgonids, another group of octocorallines, are fairly rare in Inhaca reefs; large 'sea fans' such as the many-branched, graceful *Antipathes virgata* occur where there is a steady current; long whitish branches of other species appear here and there among the hard coral boulders. They are firm, but pliable, strengthened by internal calcareous spicules in a horny support (see Fig 9.9d and Section 9.4.5).

Two species of very large anemones (Actiniaria), about 30 cm or more in diameter, may be found between the hard corals; both are greenish in colour and have tentacles over 7 cm long. One species, *Stichodactyla* has several individuals of commensal fishes, *Dascyllus trimaculatus* and *Amphiprion allardi* (Fig. 8.17b and c) nestling between the tentacles thus resembling the shorter-tentacled *Heteractis magnifica* among the sea grasses (see Section 9.4.5). The other very large anemone with long tentacles, *Radianthus* sp. has no commensal fishes.

Occasionally a spreading cluster of over 50 long, slender, cream-coloured tentacles of a burrowing anemone, *Cerianthus* sp. (Fig. 8.9e), protrudes from a wide hole in the sand between corals, about 5 cm of the polyp and its tube appearing above the surface. When disturbed it disappears extraordinarily rapidly.

Another group of compound coelenterates, the colonial anemones or zoanthids, have many tentacles in each large polyp, but there is no calcareous skeleton (see Section 7.4.2). They form firm, flat, encrusting colonies with a thick base into which tough, wide, but short, polyps can be withdrawn. A pale green species, *Zoanthus natalensis*, lies on the sand between corals and its surface is encrusted with sand (Fig. 8.9d). Although zooanthids feed in coelenterate fashion with tentacles and nematocysts, they too carry zooxanthellae which supplement their carnivorous diet by producing useful metabolites in excess.

Patches of many solitary, bluish-white, delicate cup sponges occur on dead coral. They are perforated by many ostia and have a dorsal osculum. Boring sponges are visible where they protrude at the surface as coloured patches (see Section 6.4.2).

The red encrusting sponge seen on the reef flats on coral rubble and on the vertical face of the rocks at Ponta Ponduini at extreme low tide, also occurs in the reef. Yellowish and greenish grey encrusting sponges may also be found on the old coral bases. These are all included in the Order Demospongia in which the skeleton may be a mixture of siliceous spicules

Fig. 8.9 Coral reef associates:
c. soft coral, *Sarcophyton* sp. (x 1);

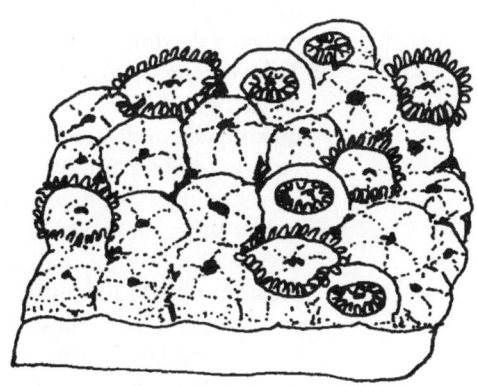

d. *Zoanthus natalensis*, opening under water, fragment of colony (x 0,5) (after Branch and Branch 1981).

e. Cerianthid anemone in burrow (x 0,5); (c and e from photographs by Robin Harris).

and spongin fibres. The canal system, through which they filter water and feed, is of the complex 'leuconoid' type (Fig. 6.7c).

Serpulid worms colonise the bases of branching corals (see Section 6.5.4); their coiled calcareous tubes are abundant. The upright bases of corals accommodate the more delicate sandy tubes of the plume worms and the crevices frequently provide space for ubiquitous sipunculid worms as on the reef flats.

8.5.2 Commensals: Xanthids and coral gall crabs; shrimps

Five species of small xanthid crabs (c. 10 mm) with smooth and shiny, flat carapaces and stout, longish legs, shelter amongst the branches of living corals (Fig. 8.10 a-c). They are quite agile and easily slip into small spaces where they cling tenaciously. The dactyls (i.e. the last joints of the legs), are broad and blunt so that the outer coral tissue is not torn as they move over the hard skeletal surface. Each dactyl has a backwardly projecting flange that fits on to the flattened edge of the penultimate segment of the leg so that it can be locked into position with the dactyl, held horizontally, grasping the coral branches. These crabs are detritus feeders, utilising detritus that has already been collected on the corals by the ciliary, cleaning currents. They scrape up the mucus (which also has food value) with special comb-like structures, made of stiff hairs, on their dactyls; the entangled detritus is cleaned off by the specialised maxillipeds and then conveyed by smaller mouthparts to the mouth. These crabs have been reported to be partially parasitic since they may scratch and poke the polyps which respond by secreting extra mucus. One pair of crabs will occupy a single coral colony for a lifetime unless the coral is very large, when there is more than one pair. If dislodged from their base they will reoccupy their territory. The larvae are free-swimming and settle on coral after metamorphosis. Juvenile crabs are however rarely found at Inhaca Island, reportedly because of the aggressive actions of the established pair.

The species are easily recognised by their shape and colour and by their specific hosts: *Trapezia* species are usually associated with the coral, *Pocillopora*; the crab's carapace is trapezoid and has a slightly toothed front and one pair of mid-lateral spines. *T. cymodoce* is orange brown (Fig. 8.10a), *T. guttata* is pinkish with tiny red dots and *T. rufopunctata*, as its name suggests, has larger red spots. *Tetralia glaberrima* (Fig 8.10b) lives on the coral, *Acropora*. The carapace has a very smooth surface and outline, the teeth on the 'front' being very small, and there is no lateral tooth. It is pinkish in colour and has a black bar across the front of the carapace. In *Quadrella coronata* (Fig. 8.10c) the front teeth are deeply incised (like a crown); unlike the other species, its dactyls are sharply pointed and can hold securely to the soft corals on which it lives.

The female of the gall-crab, *Hapalocarcinus marsupialis*, lives in a 'pouch' on any of the three genera of the pocilloporid corals whilst the smaller male is mobile. The female is the size of a small pea and almost as round, later becoming swollen with eggs (Fig. 8.10e and e'). As a post-larva it settles on the tip of a branch of *Pocillopora* or *Stylophora* (Fig. 8.10d) and lower down on *Seriatopora*, which has thinner branches. It stays in a slight depression permanently feeding on phytoplankton. The growth form of the coral around the crab changes in response to the continuous current of water impinging on its surface from the crab's respiratory current, augmented by the stronger, intermittent feeding current. Smaller, flatter and broader branches of coral develop on two sides of the crab which coalesce above and around it leaving only a few perforations on each side. The coral tissue in the walls of the gall has no zooxanthellae and, as a result of this, the slow growth of the coral walls keeps pace with that of the crab, so that the gall does not close until the crab is fully grown! The crab is thus imprisoned in a flat 'purse' where it continues to feed, but from which it cannot escape. It is said to pick at the tissues within the gall with its chelae and probably thus ensures that the inner surface is smooth. The coral resumes normal growth above and adjacent to the gall and several female crabs of different ages may live on the same coral colony.

The minute males live in open pockets between branches of the same colony but they are mobile. A male mates with a female crab before the gall is permanently closed and she is able to store the sperm for some time until the eggs are ripe for fertilisation. The female abdomen enlarges and the sides project to form a brood pouch in which the eggs are later laid and become attached to its concealed abdominal legs. Twenty five species of hapalocarcinid crabs have been found in reefs, but only one species occurs at Inhaca Island. Another gall-crab, *Cryptochirus*, which inhabits the large-polyped *Favia* and fits into cylindrical galls has only occasionally been seen on Inhaca reefs.

In coral reefs a very large number of commensal shrimp species of at least 61 genera inhabit the branches of living coral, or live with other animals, each being associated with a specific host (Bruce 1976). This is characteristic of shallow tropical seas and the hosts provide shelter rather than an intimate feeding relationship. Several palaemonid species occur in the genera: *Harpiliopsis*, *Jocaste*, *Periclimenaeus* and *Periclimenes*. Minute post-larval and juvenile shrimps, almost invisible to the naked eye, occur on a coral colony, more especially on branching species, but usually only one pair of adults remains to reproduce. Modifications of the forms of free-living shrimps enable the coral shrimps to adapt to sheltering amongst coral branches or calyces instead of burrowing. Even when fully grown they are comparatively small, the carapace is somewhat flattened, the spines are reduced and the walking legs are stout, ending in prehensile dactyls for secure attachment.

An extreme case of lateral compression occurs in *Racilius compressus*, an alpheid shrimp (Fig. 8.11a) that associates with the bright green coral, *Galaxea fascicularis* (Fig 8.5d). It is only about 15 mm long and the carapace is deeper than long, forming a wedge about 5 mm thick, so that it easily fits in between the separate septa of the calyces of the coral. Attachment is probably aided by the complex shape of the tail fan (Fig. 8.11a') (Barnard 1958). Two other larger pistol shrimps, *Alpheus lottini* and *Synalpheus charon*, usually live between the branches of various species of *Pocillopora* and they have a more normal shape. *A. lottini* (40 mm) is pale orange in colour with a dark purple, mid-dorsal stripe and orange chelae which are often red-spotted. *Synalpheus charon* (20 mm) (Fig. 8.11b) is light brown in colour and has distinctive dactyls on the legs which bear two spines. The pistol shrimps may perform a service for the coral by making a loud noise with their snapping chela, which may distract fishes intending to prey upon the polyps (see Section 4.5.3).

Some little palaemonid shrimps associated with corals are filter-feeders but they, unexpectedly, also make a similar sound to that of *Alpheus* using both chelae of the *second* pair of legs. The peg-in-socket sound mechanism is the reverse of that in the pistol shrimps! The peg is on the immovable hand and the socket into which it smacks is on the movable finger (Fig. 8.11 c and c'). A simple developmental change has thus achieved the same result in both families. These little shrimps are almost transparent under water. *Harpilius depressus* (c. 20 mm) is bluish and *Jocaste lucina* (16 mm) has red streaks and specks. One species of palaemonid shrimp, *Conchodytes tridacna*, at

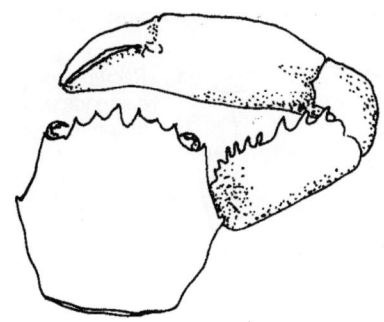

Fig. 8.10 Crabs commensal with corals:
a. *Trapezia cymodocea** (x 1);

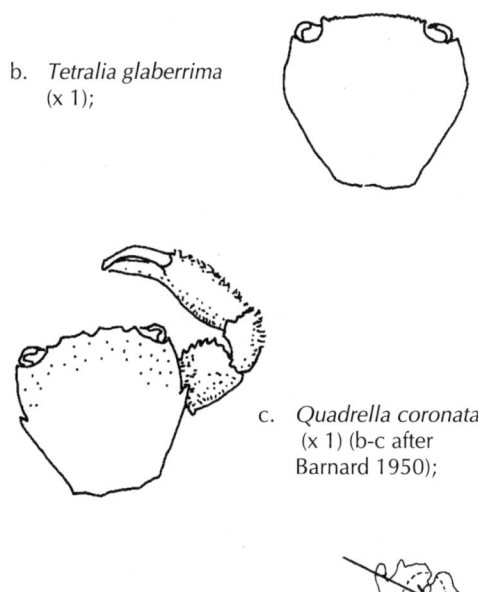

b. *Tetralia glaberrima* (x 1);

c. *Quadrella coronata* (x 1) (b-c after Barnard 1950);

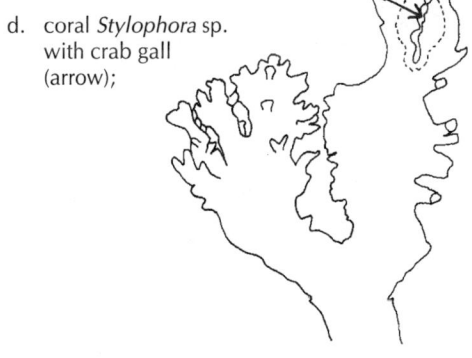

d. coral *Stylophora* sp. with crab gall (arrow);

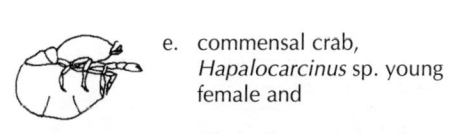

e. commensal crab, *Hapalocarcinus* sp. young female and

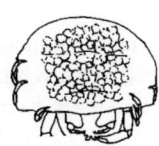

e' female carrying eggs (x 1) (after Barnard 1958).

Fig. 8.11 Commensal shrimps:

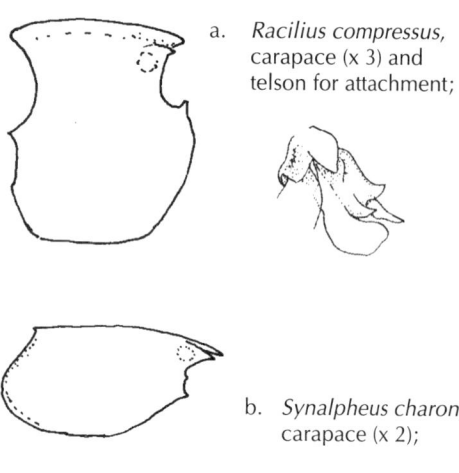

a. *Racilius compressus*, carapace (x 3) and telson for attachment;

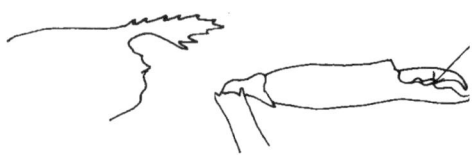

b. *Synalpheus charon*, carapace (x 2);

c. *Harpiliopsis depressus*, carapace and c' chela with peg in socket (arrow) (x 2) (a-c after Barnard 1958).

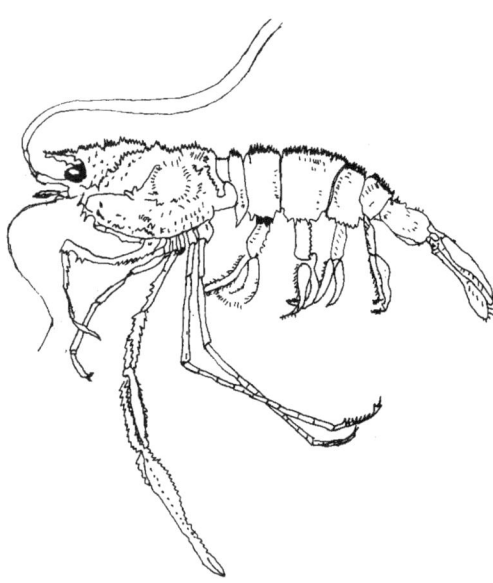

d. cleaner shrimp, *Stenopus hispidus* (x 1) (after Kensley 1972).

Inhaca Island lives in the mantle cavity of giant clams. Several other species of shrimps of the genus *Periclimenes* are adapted, by means of a variety of teeth and spines on the mouthparts, to feeding on soft food such as detritus, mucus and even coral polyps. Different species of *Periclimenes* are also associated with other members of the coral reef ecosystem such as the snake-like sea cucumber, *Synapta maculata*, the large red nudibranch, *Hexabranchus marginatus* (see Section 5.5.3), and the large sand sea anemone, *Heteractis magnifica* (Fig. 3.21c) described in Section 9.4.5. Another is associated with the feather star, *Tropriometra carinata*, described in the next section.

These little shrimps breed at all times of the year and are exceptionally fecund, egg-carrying females being always found. Larvae hatch in large numbers so that dispersal of the relatively sedentary shrimps is enhanced. An exception is found in the peculiarly flattened species, *Racilius compressus*, in which there are few eggs and they do not hatch until metamorphosis is complete. The juveniles occupy the same large colony of *Galaxea* as their parents; this coral genus is less common than other genera and has a sporadic distribution; the shrimp's survival is ensured by suppression of a motile larval stage.

The *ecological significance* of these highly specific associations is that the 'number of niches available to shrimps is increased whilst interspecific competition is prevented' (Bruce 1976).

Occasionally, the peculiar, long-armed 'cleaner-shrimp' *Stenopus hispidus* (60 mm) appears as though hovering above the corals (Fig. 8.11d). It looks more like a slender, miniature lobster as the abdomen does not have the hump, usual in shrimps. The first three pairs of legs are chelate and the third pair is much longer than the body (70 mm), stouter than the others and held laterally at right angles to the body. The carapace and abdomen are spiny and the rostrum is long. The white antennae and the alternate white and red bands on the long legs and abdomen advertise the shrimp's presence to fishes when it postures and dances, attracting attention. This shrimp has the reputation of being able to supplement its scavenger-type diet by removing fungal growths and parasites from certain species of fishes, which readily submit to its ministrations. The fish is first attracted by the shrimp's conspicuous colour pattern and then it adopts a stationary posture whilst the shrimp crawls gently over its body and removes external parasites or decaying tissue. Even predatory fishes such as eels make no attempt to eat them. Some species of the smaller palaemonid shrimps (*Periclimenes* spp.) are also 'cleaner shrimps', but these have not yet been observed at Inhaca Island.

8.5.3 Mobile crustaceans: crabs; lobsters

Many small xanthid crabs are loosely associated with coral. *Liomera*, of which there are three species, is bright red with white markings and a granulated carapace; they range from 5 mm to 20 mm in length (Fig. 8.12a). There are two species of another xanthid crab, *Lybia plumosa* (Fig. 8.12b) which has a shaggy carapace, and *L. leptochelis*, with tufts of hairs scattered on the carapace and the legs. Quite incredibly, both of these crab species

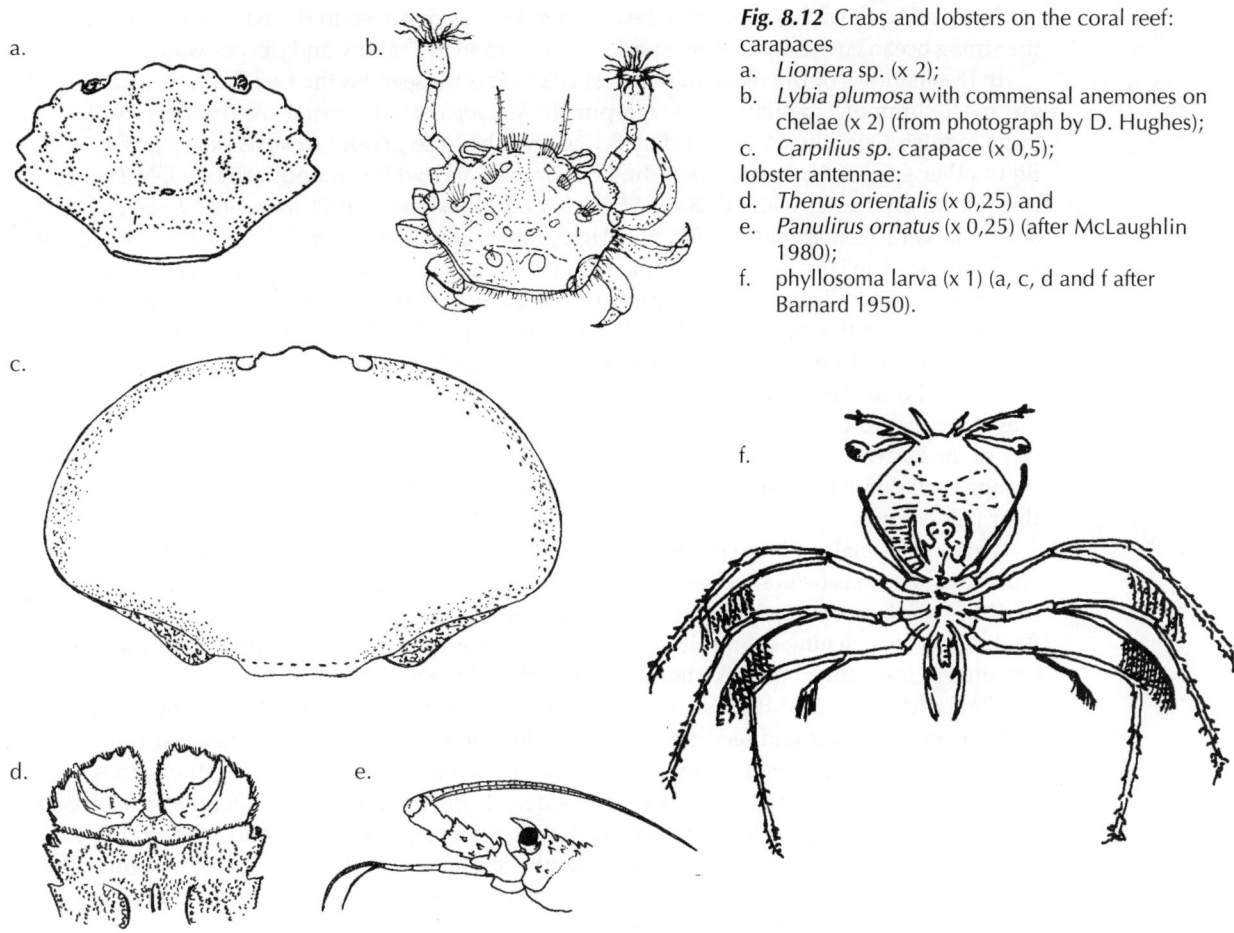

Fig. 8.12 Crabs and lobsters on the coral reef: carapaces
a. *Liomera* sp. (x 2);
b. *Lybia plumosa* with commensal anemones on chelae (x 2) (from photograph by D. Hughes);
c. *Carpilius sp.* carapace (x 0,5);
lobster antennae:
d. *Thenus orientalis* (x 0,25) and
e. *Panulirus ornatus* (x 0,25) (after McLaughlin 1980);
f. phyllosoma larva (x 1) (a, c, d and f after Barnard 1950).

carries a tiny stinging sea anemone on each chela. The anemones may sting the potential prey or intending predators of their host crabs.

In the deeper part of the reef larger carnivorous xanthid crabs such as *Carpilius* spp., (Fig. 8.12c) *Atergatis* spp., and *Menippe* spp., grow much larger than those seen on the reef flat (see Section 6.5.2). One chela in each of the species is enlarged and able to crush prey or cut it open. The red *Atergatis* spp. may be over 100 mm in diameter.

The lobsters that live in holes and crevices in the deeper parts of the reef are tropical species, beautifully coloured green, blue and white. *Thenus orientalis* (Fig. 8.12d) has a flattened carapace and enlarged, flat, plate-like second antennae, whilst *Panulirus ornatus* and *P. versicolor*, the spiny lobsters, have a more cylindrical carapace and enlarged whip-like second antennae (Fig. 8.12e). Lobsters crawl over rocks seeking food and move rapidly backwards to escape a predator by sharply flexing the broad abdomen and uropods. They are preyed upon by octopus and large fishes. The larva of a lobster is a 'phyllosoma' that has a thin, transparent, circular body and very long legs covered with feathery setae (Fig. 8.12f).

8.5.4 Echinoderms: Brittle stars; feather stars; starfish; sea urchins; sea cucumber

Brittle stars are the most abundant echinoderms in the coral reef and they are probably the most numerous of all the animal associates there. They are more easily seen at night when they emerge from concealment in the coral bases to feed. Eighteen species, which have a typical Indo-West Pacific distribution, have been found in the Inhaca reefs (see Appendix A) (Balinsky 1957). There is some difference in the fauna between the western reefs and those in the southern bay, the latter being the richer. This may be attributed to easier access to the reef for larvae carried in the strong ocean current. Different species are common in shallow and deeper water.

In the shallows of the reefs on the west coast, species seen on the reef flat (see Section 6.5.1) are most common, namely, the large purple *Macrophiothrix hirsuta cheneyi* (Fig. 6.10g), the large reddish *Ophiothrix foveolata* (Fig. 8.13a) and the little green *Ophiactis savignyi* (Fig. 6.10e). Eight other species also occur including the orange-and-white smooth-armed *Ophiolepis cincta* (Fig. 6.10h). In deeper water (7-8 m) *M. hirsuta* becomes rare, but *O. foveolata*, *O. savignyi* and a small, darker green species, *Ophiactis delagoa*, newly described from Inhaca Island (Fig. 8.13b), are common. The jaws of the last named are distinctive: a single small pair of oral papillae and broad oral shields. There are 4-5 arm spines on each segment and a pair of tentacle scales. The radial shields on the disc are shaped 'like the hoof-marks of a cow' and there are no spines on the disc surface. There is a dark spot near the distal end of each radial shield on the disc.

At the Ponta Torres reef in the southern bay the common species of brittle stars of the western reefs do occur but they are not predominant (see Section 5.5.1). *Ophiocoma erinaceus*, one of the largest and the most beautiful of brittle stars, being jet black with rows of scarlet or orange podia under the arms, climbs over living coral during the day contrasting vividly with their pale colours.

Very occasionally a large gorgonid brittle star or basket star, *Astroboa nuda* var. *nigra*, 30-40 cm in diameter has been seen here (Fig. 8.13c). It shelters under boulders by day with its arms curled inward and, being black, it is almost invisible. At night the arms are extended to form a feeding net for catching swimming prey, the mobile arms branching many times until the tiny terminal twigs form a bush-like tangle from which shrimps cannot escape.

The brittle stars owe their success to their mobility, to their ability to hide in crevices and to the variety of food and feeding habits. The long jointed arms are muscular and the animal progresses in an agile manner by putting one or two arms forward, fixing the spines on the rock; the other arms may push from the sides. Their feeding habits include filter-feeding (some *Ophiocoma* spp.), deposit feeding (*Ophiactis* spp.), browsing and scavenging (some *Ophiocoma* spp. and *Ophiotrichid* spp.), and also carnivorous feeding (*Astroboa*). Some of the brittle stars brood their young in pockets around the mouth but most of them have free-swimming pluteus larvae, which have long stiff, ciliated arms.

A single species of feather star, *Tropriometra carinata*, an unattached sea-lily or 'crinoid' occurs here (70 mm) (Fig. 8.13d). Superficially it seems to resemble a brittle star since it has a body disc and jointed arms but it is very different in the arrangement of its parts. The body disc is apparently upside down because its central mouth and eccentric anus are on the upper instead of the lower surface. The five jointed arms bifurcate at the base and each branch has pairs of side branches at intervals, giving the animal a feathery appearance. The branches have pairs of podia between the ossicles, but unlike the position in brittle stars, these are on the *upper surfaces*. There is also a shallow, ciliated, ambulacral groove on the upper surface of each arm. The arms grow from a ventral skeletal plate and curve upwards, indenting the body disc so that the ciliated grooves connect with the mouth. Underneath the disc, from the same skeletal plate, extend ten slender, flexible, jointed legs or 'cirri' which support the animal so that it is raised off the rock surface as though perching. The cirri are not otherwise used as legs since locomotion is effected by the beating of the feathery arms, adjacent ones moving up or down alternately. The cirri have evolved from the supporting stalk of ancestral sea-lilies, which were permanently attached to a sandy substratum in which the 'cirri' along the stem were partly buried.

Crinoids are suspension feeders and adopt a stationary position for this purpose. All the arms are held aloft in a current, five of the ten branches turned through an angle of 180°, so that the full force of the water flow impinges on the ambulacral grooves of all the arms. The podia in the groove are tentacle-like and secrete mucus on papillae which traps the zooplankton. There are no tentacle scales but the podia toss their catch into the amulacral groove by sudden whip-like flicks. A ciliary current carries the food to the mouth along the grooves. The feeding function of the podia is believed to be the primitive use, whilst the locomotory function in starfish and sea urchins is considered to have evolved later when the

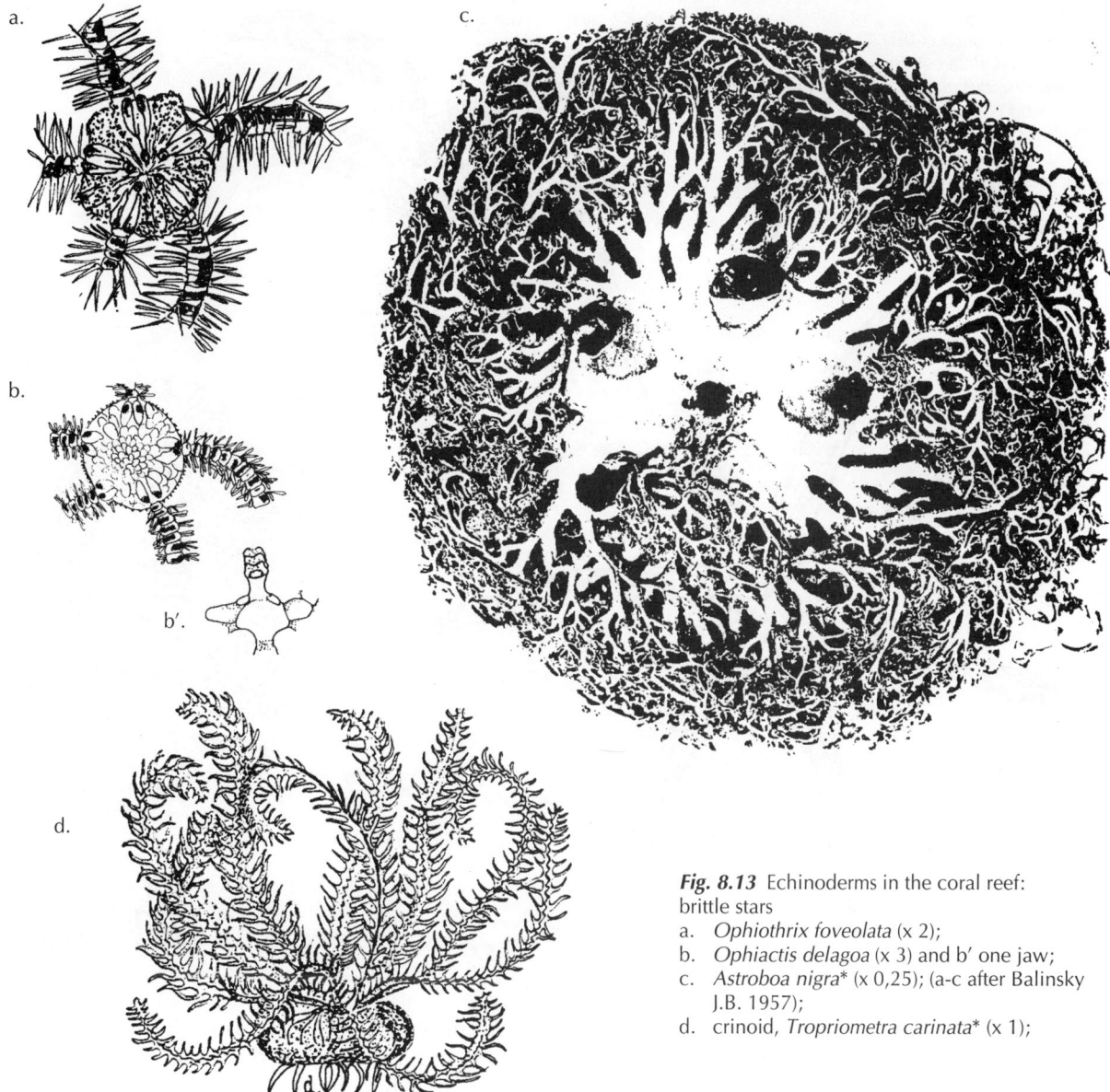

Fig. 8.13 Echinoderms in the coral reef: brittle stars
a. *Ophiothrix foveolata* (x 2);
b. *Ophiactis delagoa* (x 3) and b' one jaw;
c. *Astroboa nigra** (x 0,25); (a-c after Balinsky J.B. 1957);
d. crinoid, *Tropriometra carinata** (x 1);

disc became inverted so that the mouth became ventral. Fossil evidence confirms that the position of the disc in crinoids, with the mouth in a dorsal position, was the *primitive* position for echinoderms. Feather stars feed at night and in daylight the animals usually bend their arms inwards and conceal themselves in crevices.

In life history the crinoids resemble the sea cucumbers in having barrel-shaped larvae. The larva is called a 'vitellaria' to draw attention to the storage of yolk on which it feeds for a few days. It then settles on the bottom and fixes itself to rock by means of an adhesive gland. It develops a long stalk, like that of an attached sea-lily of deeper waters, and a disc at the top from which arms grow out. The attached stage lasts several months until the ten ventral cirri are fully formed. Then the disc breaks free and the animal becomes a free-swimming feather star. Feather stars are apt examples of the recapitulation of ancestral features during their own life history.

Unlike the brittle stars, the large, highly coloured starfishes of the coral reef ecosystem are not usually found on the coral itself nor on the dead coral rock, but lie stranded on the sandy areas when the tide recedes. An exception is *Culcita schmideliana*, the golden cushion star

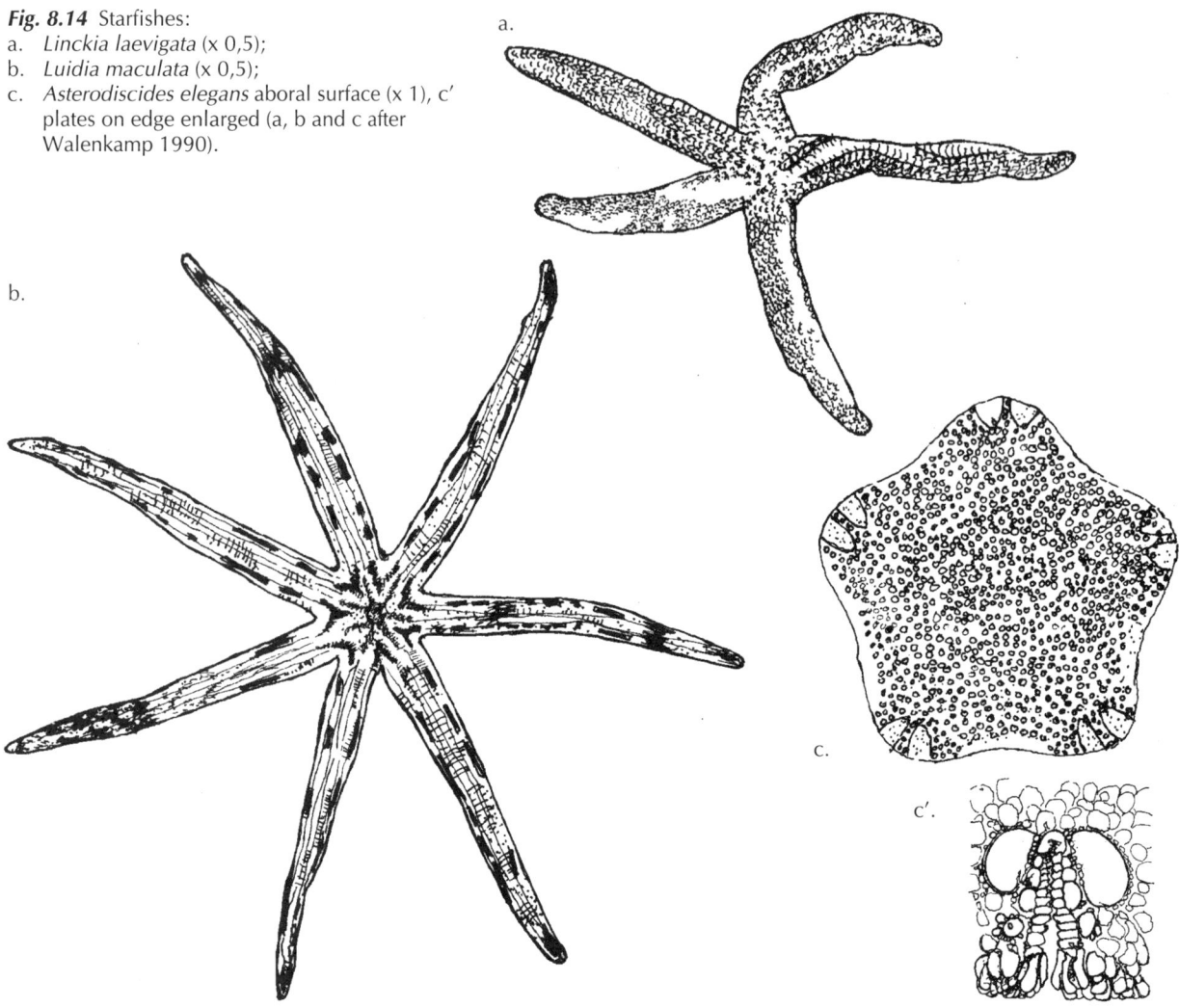

Fig. 8.14 Starfishes:
a. *Linckia laevigata* (x 0,5);
b. *Luidia maculata* (x 0,5);
c. *Asterodiscides elegans* aboral surface (x 1), c' plates on edge enlarged (a, b and c after Walenkamp 1990).

(Fig. 5.12f), which has a diameter of about 15-20 cm and is the only starfish at Inhaca known to feed on coral polyps, as mentioned in section 5.5.1. (Fortunately *Acanthaster*, the coral reef pest which did so much damage to the Australian Barrier Reef in the 1960s and 1970s does not occur here, but has once been reported from northern Mozambique (Grindley 1963).) *Linckia laevigata* (Fig. 8.14a), a sibling species of the common, little, green species, *Linckia multifora*, the six armed starfish under the reef-flat rocks, has five thick arms about 15 cm long, 5 times the diameter of the disc, but it resembles the rock-inhabiting species in having a rough granulated skin without spines. Its colour may be smoky blue, bright orange or, more often, these colours occupy large irregular patches, suggesting that the starfish may possibly have a mechanism for changing colour with age or on exposure to light. It is devoured bit by bit by a pair of small anomuran crabs, *Galathea picta* (Fig 6.12f). The greyish *Protoreaster* with large blunt red spines and the variably coloured *Pentaceraster* that live among sea grasses and sometimes near coral, are described in Section 9.4.1 (Figs. 9.3 a and b).

Two species of a large, burrowing starfish may be found, half buried in sand between coral boulders. These have fine spines along the edges of the arms and pointed tube feet for burrowing, like *Astropecten*, which burrow in bare sandbanks (see Section 9.5.1). *Luidia maculata* is yellowish in colour with dark spots and sometimes seven, eight or nine arms (Fig 8.14b). The arms are thick and about 15 cm long, smooth in texture and somewhat brittle, and bear a large number of shaggy spines along the sides. Like *Astropecten*, *Luidia* is carnivorous but feeds on brittle stars (instead of small snails), which are also swallowed whole.

On the sand at the base of the reef, 10 m deep, lies a brilliant red starfish that is sometimes tossed up and stranded on the edge of the reef. The edge of the disc is concave between extremely short protrusions representing arms and there is a pair of conspicuous smooth plates near each point on the aboral surface. This is the tropical species *Asterodiscides elegans* (Fig. 8.14c and c'). In deeper waters a large blood red starfish *Leiaster leachi* is frequently seen at the base of the reef.

One of the puzzling things about these large starfish is that smaller sizes of young stages are never found at Inhaca Island. The early life history starting with the motile bipinnaria larva with ciliated bands, and the later fixed brachiolaria larva with 10 ciliated arms and adhesive discs, in early metamorphosis are well known since the starfish will breed in an aquarium. Metamorphosis by differentiation of one end of the larva at 1 mm in size can be observed in the laboratory, yet young starfish are elusive in the field and attempts to rear them in the laboratory have generally not been successful as their specialised diets were unknown.

Some sea urchins are dangerous, in particular *Diadema* spp., the elegant, purple, long-spined sea urchin that typically inhabits coral reefs and is seen more especially in the southern bay at Inhaca Island. They shelter between rocks by day, their long protruding spines betraying the presence of crowds of twenty or more individuals in one patch. *Diadema setosum* and *D. savignyi* have dorsal spines up to 30 cm long that grow longer in sheltered waters (Fig. 5.12c). At night they emerge and walk fairly briskly (for a sea urchin) on the shorter ventral spines as though on stilts. The spines are fairly brittle and hollow, containing a *poisonous* secretion that provides protection against some predators; and the spaces between the spines form a haven for small fishes. Despite this armour they are preyed upon by many species of large fishes, helmet shells *Cypraecassis rufa* (see Section 9.5.3; Fig. 9.16e) and by lobsters.

If by chance *Diadema* is exposed on a rock by day and a possible enemy approaches, it can be seen to tilt the long spines towards the intruder in an apparently threatening attitude and it moves off in the opposite direction. The spines are attached to the skeleton of the test by ball and socket joints surrounded by two muscle sheaths. The outer sheath serves to point the spine in any direction and the inner one locks the spine into a rigid position. These sea urchins perch on the tops of coral rock at night and spread out their long spines as though on display. A beautiful iridescent, bluish-white ring appears around the dorsal anus made conspicuous by the inflation of the hind gut (see Section 5.5.3).

None of the stout-spined, slate-pencil urchins that frequent wave beaten reefs is present. The only other sea urchin in the reef itself is the oval *Echinometra matthaei* (Fig. 6.10a), described in Section 6.5.1 as an inhabitant of the reef flats.

A dark *green* cucumber, *Stichopus chloronatus* (Fig. 8.14 d-f), is quadrangular in cross section and has red-pointed papillae on the ridges; it is present in the shallow waters of the reef. Its spicules are distinctive 'tables' (Fig. 8.14f).

At Inhaca Island all the sea cucumbers are usually the size of small adults (20-30 cm) and juveniles are never found. It is presumed that these animals come inshore when almost fully grown. Life histories of tropical holothurians are not well known.

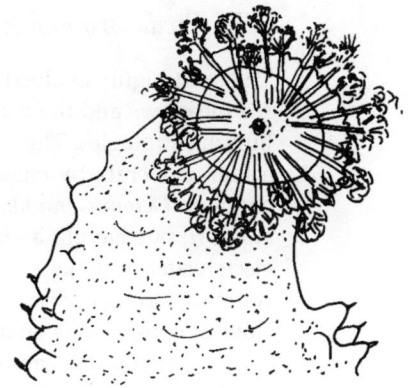

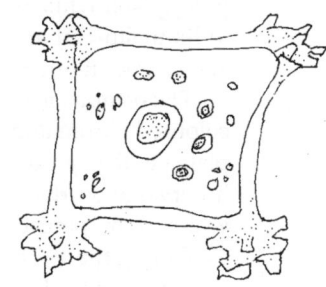

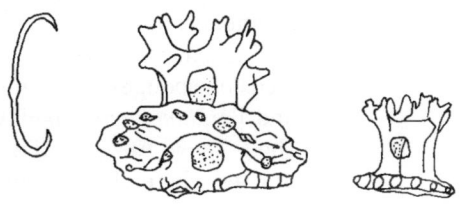

Fig. 8.14 sea cucumber, *Stichopus chloronatus*
d. head (x 1);
e. T.S. body, papillae at corners and
f. spicules (d and f after A.S. Thandar 1987).

8.5.5 Predatory molluscs in the coral reef: octopus and snails

The most highly evolved carnivorous molluscs of the coral reef ecosystem are the squid in the sea meadows and the octopus in the reef, both of the Order Cephalopoda. They differ in their hunting strategies. The squid, *Loligo*, is a stream-lined hunter that catches its swimming prey, a fish or a crab, by chasing it using rapid jet propulsion (see Section 9.4.3, Fig. 9.7o). The common *Octopus granulatus* (Fig. 8.15a) has an almost round body without an internal shell support, but having a similar muscular mantle, gills in the mantle cavity and a funnel that ejects water to make a sudden rapid movement. The octopus hides in crevices between coral colonies and, using its arms for support, it creeps slowly towards its prey, which is usually a slow-moving crab or snail. The octopus is active at night, whereas the squid hunts in daylight. The suckers on the arms of an octopus have powerful muscles so that the arms can pull apart bivalve shells or the carapace of a crab. The black beak in the centre of the buccal mass of the mouth is like an inverted parrot's beak and is retracted unless feeding. It can open shells or carapaces, while a quite large salivary gland releases a nerve toxin that has an almost instantaneous effect. The salivary gland contains digestive enzymes as well and the prey is predigested while the octopus waits a few minutes before devouring its food. In addition, true to its molluscan ancestry, the octopus has a radula with which it rasps the flesh of the prey and scrapes it out of the shell or carapace.

The octopus is notorious for its ability to change colour. In addition to possessing different coloured chromatophores it has reflecting iridocytes (like those of a giant clam) and reflector platelets that produce an iridescent shimmer. The small muscles of the chromatophores are innervated so that colour change is as rapid as in reptiles, turning from brown to grey or white, accompanied by red and blue iridescence.

During the 1950s and 1960s a long series of experiments and observations on the behaviour of the octopus in the Oceanographic Research Institute at Naples revealed an almost mammalian type of learning and memory, with neural centres located in certain anatomical regions of the complex brain (Wells 1962).

A large snail, *Rapana rapiformis* (Fig 8.15b), which feeds directly on corals, soft corals or sea anemones, lives on the reef, but is not very common. The large cowrie *Cypraea tigris* (Fig 8.15c), remarkable for its size (66-113 mm) and dark brown spots, frequents coral reefs and feeds on sponges. Large mitre shells such as *Mitra mitra* (Fig. 8.15d) prey upon smaller snails, and *Vexillum intermedium* (Costellariidae) (Fig. 8.15e) consumes small shrimps. The large Tun shell, *Tonna variegata* (Fig. 8.15f) may be found on the sandy floor between corals or sea grasses where Holothurian sea cucumbers congregate. The foot is very large and wide and can grip a sea cucumber, which is then paralysed by a salivary secretion containing sulphuric acid before being eaten. The shell is thin and light in weight, and often partly buried in sand when at rest. It is globular and light brown in colour.

8.5.6 Fish: herbivores; planktivores; omnivores; carnivores

The highly coloured fishes visible to the diver or snorkeller over the coral reefs are impressive in their number and variety. In fact, the greatest diversity of fish species of any ecosystem in the world is found among coral reefs. Three thousand species of fishes have been recorded as associated with coral reefs worldwide, many of which are localised geographically in Atlantic *or* Pacific reefs. The numbers decrease along the oceans with distance from the Indo-Malaysian reefs and in cooler seas to the north and south. Fewer species are recorded at Inhaca Island than on Tutia Reef on the East African coast near the equator (Talbot 1965), but the same 25 families that inhabit East African sheltered reefs, are represented (see Appendix A).

This greater diversity of tropical fish species in coral reefs than in temperate, shallow coastal seas has evolved for several reasons (Goldman and Talbot 1976). Extensive speciation occurs in an environment that has undergone little temperature change over a long period, that has rich and varied food resources *and* which may be subjected to temporary destruction

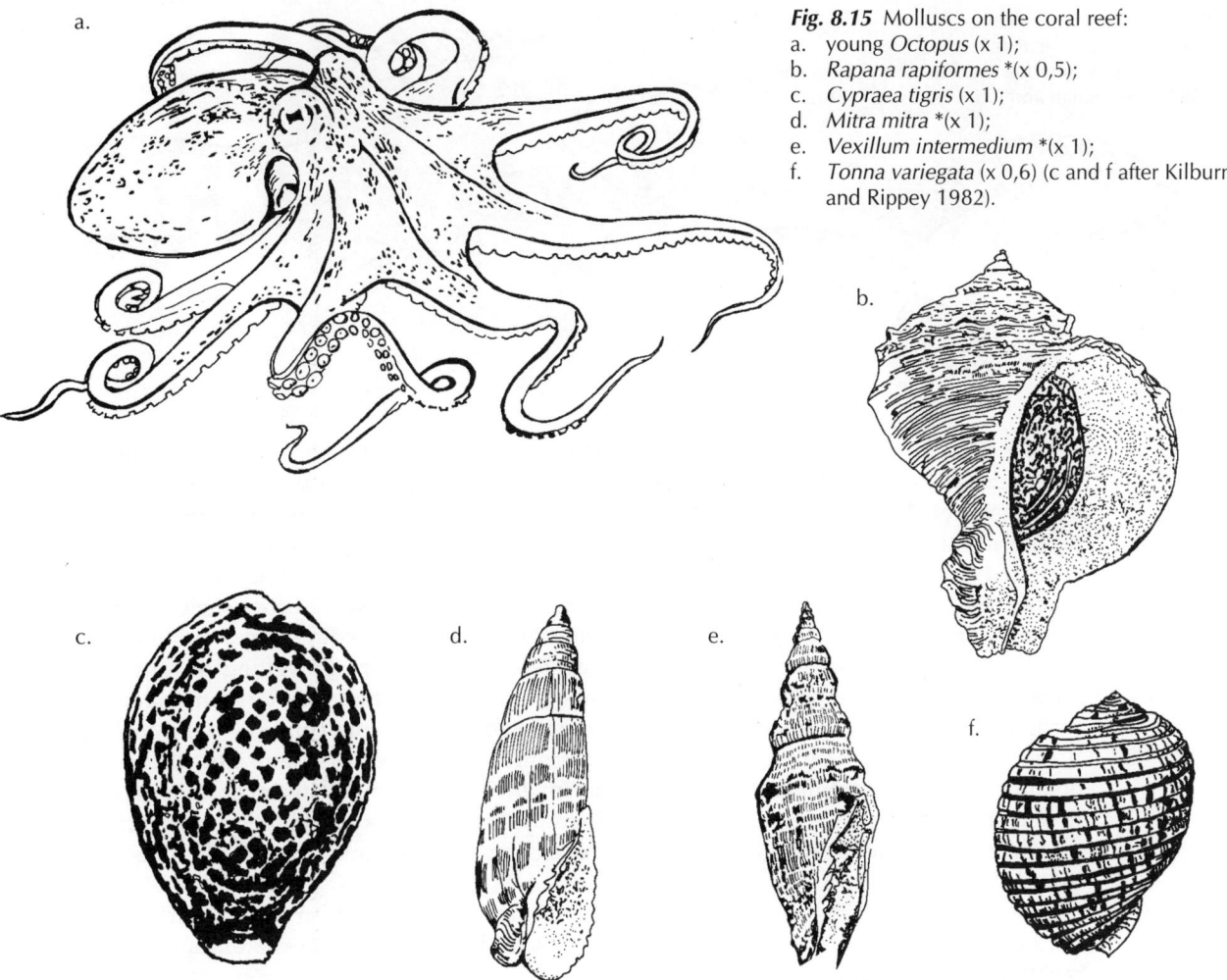

Fig. 8.15 Molluscs on the coral reef:
a. young *Octopus* (x 1);
b. *Rapana rapiformes* *(x 0,5);
c. *Cypraea tigris* (x 1);
d. *Mitra mitra* *(x 1);
e. *Vexillum intermedium* *(x 1);
f. *Tonna variegata* (x 0,6) (c and f after Kilburn and Rippey 1982).

and recovery. A mosaic of small habitats exists in a coral reef that has led to *'between-habitat' diversity*, whilst the complexity of each habitat offers opportunities for *'within-habitat' diversity*. Food resources are partitioned in a number of ways: algae and small-prey species are numerous and the times and methods of feeding by different species of fishes vary. Even when a habitat appears to be shared, the foraging methods, food-handling ability and the actual food taken differ among the fish species present, which appear to co-exist without competition. The resultant coral reef fish population is a small number of relatively small-sized individuals of many species. Population numbers are strongly regulated by predators (see below).

Secondly there are unique and varied reproductive adaptations in many species. In a few species parental care of the young in a brood pouch or by mouth-brooding protects the offspring. In some species the populations are often predominantly female with merely a low percentage of males to fertilise a large number of females externally. Should male numbers become insufficient (through predation) new males are recruited by means of *reversal of sex* of some females, thus maximising egg production. Some species spawn continuously in the equable micro-climate, where food is continuously available and not restricted to seasons. The complexity of colouration in coral reef fishes has, no doubt, enabled mates to recognise one another among the crowds of species – a case of evolution through Specific Mate Recognition. A small selection of typical coral reef fishes at Inhaca Island (from Smith and Heemstra 1986) is described below.

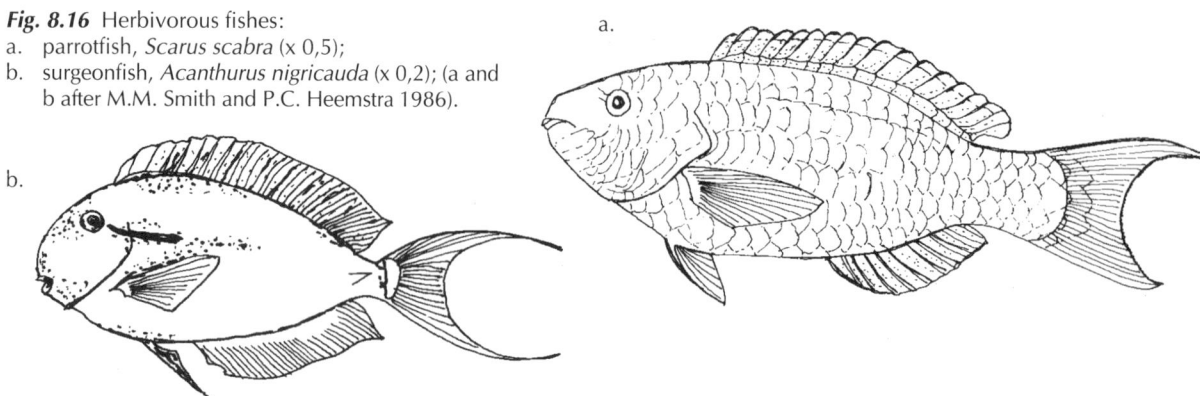

Fig. 8.16 Herbivorous fishes:
a. parrotfish, *Scarus scabra* (x 0,5);
b. surgeonfish, *Acanthurus nigricauda* (x 0,2); (a and b after M.M. Smith and P.C. Heemstra 1986).

The attached plants available to herbivores on sheltered coral reefs are the filamentous green and blue-green algae and encrusting red algae which coat the surface of dead coral skeletons, the foundation of the reefs. The growth of macro-algae is negligible, not as prolific as it is on reefs exposed to strong wave action. A more important source of algae on sheltered reefs is found *within* the living coral polyps, i.e., the large numbers of symbiotic, unicellular algae (see Section 8.2); the coral may be consumed by some fishes such as parrot fishes, largely because of its rich plant content. Phytoplankton feeding fishes are fewer in number and species in coral reefs than in cooler waters, where upwelling of nutrients enriches the plankton seasonally, and where consequently there are large numbers of sardine-like phyto-planktivores. The plankton-feeding fishes of the coral reef tend to feed primarily on zooplankton, visually selected from the water column and not filtered to obtain phytoplankton. Similarly, small benthic invertebrates, living between or under algae, may be taken by fishes grazing on rock surfaces by chance or may be chosen preferentially by some species, as in the goatfish, which uses highly tactile and movable barbels (see Section 5.6.3).

Benthic grazers are found among sixteen families of fishes in which herbivores predominate, for example among parrot fishes (scarids), surgeon fishes (acanthurids) and rabbit fishes (siganids).

The parrot fishes have fairly robust bodies covered by large scales. The mouth is small and the jaws bear coalesced teeth forming a beak with which the fish can scrape algae from rocks or break off pieces of coral. The tailfin is rounded or square in the young but crescent shaped in the adult; the bodies are often brightly coloured, more especially in mature males. *Scarus scaber* (Fig. 8.16a), for example, grows to a length of 30 cm and the body has a wide blue bar on a greenish background, the scales being outlined in red; the dorsal and anal fins have red stripes on a bluish-green background.

The surgeon fish have deep, compressed bodies, covered by a dark leathery skin in which rough scales are embedded. The young surgeon fish are, however, silvery in colour and round in outline. The head is rounded and blunt, the mouth is small and has very small incisor teeth used in scraping algae from rocks. The common name refers to the single sharply-pointed, lancet-shaped spine on each side of the body at the base of the tail. *Acanthurus nigricauda* (Fig. 8.16b), among the ten or more species of surgeon fish on Inhaca reefs, is dark brown in colour, almost black under water; it has lighter dorsal and anal fins and a black tail. This is one of a few species of surgeon fishes that ingests sand and microflora and has a gizzard-like stomach. Herbivorous rabbit fishes are more common in the sea meadows and are described in Section 9.4.7.

Schools of small fishes (10-15 cm) with small mouths hover in mid-water, feeding on plankton, preferably zooplankton. The most abundant species on the west reef appears to be the sea goldies (*Anthias squammipinnis*, Fig. 8.17a). In each small school there is one male (10 cm) and a number of smaller females and even smaller juveniles. The male will mate with each female, and if it is removed by a predator, a female will turn into a male in two weeks. These fishes are at present classified within the family Serranidae, the rock cod, although they are much smaller than most rock cod.

The damsel fishes (Pomacentridae) are small apparently timid fishes which seek concealment under rocks when disturbed. Their bodies are deep and laterally compressed and the scales are large. *Chromis* spp. occur in large schools; the body of *Chromis viridis* is suffused with a greenish-blue colour.

There is a tendency towards commensalism in this family, more especially among juveniles (see Section 9.4.7). *Dascyllus trimaculatus* (Fig. 8.17b), a jet-black little fish with three white eye-spots, and *Amphiprion allardi* (Fig. 8.17c), with three vertical dark bands, shelter among the very long tentacles of a massive greenish anemone *Stichodactyla* sp. One of the larger pomacentrids, *Abudefduf vaigensis* (Fig. 8.17d) is common.

At night the daylight species rest in the shelter of rocks and are replaced by cardinal fishes (apogonids). These are small, highly coloured fishes but they have large mouths and the eyes are much larger than those of diurnal fishes. The cardinal fishes have a dorsal fin in which the anterior spiny part is separate from the soft, posterior portion. *Apogon cooki* (Fig 8.17e) is not uncommon and is fairly typical of the characteristic bright colouration, having reddish fins and a number of bold, black longitudinal stripes and dark fins. The male cardinal fish takes fertilised eggs into its large mouth to protect them as they develop. Even after they have hatched, the young fishes shelter in its mouth in times of danger, having evolved a habit similar to that of the mouth-brooding cichlids of African lakes and rivers.

The families of benthic feeders which are broadly omnivorous include the butterfly fishes (chaetodontids), the file fishes (monocanthids), trigger fishes (ballistids), box fishes and cow fishes (ostraciontids), tobies, puffers and porcupine fishes (tetraodontids) and some of the large family of rainbow fishes (labrids).

The butterfly fishes have a much deeper body than the damsel fishes; they are extremely agile and brilliantly coloured. The teeth are bristle-like (which gives the name to the family, chaetodont) and the extremely small mouth is situated on the end of a short pointed snout. Nineteen species have been listed for Inhaca reefs (Smith and Heemstra 1986). *Chaetodon auriga*, the threadfin, for example, has a golden background colour, a large black spot at the end of the dorsal fin, and several thin, black intersecting lines on the body, a dark vertical band through the eye and another on its truncated tail (Fig. 8.18a). The outline of a butterfly fish has a distinctive squarish shape, since the dorsal and anal fins are deep posteriorly. The skin is covered by ctenoid (small) scales, which extend on to the tail fin. The butterfly fishes swim near rocks to feed, making little sallies, bumping the snout against the rocks to remove small invertebrates and coral polyps. They swim in pairs and appear to be nipping off coral polyps when they protrude.

The file-fishes and trigger-fishes have deep oval bodies and a characteristic large, anterior, dorsal spine that may be elevated and locked into position as a defence mechanism. In

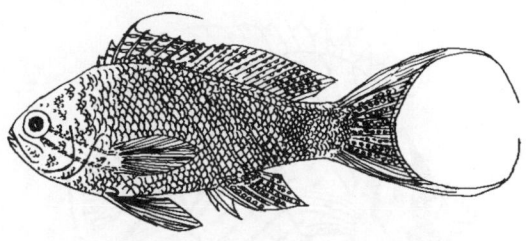

Fig. 8.17 Planktivorous fishes:
a. sea goldies: *Anthias squammipinnis* — male (x0.5)

juvenile damselfishes:

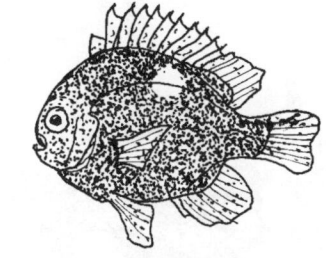

b. *Dascyllus trimaculatus* (x 1);

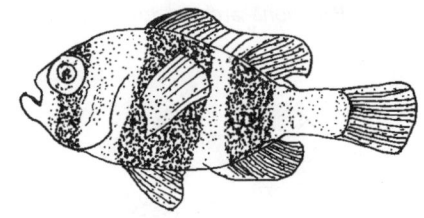

c. *Amphiprion allardi* (x 1);

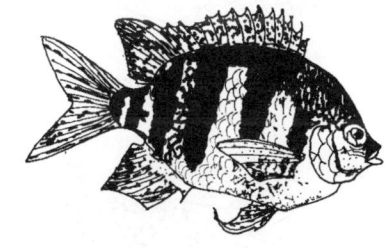

d. sergeant major, *Abudefduf vaigensis* (x 0,5);

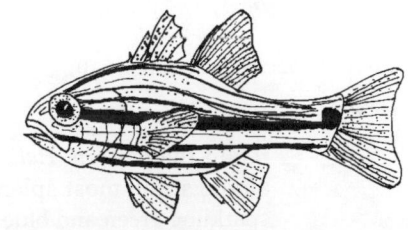

e. cardinal fish, *Apogon cooki* (x 1) (after M.M. Smith and P.C. Heemstra 1986).

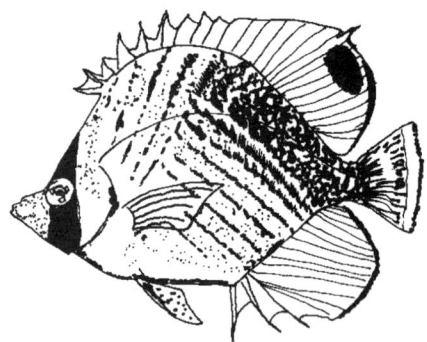

Fig. 8.18 Omnivorous fishes:
a. butterflyfish, *Chaetodon auriga* (x 0,3);

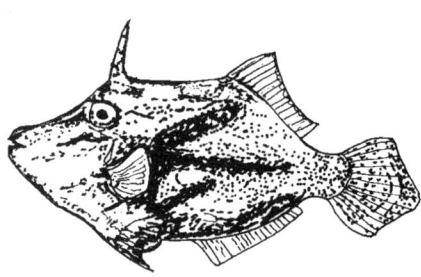

b. filefish, *Paramonacanthus barnardi* (x 0,5);

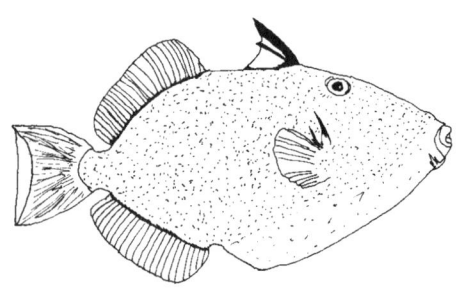

c. triggerfish, *Sufflamen chrysopterus* (x 0,3);

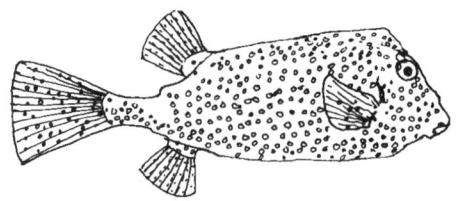

d. boxfish, *Ostracion maleagris* (x 0,3);

file fishes this spine is serrated whilst in trigger fishes it is not barbed but can be shot into an upright position by the muscles of a second spine behind it, which acts as a trigger to erect the first spine. Species of both families feed mainly on small invertebrates. They are not large (c.10-20 cm) and they are feeble swimmers. Thick leathery lips at the end of obliquely tapering snouts conceal a row of incisor teeth in each jaw. *Paramonocanthus barnardi* (Fig. 8.18b) a file fish, has an almost pentagonal body, dark in colour with lighter fins. *Sufflamen chrysopterus* (Fig. 8.18c) is a common triggerfish among the reefs in the quiet shallow water in the Bay of Maputo; the body outline is smooth, the body itself is dark and the fins and snout are golden in colour.

The fishes in the two orders Ostraciontidae (box-fishes and cow-fishes) and the Tetraodontidae (tobies, puffers and porcupine fishes) have quite different mechanisms for protection against predators. Box-fishes are encased in an armour of hexagonal bony plates under the skin, which are fused at their margins to form an external, solid box from which their fins protrude (Fig. 8.18d). On the other hand, the puffer fishes have a naked skin, without scales or bony plates. They possess the power to inflate the body with air or water (through the mouth) so as to become almost spherical and able to choke a potential predator as it tries to swallow it. When bloated, these fishes, such as *Arothron hispidus* (Fig. 8.18e) float up-side-down on the surface of the sea. This fish is dark dorsally, light ventrally and has yellow fins. The porcupine fish, *Lophodiodon calori* (Fig. 8.18f) selects sponges as food. This fish has yellow eyes and fins and a pale belly. Fishes in both these families are poisonous to man and are rejected if caught in fishing nets.

Among the families of carnivorous fishes in shallow tropical seas there are some genera that live in the shallow water permanently and others that come inshore only by night to prey upon the smaller fishes, crabs, worms and molluscs of the coral reef, since these are more active at night. Most of the carnivores have large, elongated bodies, strong tails and large fins, and are thus able to move fast; they have wide mouths and 'canine' type teeth in the jaws which are often supplemented by granular or rounded teeth on the palatine and pharyngeal bones. Besides the wrasses (labrids) there are snappers (lutjanids), scorpion fishes (scorpaenids), grunters (haemulids), eels of several families and soldier fishes (holocentrids).

The wrasses, known as rainbow fishes, are perhaps the most highly coloured of all fishes. Some genera are mainly red, blue or green, but all have intricate patterns of other bright colours. They can change colour rapidly (by means of a nervous mechanism like reptiles do, rather than a hormonal one) but retain the pattern. *Halichoeres lapillus* (10 cm), taken in shallow water at Inhaca Island, is amongst the most splendid of coral reef fishes. 'The markings on the live fish were like sparkling green and blue jewels on purple velvet' (Smith 1977) and the fins have broad orange and yellow stripes. Wrasses have lunate (rounded) tails and large mouths in which teeth are differentiated. Some species are omnivorous and others, in which the pharyngeal, molar-like teeth are well-developed, specialise in crushing molluscs. The scales of the skin are distinct; the lateral line is usually conspicuous and either bent or split into two horizontal lines, lying dorsally in the anterior part and more ventrally in the posterior part of the body. There are about 30 labrid species listed for Inhaca Island: *Cheilinus trilobatus* (Fig. 8.19a) is one example which, when adult, has a dark, red-tinged and brown body and three-lobed tail; the juveniles are brown and have a round tail (Smith and Heemstra 1986). It feeds on fishes, crabs and molluscs.

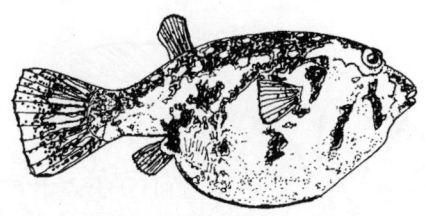

Fig. 8.18 Omnivorous fishes:
e. pufferfish, *Arothron hispidus* (x 1);

The snappers (lutanids) earned their common name from the behaviour of the fish when removed from water; the jaws are snapped shut by strong contractions of their muscles and become locked in that position. About five species are known from the sheltered coral reefs at Inhaca Island whilst different species prefer the strong waves of the east coast. The pale yellow species *Lutjanus fulviflamma* (30 cm) illustrated in Fig 8.19b occurs in the sheltered reef shallows.

Scorpionfishes (scorpaenids) have large heads armoured with bony plates bearing spines; the body is short and most are covered with scales. Glands at the bases of the numerous spines on the body and those supporting the fins contain

f. porcupinefish, *Lophodon calori* (x 0,6) (a-f after M.M. Smith and P.C. Heemstra 1986).

venom resembling that in the tail gland of the scorpion. Handling these fishes is dangerous, since the toxin is injected under the skin by sharp spines. The colouring of these fishes lends itself to concealment among stones where they lie in wait for their prey. *Scorpaenopsis gibbosa* (8 cm) (Fig. 8.19c) is fairly common and typical of the family; it is reddish brown in colour and has two tentacles on the head; the fins are large. In the firefish (devilfish), *Pterois miles* (23 cm) (Fig. 6.18b), every special feature of this family is exaggerated, making a grotesque caricature of a fish, which nevertheless is very beautiful! The pectoral fins and their spines are extended so that they resemble wings; a long spine between the eyes and short barbed spines on the snout complete the armour. The colouring is fiery red contrasting with the dark diagonal stripes on head and body and the dark spots on the tail. If by chance, the poison secreted by the skin is injected into a human hand by the spines, it is most painful for many hours (see Section 6.7).

Related to the scorpion fish is the *deadly* stonefish *Synanceia verrucosa* (25 cm), described on the reef flat in Section 6.7 (Fig. 6.18a) as a stationary fish in mud between slabs of old coral. The warning against touching this fish, even accidentally, must be repeated. It is the most poisonous creature on the shore, and the *poison may prove lethal.*

Eels (Anguilliformes) are divided into several families represented in Inhaca reefs by serpent eels (Ophichthidae), conger eels (Congridae) and moray eels (Muraenidae) (Smith and Heemstra 1986). They have no pelvic fins and in some families the pectoral fins are also absent. The body is extremely elongated, usually both the dorsal and anal fins being continuous and extending along much of the body; scales are deeply embedded in the skin and invisible to the naked eye. Eels lurk under rocks or burrow in sand and suddenly emerge to capture their prey with wide, gaping jaws, usually spiked with sharp canine-like teeth. They swim swiftly over a short distance by means of the powerful, sinuous body movements, but do not rise far from the bottom; they have no swim-bladder. The gill openings are small – another feature that points to a fairly sedentary habit. Large moray eels such as *Echidna polyzona* (50 cm) (Fig. 8.19d) are fairly common in the Bay of Maputo. They have a whitish skin marked by 25-30 vertical bars. *Conger cinereus* (Fig. 8.19e), the robust common conger eel, is dark grey in colour and may grow to over 1 m in length but the young have a golden colour. It differs from the other eels in having closely set incisor teeth that form a very powerful cutting edge to the jaws.

Soldier fishes (Holocentridae) are medium-sized fishes that feed on the larger zooplankton, benthic crustaceans and small fishes at night. The species are usually reddish in colour and have

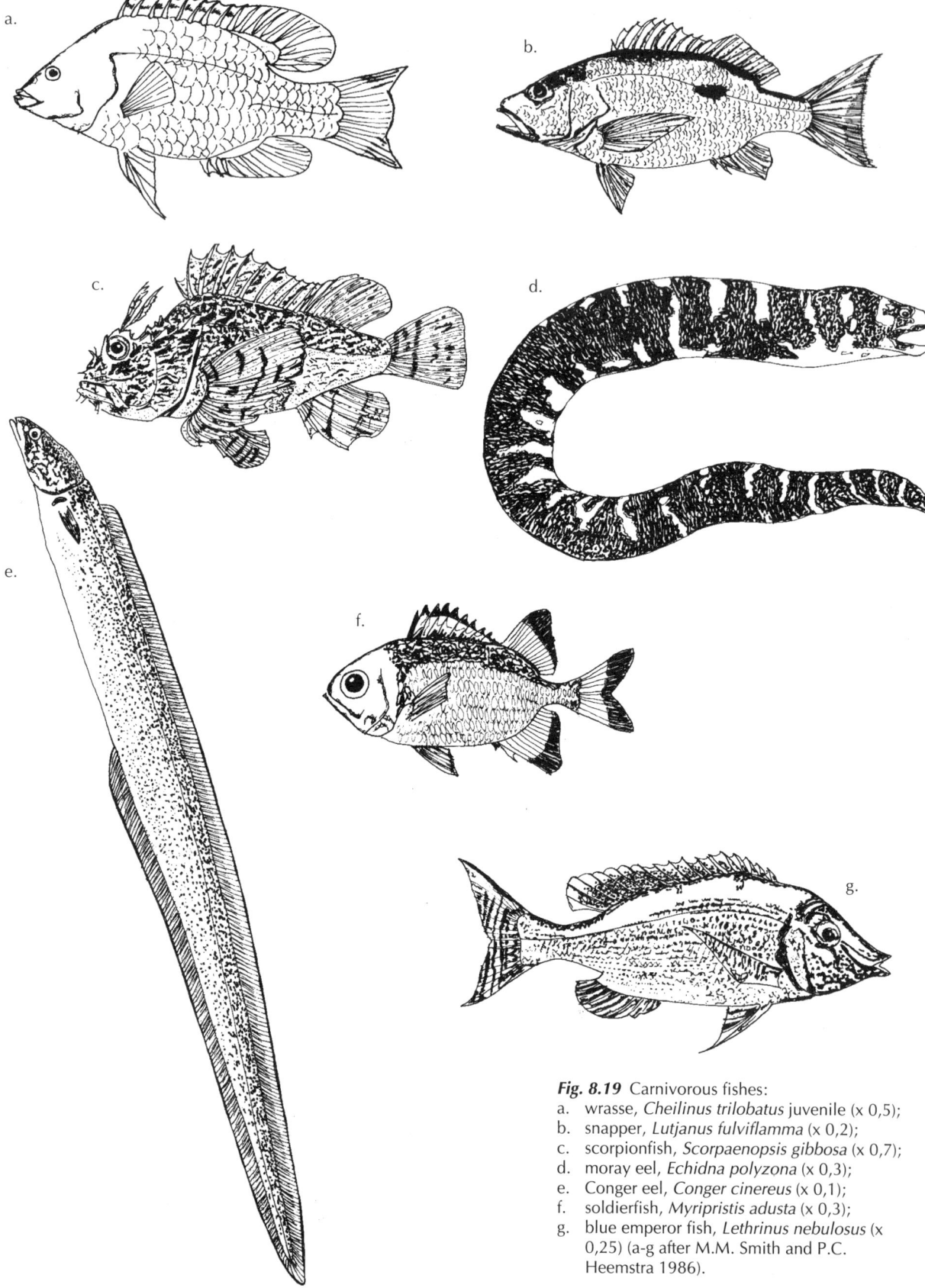

Fig. 8.19 Carnivorous fishes:
a. wrasse, *Cheilinus trilobatus* juvenile (x 0,5);
b. snapper, *Lutjanus fulviflamma* (x 0,2);
c. scorpionfish, *Scorpaenopsis gibbosa* (x 0,7);
d. moray eel, *Echidna polyzona* (x 0,3);
e. Conger eel, *Conger cinereus* (x 0,1);
f. soldierfish, *Myripristis adusta* (x 0,3);
g. blue emperor fish, *Lethrinus nebulosus* (x 0,25) (a-g after M.M. Smith and P.C. Heemstra 1986).

very distinct scales. The eyes and mouth are large and in some species such as *Myripristis adusta* (21 cm) there are a few median teeth in the lower jaw outside the gape (Fig. 8.19f).

The lethrinids are a large family of carnivorous fishes of the tropical Indo-Pacific Ocean, that are under 1 m in length. There are usually no scales on the cheeks. *Lethrinus nebulosus*, the blue emperor (Fig 8.19g), is common at Inhaca Island and grows to a length of 75 cm. It may be caught in traps among the coral reef in fairly shallow water or on lines. Its fins and mouth have a reddish tinge whilst the body is covered by minute blue spots and the head has a blue bar between the red eyes. The jaws have canine teeth in front of bands of fine teeth and lateral molars, and it feeds mainly on small molluscs.

A study of East African coral reef fishes indicated that the compositon and appearance of coral reef fish populations change by night (Talbot 1965). By day there are large numbers of active, coloured, benthic feeders and schools of plankton feeders, whilst the carnivorous fishes cruise idly by or rest on the bottom. At sunset the coral polyps emerge but the grazers disappear into their resting places under rocks or among sea grasses; the carnivorous snappers, wrasses, soldier fishes and eels become active. The schools break up as the hunters spread out singly to hunt over the reefs, sea grasses and sand flats for fishes in hiding and for the many invertebrates, which are more active at night.

At night the commonest carnivores at large are the cardinal fishes (apogonids) and the soldier fishes (holocentrids). These are fairly small fishes, highly coloured and with large mouths. Their eyes are proportionally larger than those of diurnal fishes. The cardinal fishes exhibit parental care; the male takes fertilised eggs into its mouth to protect them, and even after they have hatched the young fishes shelter in its mouth at times (see Fig. 6.18e). *Apogon cooki* (Fig. 8.17d) is not uncommon and is fairly typical in having a reddish colour and longitudinal stripes.

The soldier fishes have slightly deeper bodies than the cardinals, the dorsal fin is spiny whilst one very strong spine lies in front of the anal fin, but coloration is much the same. *Myripristis adusta* is common (Fig. 8.19f).

8.6 REEF DESTRUCTION AND RECOVERY AT INHACA ISLAND

In the first and second editions of this book (1957 and 1969) Dr P H Boshoff reported vividly on the history of the reefs since 1935, recording new colonisation and development of the reefs, including growth and extension following annihilation. More data were added in his more recent publication on the distribution and taxonomy of coral species (Boshoff 1981). His original observations are quoted here and new information is incorporated in this section.

> Between 1935 and 1938 the only coral growth known was on the *western* edge of the Inhaca Channel opposite a point midway between Ponta Raza and Ponta Punduine. The top surface of the old coral debris was covered by a field of confluent coral heads 50 m broad and 140 m long; how far it extended into the depth of the adjacent channel is unknown. In 1943 this growth had disappeared but corals started growing again seven years later on coral debris which had become exposed after being buried in sand. A typical reef was flourishing in 1958.
>
> In the latter part of 1946 large patches of coral were observed along a kilometre of the inshore *eastern* edge of the Inhaca Channel opposite the red cliff (Barreira Vermelha)... These heads grew to a height of 1,5 m before becoming confluent. Initially they grew on clumps of coral debris lying on the sand ... Additional talus formed naturally and some was added from the dead coral broken by fishing boats as they punted along. Substrata were thus provided for more coral growth. Soon there had developed an elongated coral field almost 2 km long and 65 m wide in its northern part, narrowing to 10 m towards the southern end ... Encroachment of sand later covered the landward part of the field. Coral still grows on the bank in a reef as much as 10 m wide at the northern end, but rarely more than 5 m wide over most of the length of the reef (1958). The reef extends to a depth of 3 m at the edge of the channel, giving way to a sandy slope which within 50 m reaches the floor of the channel at a depth of 20-27 m ...

>Beyond the southern tip of this now complete fringing assemblage, isolated coral heads have formed on old sandstone rocks, such as *Pocillopora, Favia, Cyphastrea* and *Stylophora* ...
>
>In 1953 a colony of *Astreopora* and another of *Goniastrea* lay on the slope below the reef. By 1956 these had increased in size and with additonal corals had enlarged the assemblage of corals downwards by a further 5 m ...

Observations at the coral reef opposite Barreira Vermehla were made in 1981/2 (Nestler *et al.* 1984). A dense growth of corals had spread on to the shoulder of the canal and down the steep eastern slope to a depth of 6 m below mean low spring tidal level. The dominant genera on the canal face were large boulder-like colonies of *Porites* and *Goniopora*. Branching genera such as *Acropora* and *Stylophora* and the bushy *Pocillopora* occupied higher positions on the shoulder of the canal bank and many other genera occurred between them (Fig. 8.1).

The other well-known reef is in the southern bay. Boshoff reported:

>It starts near the tip of Ponta Torres on its western steeper edge; it continues northwards along a channel towards the Saco da Inhaca for about 2 km. At the southern end coral grows on a vertical cliff down to a depth of 15 m from within 3,5 m from the top. The growth may however be completely annihilated from time to time, buried in silt and sand. Within a year or two the cliff has been cleared and well covered again, hence on this cliff the colonies are younger ones. The largest are 25 cm in diameter at extreme low water of spring tides. The area is one of non-confluent young coral growing against a steeply sloping cliff face ...
>
>The banks of the inshore channel leading into the southern bay has coral heads up to 3 m high and 1-2 m in diameter and the growth is luxurious, forming a field about 10 m wide. In September 1953 there was a gap in the middle reaches of this reef but in May 1954 this portion had become completely covered by a 20 cm high growth of *Pocillopora verrucosa* and *Stylophora pistillata*. This rapid growth is compatible with earlier records of as much as 12 cm increase in one month in some tropical reefs ...
>
>This southerly part of the reef has fewer species than are to be found in the reefs on the west coast and it resembles the early stages in the development of the latter. For example there are some areas with an almost pure growth of *Pavona*, areas of *Porites* only and other areas of *Montipora*.

During the 1970s the southern bay reefs were again subjected to partial destruction from which they are now recovering.

In 1974 a detailed study (Salm 1976) was made of the southern part of the Ponta Torres reef which had been destroyed by deposits of sand from a blow-out on a dune near Santa Maria, opposite the entrance to the southern bay. Several 6 m^2 areas of substratum were mapped to trace the recovery process after the dune was stabilized by planting sand-binders on the Santa Maria dune on the mainland across the strait. Recolonisation of the eastern channel bank took place in two stages. Firstly, planula larvae of the small-polyped *Porites lobata* and *P. solidula*, which are able to settle on a soft sandy base, grew into boulders on the channel floor. These were soon buried and again uncovered by the flow of the tidal water, but were dead. Their surfaces became eroded and pitted, providing a hard substratum for the settling of planulae of other species. In the secondary succession, the foliaceous species of the fast-growing *Pocillopora* and *Stylophora* species, but only the more robust species of *Acropora*, first settled and grew. They were followed by encrusting species of *Montipora* and *Astraeopora* and then by species with solid stone-like skeletons *Echinopora*, Favia, with large polyps, and the brain coral *Platygyra*, which are effective in preventing sand settling on themselves. Nearer the shore the spiny *Pavona*, the branching *Galaxaura* and stone-like *Goniopora*, which extrudes its large polyps in daylight, colonised the dead substratum of *Porites; Porites* recolonised the sandy floor of the channel.

During this study, the sandy enclave at Santa Maria was staked in rows between which sand-binding plants were grown. The species available on the dunes for the experiment were

Canavallea rosa, Cyperus maritimus, Ipomoea pescaprae, Hydrophilax carnosa, Scaevola plumeri and *Sophora tomentosa* (see Section 10.6.2).

In 1982 the Ponta Torres coral reef was studied again (Nestler *et al.* 1984). The reef had extended into the southern bay along the eastern bank of the channel, that was about 6 m deep with some deeper pools. *Porites* boulders had grown to a height of 4 m and a diameter of 3 m. Large flat areas were exposed on the landward edge of the channel at low spring tide as had been the case in the 1960s in which *Pavona* spp., *Stylophora* and *Montipora* were well developed and *Goniopora* was conspicuous. Species with large polyps such as *Favia* and *Favites* were also present. It appears that the binding of sand on the mobile dune had been successful.

In the newly formed lagoon north of Portuguese Island, protected by the westward extension of the sandbank stretching from the northern point of Inhaca Island, the sand has been completely washed away. A solid base of old coral rock, in which the imprint of calyces of *Porites* were clearly seen in 1990, has provided a hard surface for the formation of a new coral reef. Precisely the same genera as in the Ponta Torres recolonisation process are scattered 50-100 cm apart; they are forming the basis of a new confluent reef. The sea grass in the lagoon is the pioneer *Cymodocea serrulata*. In 1992, entry of sea water from the Bay of Maputo became restricted by the accumulation of sand in the narrow channel to the lagoon. The future of this coral reef is uncertain.

The incipient coral reef at the edge of the channel south of Portuguese Island seen in the 1960s had been almost obliterated by shifting sand, as the Inhaca Channel changed its course, but it appears to be growing again. A large specimen of a red gorgonid is conspicuous there. The Ponta Raza coral reef of the 1930s is now completely buried and only the large area of coral rubble remains to indicate the former reef.

Mass mortality in coral reefs is a widespread phenomenon in the tropics since they occur in regions where occasional cyclones destroy them through sedimentation or abnormal wave action. Studies on the Great Barrier Reef of Australia (Endean 1976) indicate gradual recovery, which may be almost complete in eleven years if not disturbed again. Full recovery is estimated to take 20-40 years. At Inhaca Island the long intervals between cyclones, which are needed for full recovery, rarely occur so that the reefs are in a perpetual state of rejuvenation. This is one of the reasons for the lower number of species here than is found in northern Mozambique. It makes annual studies particularly interesting because different species are frequently appearing.

The limits to reef development on the island are set by three conditions: firstly, comparatively little hard rock is available subtidally on the warm, predominantly sandy coasts; secondly, low penetration of light reduces the depth of sea in which corals can grow to about 10 m at high tide because of the high silt load of the sea in the Bay of Maputo; thirdly, cyclones in the Bay of Maputo indirectly cause a shifting of sand on the edges of the channels from time to time, which results in the partial destruction of the reefs.

8.7 BIOGEOGRAPHICAL PERSPECTIVES

The coral reefs at Inhaca Island are *not* the most southerly on the east coast of southern Africa as was written in the first edition of this book, but they are the most accessible from the shore. Subtidal coral reefs occur offshore on the east coast of the Island where wave action is strong and the sea is about 11 m deep. The nearest submerged reefs are east of Cabo Inhaca in the north, and east of Santa Maria to the south of the Island. They have not yet been thoroughly explored, but in position they are comparable with the reef which occurs 150 km to the south, off the coast of Sodwana Bay (27°30'S) in Maputaland (Schleyer, M.H 1995).

Several biogeographical questions are pertinent: what special environmental circumstances fostered the development of coral reefs close inshore at Inhaca Island? How does shelter from wave action affect coral reefs? How do the miniature reefs at Inhaca Island compare in structure and species composition with the more extensive, tropical reefs further north and with those to the south on the exposed coast of Maputaland?

Darwin (1842) wrote 'the dimensions and structure of fringing reefs depend entirely on the greater or lesser inclination of the reef slope'. Since the temperature of the sea is warm enough for corals to thrive at Inhaca Island, in the light of Darwin's observation, the reason for the occurrence of narrow reefs at Inhaca Island is that they are restricted to the only submarine slopes available near the shores, i.e. the banks of the channels.

The present reefs are formed on limestone rock of old coral beds which exist beneath the sand. At Ponta Raza a small patch of semi-fossilised coral rock lies exposed on the sand (see Section 1.3.1); the new coral reef north of Portuguese Island is forming on a flat surface of coral bedrock. Geological maps (Bosazza 1956) of the Bay of Maputo indicate several areas where old coral reefs once existed, before the bay was completely drowned in the early Holocene Period. Probably the Ponta Torres coral reef on the edge of the channel in the southern bay of Inhaca Island was formed after the isolation of the island. The sea eroded the bay and formed a deep channel which was then colonised by coral brought in by the Mozambique current from the shores of northern Mozambique (see Fig. 1.3). It is thus a younger reef than that on the west coast.

The nearest fully mature *fringing* reef is at Tulear on the southwest coast of Madagascar (23°21'S) on a shore exposed to strong wave action. It has a broad 'outer' reef in addition to an 'inner' reef. Where the rocky foundation for an outer reef is absent only an inner reef is formed. This is the case at Nossi Bé (12°S), an island off the north-west coast of Madagascar, situated on the edge of a Quaternary Plain which has been sumberged; this is similar in some respects to the position at Inhaca Isalnd. Only the 'inner' type of reef is present, sloping down to a depth of 20 m and connecting with the shore by 'reef flats' strewn with coral debris as at Inhaca Island. The Nossi Bé reef is on a sheltered shore where wave action is minimal. In these morphological features, Inhaca reefs resemble the 'inner reef' at Nossi Bé, 2 000 km to the north-east of Inhaca Island.

In both cases the coral reef ecosystem includes sea meadows, bare sand, a region of coral debris on sandy reef flats as well as assemblages of living coral. As a result of shelter from wave action there is a reduction of algae (unlike the algal crest of an 'outer reef' on an exposed shore). There is also a reduction in the amount of lithothamnion, the encrusting red algae. The coral bases are thus not well cemented together and break off in storms.

On this 'inner reef' at Nossi Bé there is a large number of coral species growing in a mosaic pattern, but there are no large tracts of a single dominant species as seen on an 'outer' reef. In this too the reefs of the two islands are similar.

Fifty-six genera are listed for Nossi Bé (Pichon 1972) and over thirty eight of these are present at Inhaca Island. This is not unexpected, since both are exposed to the Mozambique Current which distributes the larvae. At both Nossi Bé and Inhaca Island the compact and foliaceous species are more numerous than the branching *Acropora* forms, which dominate an 'outer' reef. Sheltered reefs are dominated by *Porites* spp., which are the most tolerant of raised temperatures and of both dilution and increased salt concentration of the sea. This species can clear itself of mud sediment but not sand. Corals with large polyps such as *Favia*, *Favites* and 'brain' corals are most tolerant of heavier sediment because of their efficient ciliary cleaning mechanism. These occur in deeper water while *Porites*, *Goniopora*, and *Montipora* occur in the shallows and *Pocillopora* is the furthest inshore. Species of *Fungia* are common among sea grasses, whilst *Heterospsammia* occurs in deeper water on sand.

The reefs along the coast of Maputaland are some distance offshore beyond the breaking waves in deeper water. The problem of silt does not occur and growth of branching corals is vigorous and continuous.

The animal associates of the coral reef and those on the reef flats, described in Chapter 6, together with those of the sea meadows described in the next chapter are also found in the large tropical reefs, but there the numbers of species and of individuals are larger than on Inhaca Island.

At Inhaca Island 160 species of reef-building corals have been described (Boshoff 1981); this amounts to approximately one third of the total number of species known in the Indo-Pacific region.

9

SEA MEADOWS AND SAND BANKS ON THE SUBTIDAL FRINGE

The sandy subtidal fringe between mean and extreme low spring tidal levels on sheltered shores recalls, at first sight, a field of rough pasture grasses, but the plants are not as grass-like in appearance as the narrow-leaved sea grasses on the lower shore. They are related to the pond weeds in fresh water in the families Potamogetonaceae and Hydrocharitaceae, from which they have evolved.

The sea grasses are similar to those in Kenya (Taylor and Lewis 1970) and provide shelter for a large mobile fauna, which has much in common with that of the coral reef since there is free access between these habitats. The sea grasses grow throughout the year and a vigorous growth of red coralline algae coats the stems of the dominant species, which together provide a large area of holdfast and shelter for solitary and compound sessile animals (Fig. 9.1a). The extensive system of rhizomes and fibrous roots of the sea grasses consolidate the loose, fine mud and sand in which many animals burrow. The roots remain above the deeper black mud.

The leaves of sea grasses are shed continuously and through their rapid decay they contribute hundreds of tonnes of detritus annually to both the sea and the shore. The rate of production (measured as weight of carbon per square metre) rivals that of tropical forests and exceeds that produced by seaweeds (McRoy and Mcmillan 1977). This detritus is the foundation of the food web of the sheltered sandy shore and the sea meadows (see Fig. 2.1). The plants themselves are eaten by few animals but the leaves are laden with meiofauna, epiphytic diatoms, protozoans and bacteria as well as detritus, so that food is provided for a number of animals that browse on the plant surfaces. The fauna of the sea meadows is not as rich in species as the coral reefs and many different species occur.

A completely different habitat of bare sand-banks is exposed on the subtidal fringe near Ponta Raza between branches of the tidal current which enters the Bay of Maputo through the strait at Ponta Torres (see Fig. 8.1a and c). Fine sand is deposited as this tidal current meets the south-flowing, longshore current from the mouth of the Bay of Maputo. This sand is not well consolidated and has shifting surface layers, the water table being about 10-15 cm below the surface at extreme low spring tides. The sand-banks are inhabited by a community of burrowing animals which have nothing in common with those in the sea meadows. The latter are never completely exposed to air but those associated with the clean sand-banks may remain at or near the surface for an hour or two at extreme low spring tides. Surprisingly, colonies of gaily coloured orange and purple sea pens and sea feathers, purple sand dollars, drab starfish and even amphioxus may be found there. Although the clean sand-banks do not support macroscopic plants they are rich in microscopic flora.

9.1 DISTRIBUTION OF SEA GRASSES ON THE WESTERN MEADOWS

A continuous field of the long-stemmed, broad-leaved sea grass, *Thalassodendron ciliatum* (Fig. 9.1b-e) spreads along the subtidal fringe of the sheltered western shore, parallel to the

Fig. 9.1
a. Major sea grass, *Thalassodendron ciliatum* and some associated animals*;
b. mature plant (x 0,3) and
b'. leaf tip enlarged;
c. female flower;
d. male flower;
e. pollen threads (b-e after Isaac F.M. 1968).

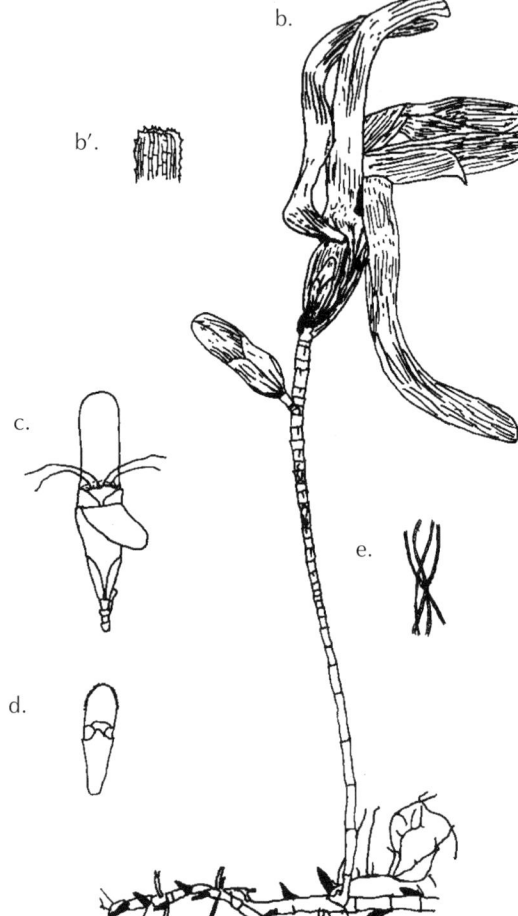

coastline from Ponta Raza to Portuguese Island and extends into the subtidal zone. This species forms an association with *Cymodocea serrulata* (Fig. 9.2a and b) below the level of average spring tide; it is limited subtidally by the Inhaca Channel in the central part of the field (see Fig. 5.1). In this area *Thalassodendron* spreads down the slope of the channel until growth stops abruptly at a depth of 5 m where light becomes insufficient for photosynthesis. The field becomes narrower where it is interrupted by the coral reef opposite Barreira Vermelha. North and south of the central part, the sea grasses abut on the bare intertidal sand-banks and subtidal sandy shoals (see Fig. 8.1).

The *Thalassodendron ciliatum/Cymodocea serrulata* association on the west shore differs from the association of *Halodule wrightii/Thalassia hemprichii* in the very sheltered north and south bays (see Sections 5.4 and 5.6). *Syringodium isoetifolia*, a sea grass with terete (almost cylindrical) narrow leaves (Fig. 9.2c and d) grows sparsely among the dominant *Thalassodendron*; some *Thalassia hemprichii* (Fig. 5.4a) intrudes from the northern bay.

9.2 BIOLOGY OF THE SEA GRASSES ON THE SUBTIDAL FRINGE

Thalassodendron ciliatum (Cymodoceaceae/Zannichelliaceae) is dioecious (Isaac 1969). The rhizomes are stout and woody, 5-10 mm in diameter, with several wiry branching roots arising at each node, to which adhere, in young plants, many small, broad, black scales (Fig. 9.1b). The woody branches are erect when supported by water and may grow 30-40 cm in height at Inhaca Island, whereas in sheltered bays on the east coast of Africa, such as Nacala Bay in Mozambique and Turtle Bay in Kenya, they may reach over a metre in height. Five to seven strap-like leaves, about 100 mm long and 10 mm broad, cluster at the ends of branches; older leaves are shed leaving conspicuous annular scars. Each leaf has a fan-shaped sheathing base, the bases of older leaves enclosing younger ones. A conspicuous ligule extends as a flap across the leaf at the junction between the sheath and the lamina; the upper part of the margin of each leaf is densely serrulated (Fig. 9.1b').

Male and female flowers (Fig. 9.1c and d) occur in the summer months, below the terminal leaf tufts on separate groups of plants. The female flowers are enclosed in three sheathing bracts which conceal a minute ovary of two carpels, each bearing a short style which bifurcates into two stigmas 30 mm long. The male flower has three smaller bracts and develops one bright pink anther, 1-2 mm long, which has eight loculi. When ripe, pollen is shed in the form of a twisted, rope-like mass which releases thread-like, sticky pollen grains (Fig. 9.1e). These float on the water surface and so become attached to stigmas of the female flowers.

Reproduction in *Thalassodendron ciliatum* shows an unusual degree of viviparity since the seedling germinates inside the fruit while it is still enveloped by the base of the third bract and it is still attached to the parent plant (Isaac 1969). The first and second bracts are shed after fertilisation, but the third bract increases in size to 50 mm in length and 15 mm in breadth; it becomes thickened, stiff and cartilaginous in texture and pink in colour. It usually remains attached to the parent plant until germination of the seedling is complete. Usually one embryo develops.

The stem grows rapidly and gives rise to more leaves and several adventitious roots, but no radicle emerges. When 5-8 leaves and 6-8 adventitious roots have developed, the sheath of the supporting bract becomes flaccid and the developing roots force it open so that the seedling is released and drops to the sea floor. Dispersal is thereby controlled and the seedling takes root in the habitat of its parent. If the bract breaks off the plant through abnormal wave action before germination is completed, it floats and may carry the seedling further afield.

Cymodocea serrulata (Cymodoceaceae/Zannichelliaceae) (Isaac 1968) is a shorter sea grass with longer leaves (about 150-200 mm), which are 10 mm broad and slightly curved (Fig. 9.2a, a' and b). The tips of the leaves are toothed and each has a ligule at its base. The shoots are short and the sheathing bases of the leaves are bright pink in colour. *Thalassia hemprichii* (Hydrocharitaceae) (Fig. 5.5a) may mingle sparsely with the other sea grasses on the west shore but is characteristic of the southern and northern bays (see Section 5.4.2). It is easily distinguished from *Cymodocea* spp. because the leaf margins are smooth and ligules are absent. Shaggy leaf bases remain on the shoot when leaves are shed (Fig. 5.5a) (Isaac 1968). Seeds have a long resting period before they germinate and the growth of the seedlings is slow. The fruit is shed and ripens below the water surface; since it has no buoyancy, dispersal by currents is very slow. The fruit is broadly conical and about 30 mm long and has a spiny surface. It bursts open and sheds many small brown seeds.

The terete leaves of *Syringodium isoetifolium* (Cymododceaceae/ Zannichelliaeae) (Isaac 1968) (Fig. 9.2c and d) are frequently found bitten off at the tips; they are always green despite a short exposure to air at extreme low tides. Two or three flowers occur in each small inflorescence.

Sea grasses are well adapted to life in the subtidal fringe where they may be 3-5 m below the surface when the tide is high. Temporarily, some are exposed to air at low tide and the

Fig. 9.2 Minor sea grasses:
a. *Cymodocea serrulata* (x 0,5) and
a'. leaf tip enlarged;
b. female flower and bracts;
c. *Syringodium isoetifolium* (x 0,5);
d. female flower (a-d after Isaac F.M. 1968).

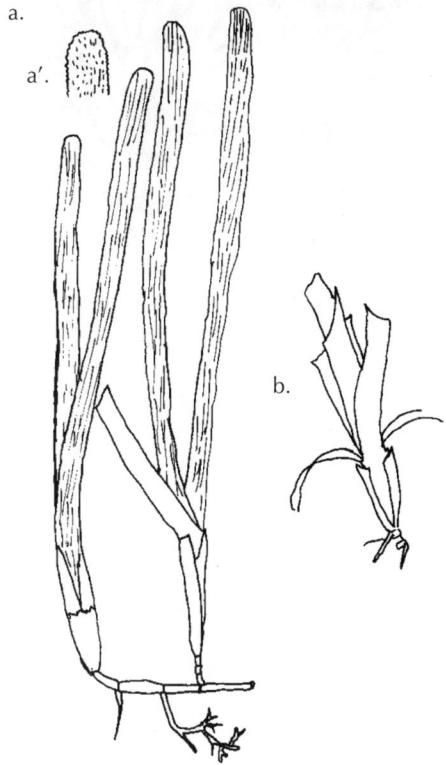
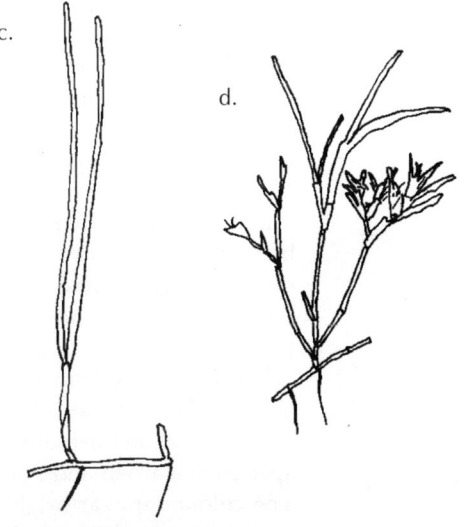

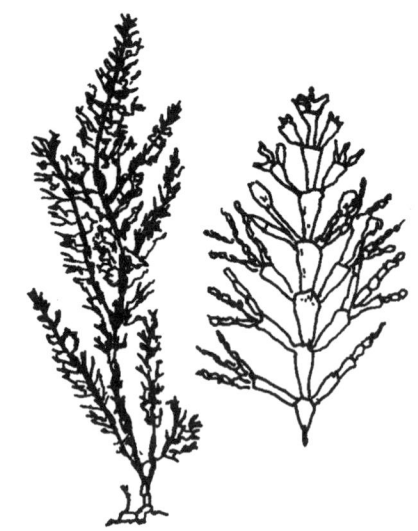

Fig. 9.2 Seaweed:
e. red alga, *Corallina cuvierii* (x 1) (after Seagrief 1980) and detail.

broad leaves *do* turn brown and die if they are exposed to air for an hour on a very hot day. There are *no stomata* on the leaves and the roots are always under water, so water loss is minimal on exposure to air. The stems and leaves float easily since the whole plant is riddled with air channels between the tissues so that all cells have easy access to gases for photosynthesis and respiration. Dissolved carbon dioxide is used in photosynthesis and sufficient light penetrates at high tide as far as the lower limit of their distribution. The cut-off point is conditioned by the amount of silt present. Nutrients in the sandy mud are abundant since bacterial cycles are at work on the massive amount of decaying vegetation (see Section 1.7.2). Dissolved salts are absorbed by the *leaves* as well as the roots. There are cell membrane mechanisms on the absorptive surfaces for inhibiting the uptake of sodium chloride; as further protection from salt, the leaves have tannin vacuoles in their cells which sequester salt that may have penetrated into the plants.

9.3 SEAWEEDS IN THE SEA MEADOWS

The only conspicuous alga in the sea meadows on the subtidal fringe is a species of articulated, coralline red alga, *Corallina cuvierii*, which grows epiphytically on the stems of the sea grass *Thalassodendron*. *Corallina* has a robust, pink, feathery form (Fig. 9.2e).

The leaves of *Thalassodendron* may be encrusted by the fine calcareous plates of a flat coralline alga, *Dermatolithon* sp., and the stems may have growths of 'bracket' coralline algae. No green, brown or other red algae have been seen in the sea meadows.

9.4 ANIMALS IN THE SEA MEADOWS

In the cartoon (Fig. 9.1a) the associates portrayed are a ghost shrimp, a brittle star, serpulid worm tubes, soft coral, a small snail and a sea horse. Burrowers in the muddy sand are few and more difficult to find except in the occasional bare patches of exposed substratum. These are completely different species from those which burrow in the bare sand-banks (see Section 9.5).

9.4.1 Echinoderms: starfish; sea urchins; brittle stars; cucumbers

The large starfish are surprisingly tough and turgid; they are supported by a slender skeleton, beneath the thin skin, composed of a continuously growing, latticed layer of calcareous plates bearing minute, fixed, projecting spines, which allow for limited bending of the arms. *Pentaceraster mammillatus* (Fig. 9.3a) is about 150 mm in diameter and variable in colour; it may be green, grey, brown or yellow and has a ring of nipple-like projections of contrasting colour on the disc. Lesser spines grow in rows on the arms and are enlarged at their tips. The other species of similar size, *Protoreaster linkii* (Fig. 9.3b) has a grey background colour on the disc and arms while the blunt spines on both are red. Why has a variety of colours evolved in one species of starfish and not in the other? A close study of their habits may solve the problem. The colours appear bright because the disc is covered by a thin secretion of mucus and is swept clean by cilia. Small animals that might settle on the surface of the starfish are nipped by its scattered pincers, less than 1 mm in size, called 'pedicellaria'. These two species may be found stranded on sand just landward of patches of sea grasses when the tide is out and they appear to be quite inert. Under water they move slowly by means of hundreds of very slender 'tube-feet' or 'podia', peculiar to echinoderms (Fig. 9.3c). Their action may be watched in an aquarium as the starfish climbs over a vertical glass pane.

During locomotion two arms pull the animal forward and three push from behind. The tube-feet are extended to about 10 mm in length and moved forward in unison according to the spread of nerve conduction down the arms. They grip the ground and start retracting so that the animal moves forward; the tube-feet are released and are stretched forward again. The grip is held by a minute retractile muscle in each podium which is attached to a tiny calcareous plate at its end; when this muscle contracts the tip of the podium is converted into a sucker that adheres to the substratum, aided by the secretion of a sticky substance. Extension and retraction are repeated rhythmically, co-ordinated by the radial branches of the nerve ring around the mouth.

The locomotory force in the podia is mediated by the sea water contained in the 'water vascular system', composed of canals radiating inside the arms from the water vascular ring around the mouth. Pressure in the system is equilibriated with the external medium through a thick, perforated 'sieve plate' on the upper surface of the starfish, known as the 'madreporite'. Each podium branches from the radial water vascular canal and bears a contractile ampulla at its base (9.3c). When the ampullae contract the podia are extended and kept distended by the closure of a tiny valve at the junction of each ampulla with the radial canal. When the ampullae relax and the valves open, the podia are withdrawn; thus the motive force in starfish is hydraulic.

The podia are thin-walled so that they also serve for respiratory exchange. In addition, there are smaller, finger-like projections all over the body surface which project from the coelomic cavity. These are called 'papulae' and are lined both inside and outside by cilia that circulate the water internally and externally, thus facilitating gas exchange between sea water and the internal coelomic fluid of the starfish.

All starfish have extensible stomachs which can be protruded through the mouth to engulf food. In these species the food has not been examined.

Larvae of starfish develop ten short ciliated 'arms' for locomotion. After the arms grow longer the brachiolaria larvae settle and develop adhesive discs anteriorally, whilst the 5-armed starfish develops on the posterior part of the larva.

Sea urchins are hemispherical spiny animals about 70-100 mm in diameter and are represented here by four species with short spines, which can be recognised by their colours, and by two species with pencil-spines (see Section 5.6.2; Fig. 5.17a). The most common sea urchin on this shore is *Tripneustes gratilla* (Fig. 9.4a), which has numerous slender white spines arising from a brown 'test', as the hard skeletal body wall is called. This sea urchin covers itself partially with dead, brown leaves of *Thalassodendron ciliatum*, which it is able to place on its upper surface with its long tube feet. This habit may be a protection from desiccation or from sunlight or from predators; it is the species that is most often exposed. *Temnopleurus toreumaticus* has dark spines on a brown test whilst *Salmacis bicolor* has alternately banded green and red spines. The slightly larger *Toxopneustes*

Fig. 9.3 Starfish in the sea meadows:
a. *Pentaceraster mammillatus** (x 0,5);

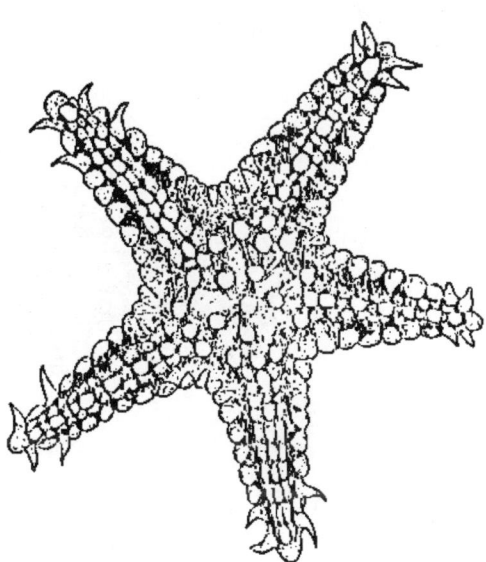

b. *Protoreaster linckii* *(x 0,5);

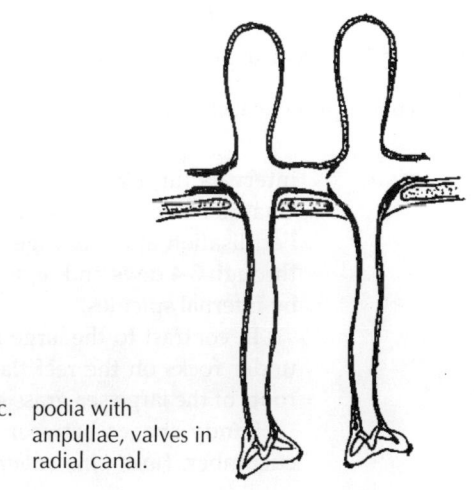

c. podia with ampullae, valves in radial canal.

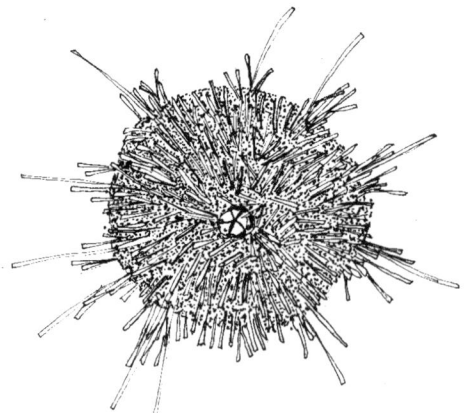

Fig. 9.4 Sea urchin in sea meadows:
a. *Tripneustes gratilla* (x 0,5);

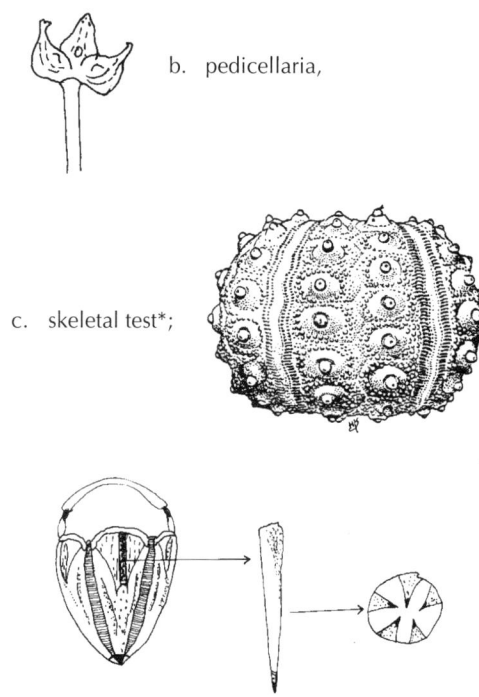

b. pedicellaria,

c. skeletal test*;

d. Aristotle's lantern; d' single tooth and d" mouth with 5 teeth (arrow to teeth) (a, b and d from Branch and Branch 1981).

pileolus is bright pink in colour; its globose pedicellaria (2 mm) contain a potent toxin which may be injected into the skin and cause intense pain as it penetrates the hand of the unwary collector. The cleaning pedicellaria of sea urchins (Fig 9.4b) have longer stalks than those of starfish and are larger.

The round shape of a sea urchin is maintained by a calcareous skeleton consisting of ten pairs of alternate wide and narrow plates in vertical rows from the upper to the lower surface (Fig. 9.4c). The podia emerge through pores in the narrow plates and work in the same way as those of starfish. They can be stretched (by water pressure in the internal water vascular canals) to several centimetres in length which helps the animal to right itself if it is overturned, whereas a starfish uses shorter tube feet to do so since it is able to bend its arms. The spines on the wider plates of the test are movable, since they articulate with knobs on the plates making ball and socket joints; they are used in locomotion, especially in the longer spined species.

Sea urchins may be herbivorous, grazing on sea grass leaves, but some prefer sessile animals, meiofauna and microflora, that coat the leaves. The grazing mechanism is a unique pentagonal system known as Aristotle's lantern (Fig. 9.4d). There are five long, sharp-pointed teeth (10 mm) (Fig. 9.4d' and d") supported by a pentagonal framework inside the mouth. This framework is composed of five triangular, calcareous pillars, called 'pyramids' and five jointed rods dorsal to the pyramids that articulate with them and hold them apart. A complex system of muscles connects the pyramids and the rods, which combine to manipulate the five pointed teeth in a grazing action. This apparatus was first described by Aristotle some two thousand years ago.

Sea urchins have been found useful in the study of early embryology. At Inhaca Island the three common short-spined species have ripe eggs and sperm throughout the year and yield gametes almost immediately after an injection of 1-2 ml of a solution of potassium chloride (0,5 M or 35 g/l) through the peri-oral membrane. Male and female animals are selected by the positive response to a drop of potassium chloride placed on the test near the genital pore. Tiny eggs are extruded by the female and the seminal fluid is milky. Unripe eggs that might contaminate the others are not shed. Eggs and sperm ooze out through the respective genital pores on the upper surfaces of the urchins and can be transferred by pipette to a petri-dish containing sea water so that development may be watched at intervals through a microscope. It is preferable to obtain fresh sea water from the Inhaca Channel to ensure that the embryos survive, as they are intolerant of diluted sea water. Fertilisation and cleavage, blastula, gastrula, and pluteus larva formation may be followed through 3-4 days and nights. The echinopluteus larva has eight short ciliated arms supported by internal spicules.

In contrast to the large number of species of brittle stars that shelter in the coral reef and under rocks on the reef flat, *few* species occur in these sea meadows. The interlaced fibrous roots of the large sea grass species may prevent brittle stars from burrowing.

Sandy spaces between the sea grasses may be occupied by burrowing species of sea cucumber, *Holothuria scabra*, the grey sea cucumber (Fig. 3.18a), *H. atra* and *H. leucopspilota*, the black species, or the brown *Holothuria* spp. of the coral reef.

9.4.2 Crustaceans: crabs; amphipods; isopods

About ten species of spider crabs are encountered in the subtidal fringe of the sandy shore since they require almost permanent contact with water. A spider crab has a triangular carapace with an anterior pointed rostrum, often bearing spines; this is quite unlike the broad carapace and front of most true crabs. Inshore species are fairly small with a lightweight carapace and slender legs ending in hooked tips, suitable for climbing and clinging to sea grasses. Camouflage seems to be the outstanding asset of these spider crabs and each genus has a different means of concealment.

Micippe thalia (35 mm) has its spiny rostrum bent downwards so that it appears blunt (Fig. 9.5a). The carapace is covered with hooked hairs to which it deliberately attaches coralline algae or bits of sea grass leaves by means of its chelae; thus it is thoroughly concealed unless it moves. *Menaethius monoceros* (26 mm) (Fig. 9.5b), when it lives among sea grasses, is the same brilliant shade of green as that of a new leaf; but when it occurs under rocks it is deep brown, matching the rock surface. Barnacles (*Balanus amphitrite*) camouflage *Schizophrys aspera* (Fig. 9.5c). The carapace of *Cyphocarcinus capreolus* (Fig. 9.5d and d') is cylindrical and overgrown by coralline algae. Its resemblance to a rhizome of *Thalassodendron ciliatum* is completed by the fixing of a real piece of the rhizome to its rostrum. Many spider crabs are carnivorous; they lie in wait for their prey, such as amphipods, which they catch with their slender pincer-like chelae. The larval life of spider crabs is short since there are only two zoea stages instead of the normal 4-6 stages in other crabs.

Sponge crabs or dromids are another family that is not encountered above low spring tide. Fossil evidence from the Jurassic period (180 million years ago) clearly shows that the ancestors of modern crabs were very similar to the dromids of today and were derived from lobsters. Among the primitive features of dromids are the filamentous gills in the branchial cavity and the vestiges of uropods (tail fan) on the abdomen. Sponge crabs have a dome-shaped carapace covered by a hairy felt. They always carry a 'pack' made of some compound type of animal such as a sponge (whence the common name), a tunicate or zoanthid anemone. A crab will pick up a small piece of a colony of its preferred animal and place it on its back where it grows to enclose the dorsal surface until only eyes and antennae are left exposed. After every moult the 'pack' is replaced and it grows larger until it covers the crab. The camouflage organism is made secure by the hairs on the carapace and is supported by the last pair of legs which are twisted dorsally and end in spines bent together to form forceps. *Dromia dormia* (60 mm) (Fig. 9.5e and f) carries a green sponge; *Dromidia unidentata* (30 mm) is covered by the brown zoanthid, *Palythoe nelliae*; *Pseudodromia* sp. (40 mm) has a

Fig. 9.5 Crabs in the sea meadows:

a. *Micippe thalia* front only (weed removed) (× 1);

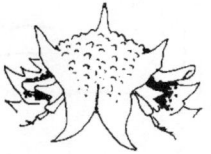

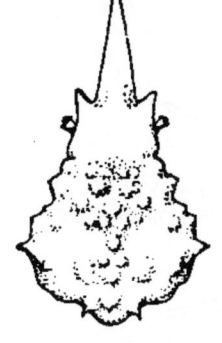

b. *Menaethius monoceros** (× 1);

c. *Schyzophrys aspera** (× 1);

d. *Cyphocarcinus capreolus** and attached alga (× 1);

d'. with alga removed

e. sponge crab, *Dromia dormia* carapace (× 0,5) with sponge removed;

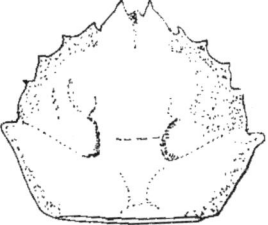

f. chela (a, e-g after Barnard 1950);

Fig. 9.5 Crabs in the sea meadows:
g. swimming crab, *Charydbis annulata* (x 0,7);

h. chela (h and i after Crosnier A. 1962).

pinkish-purple compound tunicate, *Polyandracarpa tincta*. This habit of concealment by unappetising animals makes doubly sure that predators are foiled. Most sponge crabs appear to be detritus eaters.

Swimming crabs are encountered among the sea grasses of the lower shore and their swimming adaptations are described in Section 3.6.3. Their burrows are more numerous in the subtidal fringe (and below) and may be recognised by the holes, 30-50 mm in diameter at the surface of the muddy sand, which are always marked by a ring of white shells of bivalves such as *Dosinia* sp. (Fig. 5.3d) and *Mactra* sp.(Fig. 9.7a) which are ejected when the crabs burrow. Several species of *Portunus* (Fig. 3.21a) and *Charybdis* (Fig. 9.5g and h) occur here (see Appendix A).

Empty snail shells are frequently inhabited by hermit crabs such as the large *Dardanus (Pagurus) megistos* which chooses *Tonna* shells (Fig. 5.14a-c, see Section 5.5.3). Several species of *Pagurus*, and of the smaller species of *Diogenes*, occur in the sea meadows, each in its preferred species of shell. Some of them are parasitised by a cirripede (barnacle group), *Peltogaster*, which attaches itself to the soft abdomen inside the shell. It has a life history very similar to that of *Sacculina*, which parasitises xanthid crabs (see Section 6.5.2), and when mature has the form of a shiny red swelling of an irregular shape on the crab's abdomen.

The amphipods are completely aquatic and smaller than the rock hoppers of the supratidal rocks (see Section 6.2.2). Their gills are plate-like outgrowths from the first segments of the well-developed walking legs and lie in a trough beneath the body, where they are ventilated by a current of water produced by the biramous appendages on the abdomen. Modification of the anterior appendages in different genera result in three types of feeding. *Cymadusa australis* (8 mm) (Fig. 9.6a) (Amphithoidae) uses the characteristic first two pairs of legs, known as gnathopods (since their ends are expanded into subchelae) to gather surface detritus as food and to dig a burrow in the sand. The burrow is semi-permanent since the sand is lightly cemented with a mucous secretion from the glands at the bases of the fourth and fifth pairs of legs.

Podocerus brasiliensis (Fig. 9.6b) is a member of the Corophiidae, a family of filter feeders and tube builders. It feeds on suspended particles in the respiratory current produced by the abdominal appendages. The very long second antennae are fringed with filtering setae and the second pair of gnathopods have a row of short bristles with which it removes the particles trapped on the antennae. This amphipod builds tubes of mud attached to the stems of sea grasses from which it protrudes antennae and gnathopods for feeding.

A third type of amphipod, known as a caprellid or ghost shrimp (because it is completely transparent and has a bizarre shape) is carnivorous. *Orthoptella mayeri* (Fig. 7.25b) perches among the branches of the feathery hydroid *Lytocarpus*, which grows on sea grass leaves; *Caprella scaura* (Fig. 9.6c) sits on the surface of the leaves. Caprellid amphipods are extremely thin, having an elongated thorax but an abdomen reduced to a tiny stub. The first two pairs of legs are gnathopods, the second pair having strongly hooked or relatively massive chelae; the third and fourth legs are absent, being represented only by normal, but very small, gills. In females there is a ventral brood pouch in which eggs develop and there is no larval stage in the life history. The last three pairs of legs are subchelate and thus adapted to clinging to a plant or hydroid, while the rest of the body stands erect and unsupported. Its pose recalls that of a stick insect or praying mantis as it waits for passing prey. The gnathopods of a caprellid are similarly used to catch prey and to chase away predators such as nudibranchs.

Isopods actively crawl or swim and, unlike amphipods, never feed on detritus or suspended particles but are omnivorous or scavengers. *Pontogeloides latipes* (Fig. 9.6d) may be seen at high tide in the drift line but may occur anywhere on the sandy shore. Among the sea grasses, *Paracilicea mossambicus* is fairly common and sometimes hides in shells. It is also broad and flat and is a member of the mainly herbivorous family Sphaeromidae; it rolls itself into a ball for protection as do woodlice in the soil. There are also burrowing species such as the spotted *Paranthura punctata* and *Synidotea variegata* (Fig. 9.6e) each of which has a narrow elongated shape and a loose operculum derived from the appendages covering the gills.

9.4.3 Molluscs: bivalves; snails; slugs and squid

A number of burrowing species of bivalves may be found in the sandy mud of the subtidal fringe (Kilburn and Rippey 1982) but none is as abundant as on estuarine shores. They live below the surface of the sandy mud and their density can be guessed only roughly from the relative numbers of dead shells, which is lower on the west shore than in the southern and northern bays. In addition to the cockles, ark shells and venus shells (mentioned in Section 3.6.2), which live near the surface, there are representatives of two families of the 'clam' type, which have fairly light shells. The 'trough' shells (Mactridae) lie just beneath the surface; their short siphons are *fused* together as far as the tips and extend above the surface sand for feeding on plankton. *Mactra glabrata*, perhaps the most common of these and the largest (up to 100 mm in length) has a cream or buff shell with a prominent umbo (peak near the hinge), and is covered by a brownish periostracum (Fig. 9.7a). There are several species of tellinid, some delicately coloured pink or blue, the commonest being a white species, *Tellina capsoides* (Fig. 9.7b). It is smaller than *Mactra* and less robust; it can be distinguished by the ligament of the hinge being external and by the angular posterior end of the shell, which has faintly sculptured, regular concentric lines. The tellinids burrow more deeply into the sandy mud, using a thin blade-like foot, and they have *separate* siphons. The longer inhalent siphon is protruded above the surface sand when feeding, the top being bent over the sand as it is moved around sucking up detritus in much the same way as the tube of a vacuum cleaner may be used. The food is then sorted on very large palps.

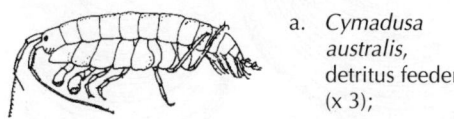

Fig. 9.6 Amphipods in the sea meadows (after Griffiths 1976):

a. *Cymadusa australis*, detritus feeder (x 3);

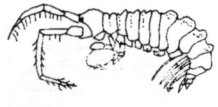

b. *Podocerus brasiliensis*, filter feeder (x 2);

c. *Caprella scaura*, carnivore (x 2);

Isopods (after Kensley 1978):

d. *Pontogeloides latipes*, scavenger (x 2);

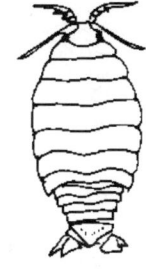

e. *Synidotea variegata*, detritivore (x 2).

All the bivalves use the filter-feeding method described for *Donax faba* in Section 3.2.2. They also have a unique system of internal sorting and digestion of food, shared only with some herbivorous snails. A slender transparent rod, known as the 'crystalline style', which is made of a mucoprotein and solid amylase, lies in a groove in the stomach and is rotated by cilia against a cuticular shield. This movement liberates the enzyme slowly and so starts digestion of plankton or detritus that has been trapped in mucus on the gills but freed by the lowered pH of the stomach, that dissolves the mucus. The rotating rod stirs the particles which then flow over the ciliated ridges and grooves of the stomach lining and are sorted; fine, partly digested organic matter is swept into openings of the digestive gland by the cilia on the ridges, and larger indigestible particles of silt and sand are carried in the grooves into the intestine for subsequent egestion.

There are very few species of herbivorous snails in the sea meadows but there are many carnivorous species and a few detritus feeders.

Among the 'top' shells, the dark red, spotted *Clanculus puniceus* with a white ridged and distorted aperture (Fig. 9.7c), and the larger brownish 'top' shell *Cantharidus suarenzis* (Fig. 9.7d) browse on the surfaces of the leaves of sea grasses.

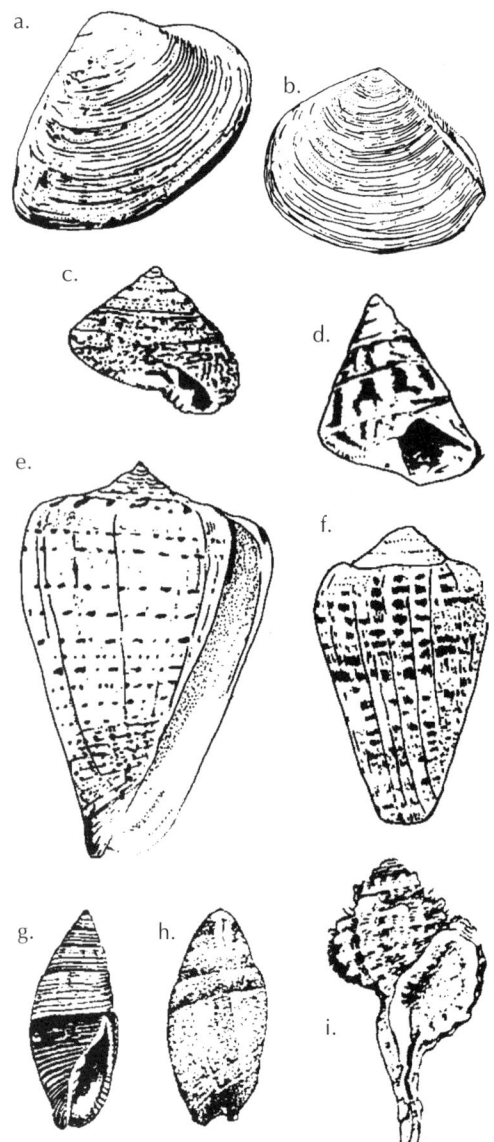

Fig. 9.7 Molluscs in the sea meadows:
a. *Mactra glabrata* (x 0,5);
b. *Tellina capsoides* (x 0,5);
c. *Clanculus puniceus** (x 1);
d. *Cantharidus suarenzis* (x 1);
e. *Conus betulinus** (x 1)
f. *C. tessulatus** (x 1);
g. *Mitra limbifera* (x 0,7);
h. *Amalda contusa** (x 1);
i. triton, *Ranularia moritincta** (x 0,5); (a, b, d and g after Kilburn and Rippey 1982).

Several species of large carnivorous 'cone' snails emerge at night from the sandy mud amongst the sea grasses to feed on worms and may still be exposed, by chance, during the morning low tide. The most common species are *Conus betulinus* (75 mm), which has a dark yellowish shell with rows of black spots, and *C. tessulatus*, (59 mm) in which the shell is lighter and has orange spots; both species feed on small polychaete worms (Fig. 9.7e & f). A smaller species *C. lividus* (Fig. 3.19e) has a dark shell with one white transverse mid-shell band; it captures acorn worms and terebellid polychaetes. A remarkable transformation of the radula teeth into sharp, barbed, poison darts has occurred in cone snails. For each 'kill' a single tooth is released from the radula sac into the mouth cavity and is propelled along the muscular proboscis to its tip; a potent nerve toxin is expelled from a large gland, that opens into the mouth cavity, and is stored in the hollow tooth. When the cone snail's prey has been located by its sharp chemical sense, it is harpooned by the armed proboscis and the worm is paralysed, then swallowed whole.

'Mitre' shells and 'Olive' shells are smaller carnivorous snails that burrow just beneath the surface sand, mainly subtidally, but some may be found in the subtidal fringe. Mitre shells are almost cylindrical with a short spire and a very short siphonal canal; characteristically the columella in the aperture has 3-5 pleats. Some of the mitre shells are highly coloured, but a common species, *Mitra limbifera* (40 mm), has a shell lighter on the upper half and brownish on the lower half (Fig. 9.7g). 'Olive' shells are subcylindrical in shape with a low spire, and a very smooth, polished surface, in which they resemble cowries since, when under water, the shell is covered by flap-like extensions of the foot which secrete enamel. The aperture has the shape of a gothic arch and there are no pleats on the columella. The shell of *Amalda contusa*, for example, has a buff background and narrow transverse, brown bands (Fig. 9.7h). 'Olives' prey upon small crustaceans and molluscs.

Species of *Tonna* have large, thin, globose, ridged shells and a low spire; they are rarely seen alive for they bury themselves by day and emerge when the tide rises, to feed on sea cucumbers. They first wrap the broad foot around the prey and paralyse it with a salivary secretion containing sulphuric acid, which is squirted through the proboscis. *Tonna costatus* (60 mm) has a buff shell with rounded spiral ridges and very slightly protruding spire; it is covered by a slightly hairy periostracum (Fig. 5.14c).

Cowries usually shelter under coral debris, but those tolerant of muddy sand may stray among the sea grasses (see Section 3.6.4). *Cypraea annulus* and *C. moneta* (Fig. 3.19a, b and c) seen among the narrow-leaved sea grasses and on the reef flat of the lower shore, may be present. Others occur singly or in pairs in the subtidal fringe sand and may be found during the midnight, low spring tide with the aid of a torch, near the coral reef.

The small triton *Ranularia moritincta* (Fig. 9.7i) has a bristly coating on the ridges of the reddish brown shell from which it protrudes a yellow, red-spotted foot. It probably feeds on echinoderms, first boring a hole in the test while the prey is held with the foot, and injecting a powerful saliva, containing sulphuric acid, which dissolves the calcareous plates in the skin. Then the proboscis is inserted to consume the soft tissues.

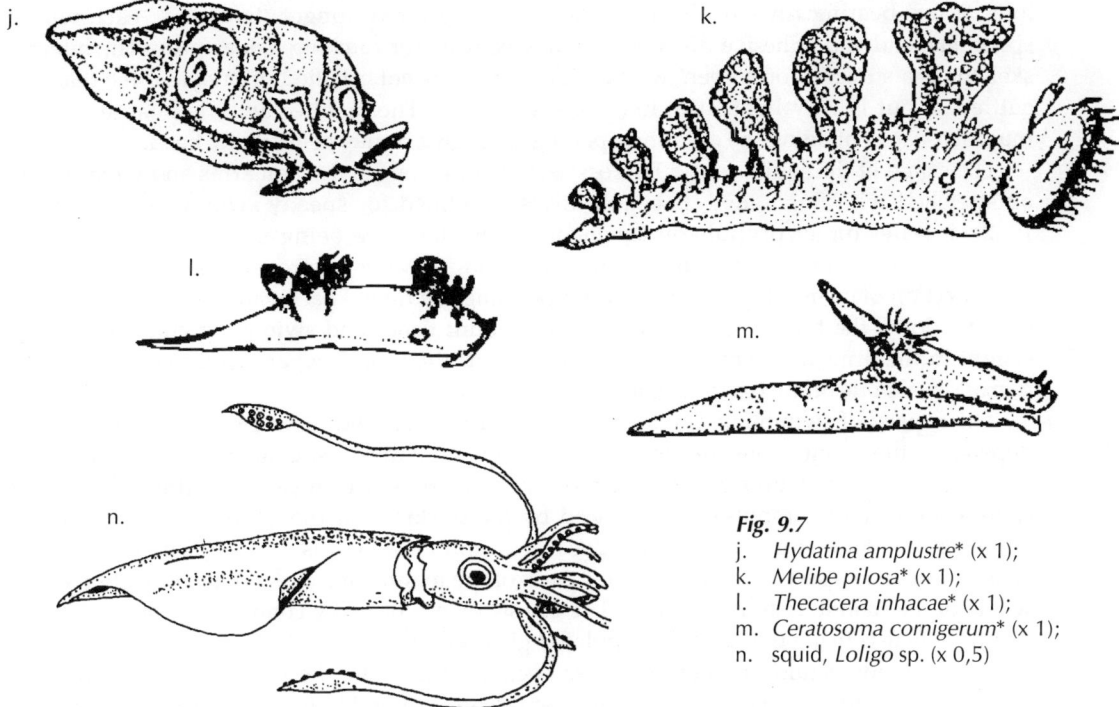

Fig. 9.7
j. *Hydatina amplustre** (x 1);
k. *Melibe pilosa** (x 1);
l. *Thecacera inhacae** (x 1);
m. *Ceratosoma cornigerum** (x 1);
n. squid, *Loligo* sp. (x 0,5)

Among the tectibranchs, i.e. those sea slugs that have shells, there are both herbivores and carnivores, but the nudibranchs (i.e. slugs without a shell), found here are carnivorous except for the large, ugly *Dolabella auricularia* (150 mm), (Fig. 5.13a) known as the sea hare because of its head appendages and squat shape. It browses on algae (see Section 5.5.3). When *Dolabella* is threatened it exudes a strong purple dye, staining the water for several metres around and making the animal quite invisible. This dye, similar to that of the whelks on a rocky shore, was the orginal source of 'Tyrian purple', exported by the Phoenicians to Greece and Rome in ancient times.

The bubble-shell, *Hydatina amplustre* (30 mm) (Fig.9.7j), a beautiful animal, has a fragile, oval shell with a very large aperture and a sunken spire. The white shell has two broad pink bands edged with black lines; the foot, body and paired head shields of the mantle are pinkish violet in colour, edged with a vivid blue. It feeds on fat cirratulid worms (Fig. 3.9c) which lie just beneath the surface sand, by hooking them with its radula teeth, sucking them into the oral tube and swallowing them whole.

Melibe pilosa, (Fig. 9.7k) one of the largest nudibranch sea slugs, is a floppy, greenish animal (100 mm) with six pairs of dorsal fleshy flaps, used in *swimming*, and a large oral 'hood' or 'cowl' bearing tentacles and a fringe. It catches isopods, amphipods and eggs of fishes in its 'cowl' which it tosses from side to side as it swims. Its own eggs are brilliant scarlet and embedded in a long string of jelly.

A notable nudibranch, first described from Inhaca Island, *Thecacera inhacae* (Fig. 9.7l), is about 20-30 mm long and a clear, translucent, orange or yellow colour, whilst its many fleshy protuberances are various shades of blue. It feeds on other nudibranchs, which are held in the jaws while the radula scrapes off pieces suitable for swallowing. *Ceratosoma cornigerum* (Fig. 9.7m) a dorid nudibranch with a cluster of external gills, is firm to the touch, bright orange in colour with purple spots.

The largest molluscan carnivores, squid (Cepalopoda), are often seen swimming swiftly through the shallow water of the sea meadows at low tide seeking prey; they evade capture by fishes or man by their speed and sudden backward spurts. *Loligo* sp. (20-25 cm) has a long cylindrical body, tapering posteriorly where a pair of large triangular, mobile muscular fins are attached and used for steering (Fig. 9.7n). The molluscan 'foot' is modified to form eight long 'arms' bearing rows of fleshy suckers. There are two longer 'tentacles', each with a spoon-like end and adhesive discs, which may be used for catching fish as prey. The internal skeleton is a straight horny 'pen' with a wide shaft; it is not calcified or buoyant like that of a cuttle fish but is very light, stiff and slightly flexible. The mantle cavity is a large, ventral muscular bag containing the gills and opening through a funnel beneath the head.

The shape of the squid, round in cross section and elongated when arms and tentacles are outstretched with fins folded, is beautifully streamlined for speedy swimming (Alexander 1982). It relies on jet propulsion for locomotion, the force being provided by the rapid contraction of the muscular mantle wall, which squirts water vigorously through the funnel. The direction of swimming is controlled by pointing the funnel backwards or forwards; when slowing down the fins act as stabilisers. Squids hunt fishes and swimming crabs, which are caught in the arms and tentacles and held close to the mouth where chitinous jaws, like a parrot's beak, bite off pieces for swallowing.

Their speed of swimming, unique among invertebrates, and the very fast escape reaction, depend on the 'giant axons' in the nerves to the mantle muscles. Unlike the fast-conducting motor nerves in vertebrates, squid nerves are not encased in myelin sheaths. The rate of conduction of nerve impulses is enhanced by the thickness of the 'giant axons', 700 μm in diameter. In swimming, a maximum speed of 20 m per second is reached. The thickest and fastest giant axons serve the more distant parts of the mantle muscle so that contraction of the whole of the mantle is simultaneous. (The classic theory of nerve conduction was developed by physiologists using the easily manipulated giant axons of squids.)

Cephalopods (squid and octopus) (see Section 8.5.5) have the largest eyes amongst the invertebrates and in gross structure they remarkably resemble the eyes of vertebrates, but their finer structure is adapted for vision under water (Alexander 1982). Squid (and fish) eyes have spherical lenses which are not homogeneous so that light can be focused on the retina. The smaller the retina cells the better the image that is formed; squids have smaller cells than those in the vertebrate retina. Nevertheless squids are nearsighted; accommodation of the lens for near and far vision is accomplished imperfectly by ciliary muscles that move the lens forward or backward rather than altering its shape, as in vertebrates. The pupil is a horizontal slit that is widened in dim light. Squid and octopus (see Fig. 8.15a) have a unique capacity for squirting black ink when attacked. This forms a large cloud in the water, confusing potential fish predators. These molluscs can also change colour rapidly.

9.4.4 Specialised tube-building and burrowing polychaete worms

Members of the detritus-feeding families of polychaetes described in Section 3.4.2, such as the spionids, bamboo worms and cirratulids, are still present in the subtidal fringe and other, much larger, worms build tubes here and there. The latter require a permanent supply of water and are more typical of the subtidal zone. A careful search of the surface sand exposed between sea grasses may reveal two openings of a U-shaped burrow about 30 cm apart, which is made by the grotesque, cosmopolitan worm, *Chaetopterus variopedatus* (150 mm). The ends of the tube are about 10 mm wide and are ivory white, smooth and polished, whilst the rest is parchment-like and buff-coloured, the whole being made out of hardened mucus. The worm (Fig. 9.8a) is soft and floppy when not supported by the water in its tube and is luminescent. The continous stream of water through the tube for feeding has the secondary effect of increasing fifteen-fold the oxygen available to the worm. The worm has neither haemoglobin nor any other respiratory pigment, unlike other tubiculous worms that suffer periodic anaerobic conditions.

Every *Chaetopterus* tube examined contains a pair of commensal porcellanid crabs, *Polyonyx biunguiculatus*, which are also filter feeders, but probably they collect coarser material from the unfiltered water that by-passes the filter in the tube.

Sometimes one may find a tough, brown, papery tube, 10 mm in diameter, standing upright 40-50 cm high, even when out of water, between sea grasses or near coral debris; this, at first glance, seems to resemble a dead, hollow woody stem. It is built by a large eunicid worm, *Eunice tubifex* (40 mm) (Eunicidae), characterised by five tentacles, eyes on the head and powerful jaws (cf. Fig. 3.10d and Section 6.5.4). The tube is made of dried mucus and is actually U-shaped, the two arms being stuck together; thus there are two openings at the top and the base is fixed to a stone buried in the sand. There are several commensals in the tube: tiny brittle stars, a small xanthid crab, *Chlorodopsis areolata*, and three species of polynoid (scale) worms; but their residence in the eunicid tube may be temporary and not obligatory. Outside the tube, on which coralline algae may be growing, a terebellid worm sometimes attaches its own tube.

Fig. 9.8 Polychaetes:

a. *Chaetopterus variopedatus* (x 1) (after Branch and Branch 1981).

9.4.5 Coelenterates: hydroids; soft corals; sea anemones; cerianthids

The classes and sub-classes of the phylum Coelenterata (Cnidaria) exhibit many different forms despite the absence of internal or external organs. The polyp type of anatomy, i.e. a tubular body and terminal mouth surrounded by tentacles, is common to all of them. The different forms of gross structure are largely dependent on the sizes of polyps and the kind of skeleton they possess, which may be external or internal, horny or calcareous or both. The skeleton may be absent altogether, support for movements being provided by the fluid in the enteron. The position and the amount of muscle and of mesogloea may allow the polyps to be *sedentary*, to *burrow* or *even to creep or swim*. In addition, polyps may be solitary, i.e. one on a single axis, or joined together to form a compound animal. All these types are represented in the sea meadows.

About ten species of hydroids with external skeletons of chitin, which are very similar to those described in the intertidal rock pools (see Section 7.5.2), grow attached to the stems or leaves of the sea grass *Thalassodendron ciliatum*. The most conspicuous is *Lytocarpus philippinus* (Plumulariidae) (Fig. 9.9a) which, in sheltered water, grows 100-150 mm high. The colony resembles a delicate, speckled, brown feather; it has a central stem and many close-set, horizontal branches which taper towards the tip. The polyps are white and set in one row on each branch, enclosed in keeled hydrothecae. It is somewhat specialised in having a division of labour between stinging and feeding polyps.

Several species of soft corals (Alcyonacea) occur in the sea meadows. They are known as the Octocorallia because the polyps have eight pinnate tentacles and eight mesenteries in the enteron instead of the six or multiples of six in true anemones and corals. In the family *Xenidae* the soft corals are delicately coloured, low, soft bushes of relatively large polyps (about 10 mm across) with tentacles about 5 mm long. These polyps do not retract (as do those of corals) and they rely for support on the slight stiffening of the gelatinous mesogloea in which numerous minute oval, calcareous discs are embedded (Tixier-Durivault 1960, Benayahu *in press*) for example, *Heteroxenia elizabethae* (Fig 9.9b).

In contrast to soft corals, the gorgonians (sea fans) have a flexible branching skeleton that is extremely strong and almost unbreakable. The main support is a central rod in each branch, made of a specific protein, gorgonin, which grows from the cells in a peg at the base of the colony. It is similar to the collagen of tendons of vertebrates but is much stronger because the collagen molecules are cross-linked by a process of quinone tanning. In addition, the mesogloea is stiffened by the presence of numerous calcareous spicules. Gorgonians usually grow where wave action or currents are strong, but at Inhaca Island one species has been seen on the sand between sea grasses, namely *Spongioderma chuni* (Fig. 9.9c and d).

Four specialised species of sea anemone may be found in the sea meadows: a sedentary type, a climber, a burrower and one which creeps over the plants.

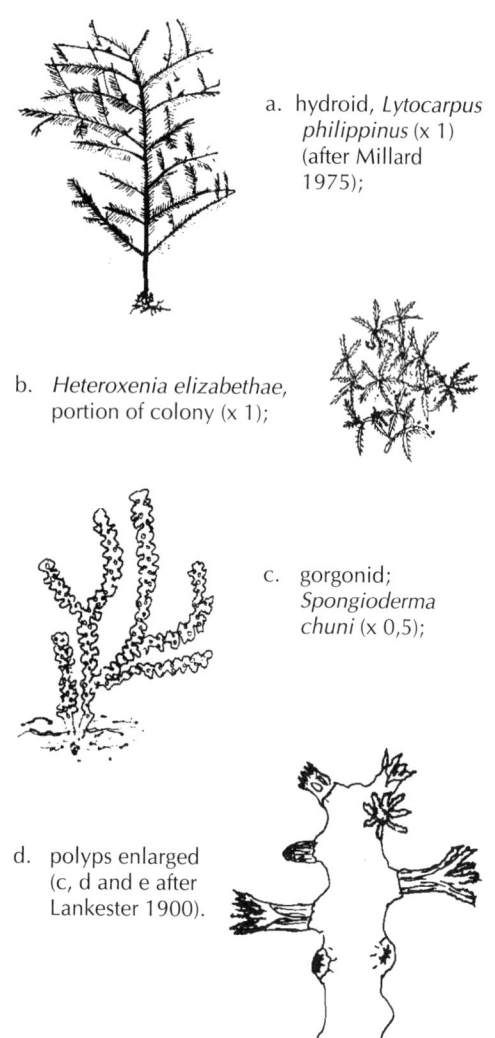

Fig. 9.9 Coelenterates among sea grasses:
a. hydroid, *Lytocarpus philippinus* (x 1) (after Millard 1975);
b. *Heteroxenia elizabethae*, portion of colony (x 1);
c. gorgonid; *Spongioderma chuni* (x 0,5);
d. polyps enlarged (c, d and e after Lankester 1900).

Heteractis (Stoichactis) magnifica (Fig. 3.21e) is the large green species of anemone that lies flat on the sand, its oral disc being up to 25 cm in diameter and covered with very short, ciliated tentacles (see Section 3.6.4).

The creeping sea anemone, *Phylactenactis tuberculata* (Actiniaria), has been found on sea grasses, unattached and moving over the supporting leaves. It is a soft, dark red anemone, covered with bladder-like vesicles. Movements would not be successful were it not for the remarkable elastic properties of the mesogloea. It is composed of long, branching, polysaccharide chains in which are embedded short lengths of protein similar in molecular structure to the collagen in vertebrate tendons. The mesogloea can however be stretched to three times its length (unlike tendons) and can quickly recover the original length when the stretching force is removed.

The cerianthid anemones are burrowers that belong to a different Order of Anthozoa since they have two concentric rows of tentacles, multiple mesenteries and one siphonoglyph instead of two. The column is narrow and may be 30 cm long when the disc has a diameter of about 10 cm (Fig. 8.9e) (see Section 3.5.3). The foot of the column becomes pointed when digging, and swollen into a bulb for anchorage, and there is a terminal pore. Cerianthids secrete a very tough, thick, gelatinous sheath, which is impregnated with mud and lines the burrow, making it permanent. Complete withdrawal into the burrow is very rapid, quite unlike the slow movements of true anemones; sensitivity to an approaching footstep is sharp and so rapid is the muscular action that very rarely is the animal seen with tentacles expanded except under water. Contraction of the column is more rapid because there are a large number of mesenteries, each bearing retractor muscles, and the volume of the sea water in the enteron is reduced by loss through the terminal pore as well as through the mouth. They also occur in sand between coral boulders.

9.4.6 Urochordata: ascidians (tunicates)

Both solitary and compound tunicates usually live attached to rocks and have been described in Section 6.4.3, but at least two species of compound tunicates at Inhaca Island prefer a sandy habitat amongst sea grasses. *Botryllus planus* consists of a soft, thick gelatinous sheet about 100 mm across or more, which rests on leaves or grows around the stems of *Thalassodendron ciliatum*, becoming securely attached. The tiny zooids are dark purple in colour and are embedded in a paler, supporting tunic, giving the whole colony a bluish tinge. The openings of the single inhalent siphons encircle a centre where there is a common exhalent aperture. This pattern is reproduced many times in the gelatinous tunic, the common openings being about a centimetre apart.

9.4.7 Herbivorous and carnivorous fishes

A number of species of small fishes, around 10 cm long or more, are encountered among sea grasses, some of which have unusual shapes and prefer sandy areas. Paradoxically, only one of these species (rabbit fishes) feeds exclusively on the plants; the others feed on plankton or small invertebrates in the sand or among the vegetation.

The most striking fishes are several species of Syngnathidae, such as the sea horse, *Hippocampus kuda* (Fig. 9.10a) and the alligator pipefish, *Syngnathoides biaculeatus* (Fig. 9.10b).

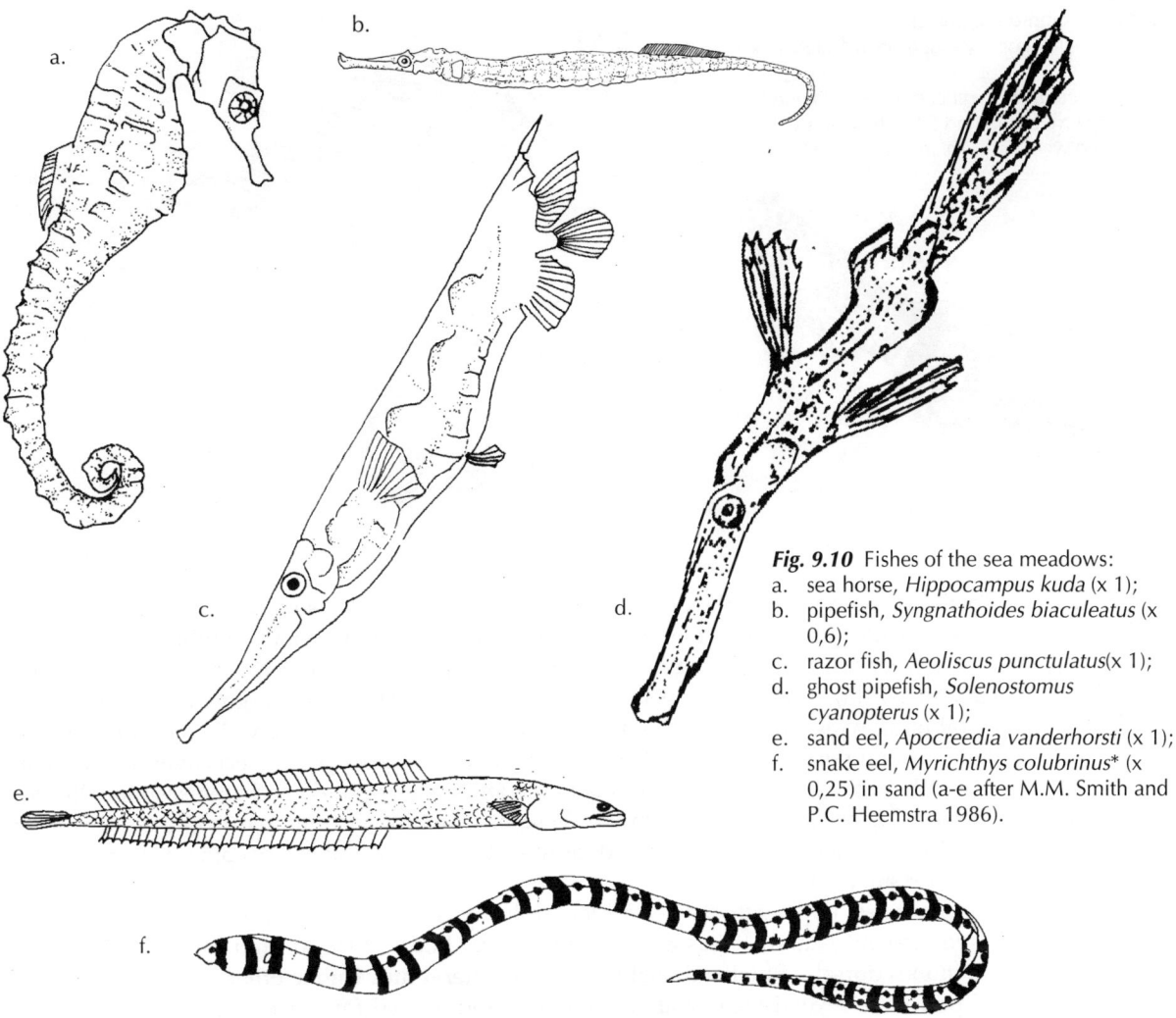

Fig. 9.10 Fishes of the sea meadows:
a. sea horse, *Hippocampus kuda* (x 1);
b. pipefish, *Syngnathoides biaculeatus* (x 0,6);
c. razor fish, *Aeoliscus punctulatus* (x 1);
d. ghost pipefish, *Solenostomus cyanopterus* (x 1);
e. sand eel, *Apocreedia vanderhorsti* (x 1);
f. snake eel, *Myrichthys colubrinus** (x 0,25) in sand (a-e after M.M. Smith and P.C. Heemstra 1986).

Both have an outer covering of large bony plates, modified scales around the body that maintain their unlikely shapes. The sea horse swims in an upright position using pectoral fins, with horse-like head bent forwards; while at rest, the coiled tail is often entwined around sea grass stems for stability and anchorage. The long, thin pipefish swims horizontally by fluttering the small dorsal fin. In both genera the fins are reduced; the mouth is an opening at the end of a narrow snout through which small animals such as amphipods are sucked in after they have first been located by sight. The males in both genera show parental care. During a nuptial dance when their bodies are intertwined the female, by means of a protruding papilla, inserts her eggs into the male brood pouch where they are fertilised. In sea horses the brood pouch is a pocket outside the abdomen and in pipefish it is under the tail.

The razor fish *Aeoliscus punctulatum* (Centriscidae) (Fig. 9.10c), and the ghost pipefish *Solenostomus cyanopterus* (Solenostomidae) (Fig. 9.10d), also have a long tube-like snout with a small mouth at the end of it. Razor fishes swim in small shoals in shallow water and ghost pipefish swim in pairs, both species with their head downwards among sea grasses in search of small crustaceans. The razor fish has large, lateral bony plates, that are extensions of the vertebrae in which it is enclosed, except for the protruding red belly; the body is compressed laterally and the edges of the plates are 'razor sharp'. The pale green 'shell' ends in a spine and the tail and anal fins are ventrally placed in front of the spine so that their vibrations, together with those of the pectoral fins, drive the fish upwards. On the other hand, the ghost

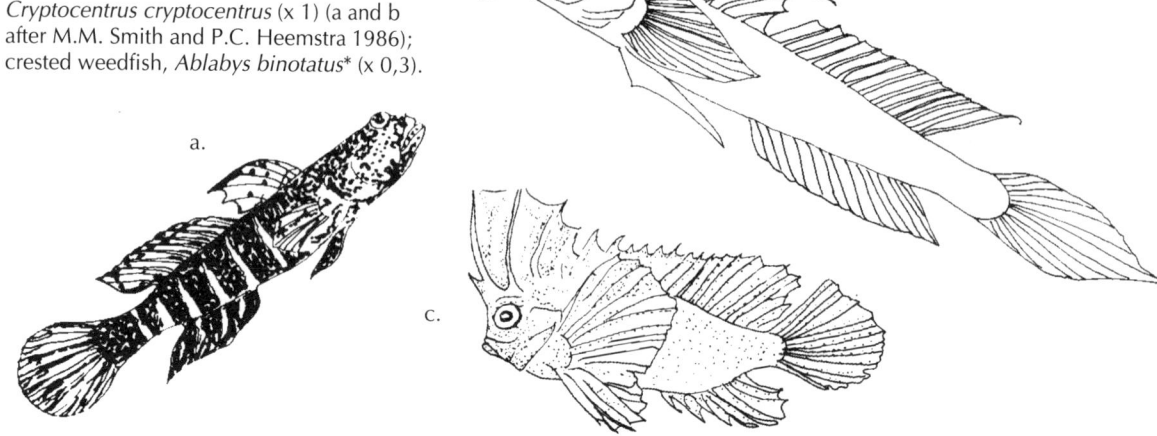

Fig. 9.11 Commensal gobies:
a. candy stick goby *Vanderhorsti delagoae* (x 1) and
b. *Cryptocentrus cryptocentrus* (x 1) (a and b after M.M. Smith and P.C. Heemstra 1986);
c. crested weedfish, *Ablabys binotatus** (x 0,3).

pipefish has brown leaf-like dorsal, pelvic and tail fins and a soft body. In this species the female carries eggs and fry in her ventral brood pouch formed from the modified pelvic fins.

A silvery white fish, resembling an eel although not a true eel, *Apocreedia vanderhorsti* (Creediidae) (Fig. 9.10e), burrows in the sand. It has conspicuous large scales, those on the lateral line having three lobes. The long dorsal and ventral fins have well-spaced rays spread along about three quarters of the length of the body. Its upper jaw projects over the lower jaw on which there are fine teeth; very small eyes are dorsally placed and are exceptionally close together. These characteristics, unlike those of true eels encountered in the coral reef, are useful for living in a burrow in sand close to the surface and for peeping out of it, on the watch for small prey.

A snake eel (Ophichthidae), *Myrichthys colubrinus* (Fig. 9.10f) is common at Inhaca Island, burrowing into sand. It has a long, cylindrical body with pointed head and tail, long continuous dorsal, caudal and anal fins and no lateral fins. Unlike other eels its stiff, pointed tail protrudes beyond the dorsal and anal fins and is used for burrowing in the soft sand. It enters its burrow tail first. It has no obvious scales; striking alternate black and yellow bands encircle the body from head to tail and older individuals have black spots between the bands. The snake eel lurks in its burrow, waiting at the entrance to waylay small fishes. It has two rows of molar-like blunt teeth on both jaws and one to two rows on the vomer. Snake eels are swallowed whole by carnivorous fishes, but the sharply pointed tail may kill the predator. It does not swim much, and has no swimbladder to regulate its buoyancy.

Several species of goby (Gobiidae) found on tropical shores have an unusual partnership with alpheid shrimps (see Section 3.6.4). *Vanderhorsti delagoae*, the candy-stick goby, which has six vertical, dark bands on the body, a spotted head and red, radially striped tail (Fig. 9.11a) is commensal with the shrimp *Alpheus rapax*; the less brightly coloured *Cryptocentrus cryptocentrus* (Fig. 9.11b) lives in the burrow of *Alpheus rapacida*.

Ablabys binotatus, the crested weed fish, or redskinfish (12 cm) (Fig. 9.11c), belongs to the family Tetrarogidae of wasp fishes and is closely related to the scorpion fishes of the coral reef (Scorpaenidae), but prefers to live amongst the sea grasses. It has poisonous spines supporting the dorsal fin which continues on to the head as a three spined crown. This fish is dark brown in colour with a distinctive white horizontal band above the pectoral fins; it has very small, deeply embedded scales and its tail is rounded. It lurks among the vegetation feeding on small, slowly moving crustaceans with its feeble teeth. It is said that if it is kept in an aquarium it learns to respond to taps on the glass in order to feed on small pieces of shrimp, which it sucks in with an audible click. The crested weed fish may grow to 15 cm in length and is often caught by chance in fishermen's drag nets.

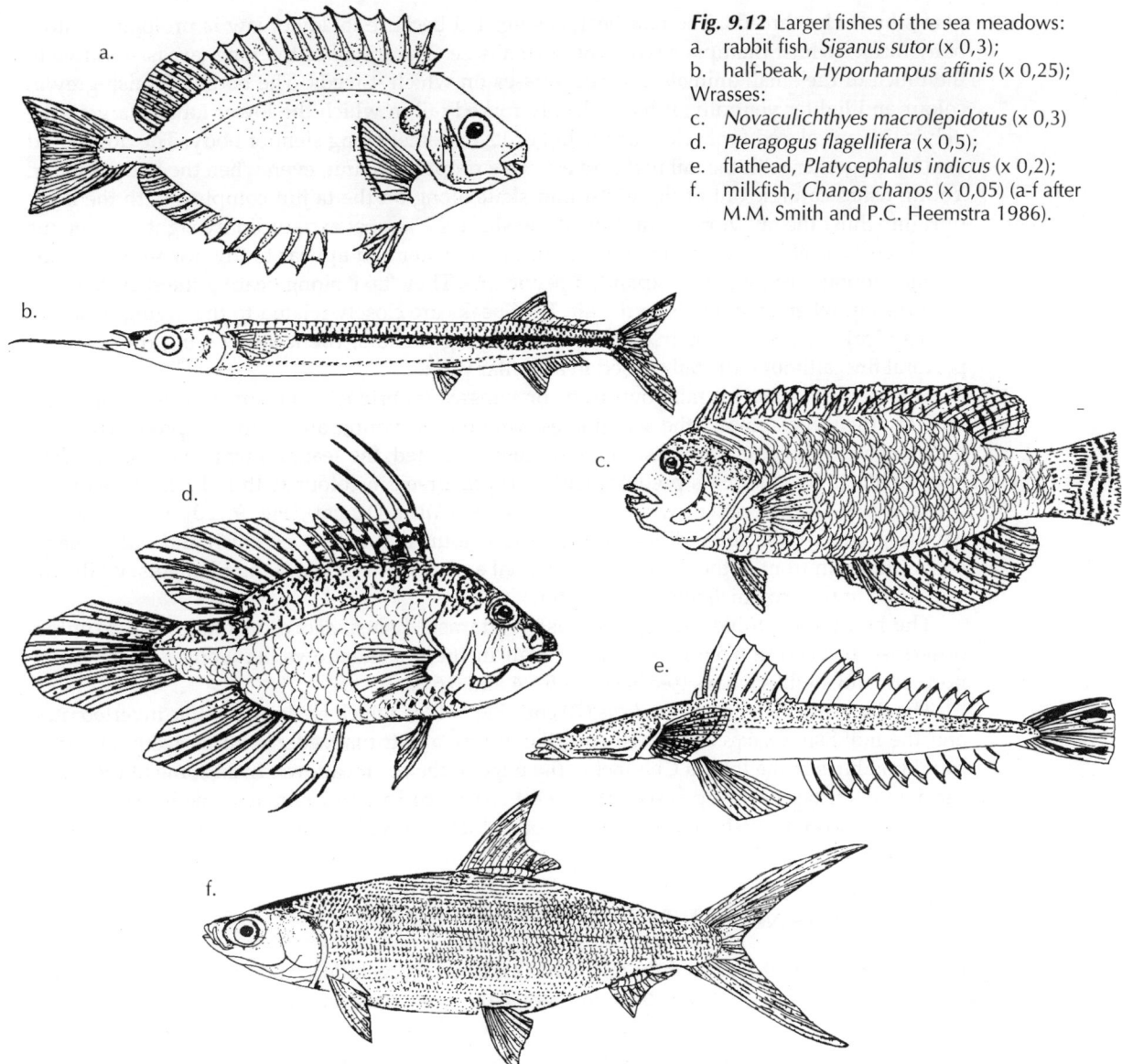

Fig. 9.12 Larger fishes of the sea meadows:
a. rabbit fish, *Siganus sutor* (x 0,3);
b. half-beak, *Hyporhampus affinis* (x 0,25);
Wrasses:
c. *Novaculichthyes macrolepidotus* (x 0,3)
d. *Pteragogus flagellifera* (x 0,5);
e. flathead, *Platycephalus indicus* (x 0,2);
f. milkfish, *Chanos chanos* (x 0,05) (a-f after M.M. Smith and P.C. Heemstra 1986).

Juveniles of the two species of damsel fishes *Dascyllus trimaculatus* (Fig 8.17b) and *Amphiprion allardi* (Fig 8.16c), which are common as adults in the coral reef, live commensally with the burrowing green sand anemone, *Heteractis magnifica* (see Section 3.5.3).

Rabbit fishes (Siganidae) are especially common in the sea grass meadows and they feed on the vegetation. Three species have been seen in the Bay of Maputo. They have very slippery, slimy, compressed oval bodies with minute scales, concealed in the skin. Strong, very sharp spines in the dorsal, anal and pelvic fins have a pair of *poison* glands at the base of each spine and are coated in toxic mucus. The mouth is small and has one row of distinct incisor-like teeth in each jaw, which has earned them their common name. They feed in small shoals, moving as a unit slowly across a patch of seagrass, grazing with heads pointing downwards. *Siganus sutor* (18 cm) (Fig. 9.12a) for example, has an olive green body, brownish above, with a large number of roundish pale spots. It becomes dark brown as it dies. It is *dangerous* to handle any species of rabbit fish.

The half-beak, or needle fish *Hyporhamphus affinis* (Hemirhampidae) (Fig. 9.12b) spawns in the sea meadows, laying thousands of comparatively large eggs (for a fish) that attach to plants by sticky threads. The half-beak is so-called because its lower jaw is prolonged into a bony 'beak', whilst the upper jaw is of normal size. The fringe of the long jaw is sensitive to the touch of the small animals and sea grasses on which it feeds. The body is bluish-grey in colour and lighter ventrally; it has a deeply forked tail in which the lower lobe is larger. The pelvic fins are placed far back near the tail of its extremely long slender body. The dorsal and anal fins together with the tail make an effective propulsion unit, even when the body is in air. Young fishes can rise out of the water and skate along on the tailfin complex, with the body upright whilst the tail vibrates rapidly at the surface – a most extraordinary sight. One genus of half-beaks is able to shoot up, almost out of the water and appear to 'fly' for some seconds in a horizontal position, with expanded pelvic fins. They 'taxi' along beating the surface water with the tail when the body is in the air. Half beaks are closely related to the 'flying fishes' of the tropical oceans, but their technique of 'flying' differs, the latter relying on enlarged pectoral fins, although the tail is used in a similar way.

Some of the colourful rainbow fishes or wrasses (Labridae), so characteristic of the coral reef, lay their eggs among the sea grasses, where their young are relatively protected when they hatch. The colours of the fishes are somewhat muted; the seagrass wrasse *Novaculichthyes macrolepidotus* (Fig. 9.12c) is predominantly cryptic green in colour with red markings on the fins and tail. In the cocktail wrasse *Pteragogus flagellifera* (10 cm) (Fig. 9.12d) another weed-haunting wrasse, the sexes differ in form and colour; the male is greenish and has a sharp-edged hump in front of the dorsal fin; both anal and dorsal fins bear long filaments, whilst the caudal fin in the brownish female is shorter.

The box fishes, tobies and puffer fishes, damsel fishes, scorpion fishes and grunters described in Section 8.5.6 may also be seen in the sea meadows, as well as those in the northern and southern bays (see Sections 5.5.4 and 5.6.3).

The flat-head, *Platycephalus indicus* (70 cm) (Fig. 9.12e), which feeds on benthic invertebrates, and the milkfish, *Chanos chanos* (70 cm) (Fig. 9.12f), a detritus feeder with no teeth, are often caught on lines in the Inhaca Channel at the edge of the sea meadows. Even some of the larger carnivorous fishes of the open sea off the east coast (see Section 7.6) may come into the Inhaca Channel and occur offshore on the west coast; they are well known to anglers in the Bay of Maputo.

9.5 SAND-BANKS OF THE SUBTIDAL FRINGE

Fine sand accumulates as slightly raised banks, bare of vegetation, and are visible at very low tides beyond the tip of Ponta Raza and along the west coast in places near the coral reef of today or where it existed in former years (see Fig. 8.1). The fauna is specialised for burrowing in fine, clean sand and is quite different from that among sea grasses at the same tidal level.

9.5.1 Burrowing echinoderms

The most abundant animal here is *Astropecten monocanthus*, a burrowing starfish (Fig. 9.13a and b), which lies beneath the surface, leaving a conspicuous outline traced in the sand. The upper surface of the starfish is sand-coloured with faint, dark radial lines and seems to be quite smooth, but through a lens a close-set pavement of ossicles may be seen. The periphery is edged with close-fitting plates and prominent spines. The ossicles, known as 'paxillae', spread parasol-like over a veritable forest of minute gills, protecting them from contact with sand or dry air. The podia of these starfish are pointed and supported by minute internal rods at the ends (instead of discs as in other starfish) and thus are adapted for burrowing. There are no pedicellaria. The main food of *Astropecten* consists of minute snails and juveniles of others such as mitres and olives, which they swallow whole. The intestinal branches inside the arms of the starfish are full of empty shells, 1-2 mm in length, from which the flesh has been digested. Another less common burrowing starfish, *A. hemprichii*, has many more spines on the margin and around the mouth; it has no dark markings on its sand-coloured surface.

Two species of 'sand-dollars' or 'pansy shells', *Echinodiscus auritus* (Fig. 9.14a) and *E. bisperforatus* (Fig. 9.14b and c) occupy a similar niche, just covered by a veneer of sand (at low tide). These sea urchins (diameter *c.* 12 cm) are brilliant purple when alive but the calcareous skeleton is soon bleached after death. They are remarkably adapted to living in moving sediments, being able to stay in one place despite strong currents, the shifting sandy surface and their light weight. The calcareous skeleton or 'test' of these sea urchins is almost circular in outline and of wafer thickness; for this reason they were given the common name of 'sand dollars'. The disc has a flat base and a very slightly domed' upper surface; the rim, where upper and lower plates join, tapers to about 2 mm thick and the plates are supported internally by struts of calcite. The periphery of *E. auritus*, the more tropical species, is indented by two radial slits posteriorly. *E. bisperforatus*, more commonly found on Natal shores, has two slit-like perforations, called 'lunules', in a similar position which are enclosed within the margin of the test.

Fig. 9.13 Animals in clean sand: burrowing starfish, *Astropecten monocanthus*
a. dorsal view* (x 0,7);

The flat 'dome' of the sand dollars reduces the 'drag effect' of the water currents and the interior struts act as ballast to lessen the 'lift' of water movement. The slits and lunules in the test also reduce the 'lift' since moving water and sand may pass through them to the upper surface. In these ways the sand dollars evade the hydrodynamic effects of flowing water up to a certain velocity, but if this point is exceeded the resistance of the sand dollar is overcome. Sand dollars are dislodged in storms and some are cast up on the drift line. Lunules are theoretically more efficient than slits in resisting tidal currents or wave action. One might verify this theory after a storm at Inhaca Island, by studying the drift line on the west shore, south of the Marine Biological Station and counting the numbers of the two species which have been washed up. It is pertinent that the species that lives in the stronger water movements on the Natal coast has lunules, whilst that with slits lives on the tropical coasts of East Africa, which are sheltered by offshore reefs. The distribution overlaps at Inhaca Island, where *E. auritus*, the tropical species, is more common on the sheltered shore.

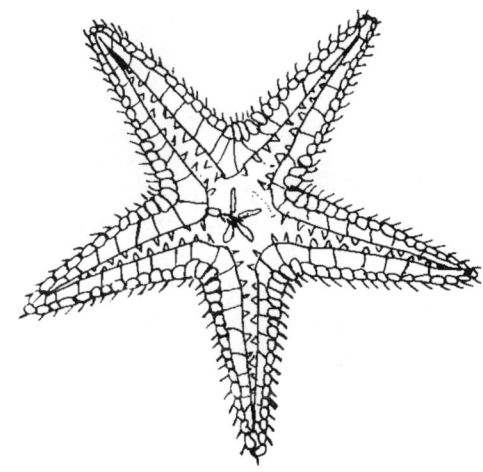

b. oral view showing pointed podia in ambulacral grooves*;

The 'test' of the sand dollar has a velvety covering of tiny spines, capable of movement in ball-and-socket joints. Some on the upper surface are club-shaped or have a prickly surface and are so close together that sand cannot lodge between them. Other spines on upper and lower surfaces are needle-shaped, those on the rim being longer. These spines are used for burrowing and for locomotion, instead of tube-feet as in regular sea urchins. The spines on the rim rock to and fro, clearing sand around the periphery of the animal; those underneath swing in synchronised action as levers to push the animal forward and downward. The movements of the spines and of the cilia on their stems stir up the sand and water so that the animal is enveloped in 'liquefied' sand, which is easier to penetrate than wet sand.

The currents generated by the cilia on the spines keep water circulating from apex to rim on the upper surface, but on the lower surface converging ciliary currents cause suspended particles to be dumped. Supplies of oxygen-laden sea water wash over the upper surface where petal-like markings on the test are perforated by large 'tube-feet' that protrude through the holes whilst the animal is under water. These are the main respiratory organs; there are no gills around the mouth as in regular sea urchins. Gas exchange occurs between

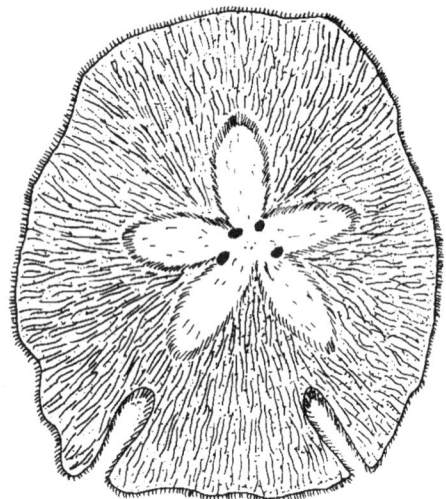

Fig. 9.14 Sand dollars in clean sand:
a. *Echinodiscus auritus** (x 1);

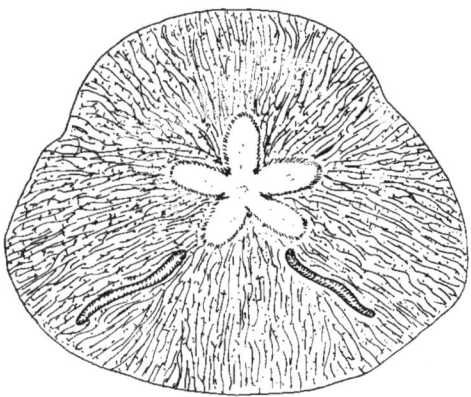

b. *E. bisperforatus** (x 1);

c. pedicellaria enlarged

the sea and the circulating water inside the water vascular system through these respiratory podia. The petal-like markings, which are visible only when the spines have been removed, suggested the name 'pansy-shell' by which these sea urchins are also known.

Sand dollars feed on both interstitial and epipsammic diatoms and probably bacteria. Particles are collected by hundreds of small, ventral podia with sticky tips and conveyed along radial lines that converge on the mouth. The hard siliceous particles of diatoms are crushed between the five jaws of the 'Aristotle's lantern', which is similar in shape to the pentagonal jaw apparatus of regular sea urchins (Fig. 9.4d) but is flattened and adapted for grinding food. The stomach of a sand dollar is full of broken diatom valves and grains of sand, which are excreted through the ventral anus after the organic material has been digested and absorbed.

Sand dollars have the usual line of defence against organisms that may settle on the test, i.e. crowds of tiny pedicellaria among the spines, which act as forceps to remove foreign bodies.

It might be expected that the dispersal of gametes for reproduction would be hindered by a covering of sand, which might prevent access to the clear water above where fertilisation should take place; however, long tubes may be extended through four genital pores on the upper surface through which gametes are ejected into clear water above. Fertilisation is rapid and the embryo soon becomes a free swimming pluteus larva similar to those of other sea urchins. Settlement occurs when the new sea urchin primordium has grown *inside* the larva. The larva is attracted to the sand banks by a chemical released by the parents and metamorphosis into an adult occurs rapidly.

The heart urchin, *Lovenia elongatun* (Fig. 9.14d-g) burrows about 10 cm deep into the sand but maintains an open chimney to the surface. It is oval, almost heart-shaped with a flattened base; it is covered with fine, light, moveable spines, two tufts of which, one anterior around the mouth and the other posterior around the anus on the ventral surface, are larger and more robust. On the upper surface of the test (best seen in dead animals in which spines have been lost) there is a five-petaloid pattern of ambulacral pores, as in the sand dollars, but the small podia of only four petals are respiratory papillae. The podia on the fifth petal are long and extensible; each looks like an old-fashioned chimney-sweep's brush with a rosette of radial, calcareous spicules at the tip and a scraper on the brush 'handle' just behind it (9.14g). These podia maintain an open chimney-like passage to the surface of the sand, sweeping it constantly and the walls are cemented with a mucous secretion. A similar petaloid arrangement on the ventral surface around the mouth supports flexible podia for feeding on fine particles in sand. Mainly foraminifera (protozoans with calcareous coiled shells) (Fig. 3.22) are selected. Heart urchins do not have an 'Aristotle's lantern' for grinding hard particles, but shells are swallowed whole and dissolved in the stomach by acid secretion.

A pair of tiny commensal brittle stars, *Amphylicus (androphorous) scripta* cling permanently to the under-surfaces of each of the two species of sand dollars. The female is the larger (disc 3 mm and arms 15 mm) and the male is about one third her size. The old specific name *androphorous* suggests that the female always carries the dwarf male but it seems to be capable of moving independently over the disc.

A very large brittle star *Ophiocentrus dilatatus* (disc 10 mm in diameter, and arms 15 cm long), makes a surprising appearance on the sand bank directly opposite the Marine Biological Station. The violet and orange disc lies about 12 cm beneath the surface of the sand in a permanent burrow; the arms, which are bright orange with mauve bands interrupted ventrally, stretch upwards through a narrow 'chimney'. The tips of two or three arms protrude from the opening and move around on the sand over a radius of 4 cm or so, collecting food particles that are passed by podia, tentacle scales and cilia down the arm to the mouth under the disc (see Fig. 3.13d).

The sea cucumber adapted to this particular environment is the large *Holothuria pervicax* (30 cm); it has a light sandy colour with two rows of large, dark brown blotches and a brown speckled pattern on its upper surface. It is a detritus feeder, which lies partly buried near the surface of the sand, feeding below the surface with its muscular, peltate tentacles.

9.5.2 Burrowing crabs

Closely associated with the sand dollars is the burrowing crab, *Dorippe dorsipes* (Fig. 9.15a), which carries on its back an individual of either of the two species of *Echinodiscus*, living or dead, held securely by the last two pairs of legs, which are twisted into a dorsal position. If a sand dollar is seen moving in jerks, the crab underneath it will be found. The crab is oval, with a broad, dentate front; the chelipeds and the first two pairs of walking legs are flattened and hairy and are used for digging backwards into the sand until the sand dollar is found and picked up.

The preferred habitat of the common sand-burrowing crabs, *Calappa* (Fig. 9.15b) and *Matuta* sp. (Fig. 9.15c) is the clean, soft sand of the subtidal fringe, although they may be encountered a short distance up the shore in muddier sand. These crabs have pairs of flattened, oar-like legs, used simultaneously as spades for digging a burrow. Their submergence in sand, literally up to the eyes, precludes the use of the route for respiratory water supply, common in most other crabs, through ventral openings at the bases of the walking legs. The respiratory current in these crabs enters the gill cavity anteriorly through the openings above the bases of the chelipeds. In both genera when at rest, the chelipeds are folded over the mouth parts in a position which has given rise to the common name 'shame-faced crabs'. The upper edge of each chela has a crest, toothed and fringed with fine hairs, which filter the inhalent respiratory currents and removes the sand. In the centre of the 'face' a single jet of exhaled water is squirted out so forcibly that the jet is often the first sign of the presence of the partly buried crab. The crabs, however, leave their burrows to hunt and feed on snails, hermit crabs and bivalves. Shells are chipped away with the large chela, which exploits the mechanical advantage of a large tooth placed near the hinge. *Matuta* can be distinguished from the dome-shaped 'box-crab', *Calappa*, by its flat carapace, prominent lateral spines and the much lighter body, which,

Fig. 9.14 Heart urchin, *Lovenia elongatum* (x 1):
d. upper surface*;

e. lower surface*;

f. animal in burrow showing chimney*;

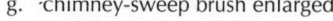

g. chimney-sweep brush enlarged.

with the more paddle-like legs, fit it for swimming rather better than the heavier, rotund body form of *Calappa*.

Two less common burrowing crabs also occur: the 'masked crab' *Gomeza bicornis* (Fig. 9.15d) and the 'frog crab' *Ranina ranina* (Fig. 9.15e). The masked crab is buried more deeply than *Calappa*, since only the tips of the antennal flagella are visible above ground. These together form a respiratory tube leading to the gill chamber centrally; the exhalent current passes out laterally beneath the second maxillae as in normal crabs. This respiratory adaptation is the reverse of that in *Calappa*. The peculiar frog crab, *Ranina*, is a primitive crab with partly extended abdomen (as in lobsters) and four pairs of flattened and expanded legs for burrowing. The carapace has a very broad, multidentate front and an elongated shape, narrowing posteriorly.

The elbow crabs, which also prefer clean sand, live in temporary burrows just beneath the surface. They have a round carapace, reinforced around the periphery by beading, and slender chelipeds in which the third segment is very long and usually bent at right angles to the fourth, so that it resembles an elbow. *Myra fugax* (40 mm) has an almost round, pinkish carapace with a posterior spine, whilst *Philyra platychira* (15 mm) is completely round and white (Figs. 9.15f and g). The respiratory current in this family is guided into openings at the base of the buccal cavity by elongated third maxillipeds. These crabs lie in wait for a passing amphipod or shrimp and snatch at it quickly with their small pincers on the ends of extended 'arms'.

9.5.3 Burrowing molluscs

The most frequently seen carnivores are the largest species of moon shell (naticids), which burrow superficially (Kilburn and Rippey 1982) (see Section 3.5.2); *Polynices didyma* (40 mm) has a buff, globular shell, an amber-coloured operculum and a brown callus on the under side of the shell (Fig. 9.16a). The contractile and flexible, white foot, with which it ploughs under the sand, has muscular flaps in front (propodium) and at the sides (mesopodia). These are inflated with sea water to augment the hydraulic role of blood sinuses during penetration into sand and for anchorage (Trueman and Brown 1992). On being disturbed, several minutes elapse while it withdraws into the shell, water being squirted out *through pores in the foot* as it does so. These features are further exaggerated in the gross, superlatively slimy *Sinum planum* (70 mm), which has a buff shell with brownish marks and similarly inflated flaps on the foot, but it is tougher; it seems that the foot cannot be completely withdrawn into the shell. Both species feed on burrowing snails and bivalves, the latter suffocating them with mucus.

Small brown 'auger' shells, *Terebra anilis* (Fig. 9.16b) (30 mm), are characteristic of the clean sand. They have long narrow shells with many whorls and a thin lip around the aperture. They burrow vertically into the sand, leaving only the tip of the shell exposed above the surface. They feed on polychaetes and acorn worm, *Ptychodera flava* which inhabits patches of the sand banks (see Section 5.4.3, Fig. 5.9c). Venom is injected through harpoon-like teeth from a gland that ends in a pressure bulb (as in cone shells). Much larger terebrids with red or yellow shells, that are sometimes stranded on these sand banks of the subtidal fringe, have come from the subtidal zone, e.g. *T. dimidiata* (Fig 9.16c).

Phalium areola (50 mm) (Fig. 9.16d) a small 'helmet' shell is smooth and globose in shape with a squat spirally marked spire and thickened aperture. It feeds on sand dollars, first squirting neurotoxic saliva over them to paralyse the 'spine' muscles. A hole in the sea urchin's test is rasped by the radula and the proboscis is inserted into the tissues. Very occasionally the 'red helmet' shell, *Cypraeacassis rufa*, a very beautiful large helmet shell (Fig. 9.16e) that has a thick reddish shell with an inner white layer, may be found stranded. This shell is used in carving cameos, outlining a white figure on a red background. It feeds on the sea urchins, *Diadema* spp.

A 'harp' shell, *Harpa ventricosa* (100 mm), has a polished, pinkish-brown shell, marked by many curved ridges and a wide aperture. From the back, it somewhat resembles a harp in shape (Fig. 9.16f). This snail glides over the surface sand with a huge, leaf-like foot and leaves a mound of sand when it burrows. It feeds on small sand crabs, first enveloping the prey with the foot and coating it with mucus.

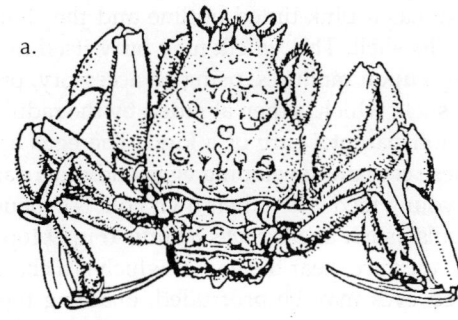

Fig. 9.15 Crabs burrowing in clean sand:
a. sand dollar carrier, *Dorippe dorsipes** (x 1);
b. box crab, *Calappa hepatica* dorsal and frontal views* (x 0,7);
c. *Matuta lunaris** (x 1);
d. masked crab, *Gomezia bicornis* (x 1) (after Barnard 1950).
e. frog crab, *Ranina ranina** (x 0,5);
elbow crabs:
f. *Myra fugax* (x 1) (after Barnard 1950);
g. *Philyra platychira** (x 1).

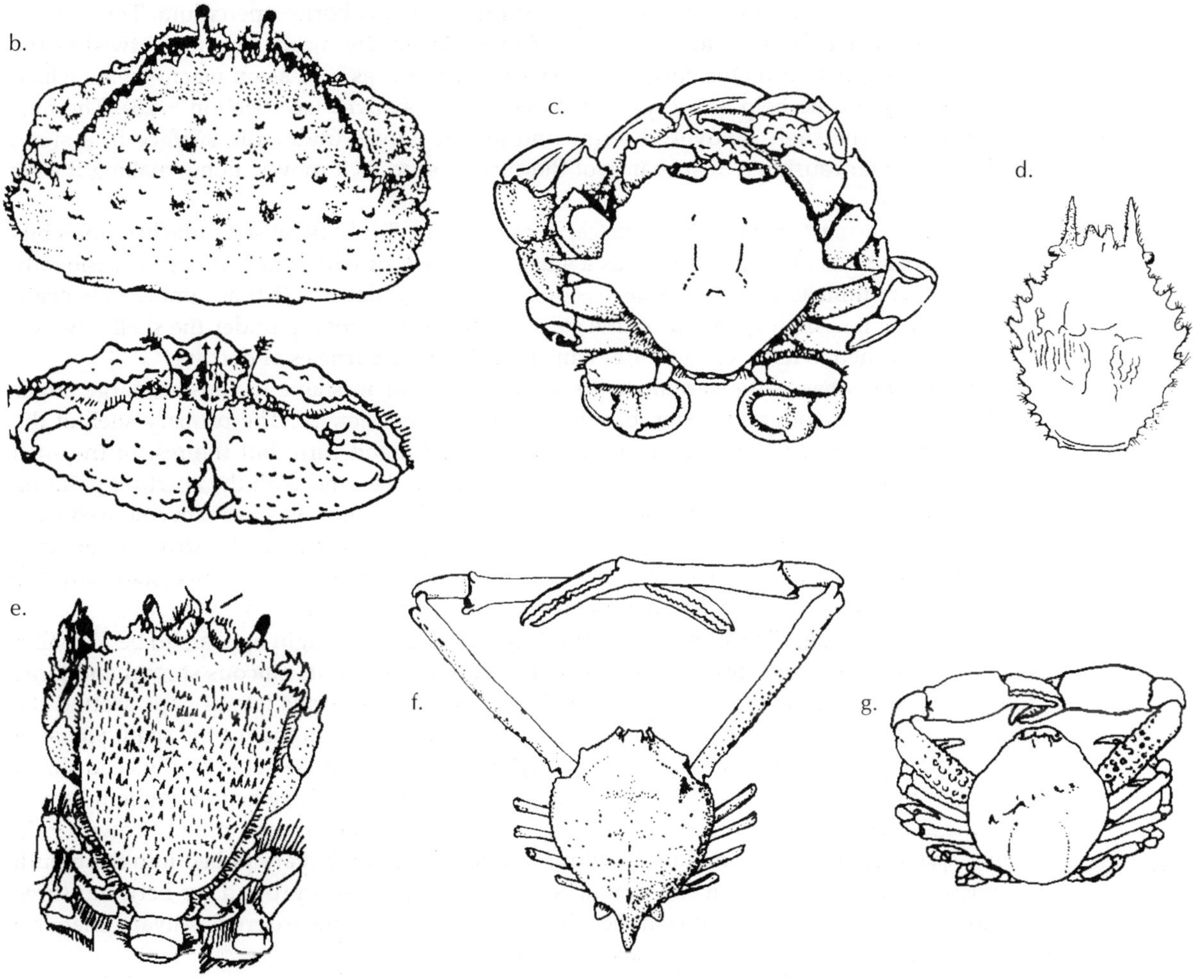

Sometimes large snails from the sea meadows are stranded on the clean sand. *Fasciolaria trapezium* (Fig. 9.16g), a 'tulip' shell (or 'horse' conch) may grow to 120 mm. Its shell is brown, it has a long siphon and many smooth, angular projections on each whorl. It preys upon molluscs, being able to open bivalves with its strong foot and to insert its proboscis into the apertures of snails. *Fusinus colus* (Fig. 9.16h), the 'spindle shell', belongs to the same family and also buries itself when not foraging. The shell is covered by a dark fibrous periostracum. It feeds on polychaete worms burrowing in muddy sand.

The robust whelk, *Chicoreus ramosus* (Fig. 9.16i) grows to 120 mm in length and has a thick heavy shell with a low spire and a large aperture, the lip of which has thick projections resembling teeth, the uppermost protruding as a 'tusk'. Rows of blunt spines extend from the apex of the shell to the siphon. The aperture has a pink-tinted outline and the stout siphon canal is at a slight angle to the main axis of the shell. This whelk may bury itself in the sand leaving only the dorsal spine exposed. It preys upon molluscs, using an accessory, proteolytic gland on the foot to make a soft place on the shell, which is then abraded by the radula.

Most of these large predatory snails exhibit parental care of a kind. The eggs are laid in masses of tough, horny capsules; some of these contain developing eggs and others are 'nurse eggs' and merely contain yolk on which the young feed when their own yolk is exhausted.

There are several species of 'hopping snails', such as *Strombus decorus* (Fig. 9.16j). It has a large thick shell (60 mm) with a long narrow aperture, near the end of which is a characteristic notch through which one of the long-stalked eyes may be protruded, enabling the snail to peep around the shell and look behind. Its foot is long and narrow, and attached at its posterior end is an extraordinarily sharp and jagged, narrow, horny operculum. This is used as a lever to enable the snail to jump forward or backward by means of a rapid twist of the foot. *Strombus* spp. feed on the surface diatoms and detritus as well as on macro-algae, which are digested with the help of a crystalline style, as in bivalves (see Section 9.4.3). The foot burrows until the shell is covered with sand and it moves under the surface in a slow, shuffling manner causing a disturbance of the surface sand as it moves along, making it easy to detect.

Xenophora corrugata is known as the 'carrier shell' (Fig. 9.16k) because it attaches small bits of other shells to its own. This snail has a small, narrow foot with which it hops occasionally using its operculum as a lever as *Strombus* does. It feeds on detritus that accumulates beneath the shell when it is slightly raised off the substratum. Water moving under the shelly 'wings' that it has attached to itself, creates 'lift', which draws in the particles of detritus.

Pyramidellids have an external shell somewhat similar in shape to those of large auger shells, although smooth and wider, but the *apex* is always *deformed*. The primary shell of the newly metamorphosed larva, the protoconch, has a left-hand turn, but the rest of the shell twists normally to the right as it grows. These snails lie in the sand with the aperture pointing upwards (unlike *Terebra*). *Pyramidella maculosa* (Fig. 9.16l) (25 mm) has a cream coloured shell with many brownish spiral markings. The proboscis ends in a sucker-like structure and the jaws are modified into a sharp stylet; there is no radula. *Pyramidella* is associated with the small acorn worm, *Ptychodera flava* and parasitises it (see Section 5.4.3).

A tectibranch slug, *Philine aperta* (30 mm) (Fig. 9.16m), at first sight appears to be a smaller edition of the slimy *Sinum planum*, as it is similar in whitish colour, mucous shroud and large foot. One portion of the foot forms a head shield and the rest forms lateral flaps around the almost cylindrical body. The shell, however, is quite different from that of the snail, *Sinum*; it is internal, fragile, flat and circular and has only one whorl. Mucus secreted by the head shield envelopes the animal and passes backwards over the body by ciliary action, while it moves over the surface like a snow plough heaping sand on each side (Trueman and Brown 1992). *Philine aperta* feeds on small molluscs, worms and foraminifera that are pulled into the mouth by means of hooked teeth on the radula, swallowed whole and ground up between three calcareous plates lining the stomach. Tectibranchs do not have the external salivary chemical secretions that the gastropod snails possess.

Bivalves with delicate shells live in the sand banks. Occasionally one may find the deep-burrowing, brown, fringed shell of a primitive protobranch, *Solemya africana* (Fig. 9.16n). It has a thick, brittle, yellowish-brown periostracum which forms a permanent fringe, like an awning, over the opening of the slightly gaping shell. This fringe can be tucked inside the shell when it is closed in its burrow at low tide. The burrow is Y-shaped and the animal retires to the deepest part at ebb tide. When the sandbank is covered with water these bivalves come up to an upper arm to feed on suspended particles collected by the gills. The razor shell, *Solen cylindraceus* (Fig. 5.3d), which also burrows deeply, may be found here too. Bivalves of the sea meadows with shallow burrows are absent, probably because the surface sand is unstable.

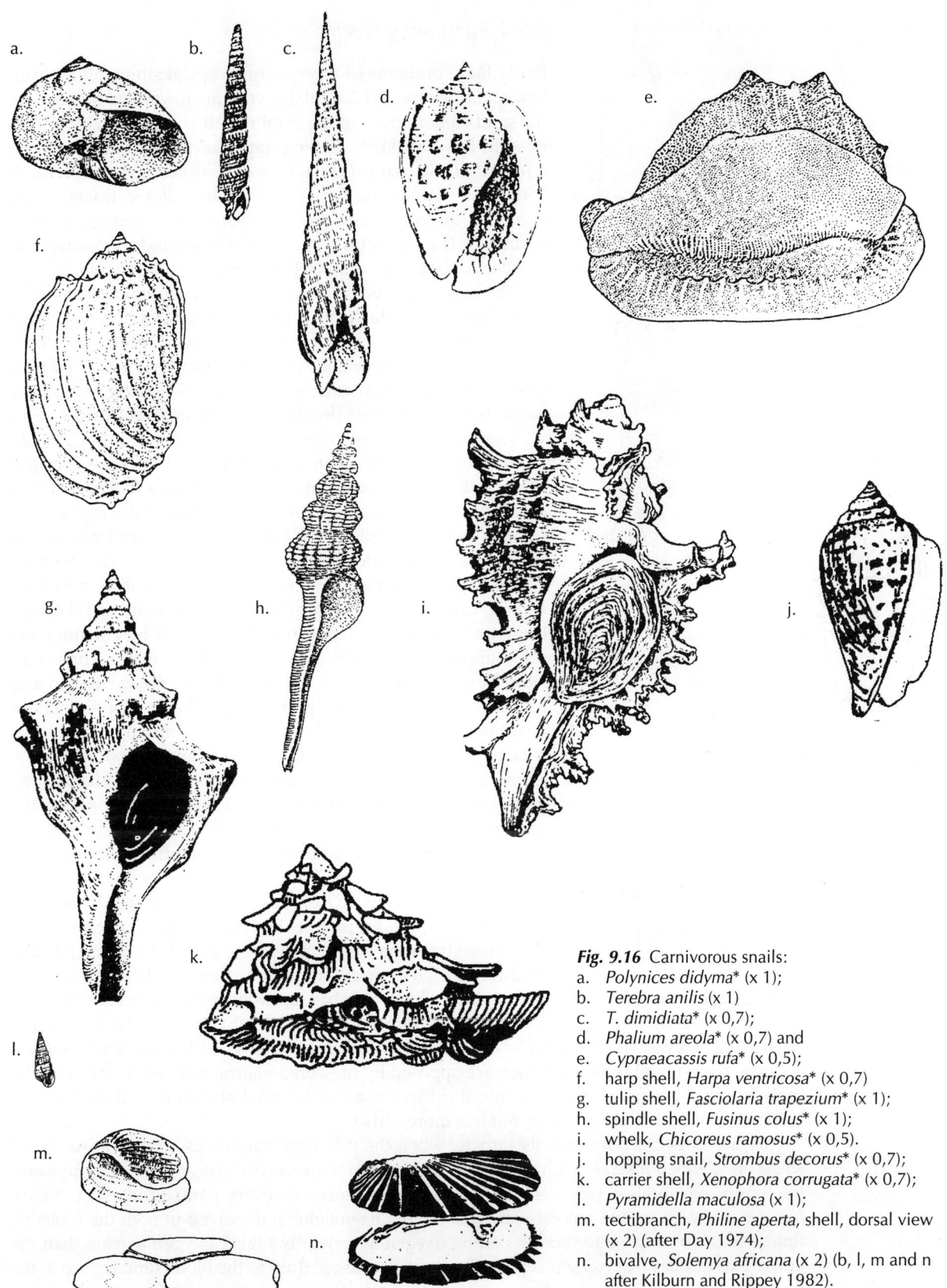

Fig. 9.16 Carnivorous snails:
a. *Polynices didyma** (x 1);
b. *Terebra anilis* (x 1)
c. *T. dimidiata** (x 0,7);
d. *Phalium areola** (x 0,7) and
e. *Cypraeacassis rufa** (x 0,5);
f. harp shell, *Harpa ventricosa** (x 0,7)
g. tulip shell, *Fasciolaria trapezium** (x 1);
h. spindle shell, *Fusinus colus** (x 1);
i. whelk, *Chicoreus ramosus** (x 0,5).
j. hopping snail, *Strombus decorus** (x 0,7);
k. carrier shell, *Xenophora corrugata** (x 0,7);
l. *Pyramidella maculosa* (x 1);
m. tectibranch, *Philine aperta*, shell, dorsal view (x 2) (after Day 1974);
n. bivalve, *Solemya africana* (x 2) (b, l, m and n after Kilburn and Rippey 1982).

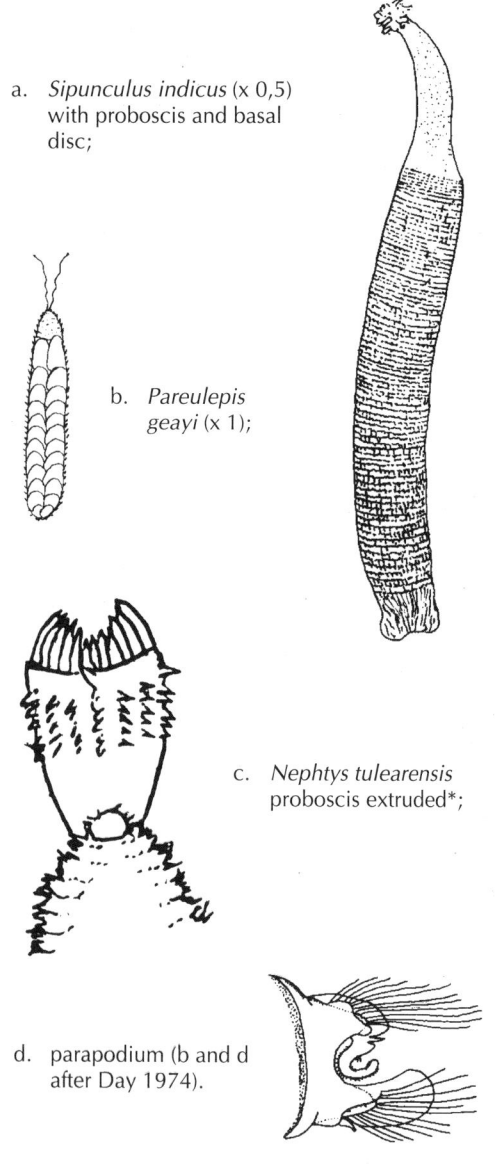

Fig. 9.17 Worms burrowing in clean sand:

a. *Sipunculus indicus* (x 0,5) with proboscis and basal disc;

b. *Pareulepis geayi* (x 1);

c. *Nephtys tulearensis* proboscis extruded*;

d. parapodium (b and d after Day 1974).

9.5.4 Burrowing worms

Fairly large numbers of a very large tropical sipunculid worm, *Sipunculus indicus* (20-25 cm) dig very deep, vertical burrows in the sand bank in front of the coral reef. It is cream in colour and its skin has wrinkled, rectangular markings. Its head can be completely introverted; when extended, it has a frill of tridentate papillae on the end, lined by ciliated tracts which enable it to collect detritus. Cautious and deep excavation results in the extraction of the whole animal, revealing the swollen bulbous end that anchors it in the sand (Fig. 9.17a).

There is one species of scale worm (polynoid), *Pareulepis geayi* (Fig. 9.17b) which, quite unusually for this family, prefers a sandy habitat to a rocky one. It is almost white in colour, small (30 mm), narrow and compact in shape; it has 12 pairs of smooth scales (elytra) each edged with 10-12 little teeth. This worm is a slow-moving predator on very small animals.

The carnivorous polychaete, *Nephtys tulearensis* is probably the most rapid swimmer of all worms and it burrows speedily using swimming movements. So active is this worm that it is easily mistaken, when in action, for the lancelet (amphioxus), which also swims and burrows in this habitat, but the latter has the benefit of muscles attached to a notochord (see Section 9.5.5). *Nephthys* is pearly white in colour, about 80 mm long and 2.5 mm wide (as is amphioxus); it has pairs of relatively large parapodia on each segment, stiffened by conspicuous setae and bearing a gill (Fig. 9.17c). The worm swims with lateral undulatory movements of the body, the waves moving forwards, aided by rowing strokes of the parapodia. These movements are similar to those of other swimming polychaetes such as *Nereis* or *Eunice*, but the exceptional speed of *Nephtys* is expedited by a *different synchronisation* of the parapodia and the body undulations (Trueman 1975). The strong strokes of the parapodia take place more rapidly and are confined to the crest of each body wave; relaxation of the parapodial muscles occurs on the inner curve of the body wave. *Nephtys* burrows as an escape reaction. It has an extremely small head and dives head-first into the sand, keeping its body at an angle to the surface. Violent extrusions of the proboscis ram into the sand and undulatory body movements continue until the worm is completely buried, extraordinarily rapidly. This worm has no circular muscle in its body wall and thus is incapable of burrowing by peristalsis like the earthworm and most marine burrowing worms. Its movements are carried out by small blocks of longitudinal muscles between the septa of the body which are opposed by the dorso-ventral muscles. It relies on the hydrostatic pressure of the coelomic fluid to act as an internal skeleton as it swims and burrows, as do other polychaetes, but it is more active.

Nephtys does not remain in its burrow when the tide rises but swims around in search of small prey, which it seizes with powerful jaws inside the proboscis. *Oxygen demand is high* and the worms have gills between the dorsal and ventral lobes on every parapodium (Fig. 9.17d). *Nephtys* is almost unique among worms in having haemoglobin dissolved in both the coelomic fluid and the blood, the former unloading oxygen at a slightly higher oxygen tension than the latter. This implies that oxygen is passed from the coelomic fluid to the blood (and thence to the muscles) during activity. This device is reminiscent of the transfer of oxygen from haemoglobin to myoglobin in vertebrate muscles or from maternal blood to foetal blood in mammals.

Balanoglossus studiosorum (Fig.3.17b), described in Section 3.6.4, is the large acorn worm that lives on the bare sand-banks; its presence is easily recognised by its large sand worm-cast. A different feeding method is used by a smaller enteropneust, *Ptychodera flava* (Fig. 5.9c), which lives in large groups in patches of soft sand. It is encountered in the warmer coves of the southern bay (see Section 5.4.3). It feeds on small particles of detritus, bacteria and diatoms on the surface sand that are collected in mucus on the proboscis by the currents produced by cilia. It swallows little sand so that no worm casts are visible on the surface sand. Associated with these acorn worms are the terebrid and pyramidellid molluscs that feed on them.

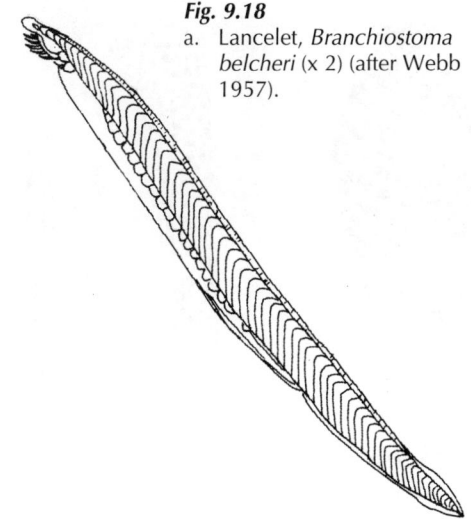

Fig. 9.18
a. Lancelet, *Branchiostoma belcheri* (x 2) (after Webb 1957).

9.5.5 Lancelets (amphioxus)

Lancelets (known by the name 'amphioxus' in older textbooks) have been identified from the subtidal sand banks beyond Ponta Raza and may be collected by dredging in shallow water or by digging at very low tide in the sand of the subtidal fringe. These are the cosmopolitan *Branchiostoma lanceolatum* and the East African *B. belcheri* (Fig. 9.18a). They are about 30 mm long, smooth and pearly white in colour, and when exposed they swim away and burrow immediately, head first or tail first, with incredible rapidity. (*Nephthys*, on the other hand, cannot reverse.)

Lancelets are classified as Protochordates: they are regarded as the 'theoretical ancestor' of vertebrates – a prototype as it were – resembling an embryonic stage in the development of a fish. They have fish-like features, such as gill slits in the pharynx, segmental muscles attached to a dorsal skeleton (notochord) and a postanal tail. The nerve cord is dorsal (instead of ventral as in worms and arthropods) but there is no brain. They differ from fishes in that the respiratory current, which passes in through the mouth and out through the gill slits of the pharynx, is also used for collecting food by means of ciliary currents. They lie in slanting burrows with the anterior end protruding for feeding on suspended particles. Sensory tentacles around the mouth enable the lancelet to test the water and so control intake of food by stopping the feeding current.

The similarity in mechanical function between amphioxus and *Nephtys* is striking (Trueman 1975). In amphioxus the body muscles are small, longitudinal blocks, attached to connective tissue of the septa (myocommata); these septa are attached to the connective tissue surrounding the flexible notochord. Efficiency in amphioxus is greater because their septa are attached at an angle to the notochord and zigzag across the body wall. This arrangement has the advantage of packing many more muscle fibres into each segment than would be the case if the septa were straight, at right angles to the skeletal rod. In addition, muscle contractions on alternate sides of the body exerts a torsional force on the notochord to bend it during swimming. The notochord is an actively flexible, 'hydrostatic' skeleton, more complex than that of the coelomic fluid of *Nephtys*. Each segment of the notochord contains horizontal, minute, transverse plates of muscle fibres composed of paramyosin and are 1,5 microns thick; they contract when the body segments contract since they have a common innervation. There is no deformation or kinking of the notochord when the body bends because the spaces between its muscle plates are *filled with fluid* that is displaced forward during contraction. This has the effect of making the flexible notochord stiffer during rapid contraction, which may reach a rate of 20 times per second.

9.5.6 Burrowing coelenterates: sea feathers and sea pens

A splendid array of highly coloured sea feathers and sea pens (Tixier-Durivault 1960, Benayahu 1994) stand 100-200 mm high above the surface on a sand-bank beyond Ponta Raza

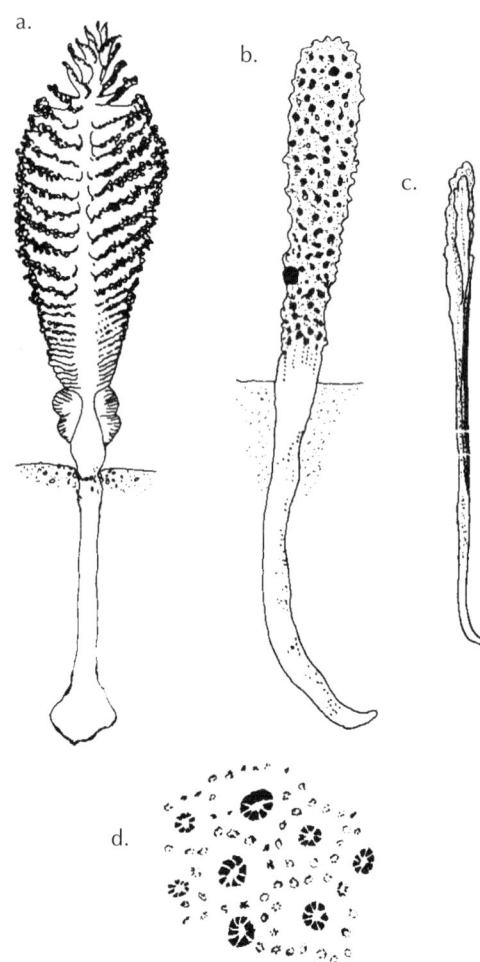

Fig. 9.19 Pennatulids in clean sand:
a. *Virgularia gustaviana** (x 1);
b. *Veretillum leloupi* (x 1) and
c. skeleton;
d. autozoids, surrounded by siphonozoids enlarged (b-d after Tixier-Durivault 1960).

on the edge of a small channel; they suddenly disappear as one approaches. They have also recently colonised the clean sand in the new lagoon north of Portuguese Island. These are pennatulids, a group of octocoralline coelenterates in which feeding polyps (autozoids) have eight pinnate tentacles, as in soft corals (see Fig. 9.9b, Section 9.4.5); a second much smaller type of polyp (siphonozoid) does not have tentacles. Each colony of polyps is partly supported by the large bulk of calcareous spicules in the mesogloea and partly by hydrostatic pressure in the enteron, maintained by the siphonozoids. As the colony emerges slowly from the sand, water is drawn into the enteron by the beating of the cilia: the siphonozoids then close and maintain the internal pressure of water. During the sudden contraction of the colony into the ground, water is expelled quickly through the opened orifices of the siphonozoids and the autozoids.

The most striking pennatulid is *Virgularia gustaviana* which has an erect, brilliant orange colony in the shape of a large feather over 200 mm in length (Fig. 9.19a). The shaft of the 'feather' is supported by a single stiff rod made of aggregated calcareous spicules; purple autozoid polyps are closely set on stiff branches of the 'feather' and are arranged in rows around which are circles of very small siphonozoids (Fig. 9.19d). The submerged part of the orange shaft ends in a bulbous swelling.

The sea pens have more compact cream or mauve bodies with purple polyps that are also supported by a stiff mass of consolidated calcareous spicules in the form of a narrow rod (Fig. 9.19c). Species differ in body form; *Veretillum leloupi*, a species described from Inhaca Island, is the most common. It grows to about 120 mm long and has a narrow cylindrical shape, 10-20 mm in diameter, slightly curved at the base (Fig. 9.19b). Autozoids are small and scattered closely over the whole surface above ground and each is surrounded by a ring of siphonozoids (Fig. 9.19d). A less common, sibling species *V. cyanomorium* has a smaller body and larger polyps, but has the same body shape. Another common species, *Cavernularia lutkeni*, has a squat club-shaped body, about 10 cm high, with a swollen flat top.

Pennatulids burrow until they are completely submerged about half an hour after the tide has left them exposed. They may use slow peristaltic waves or rapid contraction with explosive emission of water through the siphonozoids. They emerge under water as the tide rises and catch zooplankton by means of their pinnate tentacles, which have typical coelenterate nematocysts (see Section 8.2). The mucus covering them becomes luminescent in the dark, when the animal receives a stimulus, mechanical, chemical or electrical. Waves of luminescence pass over the body in concentric circles from the point of stimulation. Oxygen, calcium ions and ATP (adenosine triphosphate) are involved in an enzyme reaction when luciferyl sulphate becomes active luciferin. The reaction is carried out in discrete clusters of special cells (photocytes) in the endodermis of parts of certain autozoids and is accompanied by secretion of mucus. The cells' response to external stimuli is mediated through the nervous system. Although the mechanism of luminescence is well understood the role that it plays in the life of the pennatulids can only be guessed at this time.

A large brown cerianthid anemone (15 x 4 cm) burrows along the edges of the sand banks facing the shallow channel. They are common here but cannot easily be excavated because the leathery tube, which it constructs out of mucus and sand, penetrates 30 cm or so into the sand and the retraction of the anemone is extraordinarily rapid (see Section 9.4.5). They are best seen under water in sandy patches between coral colonies when they extend their long slender tentacles above the sand from the tube below (Fig. 8.9e).

PART 3

Land Flora and Fauna

Twenty-seven years after the declaration of Inhaca Island Nature Reserves in 1965, there are marked regeneration of disturbed bushland, a natural succession on abandoned cultivated plots on the dunes and a concomitant increase in the populations of insects, birds, reptiles and small mammals.

Chapter 10 in this edition relates the land plants to different habitats created by both natural processes and by human disturbance. The recovery of vegetation, as a result of both traditional and modern practices, is described. The system of agriculture and the reclamation of wind-eroded dunes through planting are outlined.

The account of common birds in Chapter 11 is based on five years of monthly records, 1976-1981.

Chapter 12 on amphibians, reptiles and mammals includes both old and new records. Welcome additions to the reptilian fauna are the turtles that nest again on the eastern dunes.

Representative collections of insects have been made at Universidade Eduardo Mondlane in the last five years. A selection of species from many families has been described in Chapter 13.

Chapter 14 introduces land molluscs. Snails were collected some time ago, but slugs, which leave no shell that have often been seen, still remain to be described.

In the decade 1963-1973, collections of over 300 species of diatoms were made in the freshwater swamps on the island. In Chapter 15, dominant genera are discussed in relation to the changes in area and character of the swamps.

10

TERRESTRIAL VEGETATION

Jan de Koning (ex Universidade Eduardo Mondlane) and
Kevin Balkwill (University of the Witwatersrand)

The land area of Inhaca Island is about 40 km^2 (see Fig. 1.1). It was formed as a result of sea level changes in the recent Holocene history of the large estuarine bay which it borders (see Sections 1.1 and 1.3). The oldest low dunes on the west coast are exposed to the usually mild winds of the Bay of Maputo, but occasionally there are storms. The outer dune ridge on the east coast is exposed to very strong, predominantly south-east, salty winds from the Indian Ocean. This dune ridge is continuously being built higher and higher; it shelters a series of dunes which decrease in height towards the centre of the island. The central part of the island, between east and west dune ridges, occupies over half the total area of the island and has very low dunes and valleys. About a quarter of this area is covered by freshwater swamps: one quite large swamp and several smaller ones, some of which have been drained for agriculture in the last 50 years. The soil is loose and sandy, except for areas in and around the freshwater swamps, where humus has accumulated. Intertidal mangroves occur in the north and south bays and in a creek south of Ponta Raza.

The island has been inhabited for hundreds of years (see Introduction) and the population has doubled during the last ten years (see Section 16.1). The people still depend on the natural vegetation for fuel, building materials, fruit and medicines. The best soils in and around the swamps are reserved for agriculture and horticulture. The most disturbed ground is therefore the central area and the southwest and northeast peninsulas, where the original forest has been replaced by crops and the low dunes and valleys are inhabited. Huts are surrounded by large 'mutis' (gardens), where the trees that bear edible fruit or are used as sources of medicines are conserved; grasses and herbs grow in the open areas. The result is an attractive 'parkland'.

The sheltered, inner western and north-eastern dunes have been disturbed by collection of firewood, felling of hardwood trees for timber and shifting cultivation. In order to conserve the land resources, abandoned cultivated land on the lower dunes and large areas of the decimated bush on the east and west dunes have been protected in nature reserves since 1965 (see Fig. 16.1). Regeneration of the vegetation is being studied by the biologists in Universidade Eduardo Mondlane of Maputo and conservation is supervised by rangers in its employ (see Section 16.2).

The composition of the vegetation is primarily determined by the warm, humid climate (see Section 1.4). Other factors form gradients that determine the particular vegetation in a specific area. These factors include exposure to salty wind, altitude, slope, temperature, humidity, drainage, amount of humus in the sand, light, salinity, level of water table, the degree of human disturbance and the time that has elapsed since disturbance. The vegetation on the island has now become a complex, large-scale mosaic of both original and derived types (Fig 10.1). The west dunes are largely covered by evergreen bushland and thicket and the leeward side of the east dunes by evergreen dune forest of the Tongaland-Pondoland regional mosaic (White 1983; Moll and White 1978). Secondary dune forest is well developed in areas on the east dunes which have been protected in nature reserves over the last 27 years. The gradients that affect particular areas on the island are mentioned in relation to the vegetation composition in appropriate sections of this chapter. While the initial establishment of a species in a particular area depends on its ecological tolerance, its continued existence in the community depends on successful pollination and dispersal of disseminules. A large variety of birds and insects thus play important roles in the regeneration of plants (see Chapters 11 and 13).

The people who live on the island are faced with two major erosion problems. On the west shore of the island, storms in the Bay of Maputo erode the base of the red cliff at Barreira Vermelha. Changing currents in the bay displace sand on the west shore and Portuguese Island so that the narrow belt of sand-binding plants above high tide on the west coast has been obliterated in the last ten years and the sea reaches the trees. Secondly, the sand that is blown over the crest of the eastern ridge of the dunes in certain areas is accumulating between the topmost trees of the forest, burying their trunks.

The dune ridge facing the Indian Ocean has been built up by wind-deposited sand over thousands of years and the partial burying of trees is a natural process which does not destroy the trees. The problem is, however, the threat of continued sand deposition on the agricultural land in the central 'parkland'. Stabilisation measures have thus been proposed to protect the hinterland. A study of the vegetation of Inhaca Island has therefore two aspects, ecological and economic, which are, of course, interrelated.

In this chapter the vegetation (near the Marine Biological Station) of the older, western dunes is described in the first section, where the primary vegetation is preserved as undisturbed scrub forest (White 1983). The effects of differences in light intensity, a result of human disturbance, may be observed in several places. In the second section, the vegetation on the younger dune sequence on the eastern side of the island is discussed. There, the vegetation ranges from sand-binding pioneers through dwarfed vegetation to wind resistant trees on the crest of the dune. To the west of the crest, increasing shelter from wind allows primary and secondary forest with trees taller than those on the west coast to grow on the sheltered dunes which decrease in altitude towards the central 'parkland' and the freshwater swamps, that adjoin the mangrove swamps. The last section deals with agricultural practices, recovery of abandoned cultivated land through conservation, both traditional and modern, and two cases of experimental reclamation of dunes.

10.1 SCRUB FOREST, BUSHLAND AND THICKET ON THE WEST COAST

The dune ridge on the west coast runs from the fishing harbour to Ponta Raza and follows the curves of the coastline to the south-west point, Ponta Punduine, where it is protected from wind. The highest point on the western dunes (67 m) is on a plateau at Barreira Vermelha, 2 km north of the Marine Biological Station, where red soil becomes visible as parts of the cliff break away when eroded by occasional heavy seas. The plateau occurs on the ridge behind the cliff that has long been cultivated. Now cultivation has been abandoned for about 15 years and regeneration is taking place (see Section 10.5). The coastal ridge is otherwise low and the vegetation clothes the dunes right down to the sea. The sea level in the Bay of Maputo is rising (caused by sand deposition) and in the last ten years the narrow band of stabilising plants which occurred between high tide mark and the bush has disappeared, and so have the land hermit crabs which used to rest in burrows below their leaves during the day.

The coastal bush is not uniform in character although many hundreds of years ago these western dunes were probably completely forested. Today, several larger and smaller areas of undisturbed forests exist at old burial sites. Other areas have been badly disturbed where trees were felled, or less so where firewood was collected. Since 1976, most of the coastal bush

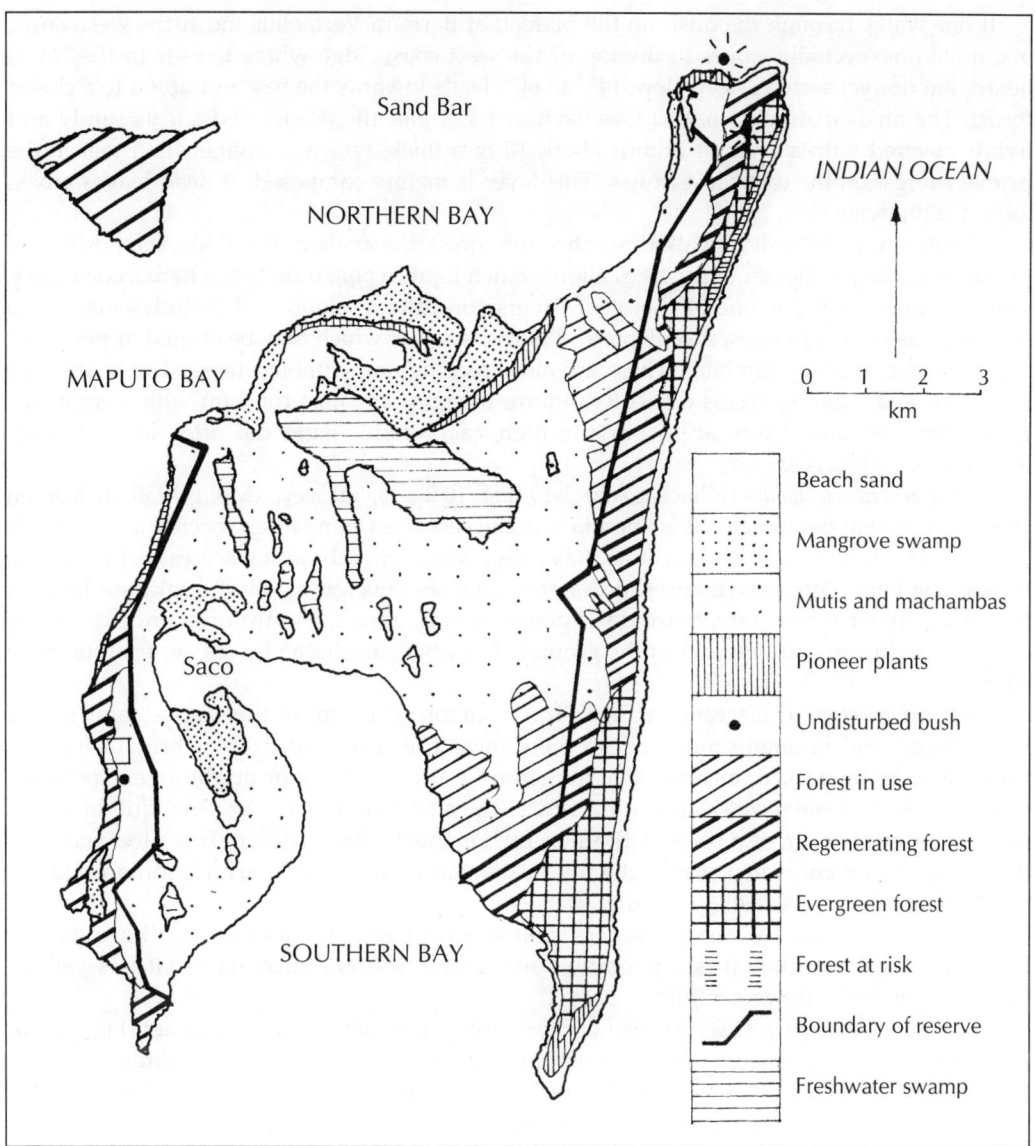

Fig. 10.1 Map of the vegetation: the distribution of the vegetation types of Inhaca Island is continually changing, because of the influence of erosion, sand deposition, agriculture and protection in Nature Reserves. This map shows the distribution recorded in 1990 (courtesy of Instituto de Desenvolvimento Rural Moçambique).

area has been better protected in nature reserves and the undisturbed forests serve as seed banks for the regeneration of scrub forest.

10.1.1 Undisturbed scrub forest on the outer dune slope

Along the west coast, the patches of undisturbed forest occur on the seaward facing ridge. These patches may be large, or as small as 20 m²; they may be surrounded by disturbed thicket or border on cultivated land. The forest can be recognised by the thick undergrowth, the closed canopy and the presence of some shade tolerant species which do not easily develop in the early successional stages. The majority of species in this forest are also found in the thicket, and like these are shorter than in the sheltered east dune forest. The stunted growth is the result of the salty winds from the Bay of Maputo.

If one walks through the bush on the plateau at Barreira Vermelha and turns westwards, one quite unexpectedly comes to the top of the west coast ridge where the sea in the bay is heard, but not yet seen. A steep slope (45° to 60°) leads towards the bay and down to a closed forest. The air is distinctly humid (see Section 1.4.1) and the orange-red soil is sandy and lightly covered with a layer of humus about 10 mm thick, which is thinner than that in the primary forest on the east-coast dunes. This layer is mainly composed of dead leaves mixed loosely with twigs.

Although it is difficult to walk through scrub forest, the angle of the slope, the position of the sun and the low height of the trees allows much light to penetrate to the herbaceous layer. The herbaceous layer is one of the most diverse on Inhaca Island and includes many tree seedlings, as well as grasses, succulents and herbs, some of which may be annual or perennial and prostrate, erect or climbing. There are also creeping and climbing ferns. The bushes are well-branched, making access difficult, and are mixed with short trees on which epiphytes grow. The trees and shrubs are up to 8 m high, cast a light shade and often have leathery leaves. Woody lianas are rare.

Sansevieria hyacinthoides (Liliaceae *sens. lat.*) (Fig. 10.2 a-c), 'xikweja' (Mogg 1958; de Koning 1993), a succulent perennial that is abundant in the east coast dune forest, occurs here in small patches. The white, smooth rhizomes have big, white, membranous scales at the orange nodes. The long roots have many short lateral branches covered by short, pink root hairs. It has erect, linear leaves which are dark green or variegated. The inflorescence is cream-coloured and grows straight out of the ground. The fibres inside the leaves are used to make string.

Aloe parvibracteata (Liliaceae), 'manga', is a common succulent in the sandy-loamy soil in shady forest and in open sandy soil on the dunes. The succulent dark or brownish green leaves may be unmarked, or may have numerous scattered, white or bright green spots; the tips droop so that the rosettes appear flattened. The aloe produces red or pink, dull or glossy flowers in midwinter; the base of the perianth is markedly swollen. This aloe was first described from a cultivated plant growing in the garden of the old fortification around the town of Maputo, where it still flourishes.

Kalanchoe paniculata (Crassulaceae) is common near the sea and occurs all around the island in forest margins. It is a perennial, evergreen, solitary succulent that has yellow, butterfly-pollinated flowers in July.

The fern, *Microsorium scolopendrium* (=*Phymatodes scolopendria*, Polypodiaceae) (Fig. 10.2 d-e) forms a mat of creeping rhizomes or becomes epiphytic with single stems climbing on the branches of bushes and trees, even 5-6 m from the ground. The well-branched creeping rhizome may be 10 mm in diameter, is greenish and bears bright brown scales and numerous dark brown rhizoids. The widely spaced (150-400 mm) fronds appear feathery, with the blades cut close to the midrib. Round sori are borne near the raised ribs. The other common fern here is *Microgramma lycopodioides* (Polypodiaceae), which differs by being always epiphytic and having undissected fronds. Creeping rhizomes bear widely-spaced fronds. The simple, stalked fronds are longer and narrower when fertile. Both ferns are typical forest species, so that their presence indicates that the vegetation on this slope of the ridge has been undisturbed for many decades.

Like the herb layer, the shrub layer of the undisturbed scrub forest is diverse. Among the most common are *Croton pseudopulchellus* and *Suregada zanzibarensis* (both Euphorbiaceae). *C. pseudopulchellus* is a low, deciduous shrub, no more than 2 m high. The leaves have silvery scales below. Male and female buds are produced on the same inflorescence and after ten months open into white to yellow flowers. The three-lobed capsules are about 10 mm in diameter and covered with silvery scales. A concoction of the root is used for curing asthma. In contrast, *S. zanzibarensis* reaches 3-5 m at Inhaca Island and is evergreen; its leaves are irregularly toothed and dotted with glands. The small, greenish flowers are solitary or in clusters in the axils of leaves; the three-lobed capsules are brownish.

The tree layer is less crowded than the herb and shrub layers, and most of the trees have *smaller leaves* than those in older forests. There are few lianas here but the number of epiphytes on the trees and shrubs is striking. Apart from the ferns, mentioned above, epiphytic orchids occur.

The orchid, *Aerangis mystacidii*, is confined to undisturbed areas at an altitude between 10 m and 15 m above sea level. Germination takes place in the fissures in bark on the stems and branches of shrubs about 5 m high, but plants continue to grow in clumps on fallen branches. During the dry season, most, or sometimes all, of the leaves are deciduous. Young leaves and roots start growing after the first rains and are a dull green. The horizontal roots may reach up to 50 cm in length. The leathery net-veined leaves are shiny, dark green above and pale and glossy below; they develop almost horizontally with the apex hanging down. In May to June, when at least five leaves have developed, plants may produce 3-4 colourful, pendulous inflorescences and are a truly beautiful sight. The slightly curved peduncle has dark brown floral bracts and 4-10 flowers, each on a rosy orange pedicel. The flowers are white with rosy orange in the centre and on the apex of the labellum; the pollinia are pale yellow.

Solenangis aphylla, another orchid, has a pale brown stem that may be several metres long. Grey-green, horizontal, branched roots with pale orange-yellow tips are produced all along the leafless stem, loosely attaching the orchid to its host. The orchid appears to hang in mid air amidst the shrubby vegetation. Tiny pink and white flowers are produced along most of the stem. *S. aphylla* thrives in deep shade under the kooboo-berry, *Cassine aethiopica* (Celastraceae). This widespread, evergreen, hardwood tree grows to 6 m tall at Inhaca Island and usually occurs in forest and woodland in southern and eastern Africa. The spirally arranged, oval leaves are broader at the base and have a finely toothed margin; they are glossy dark green above and paler below. Inconspicuous green or white flowers are arranged in small cymes. The bright red berries are ovoid and pointed at the tip.

Vanilla roscheri, a semi-epiphytic orchid, also occurs here, but is described in Section 10.1.3 on the Ponta Raza forest, where it is more abundant (see Fig. 10.3 g-i).

Many trees typical of the evergreen bushland (see Section 10.1.2 and 10.1.3) are found in the undisturbed forest patches on this western slope, but *Strychnos henningsii*, the coffee bean strychnos, is a species that occurs only in closed forest. This tree grows to about 8 m at Inhaca but is taller in the mist belt of Niassa Province and coastal forests in Mozambique. The bark is scaly, with a waxy coating on the spineless branches. The net-veined leaves taper at the apex. The purple, spherical fleshy fruit with a groove on one side is edible, but contains one or two *poisonous* seeds.

The branches of the bee-sting bush, *Azima tetracantha* (Salvadoraceae), are scrambling, square in cross section and hairy. Two pairs of spines are produced at each node; they are 50 mm long and cause an unpleasant burning sensation in scratches. The leathery leaves are pale green and hairy. Although dioecious, the male and female flowers are physically alike. The small, soft fruits are yellow to white.

The vegetation on the sandy soil on the slope of the ridge at Barreira Vermelha continues down toward the sea, where the tree layer does not produce dense shade. The floor at the fringe of the forest is exposed to light. The succulents, *Kalanchoe paniculata* and *Aloe parvibracteata* (mentioned above) and some bulbous species of Liliaceae, such as *Ornithogalum tenuifolium*, are present. The grasses are represented by species of *Panicum*.

Near the sea, the unusual *Euphorbia tirucalli* (Fig. 10.2 f-j), the rubber hedge euphorbia, grows wild; it is used as a hedge around some of the huts and gardens on Inhaca Island and in many other parts of Mozambique. It lacks spines, has smooth, cylindrical branches 5-7 mm in diameter and, like other euphorbias produces abundant latex, which is used as an insect repellent and an insecticide. The leaves are small and soon fall off the branches. The flowers are highly reduced and occur in small, yellow inflorescences in clusters at the ends of branches.

The wild date palm, *Phoenix reclinata* 'kindu' (Arecaceae) (Fig. 10.2 k-m), also occurs here; it has only *underground* stems in this habitat, which contrasts with the trunks that develop in primary forest on the eastern dunes (see Section 10.2.3). The rigid pinnate leaves form rosettes at soil level and are 1,5-3 m long, with very many leaflets of which the lowermost are reduced to spines. Male and female inflorescences are produced at soil level on separate plants and are protected by brown spathes (large leaf-like structures) that are triangular in cross section. The ripe fruits are one-seeded, orange and tasty, but the volume of pulp is less than that of the edible date, *P. dactylifera*.

Just above the high water mark, terrestrial orchids may be found; *Eulophia speciosa* (see Section 10.1.4) being quite common. This species is found near the beach, near the mangrove and in dense forest where *humidity* is always high. Here at Barreira Vermelha, it grows in sparse undergrowth under shrubs in dry sandy, white soil which has little humus. Although it flowers in deep shade it produces more of its attractive yellow flowers in sunny places from November to February, peaking in December and January.

It is very probable that there has been no cultivation in this area, which lies so close to the sea, because the traditional plan for cultivation on the island, as for other islands off the Mozambique coast, included a border of forest on the outermost dunes surrounding the farmed area. This suggests that the whole of the interior of the island, once erroneously considered to be savanna, has to be recognised as the remnant of a former forest that covered the whole of Inhaca Island. The present day distribution of plants on the west dunes is the result of more or less continuous or intermittent cultivation, while the existence of patches of undisturbed forests implies that the serious protection of well-chosen areas would result in the restoration of forested areas within a few decades (see Section 10.5).

10.1.2 Evergreen bushland south of the Marine Biological Station

Between Ponta Raza and the Marine Biological Station, the first coastal dune is low, only 1-2 m above the spring tides, and in recent years the sea has encroached on its base. Near the station, the dune is only 5-10 m wide but further south the zone of dunes broadens to 150 m. The first seaward ridge is low and the valley behind it is slowly recuperating from cultivation which ceased about 15 years ago; occasional very small patches of undisturbed scrub forest may be seen.

Inland, the second dune rises to 40 m high and small patches of *evergreen bushland* have been conserved. Here and there the vegetation may be closed and light intensity at soil level is very low. These dense patches are dominated by thick-leaved, well-branched evergreen shrubs and trees such as the bush-tick berry, the coastal red-milkwood and the Natal rhus, which grow up to 10 m high. The sandy soil is covered with a layer of dead leaves about 5 mm thick, but no major A-horizon has formed.

Chrysanthemoides monilifera ssp. *rotundata* (Asteraceae/Compositae), the bush-tick berry or 'nthosani', usually consolidates the sand just above high water level. It grows as a 3 m high, much-branched shrub and occurs in the coastal belt from Maputo southwards to Port Elizabeth. The shrub has grey bark; young leaves are covered with dense, white cobwebby hairs, while mature leaves are glabrous, glossy green and succulent. The margin is toothed and the blade tapers to the petiole. The florets are yellow and arranged in small heads. Most unusually for this family, the mature, dark purple fruits of *Chrysanthemoides* are fleshy and edible. Other parts of the plant are used in local medicine.

Young trees of *Acacia karoo* (Leguminosae – Mimosoideae) (Fig. 10.2 n-o), the sweet thorn, 'gwonya' or 'nkaya', also grow just above the high tide mark. This deciduous tree has a well-

Fig. 10.2 Species of the undisturbed scrub forest (a-m) and the evergreen bushland (n-o) (drawn by B. Pike):

a. *Sansevieria hyacinthoides* habit;
b. inflorescence with buds and flowers;
c. inflorescence with young fruits;
d. *Microsorium scolopendrium* habit showing rhizome, very unusual simple frond and sori;
e. more common dissected frond;
f. *Euphorbia tirucalli* stem with inflorescence;
g. leaf which is short-lived;
h. inflorescence while in male stage;
i. male flower and bract;
j. inflorescence while in female stage;
k. *Phoenix reclinata* upper part of leaf;
l. inflorescence;
m. young fruit;
n. *Acacia karoo* branch with thorns, leaf and inflorescences;
o. pod.

branched crown; the rough bark is reddish brown or black and the branches and branchlets are red when the bark peels off. The long sharp thorns are modified stipules; they are swollen at the base and white with black tips. The leaves are bipinnate with up to seven pairs of pinnae, each with 20 pairs of rectangular leaflets. The sweetly scented flowers form golden yellow balls, each is very small, has a small cup-shaped calyx, numerous long stamens and is rich in nectar and pollen. The woody pods are slender and sickle-shaped and are constricted between the 5-10 seeds. The wood provides good timber.

Mimusops caffra (Sapotaceae) (Fig. 10.3 a-b), the coastal red-milkwood or 'ntole', is common in regenerating areas and produces a milky latex from all parts. It forms a small bushy tree, 5 m high when growing just above high tide, but grows to 10 m inland. The tree contributes to the consolidation of the dunes. The bark is dark grey and fissured longitudinally. The hard and durable wood is close-grained and was used for making boats, so that only fairly small, young trees remain today. The leaves are stiff and leathery with light silky hairs on the under surface; the base tapers, the apex is rounded and the margins are inrolled beneath. The star-like flowers have white petals and rust-coloured hairs on the sepals. *Mimusops* is well-known for its tasty edible fruits that are grey-green when young and orange-red when ripe; they also make good jam.

Rhus natalensis (Anacardiaceae), the Natal rhus or 'mbalamuno', is a scrambling bush on the western coastline. It also grows in the northern part of the island, is widespread along the Indian Ocean coast from the Ciskei in the south to Beira, 1 000 km north of Inhaca Island and also grows in Zambia and Zimbabwe. The leaves all have three leaflets in which the terminal leaflet is 1,5 times the size of the others and is obovate. They may be light or dark green and almost membranous in texture; the margin is slightly scalloped or toothed. The small, greenish yellow flowers are produced in small sprays among the leaves.

Where the canopy is thinner and more light reaches the undergrowth, Bermuda grass and maritime sedge are dominant. Different herbs, such as *Scadoxus puniceus* and various creeping species of *Commelina*, occur in the darker patches, but the herb layer is *less diverse* than that encountered in the Barreira Vermelha scrub forest (Section 10.1.1).

Scadoxus puniceus (=*Haemanthus*) *puniceus*, Amaryllidaceae grows best in deep shade in forest, forest margins or the darker patches of vegetation near high water mark at the base of the dunes. The bulb has an elongated neck from which grow membranous, rectangular to elliptical leaves with pink-dotted sheaths at their bases. Broad, deep crimson bracts at the tip of the peduncle sheath the numerous, beautiful, orange-red flowers. The many glossy, dark red berries are much sought by birds.

In clearings amongst the trees, grasses such as *Cynodon dactylon* and *Imperata cylindrica* occur. *Cynodon dactylon* (Graminae/Poaceae), Bermuda or quick grass, is a perennial and has strong branched stolons and a well-developed root system. The stolons break off easily and rapidly produce new shoots at the nodes so that it easily colonises sandy places. It is common all over the world (except in northern Europe).

Imperata cylindrica (Graminae/Poaceae) (imperial grass; Fig. 10.5 g) is a tufted, perennial grass; new tufts are formed at the tips of underground stolons. The stiff leaves sometimes have sharp tips. The spicate inflorescences have many conspicuous, long, soft, white hairs. It is used as thatching for huts.

Cyperus crassipes (=*C. maritimus*, Cyperaceae) (Fig. 10.3 e-f), the maritime sedge, also has strong stolons, but the root system is less profuse. It is a deciduous perennial with basal greyish green, slightly succulent leaves. In summer it produces a single inflorescence on a stiff peduncle and numerous very small seeds are formed and dispersed by the wind. It is thus an important pioneer and a sand stabilizer.

Three species of *Commelina* (Commelinaceae) (Fig. 10.3 c-d) occur, all of which are herbaceous creepers that root at the nodes. Their leaves and stems are slightly succulent. Flowering branches are erect and the species differ mainly in the size and colour of the flowers.

The vegetation is not stable in this area, although it is possible for scrub forest to develop here as well as on the dunes near Barreira Vermelha (see Section 10.1.1).

10.1.3 Open evergreen bushland

A slightly different, more open type of vegetation occurs towards Ponta Raza. Although *Mimusops caffra* (see Section 10.1.1 and Fig. 10.3 a-b) and *Rhus natalensis* (see Section 10.1.2) are still dominant in the shrub layer and are slightly taller and interspersed with other species. Shrubs, such as *Chrysanthemoides monilifera* (see section 10.1.2), are less obvious and the fern *Microsorium scolopendrium* (see Section 10.1.1 and Fig. 10.2 d-e), appears in the more luxuriant undergrowth, as in the undisturbed scrub forest. On the first, most shoreward, small ridge a somewhat thicker humic layer has developed, except in places where fishermen have cut down trees or shrubs to make stakes for hanging fish nets. Because the large plants are more widely spaced and the area is less shaded, a *diverse herbaceous layer* and *climbers* occur.

Roscher's vanilla, *Vanilla roscheri* (Orchidaceae) (Fig. 10.3 g-i), seen in the scrub forest, roots in the soil and then climbs into the branches of *Strychnos spinosa* (see Section 10.5.1 and Fig. 10.6 i-j), the small-fruited olax, *Olax dissitiflora* (Olacaceae) and other trees. The long stems of *Vanilla* are bright brown, long and not thickened; they have many horizontal branches in close contact with the humic layer and numerous vertical branches that arise from the nodes and lie parallel and close to one another, along the supporting tree trunk. These dark green, succulent branches appear as two fused stems, each 10-20 mm in diameter, and are photosynthetic. The leaves are terete (round in cross-section) and look like short branches of the stem. The roots are adventitious; those from the lower nodes touch the ground, those from the upper nodes are 30-100 mm long and attach the orchid to the host plant. At about 1,5 m to 2 m from the ground the branches may bend downwards and only then are flowers produced. Numerous white, fragrant flowers are formed on each of many inflorescences (a beautiful sight), but each flower opens only for a single day. Capsules take several months to ripen and burst open to reveal numerous tiny black seeds embedded in white pulp (with the characteristic smell of vanilla).

Terrestrial orchids are also present, as in the undisturbed forest, and grow in places less than 1 m above the high levels of spring tides. The more common *Eulophia speciosa* (see Section 10.1.1) may be seen as well as the less common, violet-flowered orchid, *Bonatea steudneri*. Both have thick, succulent leaves. Numerous white roots are produced horizontally from short thick rhizomes and are covered by a velamen, (i.e. a sheath modified for absorbing moisture from the air). In contrast to the roots of *Vanilla*, these grow in the coarse sand below and not in the humic layer.

10.1.4 Vegetation on dunes with fresh water influence near the Ponta Raza mangrove

Near the creek mangrove between the Marine Biological Station and Ponta Raza which lies in the valley between the low outer and higher inner dune ridges, the water-berry, *Syzygium cordatum* (see Section 10.4.1 and Fig. 10.6 d-e), is common. This tree grows on the banks of streams and rivers; here it indicates that a good supply of fresh ground water is draining from the dunes and is reached by very deep roots. The trees grow to 12 m in height and there are open spaces around them. Fairly young trees of *Hibiscus tiliaceus* (Malvaceae), an estuarine tree, were very common on the rising land at the sides of the mangrove swamp in 1990. This deciduous tree may grow to 6 m and becomes a spreading, shady tree with large, round leaves and yellow flowers.

Evergreen bushland may be found on the outer dune beside the mangrove and, here, is dominated by *Brachylaena discolor* (Compositae), the wild silver oak, and includes *Mimusops caffra* (see Section 10.1.2 and Fig. 10.3 a-b), *Sideroxylon inerme*, the white milkwood, *Acacia karoo* (see Section 10.1.2 and Fig. 10.2 n-o) and *Phoenix reclinata* (see Section 10.1.2 and Fig. 10.2 k-m).

A variety of *climbers* grow into the canopies on these dunes. Amongst them are two species of the family Apocynaceae: *Landolphia kirkii* and *Ancylobotrys petersiana*. These *woody climbers* have branched tendrils; dark brown stems with small, prominent, bright lenticels and white, sticky latex. The white flowers are fragrant and have a corolla tube with radiating lobes. Their globular fruits have hard yellow shells, edible pulp when mature and contain numerous brown, white-streaked seeds. Both climbers are widely distributed in tropical forest, bushland and thicket from QwaZulu in South Africa to Somalia. *Landolphia kirkii* has leaves 90 x 30 mm and closed buds up to 7 mm long. *Ancylobotrys petersiana* has larger leaves (130 x 65 mm) and longer closed buds (up to 30 mm long), and also has sweeter, much sought after fruits. *Secamone filiformis* (Asclepiadaceae) is another evergreen twining woody climber with abundant latex. *Sarcostemma viminale* (also Asclepiadaceae) is common here, but differs from *Secamone* by having thicker stems, no leaves and much larger flowers.

Several species of *Asparagus* (Liliaceae *sens. lat.*) occur here. They have evergreen twining stems and modified shoots (cladodes) that photosynthesise instead of the leaves (which are modified into thorns). The African species of *Asparagus* were previously grouped in the genus *Protasparagus*. *Asparagus* has bisexual flowers, 4-12 ovules and globose seeds. *A. setaceus* has terete (rounded in cross-section) cladodes and no thorns; *A. africanus* has almost terete cladodes and black thorns. The cladodes of both *A. densiflorus* and *A. falcatus* are flattened, but *A. falcatus* has larger cladodes, larger recurved thorns and a more robust habit. The short, woody, fibrous rhizomes bear many long greyish brown roots, which grow horizontally in the humic layer.

Among the shrubs present on these dunes near the mangrove, *Euclea natalensis* ssp. *rotundifolia* (see Section 10.5.3); *Sideroxylon inerme* and *Eugenia capensis* (see Section 10.5.1) are present. They are all well-branched shrubs or trees with thick leaves and are valuable hardwoods.

Sideroxylon inerme (Sapotaceae) (Fig. 10.3 j), the white milkwood or 'ntangendi', is a large, shady evergreen tree which occurs mainly in the undisturbed dune forest, and grows to 10 m here. All parts produce a white milky latex. The branches are often twisted and young branches are covered with hairs. The flowers are arranged in clusters on older wood, as in many trees of this family, or in the leaf axils. *Sideroxylon* is distinguished from other members of the Sapotaceae, such as the species of *Mimusops*, by its floral parts being in four whorls of five to six members, rather than in five to six whorls of three to four members. The spherical, fleshy fruit is purplish black when mature and contains a very sticky milky latex. The hard and durable wood is used for construction and various parts of the tree are used in local medicine.

10.1.5 Mature trees on disturbed dunes at Ponta Punduine

South of Ponta Raza, the land curves eastward towards the southwest point, Ponta Punduine and the dunes become slightly higher, reaching 15 m at the point. The dune vegetation is similar to that north of the Marine Biological Station (see Section 10.1.1), but is less closed than that on the steep slopes near Barreira Vermelha. The effects of cultivation and of goat-keeping are conspicuous. On the top of the vertical cliff at Ponta Punduine, the vegetation in one patch only appears closed, because two arborescent species of *Euphorbia* grow in groups.

The tidal river euphorbia or 'hlohlo', *E. triangularis*, is a coastal species confined to the littoral scrub between Port Elizabeth in South Africa and Inhambane 100 km to the north of Inhaca Island. The architecture of this shrub is unusual: the cylindrical trunk has two or more branches, each bearing a rounded crown of upwardly curving branchlets, giving the appearance of a candelabra. Each branchlet is triangular in cross section or has 4 or 5 angles and is regularly constricted forming segments up to 300 mm long. The undulating margins of the segments bear tubercles with small thorns. The flower-like inflorescences are cup-shaped;

Fig. 10.3 Species of bushland (drawn by B. Pike):

a. *Mimusops caffra* branch with leaves and fruit;
b. flower;
c. *Commelina* sp. tip of shoot;
d. flower;
e. *Cyperus crassipes* rosette and underground stolons;
f. inflorescence;
g. *Vanilla roscheri* stem, leaves and aerial roots;
h. flower;
i. fruits;
j. *Sideroxylon inerme* stem, leaves and fruits.

the three-lobed capsule has a distinctly curved stalk. The other species, *E. tirucalli*, is described in section 10.1.1 (Fig. 10.2 f-j). Both species have a toxic milky latex. Young euphorbia plants are rarely seen, possibly owing to the disturbance caused when firewood is gathered.

A group of large, deciduous tropical trees has recently been discovered on the dune on the west coastline of the southern bay, a few kilometres from the Saco at its head. The pod mahogany or 'hlapfuta', *Afzelia quanzensis* (Leguminosae – Caesalpinoideae) is near the southern limit of its distribution at Inhaca Island and grows only 15 m high. It does occur further south in Maputaland but in the tropics it may be 35 m high in dry forest. These trees have a widely spreading crown, 20-25 m in diameter, and they grow in orange-coloured loamy sand, lightly covered with a 10-15 mm layer of fallen leaves. The bark is greyish brown and flakes off leaving large pale patches; the branches are twisted. The leaves have 2-5 pairs of leaflets and a terminal leaflet; the leathery, shiny leaflets are almost rectangular in shape and measure 90 x 60 mm. The red flowers are arranged in simple sprays; each flower has one large clawed petal with lighter veins; the four sepals are unequal. The hard, woody pods are dark brown and glossy; inside, the black seeds (lucky beans) are embedded in a white pith and partly surrounded by a scarlet or an orange aril. Birds are responsible for pollination and seed dispersal. Few small pods with few seeds are produced at Inhaca, so that reproduction is not efficient here. This tree was once used by men on the island to make dugout canoes for fishing.

10.2 THE EASTERN DUNES AND THE SOUTH-EASTERN PENINSULA

On much of the east coast of southern Africa, gradients in certain environmental factors have very marked influence on coastal vegetation, especially on dunes. These factors include: (a) high reflectivity of sand; (b) water stress; (c) sand abrasion; (d) sand deposition; (e) salt spray; (f) wind pruning; and (g) lack of humic layer in the soil. Many of these factors are driven by the strong prevailing south-east wind, which is largely responsible for building the dunes of this coast. The waves deposit sand on the beaches; once the sand is dry, it is blown by the wind until it reaches an area protected from the wind (amongst vegetation), until it is wet again or until it is covered by other sand. When the sea level is rising, the sea periodically erodes the base of a dune while the wind periodically deposits sand in the vegetation at its crest; this process produces a very steep slope, as is seen at Inhaca. The physical slope of the dune can determine the intensity of the gradient in environmental factors. In many areas on the African coast (Tinley 1985) there are undulating dunes, parallel to the sea, which gradually *increase* in height as one moves away from the sea – which is not the case on Inhaca Island. The vegetation nearest the beach would comprise sand-binding plants and behind this there would be wind-clipped littoral thicket, which, in turn, would be replaced by dune forest. The thicket and forest usually differ in the *structure* of the vegetation rather than in species composition.

The *outer* eastern dune at Inhaca Island is, however, very high (115 m above sea level at Monte Inhaca) and very steep; thus the gradient in environmental factors is also extremely steep. At the base of the outer dune, pioneer plants (see Section 10.2.1), such as *Scaevola* and *Sophora* grow and above this there is 50-80 m of bare sand. On the eastern dunes at Inhaca Island, the intermediate thicket and bushland are represented only by a relatively narrow band of thicket on the crest of the dunes; in one area it is continuous and impenetrable for 500 m. In the lee of the crest there is evergreen dune forest (Section 10.2.3). The south-easterly wind blows through the strait at Ponta Torres and thus the thicket that would be found on dunes with a gentler gradient (elsewhere on the east African coast) occurs on the eastern coast of the southern bay, where the dunes gradually *decrease* in height. Here, and in the region of the lighthouse, thicket is found. Vegetation on the dunes south west of the lighthouse is recovering from disturbance and superficially looks like the thicket but is in fact rejuvenating forest.

10.2.1 Pioneer sand-binding plants at Ponta Torres

Near the beach, conditions are harsh. There is little shading from the sun and the white sand reflects light and heat on to the lower surfaces of the leaves. The sand drains very quickly and the salt on the leaves and in the sand produces an osmotic gradient that causes water stress in plants. The sand that blows in the wind is very abrasive and damages soft tissues; when the wind is slowed down (usually by vegetation), the sand is dropped and buries the vegetation, but when the wind is strong, it might blow the sand away again. Plants that grow in this area show adaptations for these conditions. Leaves may be succulent with thick cuticles or dense hairs; often they have dense white hairs on the lower surface. The growing tips of the stems are protected. Many of the plants grow very fast and have vertical stems which grow above the depositing sand and horizontal stems with adventitious roots at many nodes so that they spread and anchor over a wide area and are thus less susceptible to the shifting of the sand. Pollination is by wind or by strongly flying or heavy insects such as bees, wasps and beetles. Seed dispersal is usually by wind, water or by birds, after passing through the gut (endozoochory).

One of the first plants to settle just above the high water level of spring tides is strand grass or 'txangi', *Sporobolus virginicus* (Poaceae) (Fig. 10.5 a-b), which has extremely hard, yellow rhizomes with large, protective sheaths. The bamboo-like rhizomes spread over the bare sand and produce erect, leafy branches and fibrous adventitious roots at each node.

Ipomoea brasiliensis (=*I. pes-caprae*) (Convolvulaceae) (Fig. 10.4 a-c), the sweet goat's-foot or 'nkulula', is a perennial that has large, bilobed, leathery leaves on very long prostrate branches with very long, branched roots at few nodes. The strong network of branches and the large leaves protect the sand from removal by wind and encourage deposition by locally slowing the wind. The purplish-red flowers are pollinated by bees; the hard, brown, light seeds are distributed by being rolled on the beach by the wind and by ocean currents; they germinate after being stranded above the high water mark.

An important evergreen succulent, *Scaevola plumieri* (Goodeniaceae) (Fig. 10.4 d-f) has an extensive underground stem system and numerous unbranched, erect stems. The sand forms rounded mounds around *Scaevola* just above high water mark; *Scaevola* does not grow well unless the surrounding soil level continually rises. The large, almost circular succulent leaves have a thick cuticle and are closely pressed around the growing tip of the stem. The stomata respond directly to humidity and the plant is tolerant of salt and spray (Pammenter 1985). The roots do not have a salt exclusion mechanism which may be the cause of an increase in succulence in the older leaves (Harte and Pammenter 1983). Nitrogen acquisition is very efficient and conservative mineral nutrition involves the internal recycling of nutrients that are present in the soil at very low levels. The white flowers are pollinated by bees. The spherical fruits have a soft fleshy outer layer and a hard woody inner layer; it is possible that they are dispersed first by endozoochory and then by wind or ocean currents.

Phyllohydrax carnosa (Rubiaceae) has extensive rhizomes with adventitious roots and it forms dense mats on flat mounds of sand. The opposite, succulent leaves are fused to the interpetiolar stipules to form a sheath, which protects the growing tip of the stem. The white flowers have four triangular corolla lobes and are heterostylous, which enhances outcrossing.

The coastal bean bush, *Sophora inhambanensis* (Leguminosae) (Fig. 10.4 g-i) is another pioneer and a precursor of coastal thicket. This evergreen shrub grows to 2,5 m high and the young branches, the leaves and the calyces are all densely covered with silky, silvery or golden hairs. The pinnate leaves have an odd terminal leaflet; all leaflets are elliptical. The beautiful, sweetpea-like flowers are produced in sprays and are pollinated by bees. The cylindrical pods are about 100 mm long and narrowly constricted between the hard, light brown seeds.

Gazania livingstonia (Compositae/Asteraceae) forms large clumps with strong, branched stolons and rhizomes. The numerous small, simple roots, grow vertically into the sand. The lower surface of the succulent leaves is covered with small, white hairs, except on the prominent midrib. The yellow rays and disc florets attract bees, beetles and wasps and the pappus on the small dry fruits allows wind dispersal.

Carpobrotus junodii (Mesembryanthemaceae), hottentot's fig or 'xizengani', grows at the base of the seaward dunes and has long prostrate stems that seldom branch, but have many adventitious roots. The succulent leaves of this perennial are triangular in cross-section; each pair is closely adpressed and thus protects the growing tip of the stem. The magenta flowers have many petal-like staminodes and many stamens and are visited and pollinated by wasps, bees and beetles. The fleshy fruits are edible. Some members of the genus have been cultivated as sand stabilisers, but they do not do well on road cuttings unless they periodically have new soil spread over them, as is the case here.

Cynanchum ellipticum (Asclepiadaceae), a poisonous milkweed, produces many greyish green branches, which spiral around other vegetation. The underground tubers are globose or ellipsoid and may be up to 300 mm long. The flowers are white and pollinated by bees. The dry brown follicles release flattened seeds with a parachute of fluffy white hairs for wind dispersal.

Higher up the dune face, the creepers become intertwined with the shrubs and trap dry leaves of other plants. The parasite, *Cassytha filiformis* (see Section 10.5.1 and Fig. 10.6 h) forms a network of intertwining string-like branches on the herbs and shrubs of the dune. The haustoria suck sap from the living leaves and twigs of host plants.

10.2.2 Evergreen bushland and thicket: exposed wind-clipped trees; sheltered regenerating trees; wind-trimmed thicket; regenerating disturbed forest on the inland ridges

The most influential environmental factors in this littoral vegetation are the deposition of sand and wind-pruning. The wind blowing over this vegetation carries salt and a small amount of sand. As the wind blows past leaves, it removes any water vapour that has evaporated from the leaves and thus enhances transpiration. When the wind blows over a canopy, the drag of the canopy slows the air closest to the canopy. This, and collision with the leaves will lead to the deposition of sand carried by the wind and to the amelioration of the effects of the wind within and close to the canopy, whilst the effects just above the canopy will still be strong. New shoots growing within the canopy will be protected, while those formed above the canopy will be sand-blasted and dried by the deposition of salt and the enhanced transpiration. This 'clips' the vegetation and, with the deposition of sand, produces the low, dense, spreading canopies of the trees and shrubs in the thicket. The upper surface of the canopy follows the wind, rather than the relief of the dunes below.

On the crest of the outer eastern dune, the vegetation is dominated by *Mimusops caffra* (see Section 10.1.2 and Fig. 10.3 a-b), which grows only about 0,8 m tall here. The well-developed trunk sprawls over the sand and produces many branches, which in turn produce numerous erect branchlets. The leaves are rigid and are borne at the top of the branchlets. This vegetation may be impenetrable for as much as 500 m along the dune. *Diospyros rotundifolia* and *Sideroxylon inerme* (see Section 10.1.4 and Fig. 10.3 j) are stunted and shrubby, with densely branched crowns that show the marked effect of salty winds.

There is more protection on the east coast of the southern bay, below the eastern ridge. Here, the lowest dunes are 2-8 m high and are separated from the shoreline by sandy areas or outcrops of low calcareous rocks. Under the dense 8-10 m high growth on these dune slopes, the soil is covered with decomposing leaves and a thick layer of humus has formed. *Microsorium scolopendrium* (see Section 10.1.1 and Fig. 10.2 d-e) may be seen very often with

Fig. 10.4 Pioneer sand-binding species (a-i) and species of the thicket and bushland on the eastern dunes (j-l) (drawn by B. Pike):

a. *Ipomoea brasiliensis* stem and leaves;
b. flowers;
c. inflorescence with fruit;
d. *Scaevola plumieri* tip of upright shoot;
e. inflorescence with flower and young fruit;
f. front view of corolla;
g. *Sophora inhambanensis* leaf;
h. inflorescence with flower, buds and young pod;
i. mature pod;
j. *Vepris lanceolata* leaf;
k. inflorescence with buds;
l. inflorescence with fruits.

TERRESTRIAL VEGETATION 295

rhizomes fixed in the thick humus layer as in the undisturbed forest on the west coastal dunes. Much regeneration occurs here. *Mimusops caffra* and *Sideroxylon inerme* form dense thickets and *Diospyros rotundifolia* is common here. *Strychnos henningsii* (see Section 10.1.1) is present, but rare.

The tropical dune jackal-berry or 'mbenti', *Diospyros rotundifolia* (Ebenaceae), often occurs on the outer fringes of the vegetation on the lower ridges and grows as a 6 m high, much-branched tree. The leathery leaves are broadly oval, olive green above and paler below; the margins are rolled under. The unisexual flowers occur in axils of leaves on separate trees. The round fruits are about 10 mm in diameter and clasped by the persistent calyx lobes, which form a shallow cup. They mature through red to dark purple, as the calyx lobes curl back. Inhaca Island is the southern limit of *D. rotundifolia*, which is found in coastal thicket northwards to Vilancoulos in central Mozambique.

Vepris lanceolata (=*V. undulata*, Rutaceae) (Fig. 10.4 j-l), the white ironwood or 'natani', is a common evergreen shrub in coastal thicket. The bark is grey and smooth. The trifoliate leaves have glands that produce lemon-scented oil; they are bright green above and yellowish below; the margins are undulate. The greenish-yellow flowers are produced in dense terminal inflorescences. The fruit matures to orange-red. *V. reflexa*, the bastard ironwood, is very rare at Inhaca. The leathery, gland-dotted leaves are shiny green and make the tree conspicuous. It may be distinguished from *V. lanceolata* by leaflets having flat margins. The Natal teclea, *Teclea natalensis* (also Rutaceae), which is similar to *Vepris*, but has a yellowish brown flaking (not a smooth grey) bark, also occurs here.

Along the highest dune ridge near the lighthouse (Fig. 10.1), the wind-clipped vegetation appears still lower, because the branches of *Mimusops caffra* and other trees are more rapidly covered with sand; the fully developed trunks and much of the crowns are buried and only the tips of the branches are visible.

The Zululand cycad, *Encephalartos ferox* (Zamiaceae), occurs in the coastal thicket and also has a buried trunk. *E. ferox* can be distinguished from the other species in southern Mozambique by the ovate leaflets with spinous margins. New rosettes of leaves are produced at regular intervals. The male plants produce one to three cones; the scales have hexagonal faces with a triangular deflexed head. Female plants produce single ovoid-ellipsoid cones that are 350 x 150 mm and turn orange-red. Their scales have dorsally flattened heads and have two long processes extending towards the cone axis. The seeds are bright vermilion and take several months to germinate. Besides the numerous, horizontal, branching roots, masses of white, coralloid (clustered and upward-pointing) roots are produced around the trunk.

To the west of the lighthouse (see Fig. 10.1), the vegetation on the gentle slopes of the higher eastern inland ridges is recovering from decimation, which ceased some 25 years ago. The tangled nature of the vegetation makes it similar to the wind-trimmed thicket, but as the canopy is not 'clipped', it is thicket that may regenerate to forest. Variation in the depth of the humus layer and the intensity of light at soil level determines the success of regenerating seedlings. Shrubs varied between 1 m and 5 m in height in the late 1980's.

The soil is covered with dead leaves, twigs and branches and the humic layer in this thicket is about 10 mm thick on ridges and up to 50 mm thick in valleys. The herbal layer consists mainly of the creeping fern, *Microsorium scolopendrium*. In the midst of the ferns, the thick leathery leaves and erect inflorescences of *Sansevieria hyacinthoides* (see Section 10.1.1 and Fig. 10.2 a-c and d-e) are seen. Species of *Asparagus* climb into the canopy; *A. falcatus* (see Section 10.1.4) is most common.

The most common shrub is the num-num, Y-thorned oleander or 'kekani', *Carissa bispinosa* var. *bispinosa* (Apocynaceae) (Fig. 10.5 c-e). The bark of this shrub or small tree is grey; the stout Y-shaped spines may be forked once or twice and are up to 50 mm long. The opposite leaves are broad, oval, green and about 30 mm long; the margins are rolled under. The white flowers are fragrant, white or bright pink and grow in terminal clusters; they are pollinated by moths. The ovoid fruit matures to glossy red and is edible. All parts of the plant contain milky latex.

The showy ochna or Natal plane tree, *Ochna natalitia* (Ochnaceae) may grow up to 8 m tall on Inhaca Island, and grows at the *margins* of evergreen forest. This much-branched, bushy shrub or tree has grey, fissured bark, that flakes off to reveal smooth, reddish bark beneath. The branches appear purple and have bright lenticels. The young leaves have a metallic copper-coloured sheen; mature leaves are green, oval and taper to the tip and strongly to the base; the margins are slightly toothed. The very attractive yellow flowers have conspicuous orange stamens and are arranged in terminal or axillary panicles. Two or three separate carpels develop into hard, shiny black drupelets with single seeds, which contrast with the red, inflated receptacle and persistent sepals.

The high eastern dunes protect this area so that *Mimusops caffra* (see Section 10.1.2 and Fig. 10.3 a-b) and *Sideroxylon inerme* (see Section 10.1.4 and Fig. 10.3 j) grow into tall trees here. *Acacia karoo* (see Section 10.1.2 and Fig. 10.2 n-o) also grows into a large, tall tree here, but *Vepris lanceolata* remains a shrub.

Another characteristic tree in regenerating forest is the dune myrrh, *Commiphora schlechteri* (Burseraceae), which is at the southern limit of its distribution. This spineless commiphora grows to 6 m high and has smooth grey-green bark, which peels off in papery flakes. The three pairs and terminal leaflets are narrow and toothed. The small whitish flowers are in sprays or loose heads. The narrow, oval fruit has a stone surrounded by a red, four-armed pseudo-aril.

10.2.3 Primary evergreen dune forest: wind resistant species on steep slopes; undisturbed dune forest

To the west of the crest of the outer eastern dune the effects of the easterly winds are lessened and the canopies of trees are not clipped. The topography determines whether there is a transition zone between the thicket and forest and how broad the zone is (see Fig. 10.1). *Diospyros rotundifolia* (see Section 10.2.2) and *Sideroxylon inerme* (see Section 10.1.4 and Fig. 10.3 j), flourish in the transition zone, but are wind-clipped and thus smaller, shrubbier and with tighter crowns than usual.

Forests are characterised by having closed canopies and much leaf litter on the floor (Halle *et al.* 1978). Many trees have thin, smooth bark (especially if slow-growing), fluted trunks and buttress roots. Many species have large leaves (about 100 mm long), with entire margins, a leathery texture and often with drip tips (especially while the trees are young). Many of the trees display distichous phyllotaxy with all the leaves in a horizontal plane. Many species display cauliflory and ramiflory (they produce their flowers on the trunk and branches). The roots of many species are horizontal and form buttresses at the base of the stem, while many lianes and stranglers have aerial roots. In the forest, plants which flower below the canopy are protected from the wind and they are pollinated by slow-flying insects like moths. The canopy provides a suitable habitat for sunbirds, which are involved in the pollination of some species. Much seed dispersal is by means of endozoochory (having been eaten and dropped by birds).

Primary evergreen dune forest (apparently little disturbed by man's activities) is found at the higher altitudes at Inhaca Island. It extends from just over 1 km from the tip of Ponta Torres for about 7 km northwards on ridges and in valleys until it merges with formerly cultivated ground and regenerating thicket on the west and to the north (Fig. 16.1).

Although there are several species in common with thicket, the growth pattern of the vegetation consists of remarkably, regularly spaced, tall trees, with narrow tree trunks, forming a stable, uniform stand, interrupted here and there, by older, taller trees with greyish trunks; these are mostly *Mimusops caffra* (see Section 10.1.2 and Fig. 10.3 a-b) and the very common hardwood, *Apodytes dimidiata* ssp. *dimidiata* (see Fig. 10.5f).

On the steepest slopes (about 60°) of the dune ridges near the southern end of the peninsula, the vegetation is closed as in the other parts of the forest, but the species are wind-resistant and mark the transition from thicket. These are the wild fig, *Ficus thonningii*; the white milkwood, *Sideroxylon inerme* (Sapotaceae, see Section 10.1.4 and Fig. 10.3j); 'wildekornoelie', *Kraussia floribunda* (Rubiaceae); the mitzeerie, *Bridelia micrantha* (Euphorbiaceae); the dune jackal berry, *Diospyros rotundifolia* (Ebenaceae) and the wild date palm *Phoenix reclinata* (see Section 10.1.1 and Fig. 10.2 k-m).

The seeds of the wild fig or strangling fig, *Ficus thonningii* (Moraceae), may germinate on the ground and form a free-standing tree 20 m tall. Often they germinate on the branches of other trees and grow thin aerial roots that eventually form a woody network around the trunk of the host tree. The parasite is supported by the host until it rots after being strangled; then the network of new roots can usually support the leafy crown. The deciduous leaves are elliptical with square bases and tapering tips. The figs are about 10 mm long and are borne in leaf axils; pollination is effected by highly specialised wasps. Ripe figs are eaten by birds such as barbets and thus seeds are distributed widely over the island.

At the forest margin, *Kraussia floribunda* is common. This small deciduous tree has branches that grow perpendicularly to the main stem. The opposite leaves have bacterial nodules in pockets in the axils of the lateral veins; the hairy interpetiolar stipules are lost early. The creamy white flowers are 10 mm in diameter, produced on long stalks and arranged in cymes. The stamens occur in the wide mouth of the flower and protrude for about 10 mm. The round black fruits are edible and have a sweet flavour. This tree grows inland in southern Africa, and in dune scrub and forest; it occurs from Inhambane to Port Elizabeth.

Bridelia micrantha, which also occurs on the western coastal dunes, is a small tree rarely found in open woodland, although it occurs near the coastline in the Transkei and spreads northwards to eastern Zimbabwe. The lateral veins of the oval leaves run straight to the margins without branching. The very small, yellowish flowers are clustered in the leaf axils. The small, sweet, blackish fruits are edible and produced early in December and January. The leaf sap is used to soothe sore eyes and the roots are powdered and mixed with fat to cure gastric pains.

The wild date palm, *Phoenix reclinata* (Fig. 10.2 k-m), has trunks 8 m high in this forest (cf. Section 10.1.1, underground stems). The feathery leaves are cut and used as effective brooms.

In the primary forest, the sandy soil is covered with a uniform humus layer, 10-20 mm thick, which comprises dry, decomposing leaves and twigs. Although the trees are 8-12 m tall and form a *closed canopy*, there is only *one storey* in the canopy of the forest and thus the forest floor is well lit. There are *few shrubs* on the forest floor and there are *few lianas*, but impenetrable tangles develop where fallen trees leave a gap in the canopy.

The forest floor is covered by a thick, closed layer of the fern, *Microsorium scolopendrium* (see Section 10.1.1 and Fig. 10.2 d-e), just as in the undisturbed forest on the western coastal dunes. This may be the cause of slow regeneration of forest species, in spite of the good humic layer, because of inhibition of germination.

The herbaceous layer of this primary dune forest has the fern and the other herbs present in the scrub forest (see Section 10.1.1), as well as *Dietes iridoides*. This evergreen iris (Iridaceae) has erect, bluish green leaves and a simple erect inflorescence of white flowers.

A woody climber, *Synaptolepis kirkii* (Thymelaeaceae) creeps into the canopy. The main stem has striking lenticels; in the undergrowth, opposite, small, entire leaves are produced on horizontal branches. The small, red drupes appear amongst the leaves.

The trees and shrubs in the forest include many species from the thicket: *Mimusops caffra*, *Carissa bispinosa* (see section 10.2.2; Figs. 10.3 a-b and 10.5 c-e), *Syzygium cordatum* (see Section 10.4.1 and Fig. 10.6 d-e) and *Garcinia livingstonei* (see Section 10.4.1 and Fig. 10.6 a-c). The last two are the fruit trees that are always retained on land cleared for cultivation (see Section 10.5.2).

Some trees and shrubs are now found *only* in the primary forest at Inhaca Island. The white pear or 'tsomaháte', *Apodytes dimidiata* ssp. *dimidiata* (Icacinaceae) (Fig. 10.5 f), a hardwood shrub or tree, grows to 12 m tall. It is easily recognised by its wavy trunk on which several species of lichen grow. The crown is well-branched and the alternate leaves are bright green and glossy above and paler with a yellowish midrib (tinged with pink) below. The small flowers are white and the fleshy fruits are black with red lateral appendages. This tree is the tallest and probably the oldest in the area. It is found in coastal bush, mountain woodland and grassland and is widespread in Africa from Ethiopia to the Cape in South Africa. Other characteristic trees of the evergreen dune forest are described below.

Two trees, *Psydrax locuples* and *P. obovata*, were formerly included in the well-known genus *Canthium* (Rubiaceae). Both species are shrubs in this forest, the former growing to 5 m only and the latter to 10 m. The bark in the genus is grey and rough. The leaves are opposite and the veins are prominent below; *Psydrax* means blister and both species have blister-like swellings in the axils of the secondary veins. The leaves of *P. obovata* are oval, those of *P. locuples*, the whipstick, elliptical. Flowers of both are small and greenish white; the fruits are black.

Psychotria capensis (Rubiaceae) has small cavities in the axils of some of the secondary veins below; these might house predatory mites. The wood is tough and hard and is used as timber in the construction of huts.

Diospyros inhacensis (Ebenaceae), the Zulu jackal berry, occurs alone or in small groups in this forest. It grows 12 m tall and has a straight trunk and small crown. The leathery leaves are arranged in one plane and dry blackish. The narrowly ellipsoidal fruit is supported by three calyx lobes which distinguishes it from other species of *Diospyros* in the eastern thicket.

In the canopies of some trees, the semi-parasitic shrub, *Erianthemum dregei* (=*Loranthus dregei*, Loranthaceae) can be seen. It is not host-specific and usually forms a denser area in the crown of the host. Plants grow to about 1 m tall and have thick, leathery and hairy leaves. Flowers are borne in clusters and are densely covered with white hairs. The tips of the corollas open into a cage; when a bird puts its beak through the cage to get to the nectar at the base of the corolla tube, the flower opens explosively and throws pollen on to the forehead of the bird, thus ensuring cross-pollination. The berries have large sticky seeds; birds eat the flesh and then wipe off the seeds, which stick to their bills, on to branches, where they germinate.

10.3 FRESHWATER SWAMPS

About 17,5% of the total area of the island, i.e. about 7 km^2, was covered by freshwater swamps in the 1920s. One large swamp about 5 km long and 1 km wide stretched across the northern part of the central area with long arms penetrating the shallow valleys between low dunes to the south of the main body of water. More than ten much smaller swamps occupied low ground between small dunes in the southwest peninsula (Fig. 10.1). In the 70 years since the island was first studied, a noticeable reduction of area has taken place in the largest swamp because it has been partially drained for agriculture (see Section 10.4), some of the smaller swamps have been reduced to water holes with no vegetation and a few small swamps have become dry.

In the freshwater swamps that are still viable, the species composition of the plants is characteristic, but not diverse. The centre of the swamps may still be open water or may be completely covered by the reed community that was originally around the margin (Noel 1959). There is very little floating or submerged flora. Outside the reeds a zone of ferns and erect perennial herbs formerly extended towards ditches that are normally dug around adjacent cultivated plots. Swamps today have a 2-3 m wide zone of land devoted to horticulture; vegetables grow well (see Section 10.4). The large swamp and some of the smaller swamps are adjacent to the mangrove swamps, but are separated from them by a salt marsh.

10.3.1 Emergent vegetation

Three species, the common reed or 'hlanga', *Phragmites australis* (=*P. communis*, Poaceae); the bulrush or 'papala', *Typha latifolia* ssp. *capensis* (=*T. australis*, Typhaceae) and papyrus, *Cyperus papyrus* (Cyperaceae), dominate the emergent vegetation. *Typha* and *Phragmites* occur alone or in combination in most swamps, but *Cyperus* occurs in large patches in only Mudalandala swamp in the south-west peninsula.

The three species grow to a height of 4-6 m and have long wand-like stems growing from submerged rhizomes. They may be distinguished by their inflorescences. In *Phragmites* the inflorescence is lax and composed of many-flowered hairy spikelets. In *Typha* the minute flowers are condensed on a terminal, dark, velvet-like rod about 200 mm long, which stands out above the long leaves; it bears female flowers at the base of the rod and male flowers towards the tip. Both sexes of flowers are surrounded by numerous rings of hair and when the seeds ripen, the inflorescence looks like a mass of cotton wool. These hairs act as parachutes for wind dispersal of the seeds. At the ends of its stems, *Cyperus papyrus* has characteristic, large, spherical inflorescences, which comprise over 100 slender pliable peduncles 150-250 mm long; these peduncles curve inwards to complete a sphere and terminally each bears small brown spikelets with bisexual florets. The distribution of *Cyperus papyrus* within subtropical lakes appears to correlate with slightly acid water (pH about 6). At Inhaca Island it occurs in one swamp in association with *Typha* where the soil is reddish (see Section 10.1.1). The leaves of reeds and bulrushes are long and narrow and are used in thatching; papyrus stems are used for walls of huts or split in half and dried, then used to bind thatch.

In areas where the fringe of the freshwater swamp is near a mangrove swamp, the fern, *Acrostichum aureum* (Adiantaceae), forms a hedge (see Fig. 4.7d). It has tall, stiff leaves and is fertile in April. Grasses such as *Sporobolus virginicus* (see Section 10.2.1 and Fig. 10.5 a-b) merge with *Juncus kraussii* and the pioneer seaside sedge, *Cyperus crassipes*. At the foot of the dunes, the drainage of rain water is also marked by the occurrence of scattered *Phragmites australis*. The close conjunction of *Phragmites*, which is adapted to fresh water, and the salt tolerant *Juncus* is unusual. *Phragmites* grows where fresh water drains from the dunes; *Juncus kraussii* grows where this fresh water mixes with salt water from the mangroves (Mogg 1963). *J. kraussii* and *C. crassipes* (see Section 10.1.2 and Fig. 10.3 e-f) appear to intermingle. *J. kraussii* may be distinguished by the minute tepals, which are much reduced in *C. crassipes*. *J. kraussii* is a rush; it has a prostrate rhizome that bears separated upright shoots. The outer leaves of each shoot are short and papery; the inner two or three are tall, cylindrical and have pungent tips. Inflorescences are borne on cylindrical stems and are overtopped by a pendant bract.

10.3.2 Vegetation around the swamps

On low-lying land at the foot of the dunes, where there is no cultivation, a zone of water-tolerant herbs occurs around the swamp. These include the grasses *Pentodon pentandrus* var. *minor* and *Imperata cylindrica* (see Section 10.1.2 and Fig. 10.5g) and *Conyza canadensis* (Asteraceae). At the edge of the swamps adjoining cultivated land, there is a ridge on the swamp side of the ditch separating the plots from the swamp.

On the border of the mangroves, there is strong growth of succulent, halophytic herbs. The reddish, succulent and creeping *Sesuvium portulacastrum* and clumps of *Sarcocornia perenne* or *Chenolea diffusa* grow where tidal water invades the land at high spring tides (see Section 4.5).

10.4 AGRICULTURE IN THE CENTRAL 'PARKLAND'

Most of the low-lying area of the island (including the swamps) is either under cultivation now or has been so at some time. Land is periodically abandoned and then returns to forest on the inner eastern dunes, to bushland on the western dunes or to freshwater swamp in the low-lying areas. Cultivated sandy plots thus intermingle with abandoned plots interspersed with the trees and shrubs that are of use to the population; the whole resembles a 'parkland'.

Studies carried out by the faculty of Agriculture at the Universidade Eduardo Mondlane in Maputo and observations of the practices on the island show clearly that the soil fertility does

Fig. 10.5 Species from the east coast of the southern bay (a-f) and surrounding the freshwater swamps (g) (drawn by B. Pike):

a. *Sporobolus virginicus* shoot with stolons;
b. inflorescence;
c. *Carissa bispinosa* ssp. *bispinosa* branch with forked spines;
d. flowers;
e. fruit;
f. *Apodytes dimidiata* branch, leaves, buds, flowers and fruit;
g. *Imperata cylindrica* rosette and inflorescence.

not permit a substantial agricultural contribution to the subsistence of the population on the island. Maize, flour and sugar from the mainland has to be purchased with the proceeds of the fishermen's wages. The new Integrated Development Plan of 1990 suggests that people should concentrate on growing vegetables around the swamps.

10.4.1 Preservation of fruit trees

Areas in the valleys between the low dunes are cleared of bush, and burnt to fertilise the soil, before groundnuts or even maize are planted. After one or two or more poor crops, the land is abandoned and allowed to regenerate by lying fallow; thus 80% of 'agricultural' land is always fallow. During clearing and planting, care is always taken to avoid damage to fruit and medicinally important trees.

Fruit trees (such as *Anacardium occidentale*, the cashew nut, and *Mangifera indica*, the mango) are planted around the cultivated plots, separating them from the abandoned cultivated areas. *Anacardium occidentale* nuts are used to brew a refreshing beverage. The leaves are glabrous, shiny and leathery; the blade is tinged with red and the midrib is prominent below. The inflorescence is a panicle with erect branches. This tree is a native of tropical America and is widely cultivated in southern and east Africa, where it has become naturalised. The mango does not spread easily and grows slowly in Mozambique. On Inhaca Island the tree grows up to 15 m high, but on the mainland they grow poorly and do not give good yields. The branches start about 3 m above the ground and form a closed, rounded crown. The large leaves are alternate, petiolate and have entire margins.

Four indigenous fruit trees are never cut down when ground is prepared for cultivation. The marula, *Sclerocarya birrea* ssp. *caffra* (Anacardiaceae) grows to a height of 12 m and has an open crown. The outer bark is greyish; the inner reddish brown bark is used locally to make a concoction for the treatment of dysentery and diarrhoea and also contains an antihistamine. The alternate leaves comprise 7-13 pairs of leaflets and a terminal leaflet. This dioecious tree flowers before the new leaves are formed at the beginning of the season. The yellow fruit contains a fleshy, juicy, nutritious mesocarp which makes a delicious jelly and also an alcoholic drink; it has four times as much vitamin C as oranges.

The African mangosteen, *Garcinia livingstonei* (Guttiferae or Clusiaceae) (Fig. 10.6 a-c) is usually a small solitary tree that branches from the base; the side branches grow out at an acute angle. The leathery leaves are in whorls of three and have inrolled margins. Both leaves and branches exude a sticky, yellow latex. Extracts of leaves and flowers have antibiotic properties. The abundant fleshy fruits are very refreshing and are used to make wine. These trees do not occur in the forests with *closed* canopies.

The water-berry, *Syzygium cordatum* (Myrtaceae) (Fig. 10.6 d-e), is more common than the other indigenous fruit trees and seems to reproduce more easily on the island. It has a dense crown and reaches 12 m in height; the trunk branches some distance from the ground and the bark is rough and dark brown. The cordate leaves are clustered at the ends of the branches and are bluish green above and pale green below. The flowers are sweetly scented; they produce abundant nectar and attract numerous bees and other insects. The numerous long stamens are conspicuous. Mature fruits are deep purple, edible (although with an astringent taste) and make an alcoholic drink. *Strychnos spinosa* is also preserved (see Section 10.5.1) and is useful in the regeneration of the vegetation.

10.4.2 Cultivation in and around freshwater swamps

Because the only successful agriculture relies on ground water, the swamps on the island are good for planting. Areas on the periphery of swamps are cleared, tilled and planted. For years, swamps have been drained by driving trenches, 0,5-1 m deep, radially towards the centre and adding others tangentially to these. The area is thus cut up into a number of small raised, drained plots, which are first hoed deeply to eradicate the deeply rooted rhizomes of reeds and grasses. In the larger swamps, only the fringes have been used and deep water remains in the centre. The whole of some of the smaller swamps have been used and they have become almost dry, while in some, small water holes remain. The area covered by fresh water swamps has dwindled substantially over the seventy years in which the island has been studied.

Crops include maize, cassava, rice and sweet potatoes. Coconut palms have been planted in large areas near the mangrove swamps, and in some areas, pawpaw, guavas and mangoes are grown successfully. Drying swamps are burnt annually to encourage the growth of *Imperata cylindrica*, for thatching.

In recent years, vegetables have been grown successfully in borders about 3 m wide around the margins of swamps. Some of these gardens extend for over 500 m along the valleys between low dunes. Cabbages, beans, pumpkins, marrows and tomatoes are grown.

10.5 NATURAL REGENERATION OF ABANDONED CULTIVATED LAND

Plots of land where groundnuts or sweet potatoes were grown on the western plateau have been abandoned for many years, as was the custom when the soil is exhausted. Various stages of redevelopment of thicket and bushland may be seen on the plateau, which has been incorporated into a nature reserve since 1965 (Campbell *et al.* 1990a). The rate of regeneration depends on the time that has elapsed since cultivation, the local fertility of the sandy soil, the proximity of conserved trees and on frequency of incursion by goats, which are normally kept tethered. Between the abandoned cultivated plots there are patches of forest which have remained untouched for several decades. Without the influence of goats, restoration of a reasonable forest would take about 30 years.

10.5.1 Pioneers in the regeneration on abandoned land

The most important plant on abandoned land on the plateau is *Helichrysum kraussii* (Asteraceae) (Fig. 10.6 f-g), 'xihamkelo', an erect shrub varying in height and diameter from 0,3-1,5 m. It is branched from the base with erect or slightly bent branches. The short, alternate, linear leaves are slightly succulent, fragrant and borne at the tips of twigs. The older branches often bend downwards and cover the sandy soil; these may be dead with only the remnants of old inflorescences.

A parasitic, orange creeper, *Cassytha filiformis* (Lauraceae) (Fig. 10.6 h), grows mainly on *Helichrysum*. The glabrous stems have haustoria which penetrate young leaves and twigs of the host plant and suck sap. Even when infection is extensive, it does not seem to affect *Helichrysum* significantly.

There are usually gaps between *Helichrysum* plants where other species may grow. One of the first trees to develop is *Strychnos spinosa* (Loganiaceae), the spiny monkey-orange or 'massala' (Fig. 10.6 i-j), which is rarely parasitised by *Cassytha*. In this area, *Strychnos spinosa* may be a tree or shrub that was conserved during cultivation or a small squat shrub developing from seed with, and to the same height as, *Helichrysum* (Campbell *et al.* 1990a). The tasty fruits are rich in vitamin C and popular with both young and old people. The hard shell is opened, the pulp is eaten and the seeds (which contain strychnine) are rejected and scattered, so that seeds are widely dispersed. The dried shells may be used as receptacles in homes or plant nurseries or converted into musical instruments, such as timbalas.

Several grasses populate the open spaces between the shrubs. The red-top grass, *Melinus repens* (=*Rhynchelytrum repens*, Graminae/Poaceae) (Fig. 10.6 k) has beautiful pale-rose inflorescences with a silvery sheen. *Aristida barbicollis* is an annual, tufted grass from east and southern Africa; its fruit has three sharp awns that hook onto clothes or skin of passers-by. *Hyperthelia dissoluta*, a perennial, tropical, woodland grass has piercing fruits.

The bare spaces between bushes of *Helichrysum* are also colonised by two yellow-flowered herbs, *Crotalaria monteiroi* (Leguminosae – Papilionoideae), 'lekalahumba' and, more rarely, by the evergreen, prostrate *Chamaecrista mimosoides* (*Cassia mimosoides*; Leguminosae – Caesalpinoideae). These plants fix nitrogen and thus improve the soil fertility. Later on, shrubs from the surrounding bush invade these areas.

A woodland shrub, *Hymenocardia ulmoides* (Euphorbiaceae), the lesser heart fruit or 'tatalatani', has grey striate bark. Young leaves are red and mature to dark shiny green, with glandular dots below. The pink female flowers have long thread-like styles which suggest wind pollination. The rose-coloured, rectangular fruits have wings and a narrow apical notch with persistent styles and are wind dispersed. This species normally occurs in woodland or thicket at low altitudes.

The dune myrtle or 'nkelenkele', *Eugenia capensis* (Myrtaceae) is a tree of 2 m in height in the adjacent bushland and thicket, but had grown to only 0,3 m by 1988 in these regenerating areas. The leathery leaves are almost circular with a rounded apex. The white flowers have many stamens. The fruit is edible.

The African dog rose, *Xylotheca kraussiana* (Flacourtiaceae), is a spiny sub-erect shrub with hairy, alternate leaves. The conspicuous white flowers have many central yellow anthers. The orange fruits are edible. As in the whole of southern Mozambique, *X. kraussiana* usually occurs in open regenerating grassy areas at Inhaca Island.

The dune soap-berry, *Deinbollia oblongifolia* (Sapindaceae), had grown to 0,5 m here in 1989, although it may reach 3 m in undisturbed bush. It is an erect shrub with slender woody stems and pinnate leaves. The terminal racemes have masses of white flowers.

10.5.2 Role of conserved fruit trees

The practice of leaving fruit trees untouched when preparing land for cultivation is very important in the regeneration of vegetation. The fruit trees attract birds that have been feeding on other fleshy fruits and thus many seeds fall below these trees, which offer shade and protection to the seedlings while they are young and susceptible. It is easier for forest and bushland to re-establish from a large number of these nuclei, than from open ground or under the temporary dense canopies of *Helichrysum* (Campbell *et al.* 1990b).

10.5.3 Secondary vegetation in the later phases of regeneration

Fifteen years after cultivation was abandoned, the areas were covered with shrubs 2-3 m high. *Helichrysum kraussii* was then rare, but *Acacia karoo*, *Apodytes dimidiata*, *Carissa bispinosa*, *Clerodendrum glabrum*, *Commiphora schlechteri*, *Eugenia capensis*, *Mimusops caffra*, *Ochna natalitia* and *Vepris laceolata* had invaded the abandoned land (see Section 10.1).

The white cat's whiskers or 'nunhane', *Clerodendrum glabrum* (Verbenaceae), in forests and maritime scrub at Inhaca Island, (not mentioned earlier) had grown 8 m tall. The dark bark is longitudinally fissured; branchlets have prominent lenticels. At each node two to four leaves are produced; they smell unpleasantly and are often softly hairy below. Showy flowers are produced in terminal heads and smell very sweet (almost unpleasantly so). The fleshy fruits are crowned with persistent calyx lobes. Various parts of the plants are used in local medicine.

Where cultivation has been abandoned for more than 25 years the pioneer *Helichrysum* is completely shaded out by dense patches of bushes and small trees in an open bushland. A low herbaceous layer, with the grasses *Melinus repens* (see Section 10.5.1 and Fig. 10.6k), *Aristida barbicollis*, *Hyperthelia dissoluta* (see Section 10.5.1) and *Eragrostis capensis*, establish themselves.

Fig. 10.6 Natural fruit trees (a-e) and species involved in the natural regeneration of abandoned cultivated land (f-k) (drawn by B. Pike):

a. *Garcinia livingstonei* branch with leaves;
b. flowers on stem;
c. fruit;
d. *Syzygium cordatum* branch with leaves and fruits;
e. buds and flowers;
f. *Helichrysum kraussii* tip of branch with inflorescences;
g. capitulum showing involucral bracts;
h. *Cassytha filiformis* stem twined around host and fruits;
i. *Strychnos spinosa* stem, leaf and inflorescence;
j. twig and fruit;
k. *Melinus repens* rosette and inflorescence.

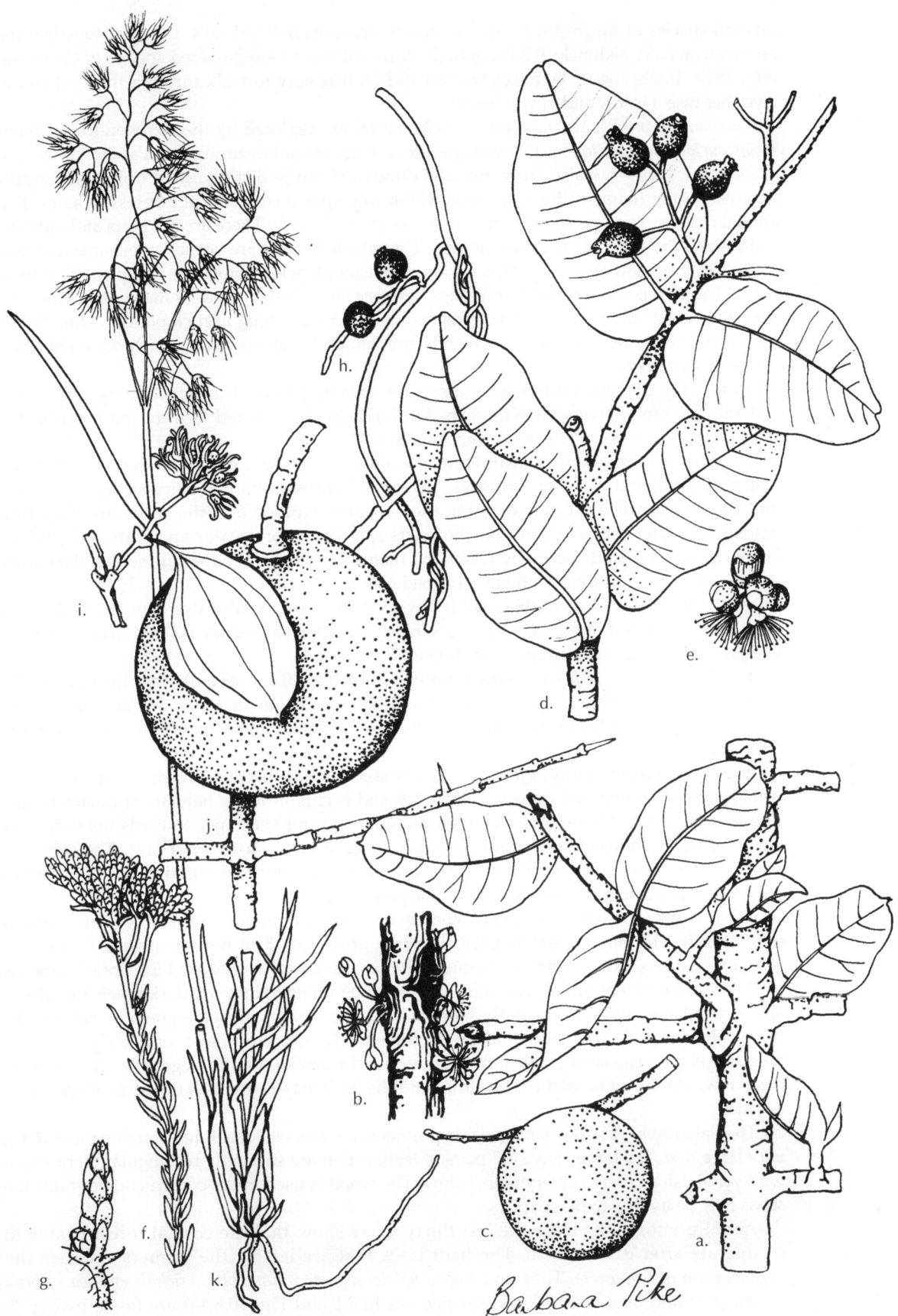

Like all species of *Eragrostis*, *E. capensis* has many-flowered spikelets, grouped together and borne on an erect peduncle, 0,2-0,5 m high. Three different blue-flowered species of *Commelina* (Fig. 10.3c-d), also occur here (see Section 10.1.2). It is easy to walk through this herbaceous layer because it is low and never dense.

Crotalaria monteiroi and *Chamaecrista mimosoides* are replaced by the nitrogen-fixing plants *Tephrosia longipes* var. *icosisperma* and species of *Indigofera* (all Leguminosae-Papilionoideae). *T. longipes* var. *icosisperma* also grows on sand dunes and can be distinguished from others by the long ascending hairs on the stem, in addition to a sparse covering of adpressed hairs. This annual or short-lived perennial sometimes has a woody base. There are 4-7 pairs and terminal leaflets and the stipules are 7-14 mm long. The purple flowers are borne on terminal and leaf-opposed racemes with long peduncles and persistent blackish bracts in conspicuous tufts at the nodes. Ripe pods may be 80 mm long. *Tephrosia forbesii* ssp. *inhacensis* has pink flowers and may be present; an infusion of the seeds may be used for fishing in rock pools to stupefy fish (it contains rotenone), although the woody species, *Tephrosia acaciifolia*, which occurs in forest fringes is more potent.

In recently disturbed areas several species of *Indigofera* occur: *I. laxiracemosa*, an annual ruderal with pink flowers, soon appears; but *I. podophylla*, with red flowers and *I. charlierana*, with pink flowers, occur after some regeneration.

Blepharis maderaspatensis (Acanthaceae) occurs everywhere on the island, especially on disturbed land, and is a semi-prostrate, perennial herb with fibrous hairy stems. The leaves and bracts are prickly. Upon wetting, the capsules explode and the seeds are dispersed ballistically. The feathery hairs on the seeds rapidly absorb water and form jelly, which enhances subsequent dispersal by water and then sticks the seed to the soil, so that the radicle can penetrate even in areas with hard, capped soil.

Agathisanthemum bojeri (Rubiaceae) is always present on uncultivated ground on the west ridge and grows best in red sandy soil; it is rarely found in shaded areas. This erect, perennial, evergreen herb has conspicuous white flowers.

In addition to the four fruit trees described in section 10.4.1, several other trees are left by farmers on the periphery of their plots. These include *Euclea natalensis*, *Trichilia emetica* and *Albizia adianthifolia*, which spread to the abandoned plots and make important contributions to their recovery.

The large-leaved ebony, *Euclea natalensis* ssp. *rotundifolia* (Ebenaceae), is a shrub that usually has spreading and drooping branches and occurs in many habitats at Inhaca Island. The leaves are 80-100 mm long, have an undulate margin and taper towards the base. The small, off-white flowers are fragrant and produced in dense axillary sprays. The spherical fruits mature through red to black. The yellow roots are commonly used as toothbrushes or may be dried, powdered and taken to relieve pain.

Trichilia emetica (Meliaceae), the mahogany tree, is common in secondary vegetation and is often planted for shade near houses. The tree grows to 15 m high at Inhaca Island and produces dense shade all the year round. The leaves are composed of 4-5 pairs of opposite leaflets and a terminal leaflet with thick petioles 100 mm long; they are dark green and glossy above and have dense, short, curly hairs below. The flowers are pale green or yellow; the three-valved capsules are abundant and the three black seeds are almost completely concealed by a scarlet aril. The seeds, called 'nkuhla' are used to make soap or vegetable oil or may be eaten raw. The wood is used for furniture and the bark may be used as the basis of an enema or emetic.

The flat-crown, *Albizia adianthifolia* (Leguminosae – Mimosoideae), has rough bark and is a very large tree. The leaves have 4-7 pairs of leaflets that are strikingly rectangular, dark green with yellowish or rusty velvety hairs below. The wood is used in the construction of huts; the seeds may be used to make soap.

Aerial photographs from the last thirty years show that the coastal forest is able to recuperate after disturbance. The fruit trees that are left by the farmers enhance the regeneration of the forest. They and *Euclea natalensis* ssp. *rotundifolia*, *Trichilia emetica*, *Albizia adianthifolia* and *Strychnos spinosa* (see Section 10.5.1 and Fig. 10.6 i-j) are fast growing. *S.*

spinosa is able to grow in the *Helichrysum* thicket, but the others require more shade and humus in order to become established. The vegetation goes through a very tangled stage, before an open forest with a diverse herbal layer is re-established. If no further disturbance occurs (as will be the case in the nature reserves), the recuperated area will eventually be almost indistinguishable from the primary forest.

10.5.4 Experimental reclamation of a mobile dune

At Santa Maria, on the mainland opposite Ponta Torres, the high dunes are exposed to south-easterly winds, which blow through the strait and deposit sand in the channel of the southern bay. There is a coral reef on the east channel bank in the southern bay and, as corals are particularly vulnerable to settling sand, this reef had been partly destroyed in the last 50 years. In the 1970s, indigenous herbs and shrubs were planted at Santa Maria in order to halt erosion of the dunes and consequent damage to the coral (Salm 1976).

The area was first stabilised by laying branches between stakes in the sand. The branches provided shade and protection from the wind for developing seedlings. Several grasses and sedges were then sown including *Cyperus crassipes* and *Sporobolus virginicus* (see Section 10.2.1 and Fig. 10.3 e-f and 10.5 a-b), which grew successfully, as did the shrub, *Sophora inhambanensis* (see Section 10.2.1 and Fig. 10.4 g-i). More recent experiments suggest that these species could be used in the rehabilitation of other blowouts in the dunes; *Tephrosia* and *Indigofera* would probably be good additions. The coral reef has made a good recovery (see Section 8.6).

10.5.5 Reclamation of a wind-eroded dune facing the Indian Ocean

The lighthouse at Inhaca island was built at the end of the 19th century but wind erosion on the south-eastern aspect on a small promontory below it, caused the loss of all vegetation so that the sand swirled upwards and hindered the work of the lighthouse staff; by the 1920s, this was an acute problem and stabilisation was necessary (Cardosa 1959). First low-growing bushes from the eastern dunes were planted, but this proved ineffective as the plants were uprooted by the wind. In the late 1930s a 600 m long double fence was erected along the shoreline above high-water mark. Parallel to this, fences (made from branches of local trees and shrubs) were erected at 10 m intervals, to the top of the dune as well as criss-cross fences for protection from cross winds. In the spaces between the fences several thousand seedlings of the tree *Casuarina equisetifolia* were planted in rich humus contained in baskets; the humus was used by the seedlings until they rooted in the sand beneath. The spaces between the seedlings were planted with sand-binding indigenous herbs. Within three years, *Casuarina* saplings formed a low thicket over the whole area; they had an extensive root system, which together with the network of shrubs and herbs stabilised the sand. *Casuarina equisetifolia* grows into a large tree. It has modified green branches that photosynthesise and are surrounded by rings of scale-like leaves. The much reduced male flowers are borne in slender catkins, while the mature fruits form cone-like structures. The seeds have a membranous wing, so that both pollination and seed dispersal is by means of wind.

By 1990 the *Casuarina* trees were slowly dying, but between them vegetation similar to thicket had developed. The herbaceous layer is composed of *Sansevieria hyacinthoides* (see Section 10.1.1 and Fig. 10.2 a-c) and *Microsorium scolopendrium* (see Section 10.1.1 and Fig. 10.2 d-e) which in some places completely cover the sand, indicating that a good humus layer can be formed within decades. This suggested that recovery from bare ground to stabilised forest was possible. A good shrub layer had not yet formed, although *Vepris lanceolata* was present. Seedlings of various trees that occur on the inner dunes (see Section 10.2.2), including *Commiphora schlechteri*, were present. Four other noteworthy species were *Zanthoxylum delagoense*, *Cassine aethiopica* (see Section 10.1.1), *Dodonea angustifolia* and *Grewia caffra*.

The dune knobwood or 'kekani', *Zanthoxylum delagoense* (Rutaceae), a climbing shrub or small tree, with recurved prickles on the stems and sometimes on the petioles and leaves, is endemic to coastal dunes in southern Mozambique. It grows to 4 m high. The pinnate leaves have elliptic leaflets, with gland dots and crenulate margins. The panicles of white flowers are clustered at the ends of the branches. The dehiscent orange-brown follicles split at maturity to reveal single shiny black seeds.

The sand olive, *Dodonea angustifolia* (=*D. viscosa*) (Sapindaceae), rarely grows taller than 4 m and is pantropical. It is very common from the Cape in South Africa to Beira in Mozambique, but is also common inland, in Zimbabwe and Malawi. The bark is dark grey. The leaves are simple, alternate, narrowly elliptical and resinous. The flowers are greenish yellow and arranged in axillary or terminal heads. The light capsule with two or three greenish, papery wings is wind dispersed. *D. angustifolia* has been cultivated in Africa for very many years because it makes a good hedge, its leaves and roots have medicinal properties and its edible seeds germinate easily. It facilitates the consolidation of sand.

The climbing grewia or 'itxubu', *Grewia caffra*, of the jute family, Tiliaceae, is a well-branched shrub which may climb into taller trees. The quadrangular stems have brown, rough bark. The simple, oval and pointed leaves bear numerous stellate hairs, especially on the veins; the margins are finely toothed. The yellow flowers are about 20 mm in diameter. The lobed berry is produced on a densely, hairy stalk (gynophore), which is inserted on a glabrous pedicel.

Amongst the dying *Casuarina* trees, there are also masses of a fleshy climber, *Cissus quadrangularis*. Tendrils support the winged stem which is 50 mm in diameter and has reddish lines on the angles. Simple, broadly ovate, dentate leaves are borne at the apex of the stems, but soon drop. Flowers are borne in small cymes. The fleshy fruit is red and smooth when ripe, but is often parasitised and then becomes woody and brown.

Near the sea, a quite dense herbal layer is present between the *Casuarina* trees. *Achyranthes aspera* (Amaranthaceae) is dominant. It is a bushy herb that grows 1 m tall. The leaves are opposite, elliptic and have white hairs on the veins when young. The spicate inflorescences are terminal or at the tips of axillary branches. The small, inconspicuous flowers point downwards at maturity and are surrounded by prickly bracts, which stick to passing animals. The soil is covered with decomposing *Casuarina* roots and branches and a whitish-grey layer of active fungi; beneath this, a deep layer of sand mixed with organic matter is present.

The planting of *Casuarina* has had a very positive effect: the trees have stabilised the dune for 50 years and have so altered the soil and microclimate, that it has become possible for a stable forest to develop. The recuperation of the forest in this area and the success of the protective planting augurs well for the growth of forests in the Inhaca Island nature reserves. In the places where sand blows over the eastern dunes (Fig. 10.1) and on to the sheltered forest on their western slopes, there is a danger that it might accumulate on the agricultural land in the central 'parkland'. This danger can be averted, however, by stabilisation of the eastern face of the dunes.

11

BIRDS OF INHACA ISLAND

J.J. Feijen and H.R. Feijen (Ex Universidade Eduardo Mondlane)

The bird population of Inhaca Island is rich and varied; about 200 species, representing 57 families, have been recorded since the first survey was organised in 1936 by the National Natural History Museum in Mozambique (da Rosa Pinto 1959). About 120 species are residents; about 65 species are migrants and 15 species are ocean vagrants seen flying over the seas around the island. The migrants come from different countries at various times of the year. Palaearctic migrants breed in the northern hemisphere in summer and fly to southern Africa for the southern summer, many resting on the island en route and some of them overwintering there, but none of them breeds on the island. Most of these are shore birds which congregate in their thousands at Ponta Raza and in the Saco around March-April and September-October. Some inland species migrate to and from tropical Africa, other birds come from Namibia, South Africa or the islands of the southern Atlantic Ocean.

An up-to-date list of birds recorded on the island by da Rosa Pinto (1959), (Clancey 1971) and others over 40 years, and some records from personal observations over five years is given in Appendix B. The bird number in Roberts' *Birds of Southern Africa* (1985) is given in the text for reference. The number and variety of birds may still be increasing since the coastal bush and forests are regenerating and the birds are relatively undisturbed there. In the garden and coastal bush surrounding the Marine Biological Station the birds are not shy. The *Casuarina* trees and the coconut palms near the fishing village, in the district of Ridjene, the mangroves and open parkland are rewarding areas for study, but in the evergreen forest on the eastern dunes it requires patience. Migrants are seen on the sand spit at Ponta Raza and in the southern bay in the evenings in season (Brooke and Tuer 1968).

A selection of birds has been chosen for comment in the various habitats, the vegetation of which has been described in Chapter 10.

As far as is known, birds do not constitute a threat to fishing around the island. The seed-eaters and the fruit-eaters play a vital role in the regeneration of trees on abandoned and cultivated land and disturbed bush.

11.1 BIRDS AMONG TREES OF THE COASTAL BUSH

The birds inhabiting the coastal bush are always encountered early in the morning and in the late afternoon around the vantage points of the Marine Biological Station and in the bush along the rough, sandy road to the hotel. Visitors from the forest on the eastern dunes may also often be seen there. In general, birds can be observed at all hours of the day.

11.1.1 Medium-sized birds: Starlings; Nightjars; Pigeons; Narina Trogon

Groups of a hundred or more tropical Blackbellied Starlings (768) (Sturnidae) may be heard in the early morning during July and August, singing noisily in the mango trees near the Marine Biological Station, their glossy, iridescent feathers glistening in the sunshine. During the day they disperse in small groups to the evergreen forest to forage for fruit and insects in the trees; in the evenings they return to the mango trees. Their flight is fast and direct with loud, swishing wing beats.

Before sunrise the common Fierynecked Nightjar (405) (Caprimulgidae) may be seen resting on the sandy road to the hotel at regular intervals, each on its own spot. It prefers to rest on bare ground. At night it hunts insects from a perch on a tree, making swift sallies to catch its prey on the wing; the fairly small beak has a wide gape, made more effective by its lining of projecting bristles. The nightjars commence calling at dawn and dusk in August, but continue all through moonlight nights. In November, eggs are laid on dried leaves and debris on the ground, hardly making a nest; usually two pale, pinkish eggs are laid.

The Green Pigeon (361) (Columbidae) feeds on berries on shrubs around the Marine Biological Station in the very early morning, but during the day it retreats into coastal bush. Near the village, however, it perches high up in the *Casuarina* trees on the cliff, fearing human predation, for its flesh is a local delicacy. It is difficult to see among green foliage, but when feeding on fruit on trees it may hang upside-down and flap its wings to keep its balance.

At the edge of the forest reserve that borders the sandy road the shy, strikingly coloured Narina Trogon (427) (Trogonidae) may sometimes be seen either solitary or in pairs, perched long periods high on the tallest trees for where they build their nests. They catch insects on the leaves and branches, sometimes flying with many twists and turns.

11.1.2 Bird parties: Finches; Flycatchers; Bulbuls; Robins; White-eyes; Shrikes; Sunbirds; Weavers

Many small birds in the bush belonging to different families may fly together from bush to bush, forming 'bird parties', which forage together. There may be an advantage to them in the greater disturbance of insects which then become available, or perhaps in their greater awareness of birds of prey. They build their nests in trees.

On the ground beneath the trees the Green Twinspot (835) one of the six species of Estrillidae (waxbills and finches) forages for seeds and aphids in the vegetation at the forest edges where grasses and other herbaceous plants grow. Its back is green and its belly is black, peppered by a multitude of white spots. They have short, stout, conical beaks adapted for eating small seeds. Both male and female participate in building the nest.

The Dusky Flycatcher (690) (Muscapidae), a resident bird, is the common species in this family. It catches moths, beetles and spiders on the ground or while hovering around branches of trees; it also eats small fruits such as those of the Ironwood tree. The Paradise Flycatcher (710), an intra-African migrant, catches butterflies in the air in undulating, graceful flight.

The Blackeyed Bulbul (568)(Pycnonotidae), a representative of these typical birds of Africa and Asia, calls in liquid, lively notes from the top of a tree, usually in pairs or in small groups. The feathers on its black head rise stiffly to form a small crest at the back. It feeds mainly on the small fruits of trees and the nectar of flowers, but it also catches insects on the ground or in flight, and sometimes it preys on small lizards.

The brightly coloured Natal Robin (600) (Turdidae) usually rests in the undergrowth in the bushland and forages on the ground for spiders, centipedes and insects, but when the trees produce fruit in winter, it moves on to the branches to feed. It sings with a rich, melodious whistle and, incredibly, can imitate the songs of about 30 other species of birds.

The tropical Yellow White-eye (797) (Zosteropidae) is so called because it has a conspicuous ring of white wattle-flesh around the eye. It has a yellow breast. Small flocks of up to 20 birds forage together in the canopy of the forest, gleaning insects from leaves, stems and flowers or catching them in mid-air, meanwhile keeping up a constant twittering.

Bush Shrikes (Malaconotidae), which are insectivorous birds, frequent the evergreen forest on the eastern dunes. The Orangebreasted Bush Shrike (748) has a call of 5-8 ringing notes, varying in pitch and tempo. It can be coaxed out of the bush by making a spishing sound, 'Psh-Psh', a technique which may be successfully used for many other shy species.

The small Collared Sunbird (793) (Nectariniidae) is unlike the rest of its family (which have long slender, recurved bills for collecting honey) in that its beak is short and only slightly curved. It hunts for spiders and insects among forest creepers, among dead leaves and in spider webs; it also probes shallow flowers for nectar and may split open tubular flowers to obtain it. It seems to prefer small berries, especially those of the Bushtick Berry, *Chrysanthemoides*, when they are ripe. This sunbird and the Yellow White-eye form small groups together and search for insects in the flowers of the mango trees in the garden of the Marine Biological Station, or in similar honey-laden flowers in the bush.

The Spottedbacked Weaver (811) (Ploceidae) is also called the Village Weaver from its habit of nesting near human habitations; it may build its oval 'hanging basket' nest on trees in the garden of the Marine Biological Station near the entrance to the laboratory from August to December. Flocks of about twenty birds look for seeds, beetles and flowers on the ground under or on the trees, being able to hold larger items in one foot whilst pecking with the bill. The other six species of weaver on the island are slightly different in colour pattern and in the shape of their nests. When the nomadic Spottedbacked Weaver arrives at its nesting site at the Marine Biological Station in August, the other weaver species such as the Spectacled Weaver (810) and the Yellow Weaver (817), which were there earlier, retreat to the forest.

11.1.3 Solitary birds: Kingfishers; Cuckoos; Warblers; Drongos; Hadeda Ibis; Owls

In the daytime at the Marine Biological Station one may be aware of the noisy call of the Brownhooded Kingfisher (435) (Alcedinidae) which lives in the coastal bush, south of the station. The colourful Mangrove Kingfisher has been recorded at Inhaca Island but it is not common. There are seven species on the island, but despite the name only two of them catch fishes: the Pied Kingfisher (428) and the rarer Giant Kingfisher (429). The Mangrove Kingfisher eats fiddler crabs as well as insects; the other species, the Pygmy (432), Brownhooded (435), Greyhooded (436) and Striped (437) Kingfishers may catch small lizards as well as insects.

On the road through the bush to the hotel the males of the Emerald Cuckoo (384) (Cuculidae) perch on the tops of tall trees calling day after day (Fig. 11.1). This cuckoo has a sweet whistle in a four-syllabled phrase which may also be heard in the evergreen forest. It is a common intra-African migrant that breeds at Inhaca Island. One white egg is laid in the nest of the Bleating Warbler (657) (Sylviidae) in dense undergrowth. The nest, near ground level, is a ball of growing leaves, still attached to the plant and sewn together with spider web which is threaded through the little holes it has made in the edges.

The black Forktailed Drongo (541) (Dicruridae) perches on a conspicuous branch of a tree, sallying forth to catch its insect prey, usually returning to the same perch. Sometimes mobs of drongos aggressively attack owls, crows or even man; they may rob other birds of their food and carry it away in the bill or the strong feet. The Squaretailed Drongo (542) is confined to the depths of the evergreen forest on the eastern dunes.

Towards evening at the Marine Biological Station the silence may be broken by a group of four to six Hadeda Ibis (94) (Threskiornithidae) shouting their raucous cry loudly and repeatedly while returning to the forest to roost, after foraging all day on the open ground probing among grasses with their long bills.

At night the Spotted Eagle Owl (401) and the Barred Owl (399) (Stringidae) can be heard hooting around the Marine Biological Station. By virtue of their very large wings and very soft feathers their flight is silent; their binocular vision, the retina rich in rods (which distinguish light and shade only) and the ability to swivel the head around 270° to compensate for their

Fig. 11.1 Portrait of an Emerald Cuckoo (384) (M.D. Feijen, Leiden).

fixed eyes, enable them to locate prey in the dark. The Spotted Eagle Owl has a varied diet of insects, crabs, toads and birds whereas the Barred Owl usually eats birds, lizards and toads. This can be checked since owls regurgitate the undigested portion of food as firm pellets which may be found near their roosting places. The Spotted Eagle Owl roosts on open ground in convenient hollows whilst the Barred Owl returns to the forest and rests on low branches; both are well camouflaged and almost invisible when at rest.

11.2 THE LOW-LYING CENTRAL AREA

The ground around freshwater swamps in the central area of the island is cultivated and people live near it in 'mutis' (huts in large gardens). A border of shrubs or tall fruit trees and trees of medicinal importance or those that give wide shade have been left standing where they grow on low dunes; grass and herbs grow underfoot, the whole is an informal type of parkland (see Sections 10.4 and 16.1). In this area many different types of birds can be seen for many reasons.

11.2.1 Birds nesting in trees: Barbets; Swifts; Herons; Crows; Sparrows; Hornbills from the forest

Near the hotel one always hears the loud calls of the Blackcollared Barbet (464) (Capitonidae), which sings from the top of a fig tree at the foot of a low hill nearby. This bird is mainly a fig-eater and is nearly always to be seen on or near a Strangling Fig Tree wherever these trees have been preserved. They sing in duet, starting with a buzzing sound which becomes a loud ringing call and answer. The birds bob up and down, opening and closing their wings as they call. Up to eleven birds may roost together on a branch; they nest in holes excavated in the trunks of trees.

In the grove of tall coconut palms nearby, the Palm Swifts (421) (Apodidae) make their nests and roost under the palm leaves. They feed as they fly, catching insects and spiders in the air and winged termites after the rains. They breed throughout the year, making nests of feathers glued to the under side of a palm leaf. Two eggs are also glued to the untidy pad of feathers and are incubated for three weeks whilst the parent clings vertically to the nest.

The Blackheaded Heron (63) and the Grey Heron (62) (Ardeidae) make nests of sticks lined with grass. Both species roost in colonies on the *Casuarina* trees and may feed on fish and crabs in the sea and on the seashore, although the Blackheaded Heron prefers to feed on land (see Section 11.5.1).

An introduced species of bird lives in the coconut groves. The black Indian House Crow (549) (Corvidae) is said to have been present here since 1940 and has multiplied although it has not been recorded on the mainland. They scavenged for food around the hotel and on the nearby beach. They used to nest in the *Casuarina* trees until about 1966 when the herons started to use these trees. The crows then made their nests in the tall coconut palms. In the last ten years the crows have spread to other parts of the island but have not been observed robbing other birds' nests as they do on the Kenya coast and in Zanzibar. It is said that they enter huts and steal food. They did not become numerous until 1957 and now they far outnumber the local birds in the garden and coconut groves. The Pied Crow (548) is also quite common on the island and nests in the coconut palms. This species is also present on the mainland.

Another introduced species is the House Sparrow (801) (Ploceidae) which was first reported in the mid-1950s and is now common around the hotel garden. It differs from the indigenous Yellowthoated Sparrow (805), which also frequents the hotel garden, in its method of foraging. The House Sparrow hops on the ground and may catch flying termites or moths on the wing; the Yellowthroated Sparrow walks on the ground or on the branches of the trees with small steps looking for insects or seeds. It flies into a tree when disturbed and flicks its wings and tail on alighting on a branch.

Sometimes a flock of 10-15 Trumpeter Hornbills (455) (Bucerotidae), birds of the Evergreen Forest that usually nest in holes in trees up to 12 m above the ground, may appear on low open ground in the central 'parkland'. They feed on fruit in the forest. They have huge, heavy black beaks that curve downwards and are topped by a high horny ridge. They make a very loud, braying call. Although this bird looks very heavy, it forages by hopping lightly among the branches of the trees and can take off almost vertically into the air in dense forest. The nests have not yet been observed on the island, although the birds are seen in the forest.

11.2.2 Birds nesting on or near the ground near fresh water swamps and cultivated ground: Bee-eaters; Wagtails; Pipits; Longclaws; Chats; Egrets; Hoopoes

There is less water in the swamps nowadays since the periphery of each swamp may be drained for agriculture and there have been several years of drought; swamp vegetation is, however, still flourishing and standing water occurs in the swamp of the northern part of the central parkland. A number of bird species prefer to live near fresh water because of the associated fauna and flora. All of them except the Bee-eater, which catches insects in flight, forage on the ground amongst grass or on bare areas, cultivated or fallow, where they select insects or seeds. They nest on or near the ground in hollows or under tufts of grass.

The Little Bee-eater (444) (Meropidae) is quite common; at night a row of up to seven birds may roost together on a branch of a tree, tightly bunched together. They remove the stings from the bees they catch by rubbing the bee's abdomen against their perch. They nest in small burrows in a low bank.

The Cape Wagtail (713) (Montacillidae) runs or flies low to catch insects and may vary its diet with tadpoles, in season. It calls frequently while foraging and wags its tail on alighting or take-off. It makes a bulky nest of grass and stems and lines the inside of the cup with fine rootlets. The Plainbacked Pipit (718) and the Yellowthroated Longclaw (728) of the same family feed and nest similarly.

The Stone Chat (596) (Turdidae), a migrant from tropical Africa, appears in summer; it perches on a bush and then drops to the ground to catch insects, flicking its tail on landing; it returns to its perch after hopping around whilst feeding.

The white Cattle Egret (71) (Ardeidae) occurs among the goats kept by villagers. It feeds on the insects on the ground disturbed by the goats; large flocks gather around the swamps in the evenings to drink before roosting. Nests are small platforms of sticks or reeds among the reeds in the swamps.

The Hoopoe (451) (Upupidae) is associated with cultivated ground around the swamps; it eats cutworms and earthworms, termites and larger insects, and even small snakes and frogs, which it finds by probing the soil with its long curved bill while walking with quick steps, nodding its head. It raises the conspicuous, pointed, red and black striped crest on its head as though to counter-balance its long bill. Hoopoes forage together and when disturbed they fly into a tree and perch for a while. The Hoopoe does not build a nest but occupies small holes in the ground without lining them.

11.2.3 Birds nesting on dry ground on the central low dunes in drier areas: Francolins; Quails; Korhaans; Dikkops

Some species of larger birds live in this fairly open area. They have long legs and forage on the ground for prey larger than that of the former group, such as lizards, frogs and toads as well as insects. They too nest on the ground or not far above it, but not preferentially near swamps.

Pairs or groups of the Rednecked Francolin (198) (Phasianidae) feed on the ground on seeds, shoots, snails and insects. They seem to fly reluctantly, running for cover rather than flying when disturbed. They make crude nests on the ground in rank vegetation. The Blue Quail (202) of the same family is smaller and not as common. It has the same horizontal posture as the Francolin when it runs; its flight is fast and direct. The Blue Quail is a migrant from tropical Africa but probably breeds on the island and seems to have become more common in recent years. Its nest is a grass-lined bowl on the ground amongst grasses and sedges. It eats insects, seeds and snails.

In the summer the Blackbellied Korhaan (238) (Otidae), which has an exceptionally long, thin neck and very long legs, occurs on the island, where it breeds; it migrates to the mainland in winter. It may be seen singly or in pairs, either standing still, relying on the camouflage of brown, buff and black, or walking with small steps and exaggerated movements of the neck. During courtship the male may fly several hundred metres with slow wing beats and then glides to the ground with wings held aloft, neck extended and black chest puffed out. This Korhaan does not build a nest; eggs are laid in a 'scrape' on the bare ground.

Very early in the morning the Spotted Dikkop (297) (Burhinidae) may be seen before it retires for the day to the shade of the bush. It does not live near water and is nocturnal, uttering a distinctive weird call. The adult bird feeds singly on intertidal crustaceans or molluscs at dusk or on very cloudy days; it also feeds on dry land near the shore where it finds grass seeds and insects. When disturbed it runs off with its head bent low and then stands still and waits. Small groups roost together during the day. It lays eggs in a shallow 'scrape' on the ground and sometimes this nest is lined or ringed with stones.

11.3 DIURNAL BIRDS OF PREY

About twelve species of birds of prey in the family Accipitridae have been recorded at Inhaca Island, including snake-eagles, goshawks and buzzards. Most of them hunt birds, reptiles and mammals on land but some dive for fish.

Fishes form the exclusive diet of the African Fish Eagle (148). Its loud four-syllabled call, dropping in pitch and intensity, is heard early in the morning and it is repeated most distinctly in the evenings. It has been sighted from most of the shores of the island and may be recognised by its large size, its white head and breast, dark wings and back and slightly forked tail. It is an aggressive, gregarious bird which swoops swiftly on its prey. Several may co-operate in seizing prey from other raptors.

The Gymnogene (169) builds its nest in a bush near the edges of a cliff and may be observed when it soars high above the forest. Gymnogenes catch small reptiles and frogs, nestling birds and large insects. They are peculiarly adapted for scooping eggs or nestling birds or other small animals from narrow holes, in that they can rotate the tarsus (ankle joint) forwards, backwards and sideways as well.

11.4 BIRDS ON THE SEASHORES

11.4.1 *Large waders: Flamingoes; Herons; Egrets*

The tallest and most extraordinary bird on the island is the Greater Flamingo (96) (Phoenicopteridae) which visits the Southern Bay. Flocks of birds stand erect at low tide, isolated from one another, spread over the vast expanse of the very sheltered sand flats in the Saco. The large peculiar bill, the legs and the feet are red, and in flight the red and black wing feathers are conspicuous. When feeding, the bird stands upright and bends its neck down so that the beak, held upside-down, may filter the muddy water which it stirs up with its feet as it shuffles around in a small circle. The mandible is deep and trough-like and the upper part of the beak forms a lid (Fig. 5.2a). The thick tongue moves laterally back and forth acting as a pump that draws water through the filtering lamellae inside the bill. It retains small crabs, polychaete worms and probably some diatoms (see Section 5.3.2). It does not breed on the island.

The large wading birds (Ardeidae) on the sand flats in the Saco of the southern bay and on the west shore of the island are represented by four species of herons and two species of egrets. The herons have an upright posture and are about 100 cm tall; they are grey or brown and have long pointed bills, long necks and legs. They are usually solitary when feeding on shore, in a freshwater swamp or on land, although they congregate when roosting and many nests are built among the branches of the Casuarina trees. The Blackheaded Heron (63), called the 'Dryland Heron' prefers to seek rodents, lizards and birds on land whilst the Grey Heron (62) more often wades in water and catches crabs and fish. In flight the herons hold the neck folded close to the body and the broad wings are bowed downwards. Both species build platforms of sticks on trees as nests. The Cattle Egret (see Section 11.3.2) prefers to feed on land but the Little Egret (64) may wade at the edge of the sea, seeking fishes or small invertebrates. It makes a platform nest of sticks in bushes or reedbeds.

11.4.2 Small waders: Sandpipers and their kin; Plovers; Crab Plover

On the western shore, sand-flats about 300-400 m from high to low tide are exposed along 7 km of the coastline and in the southern and northern bays the tide recedes for 5 km so that a rich source of invertebrate food is available to birds and is exploited by the small waders (see Section 3.5.4). They do not breed on the island but a few members of several species may overwinter there and may be seen in any month.

During the day when the tide is low the small waders spread out in small groups to feed, but when the tide rises they fly to roosting places in special areas, such as the long sand-spit at Ponta Raza, the rocks at the south-western point, Ponta Punduine and at Ponta Torres in the south-east of the island. Mixed flocks of several species of sandpipers such as the Curlew Sandpiper (272), Terek (263), Common Sandpiper (264), Marsh Sandpiper (269), Greenshank (270), Sanderling (281), Little Stint (274) and Whimbrel (290), mingling with the plovers, gather together in the evenings. The numbers reach several thousand during April, when most of them prepare to continue their journey northwards (Brooke and Tuer 1968).

Waders differ in their foraging behaviour. The Sanderling, for example, probes in wet sandy mud when a wave has just receded, and it runs upshore again before the next wave. The sandpipers walk along steadily, probing the sand for worms with slender beaks, whilst the plovers have the habit of running in short bursts, holding the body horizontally. When the plovers peck at the sand they do not penetrate as deeply as the sandpipers and are likely to obtain superficial polychaete worms or crabs.

The largest of the 'small' waders is the Curlew (55cm) (289) whose name resembles the sound of its call. It has a slender down-curved beak, longer than its legs, which is used to probe deeply in the sandy mud of the intertidal flats. The bill is twice as long as that of the Whimbrel (290) and it obviously exploits different food resources from the small waders, i.e. burrowing crabs and molluscs. Enormous flocks of Curlews congregate on the mangrove trees at high tide in preparation for their migration in April. Whimbrel are less numerous but have also been recorded *en passage* in March and October.

The most common plovers are the Ringed (245), Whitefronted (246) and Grey (254). They have short, straight bills and plain grey or brown plumage and so may be distinguished from the sandpiper family in which the bill is usually slender and curved and the plumage is patterned in brown, buff and rufous, with some white ventrally.

The Crab Plover (296) (Dromadidae) is quite different from the true plover in that it is able to deal with quite large crabs; it often feeds at night when the crabs themselves are above ground feeding on smaller crustaceans or molluscs. It has longer legs than the true plovers and runs faster with delicate steps; when it flies the long black legs trail behind. They breed on the shores of East Africa, nesting in colonies in burrows above high tide mark. A flock of 30 birds, preparing to migrate northwards, was recorded in March of one year at Inhaca Island.

11.5 SEA BIRDS

There are two styles of fishing among sea birds: some swim on the surface or fly near the surface and dive from this position, whilst others fly overhead and dive from a height to fish in deeper water.

11.5.1 Inshore birds: Cormorants; Pelicans

Groups of ten or more Whitebreasted Cormorants (55) (Phalacrocoracidae) are often observed while they are sitting erect on poles planted in the channels in the bays of Inhaca Island by fishermen, who drape their nets around the poles to catch fishes on the ebb tide. The cormorants fly low over water at high speed and dive to catch fishes. They nest in dense colonies on coarse stick platforms in trees, but although groups of 10-20 are observed in October on the Ponta Raza sand-bar, no breeding has been observed at Inhaca Island. The Cape Cormorants (56) fly low over the sea, many spaced out in a long line; they settle on the water to feed and dive from the surface. The Reed Cormorant (58), on the other hand, is solitary when fishing and roosts gregariously in trees or reeds.

The Pinkbacked Pelican (50) (Pelicanidae) occurs in small numbers from April to January, swimming at the surface of the sea while fishing singly in 5-16 m of water near the coral reef and along the Inhaca Channel, parallel to the western shore. They catch fishes of 300-400 g in the pouch of the bill and temporarily store them there as the water drains out. They do not breed at Inhaca Island; the nearest breeding site is St Lucia Bay about 100 km to the south where birds in dense colonies make platforms in tall trees. At Inhaca Island they do not come ashore and roost in trees as they do further north in Mozambique, where the human population is low.

11.5.2 Offshore birds: Terns; Gulls; Skuas

Ten species of Tern (Laridae) have been recorded flying over the seas around Inhaca Island (Brooke *et al.* 1981). The Caspian Tern (322), Lesser Crested Tern (325), Swift Tern (324) and the Common Tern (327) mingle with the sandpipers and plovers on the sand-spit at Ponta Raza in March-April. The Sooty Tern (332) and others, including the rare Blacknaped Tern (331) from the islands of the Indian and Pacific Oceans, have been seen from boats in the bay. Two or three Caspian Terns and Swift Terns, probably overwintering on the island, have been reported in July. Most Terns have large pointed wings, bent at the 'elbow', and forked tails. They fly in large flocks and individual birds plunge into the sea after hovering for a moment with bill pointing downwards. The terns disappear under water and on emerging they immediately fly away and swallow the fish while in flight. The fishes they catch are relatively small, about 20 g in weight. Two thousand Common Terns were once recorded in November off the island's shores.

Gulls (Laridae) are not very common around the island. The Greyheaded Gull (315) invades the beach near the fishing harbour to scavenge for dead fish from the nets; the Kelp Gull (312) sometimes rests on the sand-bank between the northern point of Inhaca Island and Portuguese Island. Both these species scavenge from the surface of the sea while in flight.

Three species of Skua, occasionally seen flying over the sea in the Bay of Maputo, are usually single, or sometimes in twos or threes. Both the Arctic and the Subantarctic species (307 and 310) occur as well as the Pomarine Skua (309). These are birds of the open ocean and do not come ashore; they are similar in appearance to gulls with a more pronounced hook on the beak, stronger and longer wings and darker plumage. They fly higher than the gulls and descend to the water surface to steal food, mainly fish, from the gulls and terns, attacking them singly or in pairs. The Arctic and Pomarine Skuas breed on Arctic shores and the Subantarctic Skua breeds on islands in the Southern Ocean.

11.5.3 Ocean birds: Gannets; Albatrosses; Petrels; Prions

Whilst gulls, terns and cormorants are almost always present, sightings of large birds depend very much on weather. Sometimes after a storm the larger birds may be seen from the ferry boat to Inhaca Island, on their way to the Indian Ocean. Once in November, in addition to the gulls, terns and skuas, the crowds of birds sighted included the Shy Albatross (11), Softplumaged Petrel (24), Cape Gannet (53), Sooty Tern (332) and Whitebreasted Cormorant (55). On other occasions there were, in addition, the Wandering Albatross (10), the Yellownosed Albatross (14) and the Whitechinned Petrel (32).

Cape Gannets (53) (Sulidae) usually fly within sight of land; here they are brownish, immature forms which migrate up the east coast of Africa from breeding grounds on the islands off the coast of Namibia.

The Shy Albatross (Diomedeidae) travels far from its breeding grounds around Tasmania and New Zealand; its cry is a single long-drawn-out 'waak'. The Yellownosed Albatross breeds in the Tristan da Cunha group of islands in the south Atlantic Ocean and is silent at sea, although on the breeding islands it has a high-pitched repetitive call. These large birds catch squids as well as fishes near the surface of the sea and do not dive deep.

Flocks of Pintado Petrel (or Cape Pigeon) (21) (Procellariidae) are frequently seen accompanying the pilot boat in the Bay of Maputo to the pilotage anchored just north of Portuguese Island. The birds sometimes rest on the sand-bar between the northern point of Inhaca Island and Portuguese Island. It has been reported that sometimes they come ashore and feed on pink ghost crabs (Brooke *et al.* 1981), but their usual diet consists of squids and fishes in the surface of the sea. This species is black and white like other petrels and albatrosses but it has a white chequered pattern on the dorsal surface of its outstretched, black wings. The flight alternates between flapping and gliding and it chatters as it flies. In summer they fly south to breed on subantarctic islands and on Antarctica. The Whitechinned Petrel (32) is occasionally seen from boats near Inhaca Island. It is completely black except for its white beak. This bird is almost as large as an albatross and is silent at sea, with slow graceful flight, low over the water.

Small Broadbilled Prions (29) of the same family are sometimes washed ashore during storms and have, by chance, been found dead on the shore. They are small birds that feed on plankton, filtered from the surface of the sea through the fine lamellae on the palate. They come from the southern oceans and breed on subantarctic islands. In winter they fly northwards, with fast twisting flight, and reach Mozambique.

12

AMPHIBIANS, REPTILES AND MAMMALS

D.G. Broadley (Natural History Museum, Bulawayo, Zimbabwe) and M. Kalk

12.1 AMPHIBIANS AND REPTILES (D.G. Broadley)

The varied terrain and abundant vegetation on Inhaca Island provide wide scope for reptilian habitats and freshwater swamps facilitate amphibian breeding. Nine species of amphibians and thirty-five species of reptiles have been recorded, including four species of turtles in the sea off the coast of the island (Broadley 1990).

12.2 IN AND AROUND FRESHWATER SWAMPS

A large freshwater swamp occurs in the northern part of the centre of the island and several smaller swamps lie between the low dunes on the peninsulas. The swamps support a dense growth of reeds and bulrushes and the periphery of the swamps may be cultivated or support the growth of sedges, grasses and ferns (see Section 10.3). The vegetation dies back in winter and reeds are cut on a large scale for housing; the water table drops, but the vertebrates which frequent the swamps are adapted to withstand intermittent drought.

12.2.1 Platanna; Reed Frog; Dwarf Puddle Frog; Grass Frogs; Guttural Toad

Among the frog species on the island there is a range of increasing independence of water, from completely aquatic to one completely terrestrial species which avoids water altogether, even for breeding. All the other species return from the grassland or bush to water to lay eggs; on summer nights the air is vibrant with a chorus of specific mating calls of frogs.

The Platanna, *Xenopus laevis* (Fig. 12.1a), is perfectly streamlined when swimming with legs extended and fully webbed feet; it also has lateral line sensory organs, similar to those of fishes. It lives entirely below water and rises to the surface to expel air and refill its lungs. This frog's skin is an accessory breathing organ, rich in blood capillaries. *Xenopus* hibernates in the soft wet mud in the winter when the upper layer of the swamp dries, and the skin then becomes the sole respiratory organ. In summer it feeds on insects under water; the lack of a tongue is compensated by an elongation of the fingers which are used to push food into its mouth. When there has been sufficient rain to provide free water above the ground in the swamp, the frogs breed in the water. The tadpoles have a relatively long life (6-8 weeks) and the frogs mature the following year.

Two species of Reed Frogs (Hyperoliidae), with small slender bodies, webbed feet and adhesive pads on their toes, cling to the stems of bulrushes, reeds and sedges during summer days, feeding actively on insects among the marginal plants and calling incessantly at night (Passmore and Carruthers 1979). The pale yellowish-green Tinker Reed Frog *Hyperolius tuberilinguis*, has a mating call which sounds like intermittent tapping on metal. *H. marmoratus*, at Inhaca Island, has black and white stripes with some dark mottling (Fig 12.1b); its call is a loud repetitive whistle. The latter lays its eggs on submerged plants and it may be found from the Cape (in South Africa) to the tropics, its colour pattern changing with latitude.

The smallest frog on the island, the Dwarf Puddle Frog *Phrynobatrachus mababiensis* (15 mm) (Ranidae), has a brown, warty skin, no webbing on its fingers and very little between the toes. It has a small head, squat body and very short legs. It spends little time in water and even then only at the very shallow edge where tiny black eggs are deposited in a single, thin layer of jelly which floats on the surface. Its larval life in water is short (4-5 weeks) and it becomes completely terrestrial, feeding on insects amongst the vegetation. The structure has been greatly modified from that of the ancestral Grass Frogs of the same family which have a stream-lined body and webbing between the toes, enhanced by the separation of the metatarsal bones. The Grass Frog *Ptychadaena oxyrhynchus* (Fig. 12.1c) leaps as well on land as it swims under water, but when molested it leaps into the undergrowth *away* from water.

In contrast to the Grass Frogs, the Toads (Bufonidae) are more adapted to life on land. The thick skin of the Guttural Toad, *Bufo gutturalis* (70 cm), the most widespread toad in Southern Africa, is brown, thick, rough and dry dorsally and the underside is pale and granular. It does not lose as much water on land as frogs do, because the latter have thinner skins. Two raised patches of skin just behind each tympanum, covering the parotid glands, secrete an irritant, toxic substance, defensive in function. Toads hop or run and do not leap on land; they are not strong swimmers, the webbing on the toes being stiff and minimal. They hibernate amongst the vegetation some distance from the swamp but return there to breed, travelling *en masse* from the hibernating area.

12.2.2 Eastern Hinged Terrapin; Brown water snake; Natal Green Snake

The Eastern Hinged Terrapin, *Pelusios castanoides* (Pelomedusidae), is capable of surviving in the mud of the swamp in a dry period. It is an aquatic, 'side-necked' chelonian, and it lives amongst vegetation on the fringe of the swamp. The long, smooth, brownish shell has a hinge at the front of the yellow plastron so that the head can be completely withdrawn by bending the neck sideways; the shell is then closed at the hinge, almost completely containing the animal. It feeds on aquatic insects, snails, frogs and on plants. It lays soft-shelled eggs amongst the damp vegetation on the fringe of the swamp.

The nocturnal Brown Water Snake, *Lycodonomorphus rufulus* (Colubridae), is a small constrictor (60 cm) which feeds mainly on frogs under cover of the vegetation around the swamp. The Inhaca Island record is based on a female collected at night while it was swimming among *Phragmites* reeds in a swamp; she had a Reed Frog in her stomach. This snake lays 6-10 eggs in midsummer amongst vegetation at the water's edge.

The Natal Green Snake, *Philothamnus natalensis* (Colubridae) is a harmless, slender, diurnal tree-snake that is often mistaken for a young Green Mamba; as far as it is known, no mambas occur on the island. This tree-snake is bright green above and paler below; juveniles have black cross bars anteriorly which fade out in the adult. This species may exceed a metre in

Fig. 12.1 Amphibians in or near freshwater:
a. Platanna, *Xenopus laevis* (x 0,8);

b. Painted Reed Frog, *Hyperolius marmoratus taeniatus* (x 1).

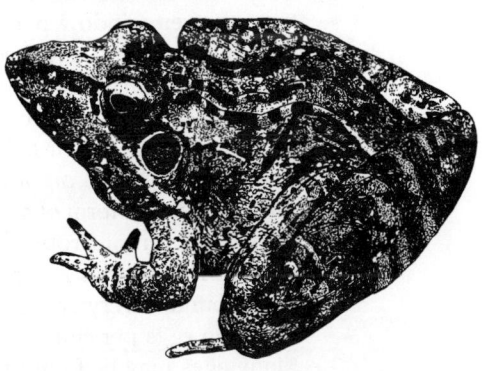

c. Grass Frog, *Ptychadaena oxyrhynchus* (x 0,8);

Fig. 12.1 Amphibians in or near freshwater:
d. Bushveld Rain Frog, *Breviceps adspersus* (x 1) in burrow; (after Passmore and Carruthers 1979).

length and is found in forests, thickets and reedbeds, feeding on both frogs and lizards.

12.3 IN THE GRASS, AMONG TREES AND ON CULTIVATED LAND

The cultivation of crops and the houses are to some degree concentrated in the low-lying central part of the island surrounding the swamps. The soil is sandy and the amount of leaf litter and humus is low; trees are scattered among long grass. Termites, grasshoppers, beetles and other insects and small rodents are more common than elsewhere and non-breeding frogs and toads are available as food for reptiles.

12.3.1 Bushveld Rain Frog

The small Bushveld Rain Frog, *Breviceps adspersus* (Microhylidae) (Fig. 12.1d) is not only completely terrestrial but lives in a subterranean burrow, appearing above the ground only during the rains when it feeds on termites, that have swarmed and shed their wings, or on ants and beetles. The female lays a few large eggs, embedded in viscous jelly, in an underground chamber and the tadpole stage is passed within the egg. *B. adspersus* is stout, shaped like a cushion with a small leg at each corner; females may exceed 50 mm in length but males are less than half the size. When molested they inflate the lungs with air, swell up and become even more rotund.

12.3.2 Geckos; Flap-necked Chameleon; skinks; Yellow-throated Plated Lizard

Two species of Gecko (Geckonidae), one diurnal and the other nocturnal, haunt the huts in the villages or rest on dead trees. Geckos are slender lizards, unusual in having on the tip of their toes scales bearing minute hairs that hook into very small pores or cracks on smooth surfaces so that they can walk on walls, upside-down on ceilings or on smooth poles and trunks of trees or even window panes (Fig. 12.2a and a'). The pupils of the large eyes dilate widely at night and close by day, becoming tiny specks. The eyelids are not movable, having become fused; they form transparent membranes, covering each eye like a contact lens; the membranes are cleaned by licking with the long fat tongue. When they live in communities, geckos often shed their tails whilst fighting but the tails can be regenerated. The Cape Dwarf Gecko, *Lygodactylus capensis* (7 cm) (Fig. 12.2a), a diurnal species with a round pupil, has a 'fifth foot' in the form of an adhesive pad on the tip of the tail. It feeds mainly on termites.

The nocturnal Tropical House Gecko, *Hemidactylus mabouia* (12 cm) feeds on moths or beetles. This species is capable of considerable colour change from very pale grey on a light background to very dark grey on a charred tree trunk. Like all African Geckos, the female lays only two eggs per clutch which are clearly visible through the abdominal skin before laying. Many eggs may be found together in nooks since several females may lay in the same place. Geckos communicate with one another by a variety of crackling sounds.

Chameleons (Chamaeleonidae) are unlike other lizards in many ways: their scales are small as in geckos and do not overlap; the head and body are laterally compressed and the neck is not defined; the toes are bound together into two opposing bundles. The tail is prehensile and cannot be shed or regenerated. The eyes protrude and can move independently, swivelling around to keep its insect prey in view. The sticky tongue can be inflated and shot out, to a length longer than the body, to catch an insect. The Flap-necked Chameleon, *Chamaeleo dilepis* (30 cm) may be found on trees or on the ground (Fig. 12.2b). Females are usually green with white lateral stripes or spots, males are orange or brown. Their proverbial change in colour is mediated by the nervous control of chromatophores.

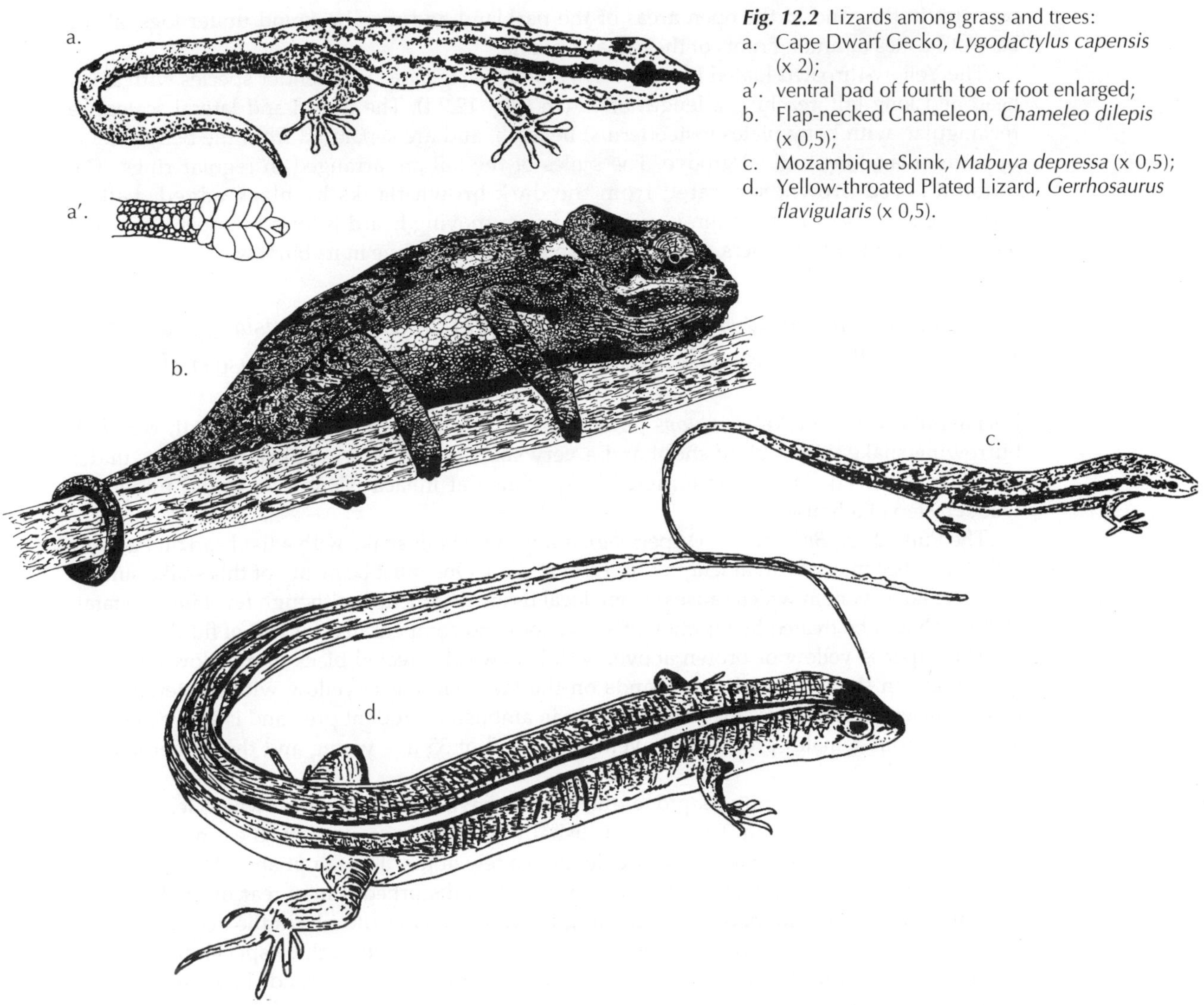

Fig. 12.2 Lizards among grass and trees:
a. Cape Dwarf Gecko, *Lygodactylus capensis* (x 2);
a'. ventral pad of fourth toe of foot enlarged;
b. Flap-necked Chameleon, *Chameleo dilepis* (x 0,5);
c. Mozambique Skink, *Mabuya depressa* (x 0,5);
d. Yellow-throated Plated Lizard, *Gerrhosaurus flavigularis* (x 0,5).

When a chameleon is disturbed, a black and yellow pattern is assumed; an angry chameleon turns black and a sleeping one becomes pale lemon yellow. The internal skin of the throat is bright orange except in breeding males which develop a pale grey throat to 'placate the female', which inflates its orange throat in threat displays. The female lays on average a clutch of 35 eggs during the rainy season in the softened soil, working through the night to excavate with her legs a hole in which to deposit her eggs, then labouring until dawn to fill it up again. The eggs do not hatch until the beginning of the next rains, when the ground is soft enough to allow the hatchlings to claw their way to the surface.

One of the most common lizards on the island is the Mozambique Skink *Mabuya depressa* (Fig 12.2c) a grey-brown, dark striped species (20 cm) which is often seen basking in the sun on tree trunks.

Wahlberg's Snake-eyed Skink, *Panaspis wahlbergii*, is a small fossorial skink which has the lower eyelid fused with the upper, the eye being visible through a transparent 'window' formed by a domed scale, as in a typical snake. Its limbs are small but usually have five digits; the ear opening is visible and the body scales are smooth. The back is pale brown, and has a blackish lateral band with a white stripe below it; the underside is white or greyish, but vermilion in breeding males. This species forages for termites beneath leaf litter and logs.

The Golden Blind Legless Skink, *Typhlosaurus auranticus*, is of moderate build and attains a length of 23 cm. It is pale orange above with traces of black stripes on the tail. This species burrows in the sand in the open areas of the parkland and may be found under logs, at the base of bushes or among roots of the cassava plants.

The Yellow-throated Plated Lizard, *Gerrhosaurus flavigularis*, is a slender species with small head and long tail, reaching a length of 45 cm (Fig. 12.2d). The dorsal and lateral scales are rectangular, with bony plates (osteoderms) beneath, and are separated from the belly shields by a granular, longitudinal groove. The scales of the tail are arranged in regular rings. The dorsum is red-brown, separated from the dark brown flanks by black-edged, yellow dorsolateral stripes; the ventrum is cream. This fast-moving lizard is found in open areas and feeds largely on grasshoppers; when it is alarmed it seeks refuge in its burrow.

12.3.3 *Fornasini's Blind snake; Puff adder; Egyptian cobra; Boomslang; Savannah Vine snake; Olive Grass snake; Herald Snake; Common Brown House snake*

Fornasini's Blind Snake, *Typhlops fornasinii* (Typhlopidae) (18 cm) is a small, blackish burrowing snake with a blunt snout and a very short tail with terminal spine. It lives under leaf litter and feeds largely on termites. A specimen at Inhaca Island was found under the wooden step of a house.

The Puff adder, *Bitis arietans* (Viperidae), is a heavily built snake with a flat head and a short tail; it may become 100 cm in length (Fig.12.3a and a'). One must be aware of this snake since it has a cytotoxic venom which causes severe local tissue destruction, although few bites are fatal. Victims should be treated by injection of antivenom and compensated for loss of fluid.

This viper is yellow or brown above, with backward-directed black and yellow chevrons (V-shapes) on the body and cross-bands on the tail; below it is yellow with scattered short black transverse bars. This sluggish snake lies in ambush for rodent prey and is active at night. The viviparous females produce on average a brood of 35 live young and then hibernate for several months.

The snakes in the family Elapidae, the cobras, have front fangs into which lead glands secreting a powerful venom. The venom is usually neurotoxic i.e. it affects the central nervous system, and an untreated bite may cause death in a few hours. The Egyptian Cobra *Naja haje* is a robust snake that can attain a length of 2.5 m. When disturbed it may rear up its head and neck and spread a broad 'hood' by extending forward the elongate ribs in the neck in a threat display. Adults are dark brown to black above, sometimes with lighter speckling or broad cross bands; the underside is yellow with brown mottling and a broad dark brown band across the neck; hatchlings are yellowish. This species inhabits open parkland and is active at night; it takes a wide variety of prey, with a preference for the toad, *Bufo gutturalis*. A victim of its bite requires large doses of antivenom. The Forest Cobra, *Naja melanoleuca* (Fig. 12.3c) has also been recorded at Inhaca Island (see Section 12.3.2). If one comes across a cobra, one should stand perfectly still until it retreats; it very rarely bites.

Among the species of Colubridae recorded from this area, two species are especially dangerous to man, whilst two others give painful bites. One rarely sees them but one should be forewarned.

The Boomslang, *Dispholidus typus* is a large venomous diurnal tree-snake, back-fanged, with a short head and large eyes, narrow, keeled scales and a long tail; it may reach 2 m in length (Fig. 12.3b). Juveniles are blackish above, with blue spots on the tips of some of the scales; they are whitish below, heavily stippled with red-brown; the iris of the eye is green and the throat bright yellow. When they reach a length of about a metre, these snakes gradually assume adult colouration, losing the green eye and yellow throat, whilst males usually become green, with or without black-edged scales, and females become grey or brown, although green females are known from Mozambique and Natal. The Boomslang preys heavily on chameleons, but also raids nests for fledgling birds, whilst some individuals prefer frogs. It is the most dangerous of the back-fanged snakes, although bites are rare, since the boomslang tries to avoid contact with man. The extremely toxic venom causes internal haemorrhage, which may be fatal within about 48 hours. A specific antivenom is required and

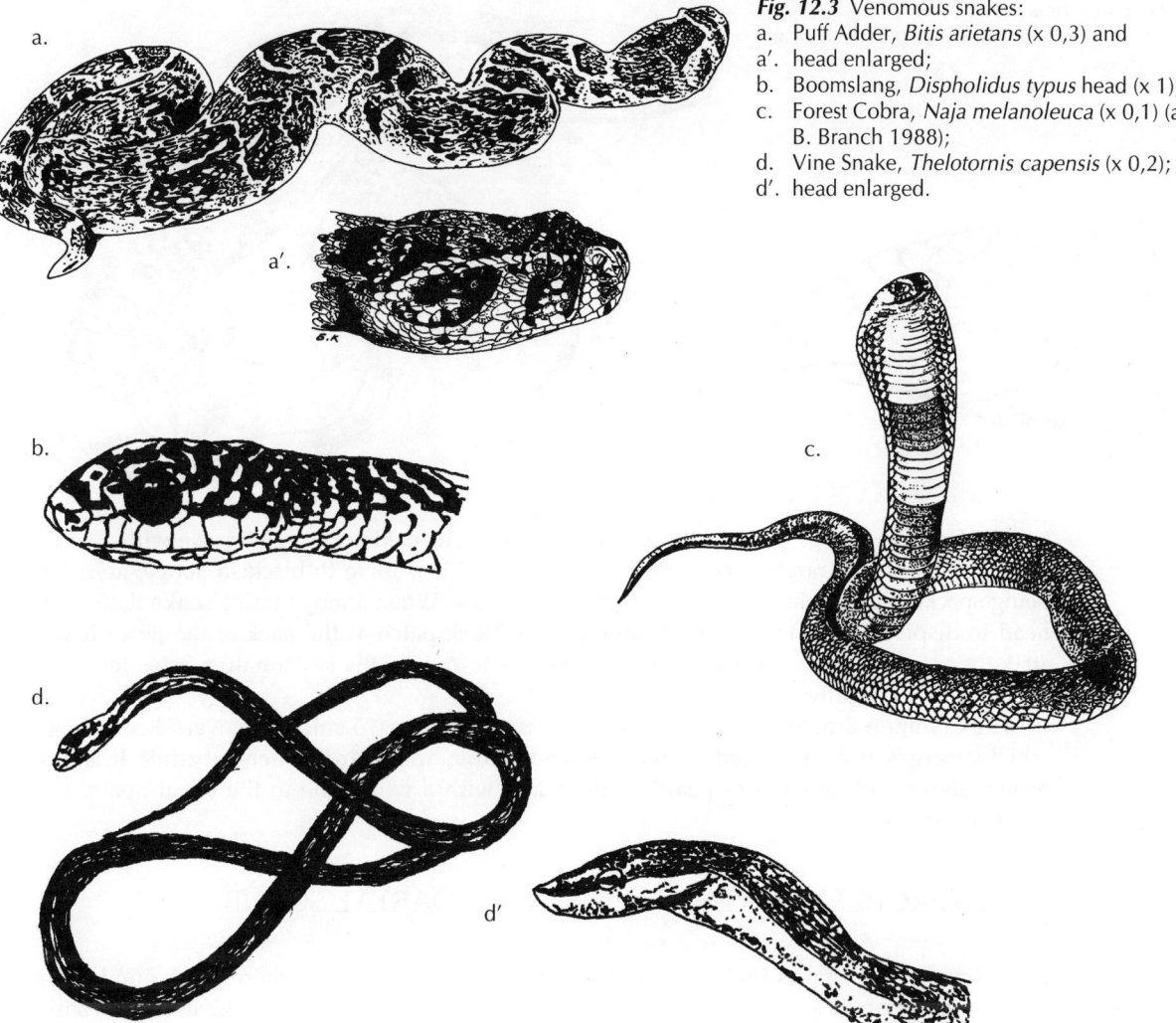

Fig. 12.3 Venomous snakes:
a. Puff Adder, *Bitis arietans* (x 0,3) and
a'. head enlarged;
b. Boomslang, *Dispholidus typus* head (x 1);
c. Forest Cobra, *Naja melanoleuca* (x 0,1) (after B. Branch 1988);
d. Vine Snake, *Thelotornis capensis* (x 0,2);
d'. head enlarged.

this is available only from the South African Institute of Medical Research in Johannesburg; it is in short supply owing to the very low venom yield of the Boomslang.

Another tree-snake, the Savanna Vine Snake, *Thelotornis capensis* has a lance-shaped head with large eyes and horizontal pupils, a very thin body and a very long tail (Fig. 12.3d and d'). The head is blue-green with a black and pink Y-shaped marking on the top, pink on the temporals (sides of the head) with black margins to the plates; the grey-brown body is mottled with ash grey, black and pink, resembling a lichen covering a branch of a tree. This back-fanged diurnal snake sits in a tree or dead bush with a commanding view. It relies on its camouflage to escape observation and remains motionless for hours, sometimes extending its orange-coloured forked tongue. When a lizard, small snake or frog comes into view it stalks the prey slowly, finally seizing it after a short rush. Birds' nests are also raided for eggs and fledglings. The venom of this snake can also cause internal haemorrhaging in man, but few bites produce serious symptoms. There is no antivenom available so treatment is restricted to blood transfusions and injection of fibrinogen and vitamin K.

The Olive Grass Snake, *Psammophis phillipsii*, is a large, robust, brown, fast-moving back-fanged snake with a blunt head and a long tail. This is a savanna species, often found in reedbeds or long grass; one was found on the edge of the mangrove on Portuguese Island. The diet consists largely of lizards and rodents, but frogs, snakes and small birds are also eaten. The bite of this species is painful, but not dangerous to man.

Fig. 12.4 Forest reptiles:
a. Giant Legless Skink, *Acontias plumbeus* head (x 1,5);
b. Tree Leguaan, *Varanus albigularis* (x 0,2).

The Herald Snake, *Crotaphopeltis hotamboeia*, is a nocturnal back-fanged snake with a broad head and short tail, rarely exceeding 75 cm in length. It is olive to blackish above, juveniles being speckled with white; it is uniformly white below. When annoyed, this snake flattens its head to display an orange upper lip and a shiny black patch at the back of the head; it will strike viciously, but its venom is not dangerous to man. This savannah species feeds on amphibians, especially toads.

The Common Brown House Snake, *Lamprophis fuliginosus* (75 cm) is a powerful constrictor which emerges at dusk to feed on rodents when adult, and lizards when subadult. It is red-brown above and 'mother- of-pearl' white below with a white line to the snout above and below the eye.

12.4 EVERGREEN FOREST AND DENSE COASTAL SCRUB

Some lizards and snakes prefer the shady, closed canopy of forest where the ground has a thick layer of humus and leaves above the sand. Insects and other arthropods, snails and slugs are available as food and some species feed on other reptiles or their eggs. Some of the species described in the previous section may also live here.

12.4.1 Burrowing Skinks; Tree Leguaan

The Mozambique Dwarf Burrowing Skink, *Scelotes mossambicus*, is a small limbless species found in the sand and humus in thickets. It is pale brown dorsally, each scale dark-centred; ventrally it is pale with only traces of spots; the ear-opening is almost hidden. Females produce one or two young. The Zulu Dwarf Burrowing Skink (*Scelotes arenicola*) is somewhat similar; it has a dark vertebral stripe that breaks into two lines of spots; the tail is bluish. The Giant Legless Skink *Acontias plumbeus* (Fig. 12.4a) is a stout species with a short tail that grows up to 56 cm. The colour varies from dark brown to jet black. It was once found on a sandy path near relict forest on the western dunes.

In contrast to these small skinks a species of Varanidae, which includes the largest lizards in the world, occurs. This is the Rock or Tree Leguaan, *Varanus albigularis* (Fig. 12.4b) which frequents the forest. This lizard may attain a length of 140 cm and has strong limbs with sharp claws. The skin is tough and covered with small bead-like scales. The tail is longer than the body, cylindrical at the base and flattened near the end, with rectangular scales. The colour pattern is variable but basically blackish above with transverse rows of pale yellow spots and broad yellow crossbands on the tail; the head is grey above, yellow laterally and the limbs are speckled black and yellow. If attacked it may bite, scratch with its long claws or lash with its muscular tail; or it may sham death. This sluggish lizard feeds largely on snails, millipedes and beetles, but it also relishes carrion and will eat any small animal that it can catch.

12.4.2 Forest Cobra; Eastern Purple-glossed Snake; Shovel Snout; Variegated Slug-eater;

The Forest Cobra (Elapidae) (Fig. 12.3c) is a relatively slender species, one caught on Portuguese Island was 2,4 m long. It is the only cobra with shiny dorsal scales and it spreads a long narrow 'hood'. This cobra is yellow-brown, blotched with black above, darkening towards the tail, which is usually uniformly black; below, it is bright yellow, heavily marked with black patches and without distinct bands on the throat. Sometimes it may be found in mangroves, for its catholic diet of small vertebrates is known to include the mudskipper, *Periophthalmus*.

Six harmless species of Colubrid snakes have been recorded in the forest.

The Eastern Purple-glossed Snake, *Amblyodipsas microphthalma*, a small, back-fanged fossorial species, preys upon legless skinks. The Mozambique Shovel Snout, *Prosymna janii*, a small pale-brown, spotted snake is nocturnal and feeds on soft-shelled reptile eggs. The Variegated Slug-eater, *Duberria variegata* is a small, stout snake with a blunt head and short tail, brown-spotted above, yellowish white below with black marks. This gentle, slow-moving, shy snake forages among grass roots and rotting timber in damp places for slugs and snails, (of which there are eight species recorded on the island (see Section 14.2). Females produce 7-20 live young.

12.5 MARINE TURTLES IN THE SEA AROUND THE ISLAND

Inhaca Island has one shore facing the Indian Ocean and others sheltered by the Bay of Maputo where sea grasses grow. As with other phyla, different species of turtles frequent the exposed and sheltered shores.

12.5.1 Loggerhead; Hawsbill; Leatherback; Green Turtle

The Loggerhead Turtle, *Caretta caretta* (Fig. 12.5c) is most often seen on the east coast of the island. It has a very large head and an elongate shell which tapers posteriorly. The head and carapace are uniformly red-brown, the plastron is yellow and the skin pale olive; it may exceed a metre in length. They are sometimes seen floating in the waves as they rise before breaking, the golden yellow plastrons reflecting the sunshine. Loggerheads nest on the beaches of Maputaland. The young turtles drift in the surface waters of the Indian Ocean for the first three years of their life; adults feed in shallow coastal waters using their powerful jaws to crush crabs, molluscs and urchins.

On the east shore of Inhaca Island from November to mid-January about 15 females come ashore, often at the very place of their birth, climb the dune and dig their nests (Costa 1989). They are usually over four years of age. Sometimes they wander around and do not lay eggs, but return to the sea. About 100 eggs are laid at one time and completely covered with sand, taking on average 59 days to hatch. This is a little longer than on warmer shores. Sometimes the number of hatchlings is decimated when they are caught by the pink ghost crab *Ocypode madagascarensis* but they are protected by rangers at Inhaca Island. Adults and newly hatched young swim northwards to the tropics, adults making about 40 km a day.

The Hawksbill Turtle, *Eretmochelys imbricata*, is usually found along coral reefs and possibly still occurs on those offshore on the east coast of the island. The thick, keeled shields of its carapace overlap like the tiles of a roof, and its long, narrow head is equipped with a hooked beak for extracting prey from crevices. The carapace shields are yellow, heavily streaked with red-brown and black; these are the source of the commercially exploited 'tortoise shell'. The adults feed on sponges; they do not breed locally.

The Leatherback Turtle, *Dermochelys coriacea* (Fig. 12.5b), is an offshore pelagic species but it breeds on the dunes of the eastern shore of Inhaca Island (see Section 3.8.1). This is the giant

Fig. 12.5 Marine Turtles:
a. Green Turtle, *Chelonia mydas* (x 0,05);
b. Leatherback Turtle, *Dermochelys coriacea* (x 0,05);
c. Loggerhead Turtle, *Caretta caretta* (x 0,07); (a-c after R.C. Stebbins 1966).

of the marine turtles, reaching a length of 1,7 m. This turtle has evolved considerable control over its body temperature, unlike most reptiles. The elongate carapace lacks shields, but the rubbery-textured skin is moulded into five oily, longitudinal ridges thus providing insulation. The head is equipped with a hooked bicuspid beak and the front flippers are long. Adults are black above, usually with scattered white spots, and the lower surfaces are white. Adult turtles feed exclusively on jellyfish and bluebottles (Fig 3.24d); they are therefore vulnerable to floating clear plastic bags which, if swallowed, can block the digestive tract. Juveniles are light grey in colour (Fig. 3.24c).

The Green Turtle, *Chelonia mydas* (Fig. 12.5a), is also a large species, attaining 1,4 m. The head is dark brown to black, usually with each shield bordered white; the carapace may be streaked like the Hawksbill's, spotted or a uniform colour, but the shields are smooth and juxtaposed, never overlapping. Adults are largely herbivorous on green algae (*Caulerpa* spp.) and red alga (*Gelidium* spp.) on rocks on the east coast, or they graze in *Zostera* beds and have been recorded at the mouth of the southern bay where there is a large bed of this sea grass. The principal breeding site for the local turtles is Europa Island in the Mozambique Channel; juveniles feed on fish and other floating animals.

12.6 MAMMALS (M. Kalk)

12.6.1 Terrestrial mammals: Fruit bat and Free-tailed bat; Yellow Golden Mole; Vlei rat; Multi-mammate mouse, and others

A survey of land mammals around the Marine Biological Station and the central area of the island in 1961 reported nine species of small mammals (Eloff and De Graaf 1963). Since then a few more nocturnal species have been seen in the forest, but some have not been positively identified, although their identity may be inferred from distribution records of mammalian species in similar forests in southern Mozambique.

Three species of bats are common: a fruit bat and two insectivorous bats. Wahlberg's Epauletted Fruit Bat, *Epomorphorus wahlbergi* (Fig. 12.6a and a') is about 12 cm long and has a wingspan of about 36 cm. Its head is like a dog's, with a snout and small ears. The wings, as in all bats, are elastic membranes each forming a web, supported by the fore-arm and elongated fingers, and also attached to the hind legs. Fruit bats have two claws on each wing, on the thumb and first finger respectively, with which they are able to manipulate the fruit that they pick from the forest trees at night. They rely on sight to find their way about except in total darkness, when they emit clicks, made with the tongue, that are heard on the rebound from objects. During the day the fruit bats hang upside down by their toes in small groups among the branches of trees and may be seen in the palm trees and the coconut groves. The Epauletted Fruit Bats are brownish in colour, lighter below, and have white patches of hair at the bases of the ears; the males have white-haired glandular patches on the shoulders resembling epaulettes.

The two insectivorous species of bats on the island are 'free tailed bats', since the tail is only partially attached to the web between the hind legs and the last section of about 3-4 cm hangs freely (Fig. 12.6b'). Insectivorous bats have only one claw on the thumb of each forelimb supporting the wings. The Angolan Free-tailed Bat, *Tadarida (Mops) condylura* is about 10 cm long, including the tail and has a wingspan of about 30 cm; the Little Free-tailed Bat, *Tadarida (Mops) pumila* (Fig. 12.6b) is slightly smaller. The head in both species has a broad pug nose with prominent open nostrils and wrinkled, muscular lips with bristles. The external ears are extraordinarily enlarged and round in outline with creases on the inner sides, and they are connected by a broad, erectile band of skin across the head. These bats catch their small insect prey on the wing by means of echo-location, emitting supersonic squeaks which are reflected back to their large ears. A variety of calls, made by the larynx, is used to estimate the speed, direction and position of the moving prey. By day, Free-tailed Bats crowd into small spaces between the rafters and roofs of buildings or in cracks in trees. They crawl into confined spaces, the tips of the narrow wings folded back and tucked underneath the rest of the wings, and they rest, huddled together. They make audible noises as they jostle for position before they emerge at dusk, one by one, and drop into flight, foraging for insects on the wing. The young are born hairless in summer, only one to each female, who may carry it about until it is weaned or may leave it behind whilst hunting. These bats frequent the buildings at the Marine Biological Station at Inhaca Island.

The Yellow Golden Mole, *Calchochloris obtusirostris* (Fig. 12.6c), has been found burrowing in the sand on the dunes among littoral scrub at Ponta Rasa, Barreira Vermelha and

Fig. 12.6 Terrestrial mammals:

a. Wahlberg's Epauletted Bat, *Epomorphorus wahlbergi* head and

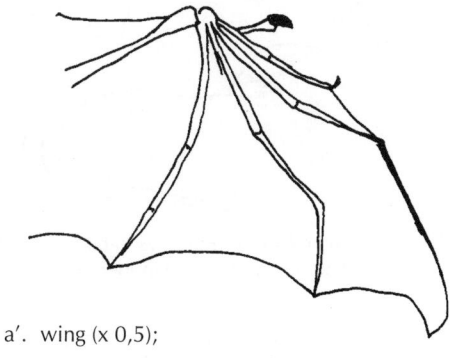

a'. wing (x 0,5);

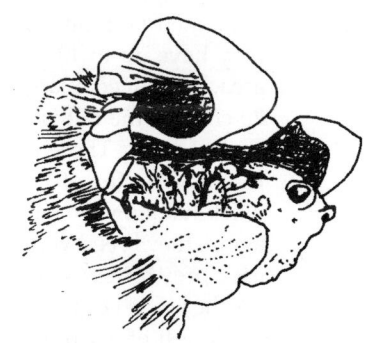

b. Little Free-tailed Bat, *Tadarida pumila*, head (x 1) and

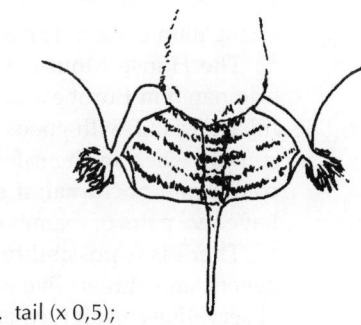

b'. tail (x 0,5);

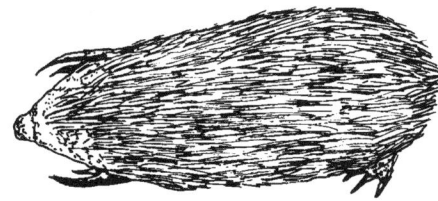

Fig. 12.6
c. Golden Mole, *Calchochloris obtusirostris* (x 0,5);

d. Multimammate Mouse, *Praomys natalensis* (x 0,3) (a-d after Skinner and Smithers 1990).

Ponta Torres, detected by the broad ridges which mark its tunnels. It lives entirely underground and is totally blind. The Golden Mole has many adaptations for burrowing: the legs are very short and strong, the claws on the forelimb are sharp, that on the third digit being the longest and shaped like a knife. The hind feet have smaller claws and a web between the toes which scoop away the sand loosened by the forelimbs. The snout has a smooth, leathery pad that aids in tunnelling. Ear openings are barely discernible beneath the fur but the internal ears are extremely sensitive to vibrations in the sand made by their insect prey in the sand or on the surface. The Yellow Golden Mole is about 10 cm long and has no tail; the long hairs are brown and glossy at the tips and the underfur is yellow. This species frequents the forested coastal dunes from KwaZulu to southern Mozambique.

Amongst the several rodent species on the island, the Vlei Rat, *Otomys irroratus*, is diurnal; it lives in damp grass near swamps and rarely enters water. It is short-legged and short-tailed, about 15 cm long, and its fur is exceptionally thick and shaggy, greyish brown in colour and darkened by a mixture of black hairs on the back. It has large yellow incisor teeth and laminate molar teeth. It is entirely herbivorous and feeds largely on grass stems, which it cuts into pieces about 5 cm long and carries to its nest on dry ground. It makes runways through the vegetation, dropping some cut grass on the way. These runways communicate with holes in the ground or hollows amongst dense vegetation, which are used as nests. They breed several times a year and produce 4-5 young each time.

The Multi-mammate Mouse, *Praomys (Mastomys) natalensis* (Fig. 12.6d) lives in or near cultivated land and has a diet of grain, seeds and insects. It is so-called because the female may have up to 12 pairs of mammae in two continuous rows from the chest to the inguinal area. It can feed a large number of offspring in one litter. The body grows to a length of about 20 cm and the tail is of similar length. They may nest in holes under trees or in the thatched roof of a hut. These nocturnal animals feed on seeds, wild fruits or maize on the cob and on insects. The fur of this mouse is fairly long, soft and pale to reddish brown, freckled with black hairs on the back and white below. After a few years of drought when the rains come again, they breed very quickly and produce very large populations in a few months, becoming easy prey for owls and snakes. They carry fleas and have, in some countries, been responsible for the spread of plague.

A nest of dried grass and leaves in the fork of branches of a mangrove tree was seen to be inhabited by a rat almost 20 cm long with a longer tail. This was probably the Tree Rat, *Thallomys paedulcus*, which is nocturnal and vegetarian. It feeds on dried seed pods (which are abundant in the evergreen forest) and usually nests in *Acacia* trees. The fur is yellowish-grey, fairly long and silky; the characteristic white underparts, white hands and feet are easily detected at night under strong illumination. Three more types of fluffy, brown rats have been captured.

The House Mouse, *Mus musculus* (75 mm) and the Black Rat, *Rattus rattus* (200 mm) originated in Europe and are now cosmopolitan; they have been inadvertently imported into Inhaca Island with goods in boats, as has happened elsewhere. They make audible twittering sounds when in the safety of their nests in houses. The Black Rat is actually dark grey in colour and nocturnal; it supplements its domestic diet with the chicks of small birds. They have five pairs of mammae and breed throughout the year.

There is a possibility that the Red Bush Squirrel, *Paraxerus palliatus*, and mongooses, genets, and shrews live on the island. The tracks of a wild pig have also been seen; but these observations await further confirmation.

12.6.2 Marine mammals: Dugong; dolphins; whales

Sometimes mammals are sighted from the lighthouse at Inhaca Island or seen from the ferry while crossing Maputo Bay.

Earlier this century dugongs, *Dugong dugon* (Fig. 12.7a), were common, grazing amongst the sea grasses that cover the sandy shoals in the Bay of Maputo, especially near Machangulo Peninsula. These mammals prefer shallow water 1-12 m deep and avoid fresh water. They are still fairly common on the coast of northern Mozambique and are now a protected species. On populous shores the increased use of fishing nets, in which they become entangled, and the delicacy of their flesh, have taken toll of their numbers. These so-called 'sea cows' (from their vegetarian habit) are dark grey mammals about 2-3 m long; they have two small fleshy anterior flippers for stability and a fleshy horizontal tail which propels them along at a leisurely pace of 6-10 km per hour. They may rest on the bottom or feed during the day if undisturbed, but have to surface every three minutes or so to breathe. The sea grasses they prefer, *Thalassodendron ciliatum*, *Syringodium isoetifolium* and *Halodule* ssp. are the common sea grasses of Inhaca Island shores and of sand-banks in the Bay of Maputo.

The head has a blunt snout and large jaws with a very thick upper lip, bearing bristles which help in tearing up the sea grasses. Incisor teeth develop into tusks in the male. The dugong cuts a swathe of sea grass a few metres long and piles it up for a time. It then shakes each mouthful before chewing it with its conical molar teeth. As in elephants, the anterior molar teeth are shed and the posterior teeth move into position for use.

About 5-6 dugongs may feed together in a group. They breed in shallow water and the young are born in summer, only one offspring being produced each season. The mother dugong suckles her offspring while lying on her back in the surface water, holding the baby with one flipper. As recently as 1976 a dugong was seen resting on its back, alongside the ferry boat in the Bay of Maputo. In June 1991 near Ponta Torres an adult over 2 m long was observed surfacing for breath several times before it swam away.

Two or more species of dolphin (Cetacea) frequent the seas around the Island. Some distance offshore in the Indian Ocean may be seen a small school of the Humpback Dolphin, *Sousa plumbea*, which grows to almost 3 m in length. This species prefers shallow coastal waters and feeds on fish which live among rocks offshore (Skinner and Smithers 1990). It has a slim body, slightly laterally compressed, bearing a dorsal fin, two small pectoral, rounded, notched flippers and a bilobed horizontal tail which propels the animal at a leisurely pace.

The Indian Ocean Bottlenosed Dolphin, *Tursiops aduncus* (Fig. 12.7b), may attain a speed of 30 km an hour, although they sometimes swim slowly as though at play, rising to the surface to breathe at intervals; seven minutes is said to be the maximum time under water when they dive. As in whales this ability depends on their having haemoglobin with a high oxygen-carrying capacity and a mechanism for shutting down the blood supply to various organs whilst maintaining that to the brain and muscles.

Dolphins and whales make sounds under water, and from the nature of the returning echo reflected back to them they can locate prey or obstacles while hunting. They also communicate with others of the same species by sounds. After a calf is born to a Bottlenosed Dolphin, the mother whistles constantly, apparently to familiarise the calf with her signature tune.

Dolphins are able to communicate information to their group about 'What? Where? and Who?' but, it is said, they cannot transmit information about 'How?' and 'Why?'. The Bottlenosed Dolphin feeds on fishes and squids and has evolved a sophisticated hunting technique. A school of some 200 dolphins may hunt together, swimming at high speed in a straight line, driving the fish ahead. The dolphins on the periphery then swim ahead and drive the fishes back again so that many of them are caught. The dolphins then reform in line and pursue the fishes in the opposite direction in the same way, reaping a rich reward. Mating is preceded by a complex courtship display. The young have a twelve-month gestation period and are suckled for between six and eleven months.

Whales are occasionally seen swimming far offshore in the Indian Ocean when the horizon is scanned from the lighthouse at the north-east point of Inhaca Island. The drastic decline in

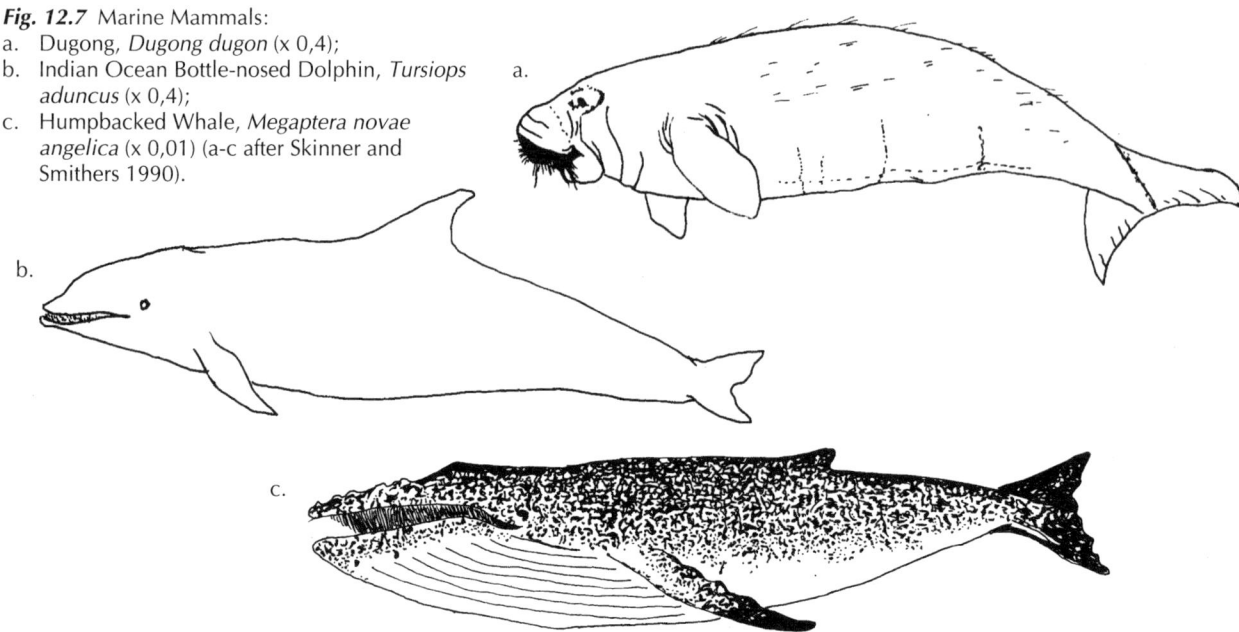

Fig. 12.7 Marine Mammals:
a. Dugong, *Dugong dugon* (x 0,4);
b. Indian Ocean Bottle-nosed Dolphin, *Tursiops aduncus* (x 0,4);
c. Humpbacked Whale, *Megaptera novae angelica* (x 0,01) (a-c after Skinner and Smithers 1990).

the populations of whales through overfishing has, however, considerably reduced this possibility.

Baleen whales feed on krill (the large shrimp-like crustacean, *Euphausia superba*) in Antarctic waters in the summer, at the estimated rate of one tonne a day. Feeding by actively filtering water through the hundreds of black, fringed whalebone plates in the mouth is a remarkably efficient way of catching euphausid shrimps, when a large concentration has been detected by means of echo location. Excess food is stored as blubber and is gradually used during the migration to tropical seas, where the planktonic shrimps and small fishes are not numerous. The whales travel north in July to tropical seas near Madagascar for calving and return to the Antarctic Ocean in September. They bring back the young whales which will grow to about 8 m long in a year. The females are the last to leave warmer waters and the first to return in the annual migrations. The gestation period is about 11,5 months and the lactating period is about 10 months, although the young are partly weaned earlier (Skinner and Smithers, 1990).

It is probable that the Humpback Whale, *Megaptera novaeangliae* (Fig. 12.7c), which is black dorsally and white ventrally, is the species occasionally seen off the coast of Inhaca Island. The flippers are unusually long, its head is covered with round knobs and it has a small dorsal fin, the 'hump'. The largest are now only about 15 m in length whereas before the recent depopulation by whale hunters they were up to 19 m long.

13

INSECTS

G. J. Ormel, Universidade Eduardo Mondlane

Insects play an essential role in the food web of terrestrial vertebrates and invertebrates, either directly as consumers or consumed, or indirectly in facilitating the reproduction of plants by participating in pollination and seed dispersal. Insect numbers are regulated by a variety of factors but mainly by predation by other insects and by parasitism.

The insect population on Inhaca Island is very similar to that of comparable habitats on the mainland, but there is more convenient access to many habitats on the island so that insect interactions and adaptations to their environment are more easily studied. A selection of some common insects on the island has been made to illustrate the taxonomy, the life styles and the distribution of insects in the major families represented; no attempt has been made to give a complete review, even of the orders; a tentative checklist of species is given in Appendix B.

13.1 DISTRIBUTION OF INSECTS ON INHACA ISLAND

The habitats of insects are very briefly as follows: coastal bush and evergreen forest on the dunes; cultivated ground with crops, scattered trees and grassland in the central open parkland; freshwater swamps; mangrove swamps; and seashores. Each of these should be subdivided into micro-habitats to accommodate the actual distribution of the insects. The flowers on the trees are frequented by insects seeking food as pollen and nectar, for example butterflies and moths, bees and bee-flies, blister beetles and flower chafers. Many insects feed on the leaves of trees as larvae or as adults but many more burrow into the wood or invade the roots; they may feed on living or dead parts or merely use the trees for shelter. Cocktail ants, tailor ants, bagworms, many kinds of beetles, termites, scale insects and shield bugs may be found on trees. On the ground beneath the trees or in the more open parkland, grasshoppers, crickets, locusts, cicadas, dung beetles, antlions, wasps and termites abound. In the freshwater swamps there are mosquito larvae, larvae of dragonflies and damselflies, whilst on the sea-shores there are few insects. On the surface of the channels in the mangrove swamps there are water-striders and in the mangrove trees, long-horned beetles, burrowing beetles, paper wasps, tailor ants and adult mosquitoes. Along the driftline there are robber flies and at high-tide mark at Ponta Mazondue, a burrowing staphylinid beetle, *Bledius* sp., which feeds on microscopic interstitial algae, has been found. Rhinoceros beetles damage the coconut palms and the tortoise-shell beetle attacks sweet potato crops or, on the dunes, the sand-binding plant *Ipomea brasiliensis* of the same family.

At different stages in the life history of an insect it may live in different habitats: some larvae are parasitic whilst the adults are not; some larvae are aquatic and the adults are terrestrial; adults may frequent flowers whilst their larvae live on leaves or are nurtured in nests in the ground. It seems, therefore, more convenient to describe the insects on Inhaca Island in taxonomic order of families rather than according to habitat.

13.2 ODONATA: dragonflies and damselflies

Adults in this order have an elongate, slender abdomen, a compact thorax, two pairs of membranous wings with complex venation and a highly mobile head with very large, conspicuous compound eyes that have very keen sight. They fly fast and far and are very skilled hunters of small insects on the wing. When resting, the dragonflies spread their wings horizontally at right angles to the body, whilst damselflies hold their wings together above the body. Both groups may be seen on the wing around the larger freshwater swamps and even on the adjacent dunes. The damselflies are smaller and more fragile than the dragonflies.

The juvenile stages (nymphs) of both groups of Odonata live in fresh water around the swamps. They have an elongated labium with a pair of pincers at the end, which is jointed to form a capturing organ, the 'mask'. This is folded beneath the head when at rest. When the nymph sees its prey it shoots out the mask with great speed and grasps it with its pincers. They feed on many kinds of animals, including small fish.

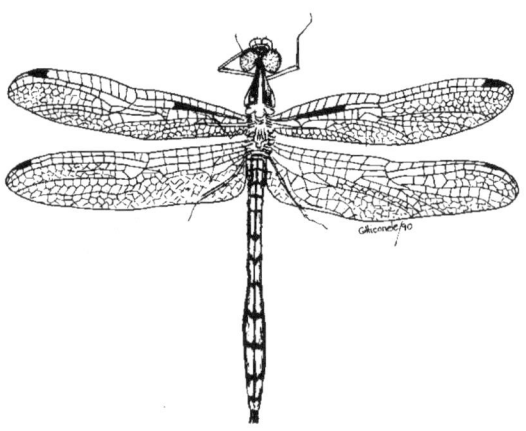

Fig. 13.1 Dragonfly, *Hemistigma albipunctata* (x 1).

Mating occurs on the wing; the male holds the female behind the head with its claspers situated at the tip of the abdomen. The female bends her abdomen forward to the accessory genitalia of the male to form the so-called 'wheel position'; transfer of sperm is effected while flying (Thompson and Dunbar 1988). The duration of copulation varies from a few seconds to several hours. Females mate with a succession of males, but the eggs are always fertilised by the last male, because he is able to remove the sperm of his predecessors in ways peculiar to each species. The male watches over the female whilst she lays her eggs; she may drop them into standing water or insert them in leaves if she has a specialised ovipositor. Males are often territorial and guard sections of the shoreline around a swamp where they attempt to attract females. Studies of mating behaviour, territoriality and the location of oviposition sites can easily be carried out on these large insects.

The most common dragonfly (Pinhey 1981) is *Palpopleura lucia* which has a wing-span of 50 mm and a body-length of 30 mm; the abdomen is light blue with black patches and the wings have large, dark areas. *Crothemis sanguinoleuta* is larger and has a red abdomen; *Orthetrum* sp. has a blue and yellow thorax and the abdomen is anteriorly blue and posteriorly black. *Hemistigma albipunctata* (Fig. 13.1) is 35 mm in length and has a green abdomen; the costal margin of the wings is yellowish and the wing-spots are creamish white.

Ceriagrion glabrum, a common dragonfly, is about 40 mm long and has a wing-span of 50 mm; the abdomen is red, the thorax is red dorsally and its sides are red, green and white.

13.3 ISOPTERA: Termites

All species of termite are social and feed on dead plant material and thus facilitate mineralisation of organic matter in the ground. Some species damage wood structures in buildings. In large colonies, workers protected by soldiers forage through extensive tunnel systems underground and make soil-covered tracks above ground and up tree trunks. Termite nests are carefully built and maintained by workers whilst they are defended by the soldier caste. The main enemies of termites are ants. Termites provide food for some species of toads, skinks and birds, especially when swarms of reproductives are produced.

Members of four families of termites occur on Inhaca Island:

(i) Dry wood termites (Kalotermitidae), which maintain nests entirely in dead wood and have small colonies in which the worker caste is absent, but the work of the colony is performed by immatures of other castes; *Psammotermes allocerus* is characteristic.

(ii) Subterranean wood-eating termites (Rhinotermitidae) in which the soldiers exude a drop of sticky liquid from a gland in the head when alarmed;
(iii) Termitidae which consume humus, grass or wood and grow fungus as food. Those that build large mounds are not present on the Island; *Odontotermes* and *Schedorhinotermes lamianus* are present.
(iv) Hodotermitidae are harvester termites which forage freely during the day, collecting twigs and grass. They make their nests underground and their presence is indicated by small heaps of soil on the surface.

Very few animals are able to make use of the cellulose in plant material as food. All metazoan animals that can digest cellulose do so with the help of some micro-organisms or fungus. In the case of termites, the primitive ones such as the hodotermitids make use of the flagellate protozoans in the hind gut to digest the plant material they eat. The more advanced Termitidae cultivate fungi on 'combs', composed of decaying vegetable matter derived from termite faeces, which form the so-called 'fungus gardens'. The enzymes of the fungus initiate the digestion of the wood into cellulose. The breakdown process is completed by bacteria in the hindgut of the termite. By eating some of the fungus as well, they also obtain vitamins and proteins. The development of sociality in termites is thought to have originated from the requirement that individuals reinfect themselves with these micro-organisms after a moult. This is made possible by living in groups in which nutritive secretions are exchanged between individuals, i.e. by trophallaxis.

13.4 MANTOIDEA: praying mantids

Mantids hunt insects by keeping the elongated body motionless whilst the head with large, bulging compound eyes swivels around to keep its prey in view. The mantid suddenly grasps its prey by extending the first pair of legs, unfolding the elongated, toothed claw on each front leg. At rest the claws are held folded and closed, the spines fitting into a groove on the toothed segment behind, an attitude reminiscent of hands folded in prayer.

The female secretes a silken foam which she kneads with her abdomen and cerci into a neat row of receptacles, into each of which she deposits an egg. The silken mass hardens on exposure to air and after a month or so the young emerge through minute valves. When they have expanded and dried they are black and characteristically curl the abdomen over the body as they walk. The large green mantid, *Sphodromantis gastrica* is found on bushes and trees among the foliage. It feeds on caterpillars.

13.5 ORTHOPTERA: crickets; locusts; grasshoppers

In the Orthoptera the males produce sound by stridulation, i.e. rubbing parts of the body against one another. In crickets the forewings are used; in grasshoppers and locusts the femurs of the hind legs are rubbed against the forewings. The auditory organs (tympana) in the grasshoppers and locusts are on the sides of the first abdominal segment, whilst those in the crickets are at the base of each tibia of the front legs, just below the joint with the femur. Crickets produce many kinds of sounds, such as calling sounds to attract females, aggressive sounds to repel competing males. Locusts differ from grasshoppers in possessing a swarming phase in their life-history but this does not occur on Inhaca Island.

The females of locusts and grasshoppers deposit their eggs in batches enclosed in pods in the ground whilst the crickets lay them singly either in the ground or in plant tissues. The nymphs resemble the adults in many ways, but do not have the typical adult features such as ocelli and functional wings. After several moults they develop into adults and there is no pupal stage.

Crickets (Gryllidae) are able to stridulate loudly but they are not easily located because these insects can muffle the sound by lowering the forewings which are normally raised during stridulation. The common garden cricket, *Gryllus bimaculatus* is dark brown in colour and has a pale spot on the posterior part of each forewing.

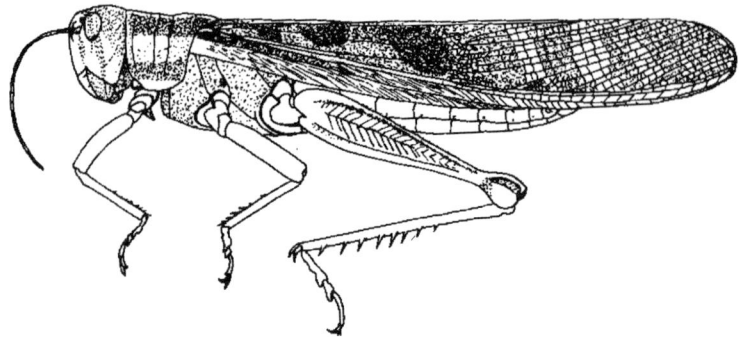

Fig. 13.2 Red locust, *Cyrtacanthacris septemfasciata* (x 1).

Locusts (Acrididae) exhibit 'phase polymorphism', i.e. they may be solitary, in transient phase, or gregarious. In these phases they differ in colour, morphology and behaviour, sometimes to such a large extent that in the past members of the same species but in a different phase have been considered different species. On Inhaca Island the gregarious phase does not seem to have been reported, although locusts on the island do change colour during the year. The red locust, *Cyrtacanthacris septemfasciata* (Fig. 13.2) is brown at first, later it becomes reddish and at the end of the year, in summer, it is yellowish, but does not swarm.

Toad grasshoppers (Pamphagidae) are stout grasshoppers with sword-shaped antennae and a distinct crest on the pronotum in many species. *Lamarckiana bolivariana* is common around the northern swamp, Inhanguene. The females are wingless but the males have wings and they stridulate while sitting in trees or bushes.

Zonocerus elegans, the Elegant grasshopper (Pyrgomorphidae), can be found all over the island in various stages of development. Adults are about 50 mm long, brightly coloured in black and yellow, with red eyes and black and red banded antennae. Wings are short and non-functional. These grasshoppers have a foul smell. The males stridulate by the vibration of the thoracic segments with the aid of special muscles. The hoppers often gather together in clusters at night and migrate together to crops during the day.

Numerous grasshoppers of the genus *Chrotogonus* of the same family are easily found. They dig into loose, sandy soil leaving only the eyes and antenna visible.

13.6 HEMIPTERA: bugs

Bugs have characteristic piercing/sucking mouth parts in the form of stylets, used for sucking blood or plant juices, depending on the species. The stylets are situated in a slender, segmented beak, the rostrum.

Hemiptera are divided into the suborders Heteroptera and Homoptera. Homoptera are terrestrial; both forewings and hindwings are completely membranous. In Heteroptera the basal part of the forewing is leathery and the distal part is membranous, as are the hindwings. Many Heteroptera are aquatic.

13.6.1 Heteroptera: twig-wilter bugs; cotton stainers; stink bugs; water striders

Twig-wilter bugs (Coreidae) feed on plants by piercing the plant with their mouthparts. Saliva is injected whilst sucking and causes wilting of the plant. They also secrete a rank-smelling fluid from their thoracic glands, the openings of which are visible near the bases of the hind legs. *Anoplecnemis curvipes* (Fig. 13.3) is common; it is about 25 mm long and dark brown in colour, the last antennal segment is reddish and the openings of the stink glands are orange. The pronotum is pointed laterally and each hind leg is markedly curved and bears a large spine in the male. It feeds on several plant species.

Some species of Cotton Stainers (Pyrrhocoridae) are found all over the island, for example, the Red Cotton Stainer, *Dysdercus fasciatus*. Adults have red wings with a black bar near the middle and on the pronotum; the body has several black lateral stripes. It lives on various plant species although its main host is cotton.

Stink bugs (Pentatomidae) possess well-developed glands that secrete a fluid with an unpleasant odour. *Nezara viridula* (Fig. 13.4) the green Stink bug, is common. It feeds on a large variety of plants, and is particularly damaging when it feeds on developing fruits which become spotted and deformed or may be shed prematurely.

The water strider (Gerridae) *Halobates* sp. (Fig. 4.16c) occurs on the surface of the water in the channels of the mangrove swamps near their exits into the bays (see Section 4.5.3). The under surface of the body is covered with a dense coat of white scale-like hairs which prevent it from getting wet. The second and third pairs of legs have thick brushes of water-repellant hairs below the knee-joint. They skate over the water surface and detect their insect prey by sensing the ripples made when they fall into the water and they dive to capture them. Courtship signals between male and female water striders are produced by making special ripples on the surface of the water.

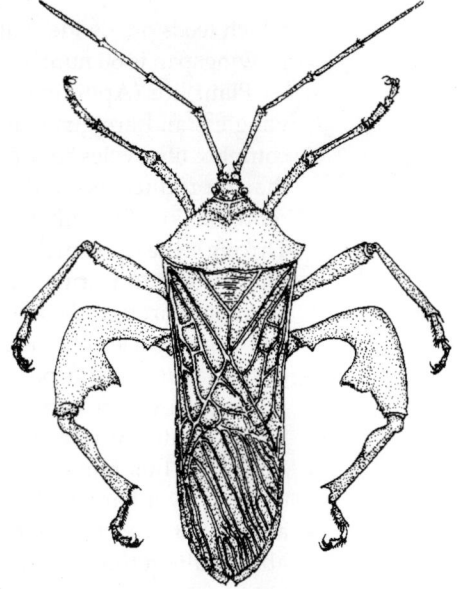

Fig. 13.3 Twig wilter bug, *Anoplecnemis curvipes* (× 2) (after Annecke and Moran 1982).

13.6.2 Homoptera: spittle bugs; cicadas; plant-lice; scale insects; mealy bugs

Spittle Bugs (Cercopidae) are also called froghoppers because the adults can jump, whilst nymphs live enclosed in a mass of foam which is produced as protection against predators and dehydration. Nymphs feed on plant sap and much of the fluid is excreted, mixed with a waxy substance from glands under the abdomen, secreted into a chamber made by membranous flaps which meet in the midline ventrally. The insects' tracheae open inside this chamber which has a valve near the anus. The Froghopper forces air through the secretion by working the terminal segments of the abdomen in and out telescopically and in this way the foam is created. Nymphs complete their development within the foam. After the last moult adults leave the foam, their wing buds expanding into stiff wings, held roof-like over the abdomen. *Locris* sp. is a common Spittle Bug on Inhaca Island. Adults are about 12 mm long and have red wings; the pronotum is red and black whilst the scutellum is black.

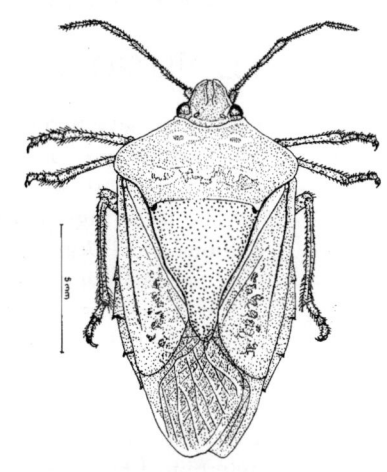

Fig. 13.4 Green stink bug, *Nezara viridula* (× 3) (after Annecke and Moran 1982).

Cicadas (Cicadidae) are well known for the loud, shrill noise made by the males. The sound is produced by special organs located on the dorsolateral surface of the first abdominal segment. These organs or 'timbals' are tightly stretched drum-like membranes to which strong muscles are attached, which enable them to vibrate rapidly to produce the sound. Folded membranes on each side of the drums act as sounding boards to increase the intensity of the sounds. On the ventral surface of the first abdominal segment there is a small round, shining plate, the mirror or 'tympanum', which acts as a hearing organ. Both males and females possess tympana at the base of the ventral surface of the abdomen and hence are capable of hearing the sounds produced by the males. Nymphs of cicadas live in the soil, feeding on roots of plants; some species have an extremely long period of development, lasting several years. Adults sit close together on a branch of a tree with their beaks embedded in the plant, feeding on the sap. *Oxypleura lenihani* is a large cicada, 28 mm in length with a wing-span of 90 mm, which feeds on a wide range of trees. *Brevispana brevis* is smaller; its body length is 15 mm and its wingspan is 60 mm; it feeds only on *Acacia karoo*.

Plant lice (Aphididae) are small insects that live on plant sap. They may be winged or wingless and are generally found in groups on new growth, flowers and fruits. Many have complex life cycles in which individuals that reproduce sexually alternate with individuals that reproduce asexually. The winged individuals are usually those that practise sexual reproduction. Dorsally they have a pair of tubes called cornicles near the end of the abdomen. Through these cornicles waxy substances may be secreted to repel enemies or as alarm pheromones to warn members of the colony about the presence of predators. Excess plant sap is voided through the anus and is called honeydew. The honeydew is often collected by ants and some species of ants look after the aphids from which they obtain honeydew. A number of species are serious pests of agricultural crops.

Scale insects (Coccidae) are extremely small and adhere to fruit, leaves or twigs for their whole adult life without moving, their slender lancet-like beaks thrust into the plant. The 'scales' in this family are made of dried-out skins, which they have cast whilst moulting, covered by a hardened waxy layer. The first instar nymphs, the crawlers, have legs and are active, but after the first moult they do not form legs and become sessile. Often males are absent altogether but if they are present they are winged. Many scale insect species are notorious agricultural pests. In Mozambique large areas of manioc have been infested with *Phenacoccus manhiota*, a mealy bug (Pseudococcidae) which has caused considerable damage. It is also present on Inhaca Island and is found on the host plant manioc, wherever it is grown. It can be recognised by its white, waxy secretion, which has given the family the common name of mealy bugs.

13.7 NEUROPTERA: lacewings and antlions

Lacewings have two pairs of similar membranous wings with a conspicuous network of veins. Most adults prey on insects and have biting and chewing mouthparts. The larvae are also predacious. Of these the most common family is that in which the larvae are known as antlions.

The lacewings or short-horned antlions (Myrmeleontidae) form the largest family of antlions and *Palpares libelloides* is most common on Inhaca Island. The body is about 80 mm long and the wingspan may be 150 mm; it is dark brown in colour and the thorax is hairy. Adults of antlions (lacewings) resemble dragonflies to some extent but their antennae are clubbed at the tip instead of being thread-like. The wings are beautifully patterned with large dark areas on a transparent background, but the insect's flight is clumsy and they fly at dusk or during the night. Eggs are laid singly in the sand and the larva lives in the sand lying in wait for its insect prey (Youthed and Morgan 1969). The position of their antlion larvae can be determined by the small pits with sloping sides that they dig. Insects that fall into a pit are captured by the larva, which is lying in wait at the bottom. The larva has no mouth but the sickle-like mandibles, pierce the prey and juices are sucked in along the grooves of the mandibles which are roofed by the maxillae. The larval life may be as long as three years but adults are short-lived.

Owlflies (Ascalaphidae) are antlions with knobbed, long antennae, sometimes as long as the forewings (Ferreira 1964). They are much better at flying than the short-horned antlions. They live in trees and catch their prey on the wing. Females lay white, oval eggs under rocks or stones; the larvae are black when newly hatched. They hunt for ants and other insects under stones or leaves and do not construct pits. Species of the genus *Disparomitus* are common on Inhaca Island.

13.8 COLEOPTERA: beetles

The forewings of beetles, the elytra, are horny or leathery and are not used in flight although they do provide lift in a breeze. The membranous wings beneath them are folded at rest and are used in flying. Beetles have biting mouthparts and metamorphosis is complete.

Ground beetles (Carabidae) are a large family of terrestrial beetles, usually black or brown in colour. Many species actively prey upon many kinds of insects and other animals, such as snails; others are herbivorous. They have thread-like antennae; the mouthparts protrude in front of the head, i.e. they are prognathous; the tarsus is five-jointed. Many carabids have no hind wings and the elytra are fused so that they are unable to fly. When disturbed these beetles secrete a volatile, pungent acid from the abdomen. The larvae of carabids also hunt insect prey. They are found in decaying logs or under leaf debris. They may be recognised by their sickle-shaped jaws and six simple eyes on each side of the head. A large species of the genus *Thermophilium* (Fig. 13.5) and the smaller *Cypholoba* are common.

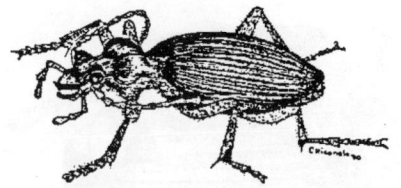

Fig. 13.5 Ground beetle, *Thermophilium* sp. (x 1).

Tiger beetles (Cicindellidae) are voracious insects both as larvae and as adults. They have long, slender legs, conspicuous bulbous eyes and long antennae. They are also prognathous. They are usually found in open, sandy areas for they prefer sunshine; at night they rest in sand or under stones. *Lophyra brevicollis* is frequently seen running over sand tracks near habitations, sometimes flying a short distance and then running again. It is about 13 mm in length, the elytra are light in colour with a dark pattern, but the body is metallic blue-green underneath and has silvery hairs along the margins. *Lophyra barbifrons*, which has uniformly light elytra is also common. Larvae of Tiger beetles live in burrows in the ground where they block the entrance with the broad head and thorax. The mandibles have long, sharp teeth with which they seize insects near the mouth of the burrow and then draw them down to devour them.

Rove beetles (Staphylinidae) may be recognised by their very short elytra and relatively large membranous wings on a straight-sided, elongated body. They may be mistaken for flies by the novice unless examined closely. The genus *Bledius* (8 mm) (Fig. 3.25c) lives in colonies, making raised tunnels of pencil thickness in the surface sand at high-tide mark, parallel to the driftline. They feed on micro-algae in the sand instead of being carnivorous like most other beetles. In one species the female exhibits parental care, but this has not been ascertained for the Inhaca beetle which has been observed only at Ponta Mazondue. Staphylinid beetle larvae have also been found in mangrove mud, eaten by the mudskipping goby, *Periophthalmus kolreuteri* (see Section 4.6).

The poisonous blister beetles (Cantharidae/Meloidae) are very common, black and yellow-orange insects which, as adults, feed on the flowers of leguminous plants. The larvae are predators or parasites on the eggs of grasshoppers. These larvae have a complex development, known as hypermetamorphosis. When they hatch, the larvae are at first tiny insects with a large head and two stiff bristles on the tail, like lice at first glance. They run about seeking the spots where grasshoppers or locusts have laid their eggpods. They burrow through the soil and bore into the eggpods to find the eggs. Then they cast the first larval skin and become fat, white maggots with short legs. These larvae feed on the eggs and grow rapidly until, at the end of summer, they change into a resting 'false' pupa, with a tough white skin. At the beginning of the next summer, the larva casts this skin and feeds for a short time. At first they are white, then they darken and after 3-4 weeks they shed the skin for the last time and the adult emerges. The adult beetles eat the petals of flowers. They are completely shunned by birds and insects because they contain an extremely potent poison, cantharidin. An extract of the beetles is sometimes included in local medicine, but may be fatal; it is supposed to have aphrodisiac properties and the beetle was known in Europe as 'Spanish fly'. A very common blister beetle on Inhaca Island, *Mylabris oculata*, has elytra with the typical 'warning' black and yellow colours.

Most species of the false ground beetles in the large and varied family (Tenebrionidae) live on the ground, but some live under bark or in stored grain. The slender larvae are called 'false' wire-worms; they have six legs and live in the soil on roots of plants and detritus.

Psammodes spp. are known as 'Tok-tokkies'. They are fairly large insects (55 mm), convex in shape, totally black and wingless; their elytra are fused. The anterior part of the back is

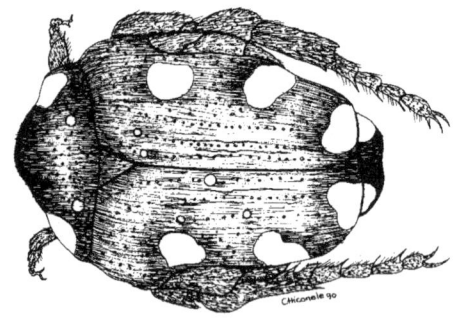

Fig. 13.6 Fruit chafer, *Mausoleopsis amabilis* (x 4).

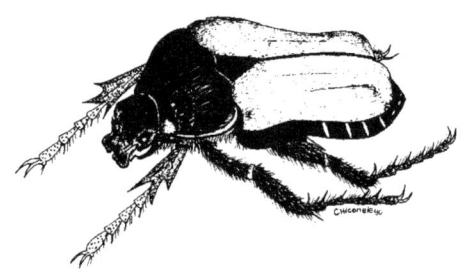

Fig. 13.7 Flower chafer, *Atrichia placida* (x 2).

smooth and shiny but the elytra have sculptured surfaces. *Psammodes vialis* is more common; *P. bertolini* is larger and differs in having a sculptured pronotum. These beetles live in leaf litter and are scavengers. The common name is derived from their habit of knocking on the ground with the abdomen during courtship; they raise the abdomen and drop it several times in succession. These beetles are well-adapted to hot, arid and sandy conditions.

Dung beetles, fruit, flower and leaf chafers and rhinoceros beetles are members of the Scarabaeidae.

Dung beetles are so-called because they feed on dung. Some species first roll it into a ball and bury it for their own consumption or for their larvae; other species do not bury the dung and some eat carrion or even decaying vegetation. On Inhaca Island there are no longer any cattle to supply dung, but the dung beetles use goat faecal pellets. Pasture land is much improved by the activities of these scarabs and they have been introduced into Australia, where they did not originally occur, with rewarding results.

Two species of the typical dung beetle, *Catharsius pseudoopacus* and the larger *C. troglodytes* (10-12 mm) are found on the island. The antennae terminate in a fan of three yellow, flattened joints that can be folded over one another; these are the organs of smell. The front legs have no tarsi but are stout and armed with four teeth which are used in digging the hole in the soil and gathering the dung into a ball. These species prefer fresh dung and bury it after digging a hole. They are rarely seen at work since they are nocturnal.

Fruit and flower chafers are often brightly coloured and very common in the flowers of shrubs and trees such as the Waterberry, *Syzygium cordatum*, and the White Ironwood, *Vepris undulata* (see Section 10.3). These beetles do not raise their elytra for flight but unfold their hind wings from a depression on the sides of the elytra. *Mausoleopsis amabilis* (Fig. 13.6) is black with white spots on the elytra and about 12 mm in length. *Atrichia placida* is bigger (20-25 mm) (Fig. 13.7) and although the head is always black, the elytra vary from dark brown to light brown with darker spots. All species have strongly pubescent elytras. *Micrelaphinis* spp. are somewhat smaller and variable in colour with both dark and light spots. Nocturnal leaf chafers are brownish-grey, but the diurnal species may be brightly coloured. *Anomala* sp., a nocturnal species, can easily be captured since it is attracted to light sources. It is about 15 mm long and has well-developed tarsal claws of unequal length. Females lay eggs in soil or leaf litter, sometimes at a considerable depth. The larvae are commonly called 'white grubs' and feed on plant roots or litter.

Rhinoceros beetles include some of the largest of all Coleoptera. Most species are dark brown and nocturnal. Males have a long horn on the head or thorax which is used in combat. Larvae of several species attack the roots or leaves of trees or other plants. *Oryctes boas* (40-50 mm) attacks the foliage of palm trees and cannot easily be found among the young leaves. The beetles are hidden between the leaf bases or under the bark of sick or dead palm trees where the larvae live and pupate.

Longhorn beetles (Cerambycidae) have very long antennae and long legs. Many diurnal species have beautiful colours in contrast to nocturnal species which have dull colours. In the tropics the larvae of many species feed upon the solid tissues of living trees, whilst others attack sick trees or dead wood. The preference for living trees is probably the result of competition with termites for dead wood. In the mangrove swamps there are few termite colonies, but the longhorns are common, such as *Dirphya gigantea* (Fig. 13.8) and *Philematium virens* var. *natalense*, which has a metallic blue-green lustre on the body both dorsally and ventrally. Adults are found on buds or stems. Eggs are laid in a crack in the bark of a tree and a number of young larvae emerge as fat, white grubs. An unusual method of oviposition is adopted by *Tragocephala variegata* (Fig. 13.9). The females chew a ring around a branch with their mandibles before laying their eggs beyond the girdle. This arrests the flow of sap and the branch dies and falls off, providing dead wood for larval development. The larvae tunnel through the wood as they feed on it. Some larvae are cannibalistic and will devour others that they meet in the wood. The larvae possess chordotonal organs along their sides which may enable them to detect others close by in the wood and avoid them. Larvae of longhorns take two or three years to mature, after which they move to the surface of the wood and hollow out holes in which to pupate. At first the pupa is white and then it slowly darkens; the adult emerges and makes a tunnel to the exterior to escape.

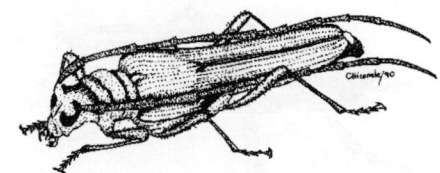

Fig. 13.8 Long horned beetle: *Dirphya gigantea* (x 2).

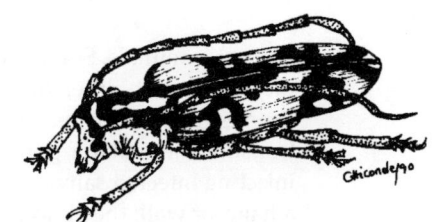

Fig. 13.9 Long horned beetle, *Tragocephala variegata* (x 2).

Leaf beetles (Chrysomelidae) are rounded and smooth and often have a metallic lustre. Their small heads are sunk into the thorax; they have short antennae and short legs with four-segmented tarsi. Most of them prefer sunshine and when disturbed they draw in their legs and drop to the ground from the tree. Tortoise beetles in the same family are broad and flat and the head is covered by the pronotum; the dorsal surface has a brownish tortoise-shell pattern. *Aspidomorpha puncticosta* is found on plants of the family Convolvulaceae which includes sweet potatoes, *Ipomoea batatus*, and the dune plant, *Ipomoea brasiliensis*, which binds the sand and is sometimes used on beaches to contain erosion. Other species of tortoise beetles attack a variety of plant species. Eggs are laid on leaves and hatch into tiny larvae which feed on them. The larvae have a row of quite large, flattened spines on each side and two prongs on the tail on which excreta become tangled. Each time it moults the larval skin is retained on the prongs.

Wood-boring beetles (Buprestidae) are torpedo-shaped and have a bright metallic lustre. Adults feed on pollen or foliage and are short-lived. *Evides pubiventris* is common on Inhaca Island. It is about 22 mm long and has a light-green metallic lustre and conspicuous black eyes. The antennae are reddish green and the body has white hairs underneath. The elytra have long lines of black hairs and their posterior margins are saw-like. The head has a reddish, metallic sheen. The larvae make long, oval-shaped galleries in moribund trees and some live for many years.

Net-winged beetles (Lycidae) are related to the glow worms and fireflies but do not have light-producing organs. They are black and yellow, elongated and soft-winged insects with a peculiar network of longitudinal ridges on the elytra. The adults live on shrubs, but the larvae live in rotting wood, probably feeding on the larvae of other insects. *Lycus* species may be seen on foliage and flowers, sometimes in large numbers.

Weevils (Curculionidae) or snout-beetles are the largest insect family. The head is produced into a beak with two pincer-like mandibles; the antennae are clubbed. Females use the snout to bore holes in which to lay eggs. The adults usually have dull colours and when disturbed they feign death. They feed on all parts of plants when adult and the legless larvae feed on the internal tissues of stems and roots. The sweet-potato beetle, *Cylas formicarius*, is common. It is a slender, ant-like weevil, about 5 mm in length, and it is a serious pest of sweet potatoes since the larvae bore into the stems, roots and tubers and often kill the plants.

13.9 DIPTERA: flies and mosquitoes

Mosquitoes (Culicidae) are of medical importance because they are the intermediate hosts of the causal agents of various tropical diseases such as malaria, dengue and elephantiasis

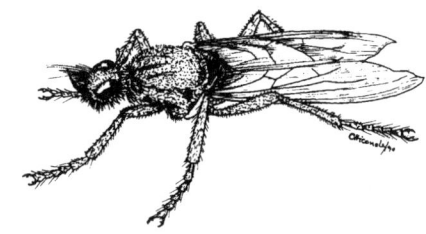

Fig. 13.10 Robber fly, *Proagonistes praeceps* (x 1,5).

(W.H.O. Report 1989). Mosquitoes have specialised mouthparts adapted for piercing the skin of their hosts and sucking blood. Only females do so since they need the protein content for the maturation of their eggs. Males live on sugars obtained from over-ripe fruits and the nectar of flowers. Eggs of mosquitoes are always deposited on or near water; the larvae complete their development in water, feeding on algae and detritus. Some species lay eggs in lakes or swamps, whilst others use temporary pools or can utilise the water in the axils of leaves, holes in trees or water left in kitchen utensils.

Some species of mosquito do not bite man but prefer domestic animals. Some bite man at night, but others do so throughout the day. After a blood meal they rest either inside or outside houses before laying their eggs, depending on the species. Species of the *Anopheles gambiae* complex (Fig. 4.14d) occur on Inhaca Island and transmit the malaria parasite *Plasmodium falciparum* from a sick person to a healthy one by injecting infected saliva. *Anopheles* spp. can be recognised by their posture when they alight on a hand or wall; the abdomen is raised at an angle and is not parallel to the surface on which it is standing; hind legs are raised as well.

Twelve species of mosquitoes were reported to be present on the island (Perreira 1959) and are listed in Appendix B. Six species of *Aedes* are aggressive in daytime as well as at night, but they do not carry yellow fever. It would be interesting to find out whether the density of populations has changed on the island in the last thirty years because the areas of fresh water have declined, restricting the breeding places of vectors.

Robber flies (or assassin flies) (Asilidae) are encountered all over the island and especially on the beaches, near the drift line. The body is covered with bristles, the proboscis is firm and horny and the eyes are large and separated from each other by a deep groove. The head is mobile and turns continuously to watch the prey. The face has a prominent tuft of hairs like a beard. Robber flies hunt other insects on the wing and will attack insects larger than themselves. Victims are killed by piercing the body with the rigid proboscis and sucking up the contents. The larvae live in the soil or in decaying wood and prey upon other insects. They develop into pupae covered with spines. The most common species of asilid is *Philodocus tenuipes* which is grey in colour, 20 mm in length and has a long, slender abdomen. A larger species, *Proagonistes praeceps* (Fig. 13.10), with a black body and brown legs, is common in the hot season.

Bee flies (Bombyliidae) have a preference for semi-arid regions with sclerophyllous vegetation such as that on the dunes at Inhaca Island. These plants flower profusely after rain and it is then that the bee flies are numerous all over the island, especially during the hottest hours of the day, for they are heat-loving and cannot tolerate cold and wet periods. They resemble bees in shape and hairiness but they have only one pair of wings. Their flight is rapid and they are able to hover in front of flowers, sucking nectar with their long proboscis. They settle on the ground between feeding periods with wings stretched out, showing dark-coloured patches in the membranous, transparent area. Bee flies also differ from bees in larval life. The female lays eggs near the eggs of locusts, grasshoppers and moths; when the larvae emerge they seek out insect eggs and live on them until they pupate in the ground. Common bee flies on the island belong to the genus *Exoprosopa*. Some species may have larvae that attack egg-pods of locusts or destructive grasshoppers or army worm caterpillars, and so play an important part in the control of insect pests in Africa.

The long-legged flies (Dolichopodidae) are frequently seen running over the mud in the mangroves at low tide. Adults prey on minute soft-bodied insects. The larvae live in humus or rotten wood, or they may be aquatic. *Thinophilus* and *Tachytrechus* are common genera. The flies are black in colour, have very slender bodies (±8 mm) and the thin legs are three times as long as the body.

The Syrphidae is a large family of hover flies, many species of which have characteristic yellow markings on the body, mimicking bees or wasps, but as they have only one pair of wings they can easily be recognised as flies.

Adults are swift fliers. Many species hover apparently motionless in the air, hence the common name of 'hover flies'. In summer they are a common sight visiting flowers, collecting nectar. They may be of importance in pollination.

The habits of hover-fly larvae are very varied. Some prey upon aphids, others are phytophagous. Some live in polluted water, for example, the rat-tailed maggot of the genus *Eristalis*, the so-called drone fly. Larvae of other species live in the nests of ants, wasps or bees. *Mesembrius* spp. are well represented on Inhaca Island.

Blowflies (Calliphoridae) may be blue or green with a metallic lustre; the body is smooth with few hairs and there are no hairs on the dorsal surface of the abdomen. Most blowflies are oviparous. A common genus is the green blowfly *Phaenicia* (8 mm) which lays its eggs in dead animals. White, legless maggots emerge and feed by secreting copious saliva from their mouths and pecking at the meat with their mouth-hooks. Digestion thus takes place outside the body of the fly and it sucks up the broth it has made. *Chrysomya* sp., another carrion breeder, is also common.

Flesh flies (Sarcophagidae) resemble house flies superficially but they are larger and have a black-striped thorax. They do not lay eggs but give birth to first instar larvae, which are laid on decaying organic matter, excreta or on meat. In this way they may transfer bacteria to food and endanger public health. They may also parasitise other insects. *Sarcophaga haemorrhoidalis* is common. It is a grey fly with three longitudinal black stripes on the thorax; the abdomen has a red tip and a checkerboard pattern.

Adult tachinid flies (Tachinidae) live in flowers but the larvae are maggots which parasitise the larvae of butterflies, beetles, wasps and grasshoppers or of spiders or centipedes, completely destroying the host's tissues. They may lay their eggs on the leaves of plants which are consumed by caterpillars. The tachinid maggots pierce the host's trachea, interfering with its supply of air, and consume its tissues. Adult tachinid flies may haunt beehives and parasitise bees, especially in summer. Female flies give birth to first instar larvae which are deposited on and adhere to the bee's abdomen. The larva enters between the segments of the bee's abdomen and lives on fluids until it is fully grown, becoming a large, white maggot with two conspicuous black, circular plates around the spiracles at the tip of its abdomen. When the bee dies, weeks later, the larva breaks its way out and pupates. Two months later the fly emerges.

13.10 LEPIDOPTERA: butterflies and moths

The larval stages, caterpillars, are often of economic importance since they attack foliage and the shoots of crops and trees.

13.10.1 Butterflies: Blues; Whites; Monarchs; Commodores; Swallow tails; Browns; Forest butterflies; Distasteful butterflies; Skippers

Blues (Lycaenidae) are relatively small and often brightly coloured butterflies. Many have flimsy, slender tails on the hind wings. Some are slow fliers and others fly rapidly. Larvae of many species are associated with ants; they have a dorsal honey gland that secretes a sweet fluid that is immediately consumed by ants. A common Blue on Inhaca Island is *Myricna dermaptera*, which is about 10 mm long and has a wingspan of 30 mm. The wings are typically metallic blue, the hindwings being short but with relatively thick tails. *Iolaus silas* is larger with a wingspan of 45 mm. The wings are violet-blue with a dark apical patch and a black margin; on the posterior angles of the hindwings there are two orange-red spots. The tails of the wings are black but the undersides are white with red markings.

Whites (Pieridae) have white or yellow as a basic colour. They vary considerably in size; males and females differ in colour pattern. The wingspan varies from 30-65 mm and many species are strong fliers. The common Whites are species of the genera *Colotis*, *Catopsilia* and *Eurema*. *Colotis evagore* males have white wings with an orange-red apical spot and a black notch on the inner side of each wing. Females may be white or yellow, and the apical spot may be red or white. *Catopsilia florella* has a wingspan of 65 mm and white wings in the male, yellow in the female. They have narrow orange-brown borders on the forewings and orange-brown dots along their outer margins as well as on the hindwings. In *Dixeia spilleri* (Fig. 13.11) the wings have a dark margin, which is broader on the anterior pair. Several species of Whites exhibit seasonal dimorphism, in which the wet season forms have darker patterns than the dry season forms.

Monarchs (Danaidae) are medium to large butterflies with black or brown wings and yellow or white spots. The head and thorax are marked by white dots. All monarchs are unpalatable or even toxic to predators. The poison is derived from the plants on which the caterpillars feed, namely milkweeds (Asclepiadaceae) such as *Cynanchus* spp. These plants contain toxic cardiac glycosides which accumulate in the caterpillars and are conserved in the butterflies. The distinctive patterns of the butterflies' wings warn potential predators that they are harmful, i.e. the colouring is said to be aposematic. The caterpillars are banded black, yellow, red or green.

On Inhaca Island *Danaus chrysippus*, African monarch, (Fig. 13.12) is common. With a wingspan of 80-100 mm, it has reddish-brown wings, the forewings having dark margins and white patches. The male butterfly releases strongly scented black dust from the black spots in the middle of each wing during courtship. Species of *Amauris* are very common too. *A. niavius*, Friar, has a wingspan of about 75 mm; the wings are dark brown with a large white central area and large, white patches in the apical region of the forewings. *Amauris ochlea*, Novice, (Fig. 13.13) has a similar pattern, but is smaller.

Commodores (Nymphalidae) have strikingly coloured 'uniforms'. *Junonia oenone* (Fig. 13.14), Blue pansy, has a large violet spot on the brown hindwings which have white margins. *J. tugela* is similar but has a broad pale orange band across the brown hindwings. *Vanessa cardui*, the 'painted lady', is not so common. It has rich, tawny red wings with black markings and some dorsal white spots, but underneath, the wings are mottled yellow, white and brown. *Phalanta* spp. have purple markings on the undersides of their brownish orange wings; they occur mainly in evergreen forests. *Byblia ilithyia*, Joker, (Fig. 13.15) has orange-red wings with black markings.

Swallowtails (Papilionidae) are large, colourful butterflies, many of which have tails on the hindwings. Many butterflies in this family mimic the colouring of poisonous Monarchs, for example *Princeps dardanus*, the mocker swallowtail, which is very variable in appearance. *Princeps demodocus*, Citrus swallowtail, (Fig. 13.16) has a wingspan of 100 mm; the wings are blackish, heavily marked with yellow patches and orange and blue 'eyespots' on the hind wings.

Browns (Satyridae) are of medium size and have orange-red markings; most species have 'eyespots' on the posterior wings. They prefer shade and fly in the undergrowth of forests. The caterpillars feed on grasses. *Melanitis leda*, known as the Dead leaf, rests by day amongst dead leaves under bushes; the colouring of its underside blends perfectly with its background, making it difficult to spot.

Forest butterflies (Charaxidae) are medium to large in size and have a big head as broad as the thorax. They fly rapidly among the tree tops in the forests but descend to the ground to settle on muddy patches, on animal droppings or on decaying fruit. *Charaxes brutus* is common in forest on Inhaca Island; its wings are tailed, brown with a light longitudinal band.

The caterpillars of Distasteful butterflies (Acraeidae) feed on plants toxic to vertebrates. They have narrow, rounded wings which are often black and red, the common warning colours among insects. *Acraea zetes* (Fig. 13.17) has black patches on red wings and orange apices. There are many, more or less, similar species of *Acraea* on Inhaca Island.

Skippers (Hesperiidae) are the most primitive of butterflies; they are dull-coloured and almost moth-like with a large head, antennae widely separated at the base and dilated terminally to form clubs, often ending in a hook. Skippers derive their name from their erratic, darting flight. *Metissella metis* is a small skipper with brown, yellow-spotted wings. The caterpillars feed on grasses. These butterflies form a connecting link between the moths and the higher butterflies.

Fig. 13.11 White butterfly, *Dixeia spilleri* (x 1);
Fig. 13.12 African monarch, *Danaus chrysippus* (x 1).
Fig. 13.13 Novice, *Amauris ochlea* (x 1);
Fig. 13.14 Blue pansy, *Junonia oenone* (x 1);
Fig. 13.15 Joker, *Byblia elithya* (x 1).
Fig. 13.16 Citrus swallowtail, *Princeps demodocus* (x 0,8);
Fig. 13.17 Distasteful butterfly, *Acrea zetes* (x 1).
Fig. 13.18 Cutworm moth; *Egybolius vaillantina* (x 1).

13.10.2 Moths: cutworms; bagworms

Some moths are nocturnal, others fly at dawn. The former may have dull colouring but those active in daylight are often brightly coloured.

The moths in the cutworm family (Noctuidae) fly mostly at night. *Egybolius vaillantina* (Fig. 13.18) has a wingspan of 55 mm; the wings are a metallic blue with clear orange spots. Antennae and palps are yellow and the legs have yellow patches. Members of the Noctuidae have larvae called 'cutworms' that may damage root crops by beheading the seedlings as though cut with a knife, although they eat very little of them. Army worms do much the same damage to tropical crops and the Boll worms damage cotton crops. Both belong to this family.

Bagworms (Psychidae) are named after the bags constructed by their larvae in which they live concealed, carrying them about for protection. The bags are made by the larvae from pieces of leaves from their food plants, from short twigs or thorns or other plant debris and sometimes they use sand grains. These are woven together by silk from silk glands. Species may be characterised according to the materials used and the way in which the bags are made.

The larvae remain in the bag during their whole lifetime, continuously extending the bag as they grow. Excrement is expelled and the interior of the bag is kept clean. In the prepupal stage the larva searches for a sheltered, secure place and the bags are firmly attached to a plant, to walls or roofs of houses. The lower part of the bags have a small opening to allow the adult males to escape. The males are winged but the females are vermiform, wingless creatures and confined to the bag. They are fertilised when the male places its abdomen through the opening in the bag of the female. Eggs are laid in the bag in large numbers and the female dies after laying eggs.

On Inhaca Island the bags of species of *Eumeta* spp. are found. *Eumeta cervina*, the Lictor Bagworm, constructs the bag of small sticks of about 6 cm in length, rounded off at both ends and attached lengthwise to the bag.

13.11 HYMENOPTERA: ants, wasps and bees

Many of the best known species of insects in this order live together in colonies; however, a large number are solitary. Amongst the species that live in groups or colonies there are grades of social behaviour from simple communal nesting to complex societies with many individuals (Wilson 1971). The species with complex societies consist of groups of females among which one or a few (the queens) lay eggs. The remaining more or less sterile, non-reproductive, individuals form a 'worker' caste that rears the larvae that emerge from the queen's eggs and perform all the work necessary to maintain the colony. Males are produced seasonally and are necessary to mate with the virgin queens that are also seasonally produced.

Hymenoptera are characterised by the following structures: two pairs of membranous wings which are not identical; biting and chewing mouthparts modified for licking or sucking; usually a basally constricted abdomen in which the first abdominal segment is fused with the last segment of the thorax forming a 'petiole' or 'wasp-tail'. The queens lay fertilised eggs (i.e. diploid) which develop into females, or unfertilised eggs (i.e. haploid) which develop into males. Metamorphosis is complete.

Hymenoptera are of great economic importance to man; bees because of their role in the pollination of crops and wild flowers and in the production of wax and honey. Wasps are equally more important because they eat or parasitise insects which are harmful to crops.

All ants (Formicidae) have the physical characteristics of the Hymenoptera and most are highly social. After the nuptial flight during which mating occurs, the males die and the females shed their wings and start a new colony. In a number of species 'soldiers' occur; these are the large worker ants with big heads and strongly developed mandibles with which they defend the colony. Communication between the large number of ants within the colony is by means of chemical signals called pheromones. Ants are probably the most successful of all insect groups; they occur in enormous numbers in all kinds of terrestrial habitats. Their feeding habits are varied; some prefer the nectar of plants, some feed on fungi and others are carnivorous. Some may feed on the excretions of other individuals in the nest.

The most prominent ant species, and probably the most abundant insect species on Inhaca Island, is the Tailor Ant, *Oecophylla longinoda* (Fig 13.19) (Way 1954). It is a tree-dwelling species which makes its nest by binding together the edges of leaves with silk secreted from the silk glands of larvae which are manipulated by the workers (see Section 4.5.2). Tailor ants are very aggressive; when disturbed the reddish brown workers rush outside the nest to confront the possible aggressor. They can pierce human skin with their strong mandibles and spray formic acid into the wound from the poison gland in the abdomen. In a colony of several nests, only one of them contains the queen, from which other worker ants carry batches of eggs and pupae to found the other nests of the colony.

Tailor ants feed mainly on other insect species and their predation is so efficient as to lead to a scarcity of the insects they feed upon in the trees they occupy. They also tend a wide range of Homoptera, including some scale insects, because of the honeydew, a sugary solution which these insects secrete from the abdomen. In certain periods of the year winged males and virgin females are released. After mating, new nests are initiated by a single fertilised queen who uses her flight muscles as an internal food reserve to bring up the first batch of larvae to maturity. Nests are populated by workers in a range of sizes. The large workers (10 mm) are said to be involved in searching for food, in nest building and in the defence of the nest. The smaller workers (5 mm) are less numerous and may be involved in tending the brood and the queen. In nests without queens some of the workers lay eggs which, because they are not fertilised, develop into males.

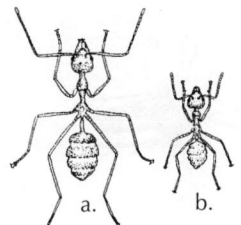

Fig. 13.19 Tailor ant, *Oecophylla longinoda*
a. worker (x 4);
b. larva (x 4);

Nests of another arboreal ant, *Crematogaster castanea*, are frequently found. The nests are attached to branches of trees. They are more or less spherical in shape and range in size from 20-50 cm in diameter. They are made from a mixture of chewed plant material and faeces called carton. *Crematogaster* spp. are called Cocktail Ants because of their habit of raising the abdomen and bending it over the back when disturbed. This is made possible by the attachment of the petiole to the dorsal surface of the abdomen. At the same time, glands at the tip of the abdomen secrete a sticky, whitish fluid with an unpleasant odour. The workers are small (3-5 mm) and are dark brown in colour; the abdomen is heart-shaped. Cocktail ants are aggressive; since the sting is modified they cannot use it offensively and so bite when disturbed. They attend aphids and coccids and feed on the sweet honeydew these insects secrete.

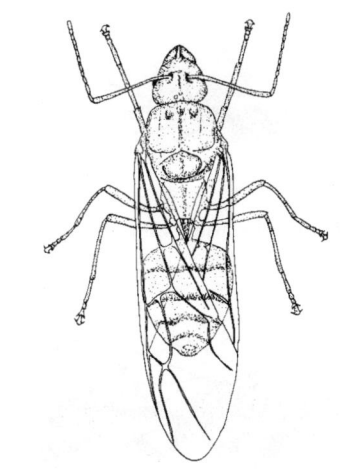

c. adult (x 8) (after M.J. Way 1954).

It would be interesting to find out what makes Inhaca Island such a favourable place for these ants. On the mainland of Mozambique, at least around Maputo, they are much less abundant.

The wasps are a very large and diverse group of insects, both in structure and behaviour. Some are only a few millimetres in size, others are over 70 mm long. Many species are solitary, others are social and live in colonies, sometimes building nests that exhibit a complicated architecture. Although unpopular because of their sting, wasps are extremely beneficial to man because they prey upon almost every kind of insect, as well as on spiders.

Velvet ants (Mutillidae) are so-called because of their ant-like appearance and the dense covering of hairs, but they are really wasps. Velvet ants are found on the beaches and other sandy places. Females are wingless and have a long sting which can inflict sharp pain when stepped upon barefoot or manipulated carelessly. Males are winged and slightly bigger than females. They are parasitoids: females search for pupae of other insect species, often those that pupate on the ground such as solitary wasps, bees and flies. They chew a hole through the wall of the cocoon and lay an egg on the pupa. After emergence, the velvet ant larva gradually consumes its host, eventually killing it. *Dolichomutilla* sp. can be found throughout the year. It is about 14 mm long and has the typical mutillid colouring of a red-brown thorax and dark abdomen with two white spots on each side.

There are many families of parasitic wasps. Chalcidid wasps can easily be recognised by the large, swollen femora of the hind legs which are ventrally toothed. Many species are fairly large (up to 7 mm), and are usually black with a sculptured integument. Larvae of chalcidids are parasites of immature stages of a great variety of insects and in some countries in East Africa they control the fly pest *Diopsis macrophthalma* in rice crops. *Chalcis* sp. has been recorded from Inhaca, but many more species of other genera are bound to exist on the island.

Hairy wasps (Scoliidae) are also very common on Inhaca Island. They may be found on rice crops. They are black and often have yellow or red bands on the abdomen. Females are larger than males. Wings are also black and have a metallic iridescence. Females have stout, spiny but relatively short legs with which they burrow in search of scarab beetle larvae on which to lay eggs. These beetle larvae are white grubs which live in soil or compost heaps. The wasp stings the grub but does not kill it. When the grub is paralysed, the wasp lays an egg in it. The egg hatches into a legless white maggot that feeds on the beetle larva. When the wasp larva is fully grown, it spins a dense oval cocoon of silk beside the shrivelled skin of the beetle larval host and pupates inside it. Scoliid wasps can be recognised by the deeply notched inner margin of the eyes and the three straight spines at the end of the male abdomen. Species belonging to the genus *Scolia* (Fig. 13.20) are frequently found on the island.

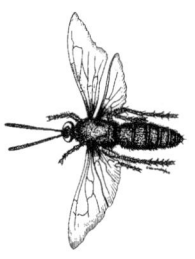

Fig. 13.20 Hairy wasp, *Scolia* sp. (x 2).

Digger and Thread-waisted wasps (Sphecidae) are solitary wasps which provide their larvae with a wide range of other insects as hosts. The females dig tunnels in sand or plant material but some species make mud nests. They have a long, slender wasp tail; the legs have two mid-tibial spurs.

Fig. 13.21 Thread-waisted wasp, *Sphex* sp. (x 2).

The sand wasp, *Bembix* sp., has yellow and black bands on the abdomen and a long, pointed labrum. It digs a slanting burrow with its mandibles, which break up the soil, and the front legs, which bear a series of comb-like spines on the tarsus, act as a rake. A 'cell' is made at the end of the 15 cm deep burrow in which its fly prey, paralysed by the sting, is laid. The wasp larva starts feeding on its host but is continually fed with fresh prey until it is fully grown. The female wasp is able to return swiftly to its concealed nest, probably using some distance-orientation cue to guide it effectively.

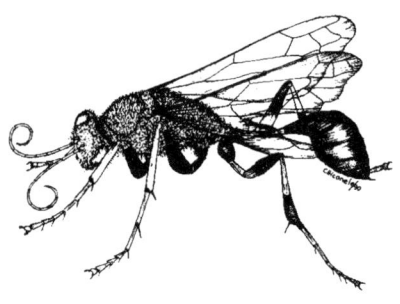

Fig. 13.22 Spider-feeding wasp, *Sceliphron* sp. (x 1,5).

Thread-waisted wasps have a thin petiole as the name implies. They make their nests close to human habitations. Two species of *Ammophila* are quite common; they search for smooth caterpillars among leaves on the ground to be used as hosts for their larvae. *Sphex* sp. (Fig. 13.21) digs a burrow and stocks it with nymphs of grasshoppers. Other species, such as *Sceliphron* sp. (Fig. 13.22), a slender, black wasp with a long yellow petiole and long black legs, collects damp mud and first makes nests consisting of several cells into each of which it puts a paralysed spider for each larva that develops from the eggs it lays. The spider supplies sufficient food to raise the larva to maturity.

Fig. 13.23 Spider-hunting wasp, *Hemipepsis* sp. (x 1).

Spider-hunting wasps (Pompilidae) may often be seen running over the ground, vibrating their wings and hunting for prey. They have curled antennae and long legs. The big, black Pompilid wasp, *Hemipepsis* (Fig. 13.23) hunts the large baboon spider. The female first stings the spider in the front of the head, thereby rendering the poison fangs harmless and paralysing its prey. She looks for a ready-made hole in which to store the spider and makes several cells in a dry, sheltered spot. She flies back to the paralysed spider, obviously remembering where she left it, and then drags it to the hole she has made. Although it may be many metres distant she remembers where it is located. She drags the spider into the hole and lays one large, white egg on its abdomen. Then she scrapes soil and twigs over the opening as though to camouflage it. The egg hatches in about ten days and the larva feeds on the still paralysed but living spider until it is fully grown. The larva spins a cocoon and cleanses its alimentary canal of waste matter for the first time. It remains in the cocoon all through the winter until warmer weather triggers pupation. After two to three weeks the adult emerges from the pupa.

Paper wasps (Vespidae) build their nests of pulp, prepared by chewing plant fibres and manipulating them into thin sheets which are allowed to dry. Nests are inhabited by several individuals which form a small society. The nests of these primitively social wasps consist of about 50 cells and are located in a sheltered place, attached by a short stalk to a tough plant stem, to branches of trees or walls of houses.

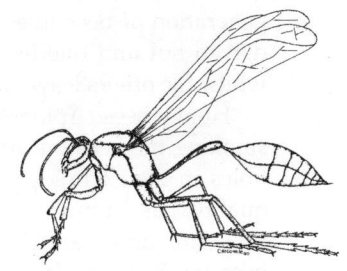

Fig. 13.24 Paper wasp, *Belonogaster* sp. (x 1).

Each cell has a fairly wide opening, pointing downwards. New nests are founded by a single fertile female, although she may be joined later by other females. Eventually one female will become dominant and lay eggs while the others become workers and help raise the larvae which are fed on chewed caterpillars. The larvae produce a droplet of sweet saliva when their mouthparts are touched, which is eagerly licked by the attendant females. At Inhaca Island, two species of *Belonogaster* are common in the mangrove swamps (Fig. 13.24), one larger than the other. The resting wasps point their wings backwards in a longitudinal position. They behave aggressively when their nests are approached too closely. In certain periods of the year, when they are quite abundant, local people complain about being attacked when cutting wood in the mangrove swamps.

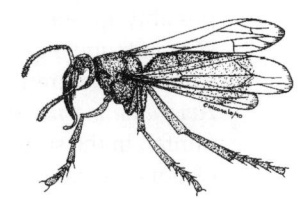

Fig. 13.25 Mason wasp, *Synacris* sp. (x 1).

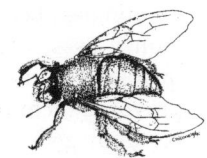

Fig. 13.26 Carpenter bee, *Xylocarpa caffra* (x 1).

Mason wasps (Eumenidae) are a diverse family of solitary wasps. Some make nests in holes in the ground or in hollow twigs; others make nests of mud, moistened with water from the crop, on twigs or on walls. A common genus is *Synacris* sp. (Fig. 13.25), a large, black wasp with the tip of the abdomen orange or white. The inner margins of the eyes are notched just where the antennae are inserted. The face (area between the eyes) is orange brown. The mandibles are long and slender, crossing each other whilst the insect is at rest. Some Eumenidae are, however, quite similar to big paper wasps but *Belonogaster* is brown whilst the Mason wasps are black. The Mason wasps feed their brood on larvae of butterflies and beetles.

The superfamily Apoidea includes both solitary and social species of bees. The food of both adults and larvae consists of nectar and pollen. Most individuals have a structure on either the ventral part of the abdomen or on the hind legs in which pollen is carried from flowers to the nest. Some species do not build nests of their own but lay their eggs in the nests of other bees; these are the 'cuckoo' bees.

Carpenter bees (Anthophoridae) are medium sized to big, black, hairy bees, sometimes with tufts of white or reddish hairs on the sides of the thorax and abdomen, on the face and forelegs. Some species have one or two yellow, sometimes white bands on the abdomen. Males are always yellow. On Inhaca Island several species of carpenter bees occur, e.g. *Xylocopa caffra* (Fig. 13.26) and *X. nigrita*. They are solitary bees; the female makes a tunnel in dead wood by using her powerful mandibles. She brings back pollen from leguminous plants and mixes it with nectar regurgitated from her crop, to form a thick paste. She lays an egg on top of this food and seals it off with a partition made of sawdust that she has chewed from the sides of her tunnel and mixed with saliva. She constructs seven or eight such cells, each containing a developing egg. The eggs hatch into larvae and the larvae grow and pupate in their cells; the adults emerge from the tunnel by breaking down the partitions. The new generation of bees lives communally in the tunnel until the next spring when the males are driven out and one female becomes dominant and retains possession of the original nest, whilst the others leave the tunnel and make new ones on their own.

Honey bees (Apidae) make their nests in trees from which the islanders extract honey by smoking the bees out of the nest. All over the island, dead trees may be seen with scorched holes in their trunks, which have been burned to obtain the honey. This honey is of poor quality, has a burnt taste and often contains dead bees or larvae! When shrubs and trees are flowering abundantly following the summer rains, the industrious buzzing of hundreds of bees can be heard from dawn to sunset, except at the hottest hours of the day. Little activity occurs in the drier winter.

The species of Honey bee on Inhaca Island is probably *Apis mellifera* var. *scutellata*, or possibly the race *littorea* which occurs in the coastal regions of East Africa from Kenya to Mozambique. Apiculture (bee-keeping) on the island is rare, although some families do have 'corticos', i.e. traditional beehives, made from the hollow trunks of trees, especially those of *Avicennia marina* in the mangroves. Honey production depends heavily on the amount of rainfall in the summer, but no reliable figures exist. Some islanders state that they obtain 25kg of honey from one swarm in a season, but this may be exaggerated. The principal nectar-producing flowers occur on the Waterberry, Milkberry, White Pear and White Mangrove (see Section 10.3). Bee-keeping is being upgraded by the National Programme of Apiculture according to the Integrated Development Plan for Inhaca Island, 1990.

Stingless bees (Apidae) are also present on the island but less common than the ordinary Honey bees. These are much smaller than Honey bees and, unlike *Apis mellifera*, they are unable to regulate the temperature of the colony efficiently. This is why a hot day frequently provokes them to abscond when their nests have been made in insufficiently shaded places. Nests are generally made either in cavities in trees or in holes in the ground.

A large number of bees from the Southern African region, belonging to various families, have been recorded; most of them are solitary bees. The species which have been collected on Inhaca Island undoubtedly represent only a fraction of the total number of species present. Some of the species recorded on the island are listed in Appendix A.

13.12 SPIDERS OF INHACA ISLAND (T. Steyn and A.S. Dippenaar-Schoeman)

Very little is known about the spider fauna of Inhaca Island. During December 1992 and January 1993, spiders were collected by hand from five different localities on the island namely mangrove swamps, the hotel area, coastal bush in the north-eastern part of the island, around a village at Monte Inhaca, and coastal bush on the southern peninsula of Ponta Torres. A total of 19 families represented by 40 genera and 44 species have so far been recorded. These are listed in Appendix B. Collecting on the island is to be continued during 1993.

14

NON-MARINE MOLLUSCS

A.C. van Bruggen (Leiden University) and M. Kalk

Seven species of land snails have been recorded from Inhaca Island but comparison with similar areas on the mainland indicates that a much larger cryptofauna exists, hidden in the undergrowth and leaf litter on the forest floor (Connolly 1925). The comparatively recent age of the island will probably not have led to impoverishment of the fauna and it may be assumed that it reflects that of the adjoining mainland. This small number of species, however, illustrates several aspects of molluscan biology: the preference of snails for sheltered habitats, their mixed biogeographical affinities (as in the other phyla on the island), different specific diets, evolution from marine ancestors and speciation in successful genera.

Land snails belong to the two subclasses of the class Gastropoda: Prosobranchia and Pulmonata. The third subclass Opisthobranchia (sea slugs) has no terrestrial representative. The majority of the Prosobranchia are marine snails; a few families (only four in Africa) with limited diversity have adapted to a fully terrestrial existence. Most land snails belong to the Pulmonata or air-breathing snails with very few species on rocky shores and mangroves and some in fresh water.

Both a land prosobranch and pulmonates occur on Inhaca Island. Prosobranchs are easily recognised by the possession of an operculum as in sea snails; it is a small calcareous or horny plate carried on the foot so that upon withdrawal of the snail into the shell the aperture will become hermetically sealed. There is no permanent operculum in pulmonates although they may seal their shells by the secretion of a mucous film, which hardens when it dries, forming the epiphragm. Several families of pulmonates have given rise to slugs in which the outer shell has been completely reduced. No slugs have been recorded from Inhaca Island as yet, although at least a few species belonging to two families are likely to occur. Some have been seen but not identified. The Slug-eating Snake has been recorded and a black slug, *Laevicaulis* sp., has been reported from the environs of Maputo (Connolly 1925).

14.1 SNAILS AND SLUGS ON LAND (A.C. van Bruggen)

14.1.1 Prosobranchia

Tropidophora (Ligatella) ligata (Pomatiasidae), one of the common species, was reported earlier from Inhaca Island (Franca de Lourdes 1960). The shell is similar in shape to the top shells on the shore, the height and the major diameter being approximately equal; the aperture is large and rounded and is closed by a calcareous operculum (like that of *Turbo* on the shores). The umbilicus (a perforation in the columella) is narrow but deep. Inhaca shells are slightly smaller than usual: 12,4 x 12,2 mm (Fig. 14.1). They have few narrow dark bands, but these are variable in scattered populations, the intensity of colour probably being correlated with

Fig. 14.1 Prosobranch, *Tropidophora ligata* (x 1,7).

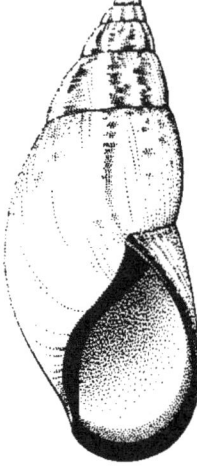

Fig. 14.2 *Metachatina kraussi*, harbivore (x 0,5) (drawn by A.C. van Bruggen).

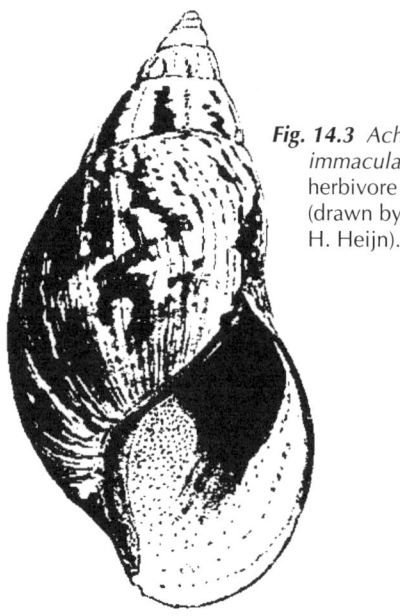

Fig. 14.3 *Achatina immaculata*, herbivore (x 0,4) (drawn by H. Heijn).

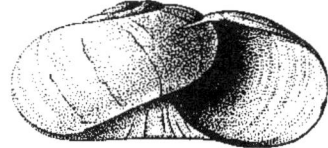

Fig. 14.4 *Nata vernicosa*, carnivore (x 2,5) (drawn by H. Heijn).

rainfall; the operculum is shown in place. This species may be found in the shelter of fallen logs in the sandy soil of the open parkland or in the humus of the closed forest. *T. ligata* is widely distributed from the Cape Peninsula in South Africa northward to the Zambesi River, although never very far inland (Van Bruggen 1978). It probably feeds on fungi and algae.

14.1.2 Pulmonata

Two species of a herbivorous family of large snails (Achatinidae) occur on the island, the larger *Metachatina kraussi* lives in sheltered forest or dense bush, since it is killed by direct sunshine. The other, *Achatina immaculata*, occurs on low-lying ground in open parkland adjacent to the freshwater swamps as well as in coastal bush. Their geographic distribution mirrors their habitat selection in so far as *M. kraussi* is endemic to a limited area of coastal bush from the Natal south coast to just north of Maputo and inland for about 80 km. *Achatina* is widely distributed in south-eastern Africa (Van Bruggen 1966).

Metachatina kraussi, the largest land snail in southern Africa (Fig. 14.2), grows to a maximum shell length of 159 mm and a major diameter of 68,5 mm. The shell tapers gradually from the aperture to the point and a conspicuous narrow brown band marks the outline of the aperture of the shell. A thin brown, deciduous periostracum covers the shell, which is dirty white except for a purple tinge and chestnut coloured markings on the early whorls. The snail's body and tail are yellowish grey, the head and neck are dark grey with a broad pale streak on each side. The broad radula has 130-150 rows of teeth, each row having a very small central tooth and over 60 broad teeth on each side, diminishing in size from the centre. This is a radula which is suitable for grazing on leaves of higher plants and large amounts are consumed.

The shell of *Achatina immaculata* has the same shape but is somewhat smaller and has a brown flame-like pattern; the columella and margin of the aperture are pink, which tends to fade (Fig. 14.3). The animal is grey with a dark-brown, mid-dorsal stripe on the neck, showing cryptic colouration like the other species. Both species play a small part in the conversion of leaves into soil and in the aeration of the soil. They are preyed upon when young by a carnivorous snail, *Nata vernicosa* (Rhytididae).

Nata vernicosa (Figs. 14.4), has a much flattened, globose, coiled shell which is yellow and glossy and measures about 10 x 17 mm; the umbilicus is wide and deep. It may be found on leaves of shrubs and herbs in the shelter of the forest. The radula has 10-30 teeth, fewer than in a herbivorous snail; they are conical in shape, larger canines occurring towards the centre, in this way being adapted to consuming small, soft invertebrates. This species has a wide distribution from the Cape Peninsula in South Africa along the eastern coastal bush (Van Bruggen 1978). Its occurrence on Inhaca Island is a first record for Mozambique.

The Streptaxidae is another family of small, carnivorous snails with shells often less than 10 mm in length. Shells of the genus *Gulella are* pale, glossy and cylindrical in shape; the margin of the aperture is usually covered by large processes of the shell (denticles) which vary in number and pattern in each of the 130 species in southern Africa. *G. triglochis* is one of the larger species (14,1 x 7,5 mm); the surface of the shell may be weakly ribbed or smooth and the aperture has only three teeth (Fig. 14.5a). It is an ecologically widely tolerant species inhabiting both forest and dune bush in Natal and Swaziland; Inhaca Island appears to be the northernmost record. *G. perissodonta* (Fig. 14.5b) is a very small species, about 4 x 2 mm, in which all the whorls of the shell are ribbed and the aperture has nine denticles, but its size and dentition are variable. The specimens have come from open vegetation at Ponta Torres and some of the processes are reduced in size. This species lives in comparatively dry situations with sparse vegetation, from the eastern Transvaal to southern Mozambique (Van Bruggen 1966). The third species of *Gulella, G. daedala* (6 x 3 mm), has a smooth shell and complex aperture denticles (Fig. 14.5c). It is found in forest and coastal bush in Maputaland (Van Bruggen 1966); Inhaca Island is the northernmost record.

From the evidence of two colour slides of a living snail it seems that *Trochanina mozambicensis* (Urocyclidae) (Fig. 14.6) occurs in the forest; it is herbivorous and often hides under the bark of trees. The species in the family Urocyclidae show various stages of reduction of the shell from a slightly depressed cone-like shape to smaller shells and finally to a small, shield-like shell, hidden under the mantle in the genus *Urocyclus* and allied forms. A caudal mucus gland beneath a short appendage at the end of the foot which wags like a dog's tail is characteristic of the family. The shell of *T. mozambicensis* is flattened and keeled at the periphery of the whorls and reaches a maximum of 11,3 x 18,5 mm. The apex shows fine spiral engraving, the rest of the shell exhibits close, regularly oblique coarse markings. The body is yellowish and the radula has tricuspid teeth and this snail is carnivorous. It occurs in open woodland from Swaziland to East and Central Africa (Van Bruggen 1978).

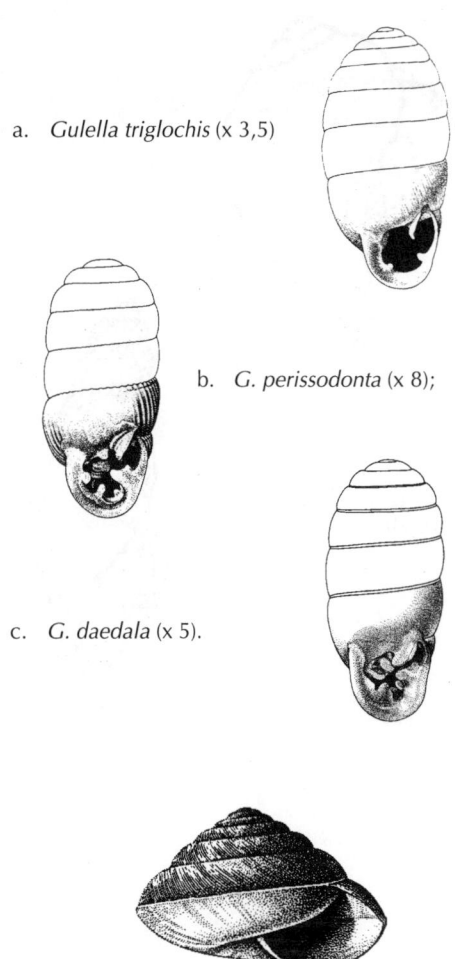

Fig. 14.5 Carnivorous snails (drawn by H. Heijn):

a. *Gulella triglochis* (x 3,5)

b. *G. perissodonta* (x 8);

c. *G. daedala* (x 5).

Fig. 14.6 Herbivorous snail, *Trochanina mozambicensis* (x 1,5) (drawn by H. Heijn).

This is a meagre collection of snails from such a rich and varied vegetation as Inhaca Island offers, and indicates that a land mollusc survey is overdue.* More representatives of Prosobranchs, slugs of the family Veronicellidae (which are large and dorso-ventrally flattened) and various snails of other families are sure to occur. A survey of a comparable, limited area in neighbouring Maputaland has resulted in a collection of about 50 species of land molluscs.

14.2 SNAILS IN FRESHWATER SWAMPS (M. Kalk)

An early medical study on the prevalence of human parasites on Inhaca Island (De Morais 1959), reported a survey of the molluscan fauna in the freshwater swamps, undertaken because 40% of the people were infected with bilharzia. *Bulinus globosus* (Fig 14.7a), the snail host of the larva of *Schistosoma haematobium* that causes the debilitating disease, urinary bilharzia, was present in two of the larger swamps. It is well known that the snail *Bulinus* survives the superficial drying of the swamps in winter. In summer its numbers are largest

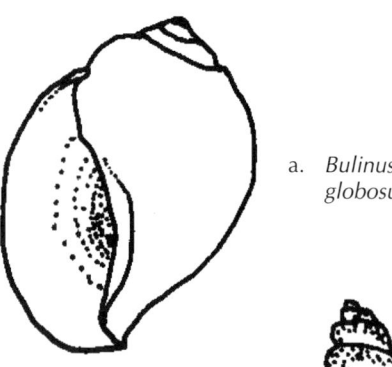

Fig. 14.7 Freshwater snails:

a. *Bulinus globosus* (x 2)

b. *Pyrgophysa* sp. (x 5) (a and b after J.A. van Eeden 1960).

at the water's edge, and decline rapidly towards the interior of the swamp where the density of bulrushes or reeds excludes light and precludes the growth of epiphytic algae on which the snails feed (Cantrell 1979).

These snails lay their eggs preferentially on water lily leaves in open water, but there are very few water lilies growing today. Intensive agriculture in and around the swamps, whilst initially exposing women to infection by the *Schistosoma* larvae may have destroyed the habitat of the snails, or alternatively, may have facilitated their spread on the swamp periphery.

A smaller number of the elongated spiral shells of *Pyrgosoma forkali* were present in 1958. These shells have a very narrow aperture and are about 10 mm long (Fig. 14.7b).

* Material for study should be sent directly to the author of this note:
Dr A.C. van Bruggen,
Systematic Zoology Section of Leiden University,
c/o National Museum of Natural History,
P.O. Box 9517,
2300 RA LEIDEN,
Holland.

15

Changes in the Diatom Flora in the Freshwater Swamps of Inhaca Island

F. D. Hancock, University of the Witwatersrand

Diatoms are ubiquitous, unicellular or colonial algae that, by their great numbers and physiological activity, play a very important role in aquatic environments. They are primary producers of food supplies for both animal and plant life in the water (see Sections 3.3 and 4.6.1) and they maintain a viable level of dissolved oxygen through their photosynthetic activity.

Research workers in this field of botanical studies have found that some diatom species are tolerant or intolerant of certain chemical or physical conditions of the water (Hancock 1973). The degree of salinity is the principal ecological factor responsible for the distribution of diatoms into marine, estuarine and freshwater inhabitants (cf. Fig. 4.23a-f); other limiting factors are the quantity of dissolved oxygen, the temperature of the water, the reduction in volume of water and competition between species (Cholnoky 1960).

When conditions are favourable for one or more species, they multiply rapidly and so form dominant associations, whilst for other species chemical or physical conditions become unfavourable. The latter includes crowding out by the dominant species, with a resultant diminishing of others to an occasional or even a rare occurrence. Thus, by classifying species and determining the percentage occurrence of the main species and their associates it is possible to tell, for example, whether the quality of the water is suitable for human consumption.

Samples were taken periodically from three of the Inhaca swamps (see Fig. 15.1) during the winter months, May and August, during the decade 1963 to 1973. The springs originating in each swamp were densely covered by a rich growth of reeds and bulrushes; this area, termed Zone I, is described in Section 10.3.1. The sandy soil was buried by decaying material derived from the vegetation, making the water slightly brown; light penetration was limited, reducing photosynthesis, which in turn resulted in a low level of dissolved oxygen and an increase of carbon dioxide and carbonates (see Table 1, Zone I).

In the lower regions of each swamp, termed Zone II, streamlets emerged from the springs of Zone I, in which small pools occurred. Here hydrophytic higher plants grew mostly along the margins of the water, thus sunlight was not impeded (see Section 10.3.2). Much of the water from Swamps 1 and 3 was drained off through furrows, dug by the local people for irrigating their crops. The different physical and chemical conditions of the water in Zone I and II in the three swamps are shown in Table 1.

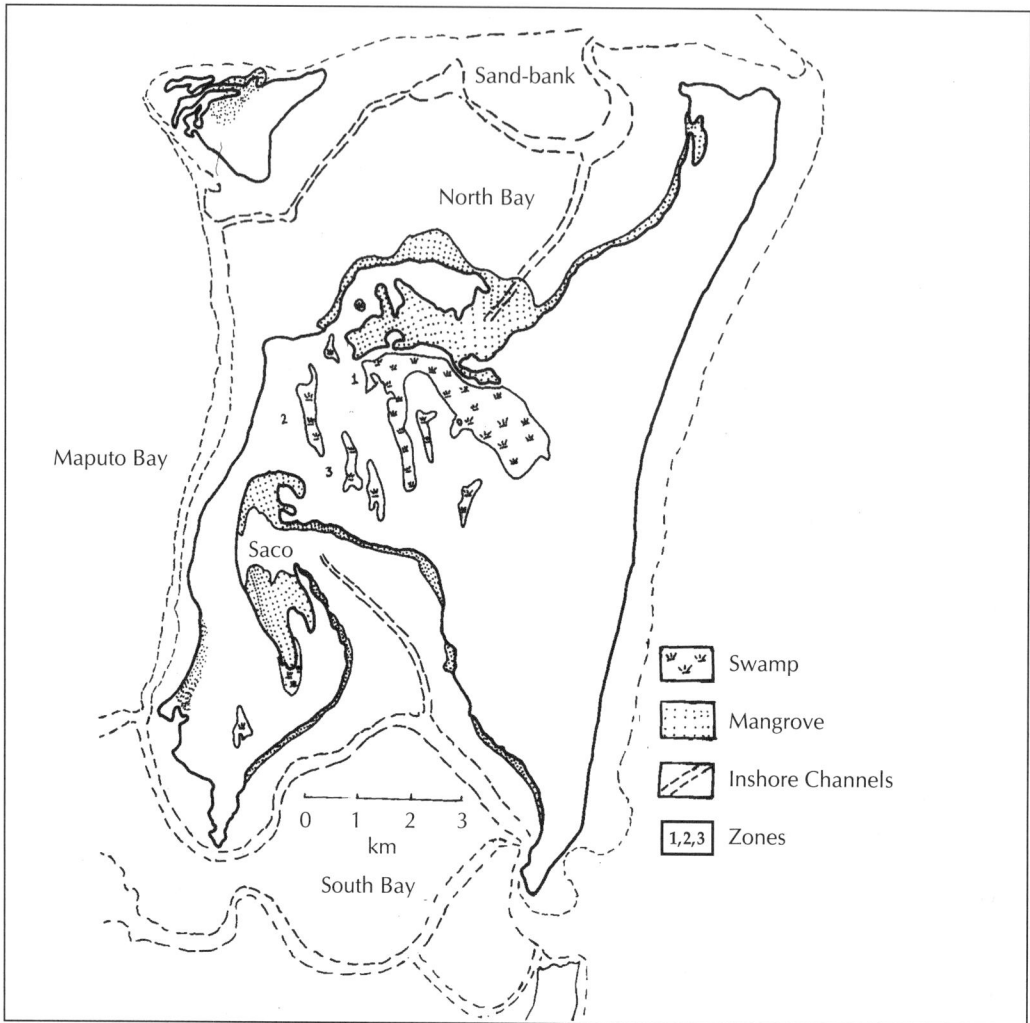

Fig. 15.1 Map of Inhaca Island (1958)*, showing sampling sites 1, 2 and 3 in the Tivanini freshwater swamps.

Table 1: Comparison of the environmental factors in Zones I and II during winter months 1963-1973

	Zone I	Zone II
Temperature (°C)	20,1	20-23
pH	5,5	6,0-7,2
Conductivity µS cm^{-1}	136,00	160
Oxygen mg/l^{-2}	4,0	6,6

Note that values in all four factors were higher in Zone II than in Zone I.

The diatom survey of the three swamps yielded 35 genera and 325 species. Details of the genera and species are listed in Appendix B. Slides of samples are housed in the Botany Department of the University of the Witwatersrand.

The species, in general, were widespread amongst the three swamps, especially in Zone I. Five genera in descending order of density in Zone I were *Nitzschia, Navicula, Gomphonema, Eunotia* and *Pinnularia* (Fig. 15.2a-e); this order differed, however, in Zone II — as shown in Table 2.

Table 2: Average percentage occurrence of diatoms of the five leading genera in the three swamps

Zone I		Zone II	
Nitzchia	13,8	*Nitzchia*	16,0
Gomphonema	20,5	*Eunotia*	13,9
Navicula	19,5	*Navicula*	12,2
Pinnularia	9,9	*Gomphonema*	9,5
Eunotia	7,4	*Pinnularia*	2,8

The percentage occurrence of four major genera was higher in Zone I than in Zone II, with the exception of *Eunotia* which was almost twice as frequent in Zone II when compared with Zone I. *Nitzchia* species dominated throughout Zone I; this would be expected in an acid habitat in which a great deal of decaying organic matter occurred. The species *N. palea* and *N. perminuta*, which are known to be indicators of low pH, were abundant here. *Navicula* and *Gomphonema* were closely associated as sub-dominants of Zone I. *Navicula cryptocephala* var. *veneta* was the most abundant species in this genus, members of which are known to be tolerant of varying pH values but sensitive to the ionic content.

Gomphonema gracile, *G. lanceolatum* var. *insignis* and *G. longiceps* var. *subclavata* all attained lower percentage values in Zone II; they have been shown to be sensitive to the ionic content of water but insensitive to pH values (Cholnoky 1960).

Pinnularia spp. and *Eunotia* spp. formed a weak association in Zone I which disintegrated in Zone II, where *Eunotia* species were more than twice as frequent as in Zone I. The increase in *Eunotia* spp. in Zone II was expected as the pH still remained fairly low, however, the inorganic content had increased while the degree of shade had decreased. *Eunotia subaequalis* was most abundant, while the frequency of *E. flexuosa* increased considerably by 1973. *Pinnularia* was of low abundance throughout Zone I, although the occurrence of this genus is generally favoured by acid conditions. It is possible that low values of oxygen and nitrates inhibited its development or it was affected by the environment where water was being reduced in Swamps 2 and 3 for crop irrigation. The higher density of key species in Zone I indicated the instability of the habitat. A very low frequency of *Acnanthes* was observed, reflecting the low oxygen content of the water.

Relative diatom frequencies vary from year to year in any given type of habitat; this holds true for individual species, especially in the unstable habitats in Zone I, caused by the increased water drainage by the growing human population. The rate of destruction of the water habitat is reflected in Table I and in the reduced areas of swamps after a lapse of thirty years.

A study of the diatoms is not only interesting in itself but it enables those in charge of water affairs to assess the state of the swamp ecosystem and to intervene for its preservation.

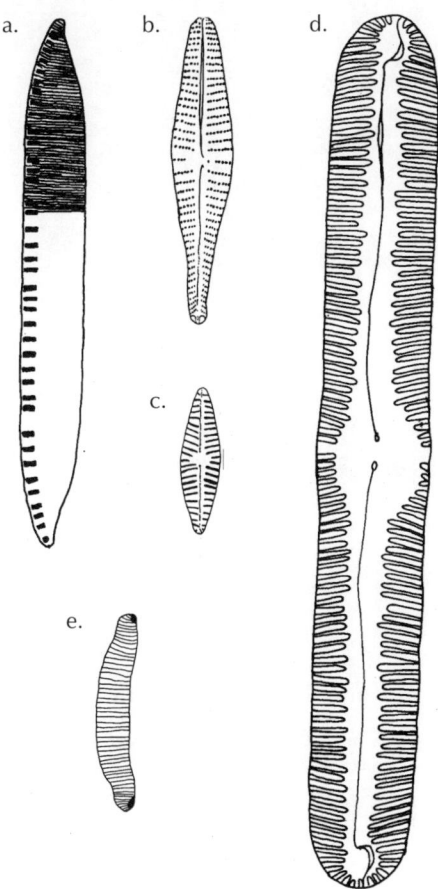

Fig. 15.2 Dominant genera of diatoms in the sampling sites of freshwater swamps.
a. *Nitzschia*;
b. *Gomphonema*;
c. *Navicula*;
d. *Pinnularia*;
e. *Eunotia*
(all approximately x 1 000)
(from photographs by F.H. Hancock and after B.I. Cholnoky 1961).

PART 4

The Future of Inhaca Island

The way of life of the people who live on Inhaca Island has changed little in the last fifty years (see Introduction), although it now appears more prosperous. The island is still a haven of tranquillity for tourists and a focus of interest for naturalists because of its tropical seashore life and numerous birds. It also has an attraction for scuba divers, snorkelers and anglers.

Nevertheless the increasing population on the island, confined to the limited area of about 40 km^2, has created problems. It has now become urgently necessary to diversify the economy in order to improve the quality of life of the local people. Simultaneously, the equilibrium between the people and their environment must be restored to prevent further degradation of resources.

It is expected that these two objectives will be achieved as a result of several factors. The policy of nature conservation initiated in 1965 is reinforced by long-standing traditional practices. The participation in the cash economy of women, who will sell their produce to an increasing number of tourists when more accommodation is built, will raise their purchasing power. The fishing industry has been steadily modernised since independence and the process will continue. Relevant research, carried out by biologists at the Universidade Eduardo Mondlane in Maputo, has pin-pointed the problems and laid a scientific foundation for the changes that must take place. The Marine Biological Station, under the control of the Faculty of Biology at the University, plays a key role in conservation in both forest and marine reserves. The increased literacy of the people since independence will facilitate their participation in the new United Nations Integrated Development Plan of 1990.

16

CONSERVATION AND DEVELOPMENT
M. Kalk and F. Costa

16.1 THE PRESENT STATUS OF THE PEOPLE AND THEIR PROBLEMS

The 1989 census recorded a population of approximately 10 000 persons on the island of whom 5 000 were permanent residents (Plano de Desenvolvemento Integrada Ilha da Inhaca 1990). The others were families who had migrated there from the neighbouring Machangulo peninsula in the previous seven years to escape the war.

The society is mainly patrilineal in character and there are about 750 families in the permanent population. Each extended family has a 'muti' or garden of 1 300-3 000 m^2 in area, in which a house is built for the family. There are separate huts for kitchen, bathroom and food store, with smaller sheds for chickens, pigs and goats. Houses for sons and their wives are added later. Chickens have the free run of the muti, goats are tethered outside in the forest amongst the vegetation. Indigenous fruit trees are incorporated in the muti and trees such as mango, pawpaw or cashew nut may be cultivated.

Outside the boundary there may be sacred trees and a cemetery. There is a footpath leading to a well, which may be up to 500 m away, and to the family's agricultural land, the 'machamba'. There are about 83 wells on the island near the small freshwater swamps between the dunes. The habitations are spread over about 47% of the land area and a further 16% is divided into machambas for crops.

The houses are built of dried reeds from the renewable supply in the freshwater swamps. The roofs may be thatched with local dried grass or, more recently, a few are made of galvanised iron. The houses are waterproof and may last for 15 years or more, although some are in disrepair because the owners are migrant workers. The displaced families build reed houses in a similar way in the southwest peninsula.

Most of the remaining 37% of the land is covered by natural vegetation such as coastal bush or evergreen forest, by freshwater swamps and reeds and, in some areas, there are mangrove swamps on the upper shores. In 1965 about 30% of the land area was declared a Nature Reserve in order to ensure regrowth of the vegetation. Various parts of it are in stages of regeneration. The people rely on the bushland for firewood and timber (Bandeira 1992).

Traditionally, space is reserved within the family 'muti' for the extended family, but the increasing population requires more and more land so that new homes were built on the fringes of the bush. The land may first be cleared for agriculture, but it is soon exhausted because the soil is poor, then houses are built. This encroachment on the forest has now stopped since the conservation plan is more strictly adhered to. The forest and bush are needed for firewood and timber so that regrowth is imperative.

There are three administrative districts on the island. Ridjene, which includes the fishing village, the airstrip and the hotel and part of the central 'parkland' of mutis and machambas (cultivated areas), is the most urbanised. Inguane, which includes the north-east and south-east peninsulas, and Nhaquene, the south-west peninsula, have larger machambas and are less sophisticated.

The district of Ridjene has acquired some of the amenities of an urban centre. There is one tourist hotel, several reed huts for tourists, small restaurants and camping sites. Facilities include one shop with a bar, a post office and a public telephone connected to the mainland at Maputo, an administrative office, a bank, a generator (for the hotel), a bakery, an open market place and a recreation field. The refrigeration and packing plant, which since 1977 has belonged to the Fishermen's Combine, stands near the shore, opposite the landing area for boats. This ensures that the daily fish catch reaches markets on the mainland in good condition. The ferry boat to and from Maputo, which may carry over 100 people as well as goods, anchors offshore but there is no formal quay. Passengers embark from small boats; goods may be carried as head-loads by men wading at low tide. On the cliff above the hotel there is a clinic with a maternity ward, manned by a medical assistant and his small staff. A male ward has recently been built. It is not well-stocked with medicines. Since independence three primary schools have been built and people under 23 years old are literate.

The cash economy of the people on the island still depends partly on migratory labour and partly on fishing (Plano de Desenvolvemento Integrada Ilha da Inhaca 1990). About 47% of the men were employed in the South African gold mines in 1989 and some men work on the mainland of Mozambique. Variable proportions of their low wages reach their families. Herein lies another problem since the South African gold mines are retrenching labour on quite a large scale. The fisheries industry accounts for the employment of 52% of the male population. This bolsters the inadequate semi-subsistence economy. There are a few self-employed men who construct boats from local timber, make sails, clothing and shoes or baskets for sale and in their spare time may be barbers.

Cash earnings from fishing depend on the relative sophistication of boats and the fishing equipment. About 1 000 men are engaged in fishing from boats for a living. They catch 400-500 tonnes of fish per year, most of which is sold on the mainland. The Fishermen's Combine runs one large ocean-going vessel and 80 registered small fishing boats of which (in 1989) 48 had petrol engines and a crew of five or six men; 32 boats had sails only. The motor boats can go further afield and fish in deeper water; they catch bigger fish that fetch a higher price. Some men supplement their earnings by fishing from the shore and selling on the local market. Women still collect shellfish and crabs from the sandflats at low spring tides for home consumption.

The fish population in the Bay of Maputo appears to be declining. A viable industry requires that all the boats be motorised. An uncontrolled collection of seashore animals will sooner or later lead to a scarcity of edible species.

There is no paid employment for women; they are totally responsible for agriculture. The women grow small quantities of cassava, sweet potatoes, rice, maize, sugar-cane and ground-nuts on the most fertile soil around the freshwater swamps, which are drained for cultivation (see Section 10.4.2). More recently vegetables, such as beans, pumpkin, cabbage and tomatoes, are cultivated for sale on fertile ground around the swamps.

The planting of cereals yields poor crops. The soil is not only sandy and of poor quality, tending to be saline, but the areas available are too small to supply the needs of the present population so that supplementary maize meal, flour and sugar are imported in bulk from Maputo and sold in the shop. Some 80% of the agricultural land has to lie fallow, intermittently, for long periods after it has been used. This soil is exhausted and there is no more land available for agricultural use. There is really not enough food on the island, even with imported maize meal, for everyone to have regular meals because of the low purchasing power of the people. However local fruit is available and a recent survey showed that overt malnutrition amongst the children was rare.

Dry wood for fuel is collected in the bush. Recently, parts of the Reserve in the south-west and south-east peninsulas have been released for the collection of firewood. Preliminary estimates indicate that the consumption of wood is 20% in excess of the annual production of the forests and mangroves. Regeneration in the Nature Reserves is not yet fast enough to ensure a constant supply. Natural regeneration of the mangrove trees does occur annually but the trees take some years to mature (Hatton and Couto 1991).

The availability of fresh water now appears to be reaching a limit; some wells are slowly drying and the lack of any protection constitutes a health hazard. Rainwater is collected near the Biological Station; boreholes supply water to the hotel and the Biological Station *and* to villages in their environs.

There are two long-term environmental problems in addition to the economic problems outlined above. The west shore is subject to erosion by the sea and should be stabilised when the time is opportune. The east shore is subjected to deposits of wind-blown sand which accumulate over the crest of the dune and continuously bury trees in certain places. In the long term the sand threatens agricultural land in the interior. This problem has to be tackled by planting *Casuarina* trees, as was done in the 1930s, below the lighthouse (see Section 10.5.5) (Cardosa 1959). This has recently been done at Ponta Torres.

In order to solve the socio-economic problems and at the same time to protect the island's shores, an Integrated Development Plan for Inhaca Island under the aegis of the United Nations was devised in 1990, with the co-operation of the people (Plano de Desenvolvemento Integrada Ilha da Inhaca 1990). The plan envisages a change in the economy through the increase of maritime and rustic tourism in which the women may be employed in catering, and the modernising of fishing methods. Social amenities will be upgraded and conservation measures to protect vulnerable areas will be extended (see Section 16.3). The plan is largely based on the research of biologists at the Universidade Eduardo Mondlane in recent years.

16.2 THE ROLE OF THE BIOLOGISTS AT THE UNIVERSIDADE EDUARDO MONDLANE

16.2.1 Research on Inhaca Island

In the last decade, a detailed survey of the vegetation of the whole island has been carried out (see Sections 10.1-10.3). It may already be concluded that forest regeneration is taking place *faster* than expected, in 50 years perhaps, rather than the 100 years often quoted in the literature. This may be explained by the positive role of traditional practices on the island. Reservoirs of seeds exist in the small, sacred patches of forest which have never been disturbed by agriculture. Similarly, trees with edible fruit or medicinal use surround the homesteads and plots of cultivated ground on the low dunes and these facilitate the spread of seeds, largely through birds.

The results of experimental reclamation of two enclaves on the dunes exposed to the south-east winds on the east coast show that it is possible to reduce the removal of wind-blown sand in certain vulnerable areas. In one case the vicinity of the lighthouse was threatened (Cardosa 1959) and in the other case the coral reef at Ponta Torres was being smothered (see Section 10.5.4 and 5). The *Casuarina* trees planted 50 years ago below the lighthouse are now dying, but in their stead a herbal layer, seedlings and saplings of littoral scrub have grown. The coral reef became rejuvenated after the sand had been stabilised by planting local sand-binding herbs and bushes (Salm 1976) (see Section 10.5.4 and 5).

In the protected mangrove swamps and forests in the nature reserves the rate of new growths from seedlings has been monitored. The rate of regeneration will help determine the quantities available as future sources of timber. The annual consumption of firewood has been estimated (with the co-operation of the people on the island), so that the conservation policy may be based on a balance between the rate of regrowth and the rate of consumption of trees. Planting of woodlots in selected places has been recommended.

A large number of tree species which have flowers pollinated by bees have been located on the island, so that the development of a small industry for the production and sale of honey is assured. A manual of apiculture has been produced and help is available from the National Apiculture Programme.

The potential for tourism is supported by the studies of the seashore and flourishing coral reefs. The distribution of the sea grass associations, which stabilise the sand flats, has been mapped (see Section 5.2). The reefs have been monitored from time to time by foreign scientists, resident for some months at the Marine Biological Station. New coral growths have been mapped at the Barreira Vermelha reef on the edge of the Inhaca Channel. The recovery and continued growth of the reef at Ponta Torres and the formation of a new reef in the lagoon adjoining the northern shore of Portuguese Island will attract tourists. Several aspects of coral reef associates and the plankton above the reefs on which coral polyps feed have been investigated by local and foreign biologists (see Chapter 8). Large assemblages of coral in several reefs off the eastern shore of the island (see Fig. 1.1) and the subtidal rocks around the old wreck at the north-east point have become popular with anglers, scuba divers and snorkelers.

The impact of the fishing industry in various parts of the Bay of Maputo and in the open ocean is being evaluated by the Fisheries Research Institute. A decision was taken to fish further afield using motor boats, to avoid the continued depletion of the local fish populations and to allow for their recovery. The usefulness of both medicinal and non-medicinal (timber) trees has been studied.

16.2.2 Conservation and Management

The Marine Biological Station is run by a resident biologist and a team of ten rangers who manage the protection of the environment and, when required, participate in research. An assistant curator of the museum and herbarium at the station keeps the taxonomy up to date, and other members of staff manage the electricity supply, the motor boats and land transport. The museum has on display members of almost every class of animal which occurs on the seashore, the common insects on land, the snakes and lizards, turtles and mammals. The herbarium has a complete collection of dried plants with names updated to the present time. The laboratories have running sea water and fresh water, electricity, and separate rooms for student classes and research workers. The library has copies of research papers and books relevant to the island. The dining room is used as a lecture room. The station has the potential for training study leaders who will spread environmental education among the residents of the island, as well as for the instruction of school pupils and students. Scientists from several countries in Europe and South Africa pay research visits to the island in turn.

A programme of conservation and management for Inhaca Island and Portuguese Island has recently been revised. Scientific studies clearly showed that the islands form a fragile system that could support the resident population, fulfil scientific expectations and allow for the growth of the expanding tourist industry *only* if some form of environmental control was put into operation.

It was already apparent in 1965 that several factors were contributing to endangering some of the sensitive components of the island's environment. This led to the proclamation of several forest reserves on the islands and marine reserves in the surrounding waters. These were later extended (Fig 16.1) and rules were formulated for the conservation of vulnerable areas.

The objectives were basically the following:

- to ensure that viable portions of protected dune forest and mangrove swamps remained intact, in view of the increasing and indiscriminate felling of trees;
- to ensure that, by keeping enough forested areas, the resident population would have a sufficient basis to cater for the daily needs of fuel-wood, building materials, medicines and other forest products, utilising the forests as a renewable and sustainable resource;
- to contain and reverse the deleterious trend in erosion-prone sites by the establishment of a stable forest cover;
- to protect corals and associated marine life from misuse by fishermen, beachcombers and tourists; and

- overall, to maintain the ecological integrity of the islands and surrounding water in such a way as to support the local community and to continue fulfilling the traditional role of the island as a place for study, scientific research and use by tourists.

The Yingwani Forest Reserve was established mainly for the protection of the eastern dune belt. It covers an area of approximately 1 100 ha of which roughly 60% is dune forest, 20% is secondary vegetation recovering from previous cultivation and the remainder (20%) is mobile, bare and scarcely covered sand. Sand encroachment of the highest forest over the brow of the dunes is critical in many spots owing to the strong south-easterly winds (Fig. 16.1). It is intended that planting with *Casuarina* trees will eventually remedy this, since they will be replaced naturally by indigenous vegetation. This reserve also includes the Tivanini mangrove area in the north-west, which is expanding seawards.

The Nyakeni Forest Reserve protects the western dune belt and extends from just south of the main village (Ridjene District) to Ponta Punduine in the south-west, covering an area of approximately 260 ha, of which only a little over 20% is true forest; the remainder, showing secondary stages of vegetation, is recovering from previous cultivation. This reserve also includes the Ponta Raza mangrove area, the only creek mangrove on the island.

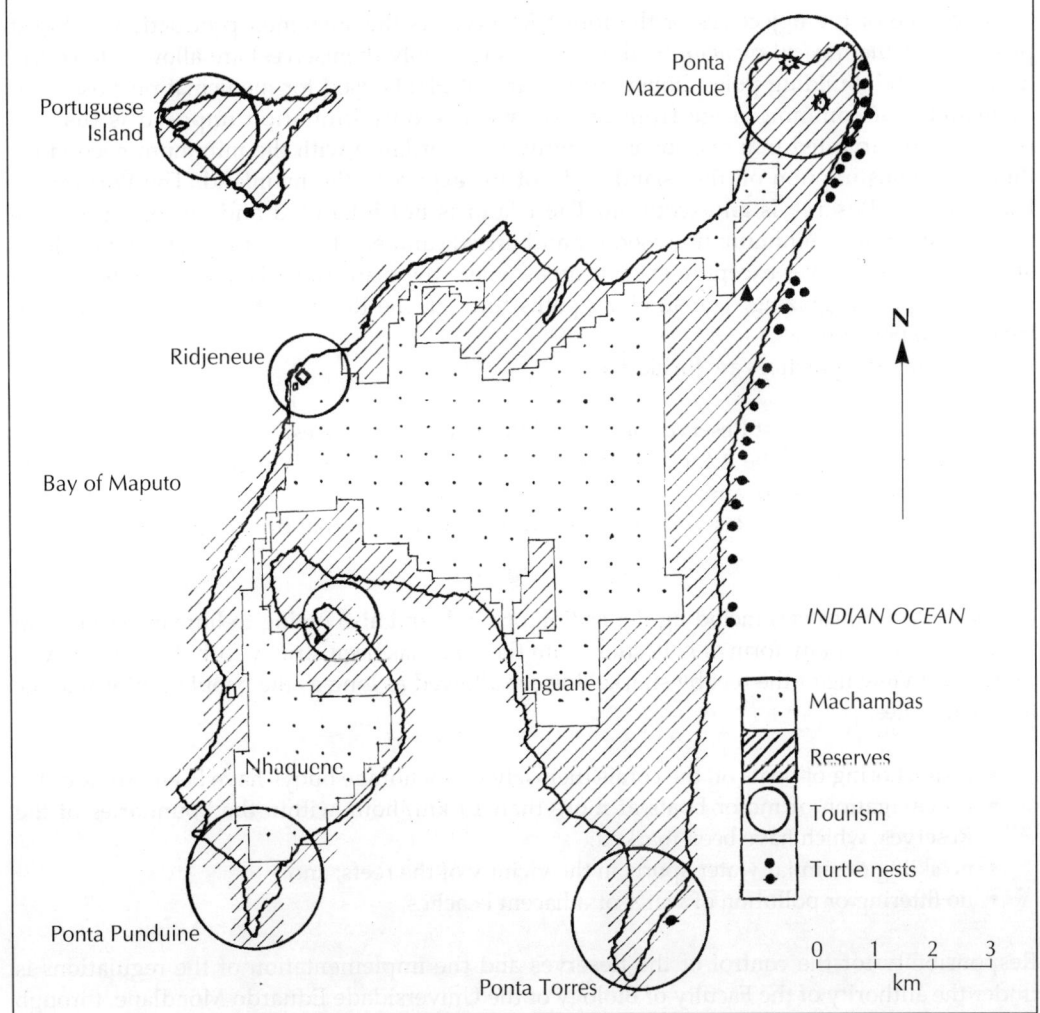

Fig. 16.1 Map of Inhaca Island: Allocation of areas for Housing and Agriculture, Nature Reserves and Tourism; sites of breeding sea turtles (*United Nations Integrated Development Plan for Inhaca Island 1990*, courtesy of the Instituto de Desenvolvimento Rural, Maputo).

In the western areas there are no mobile sands, but erosion from natural causes (tides, currents, cyclone action) has eaten into the base of the red cliff, Barreira Vermelha facing Maputo Bay, causing the upper layers to slide down. As a result the cliff shows several patches denuded of vegetation and destruction of Kingfisher nests. Deposition of sand by currents linked with the inundation of Portuguese Island has moved the limit of high spring tides to the edge of the dune bushes.

The whole of Portuguese Island and its shores constitute another reserve. Between the northern edge and the encircling sand-bar a lagoon has formed which encloses the new coral reef on a solid coral rock base. For this reason the lagoon has been awarded nature reserve status. However, there is no prohibition of fishing and collecting activities along the remainder of Portuguese Island's coastline.

The Barreira Vermelha Marine Reserve was established to protect the coral reef adjacent to the northern section of the Nyakeni Forest Reserve. The reef varies from 10-15 m in width on the edge of the deep Inhaca Channel and is approximately 1 000 m long, but extra buffer zones of 200-300 m long have been included at the north and south ends.

The Ponta Torres Marine Reserve, located in the extreme south-east of the island, is intended to protect the local, shallow coral reef and the deeper rock, a wall-like cliff, that joins the reef at its southern end. The size of this reserve is comparable with the west-coast marine reserve.

Since one of the objectives of the forest Reserves is the sustained production of forest products for the *resident community*, the islanders (but only themselves) are allowed to collect dead firewood, building poles, honey, fruit, parts of plants used for medicinal purposes and other items of common usage from the forests. The only limitation imposed is that the products be harvested in a reasonable quantity in accordance with the minimum needed for their own consumption on the island and not for export to the mainland. The Portuguese Island Forest Reserve is an exception. The island is not inhabited and *no* collections are allowed other than minimum firewood for authorised campers. Pasture for goats (not cattle) is allowed in the reserves except in specific areas where the plant cover is not in a state healthy enough to permit grazing. Transit along paths through the Reserves to walk to the sea and back is also permitted.

The following practices are forbidden:

- to settle in the forest Reserves or to build permanent structures;
- to clear the forest, be it for cultivation or other purposes (and, in general, there is to be no tilling of the soil); and
- to light a fire in the Reserves or in any way to allow external fires to penetrate into the Reserves.

Owing to natural causes including deposition of sand, and also owing to human activities in the past, fishing or any form of collection from the reefs has been strictly forbidden. However, diving and viewing of the reef by snorkelling are allowed as long as the people conform to the following rules:

- no anchoring of boats on the corals themselves (anchorage buoys have been provided);
- no navigation of motor boats at more than 10 km/hour within the boundaries of the Reserves, which have been marked;
- no skiing or similar water sports in the vicinity of the reefs; and
- no littering or pollution of water or adjacent beaches.

Responsibility for the control of the Reserves and the implementation of the regulations is under the authority of the Faculty of Biology of the Universidade Eduardo Mondlane, through the staff of the Marine Biological Station, and has been so since 1980. A small ranger force, which operates from outposts located at six strategic points, and a mobile patrol around the islands constantly monitor the Reserves. This has proved especially useful in the protection of nests, eggs and young of the Leather-backed Turtle and the Loggerhead Turtle on the eastern dunes. It was estimated that in 1989, 2 000 young turtles succeeded in reaching the sea after hatching (Costa 1989).

Rangers at Portuguese Island and Ponte Torres also supervise the camping sites.

16.3 THE INTEGRATED DEVELOPMENT PLAN, 1990

The plan (Plano de Desenvolvemento Integrada Ilha da Inhaca 1990) for the immediate future of the islanders has been drawn up by the National Institute of Physical Planning in Mozambique. It was formulated in consultation with the people of the island, with representatives from the Institutes of Meteorology, Hydrology, Fisheries and Agronomy, the Faculty of Biology at the Universidade Eduardo Mondlane in Maputo, the Fishermen's Combine on Inhaca Island and the hotel industry. The Development Plan will be carried out under the aegis of the Maputo City Council. Finance to implement the plan will be raised from foreign development aid.

One of the first tasks agreed upon was an extension of education on environmental objectives and conservation for both adults and children on the island. The three schools have already been upgraded to Standard 7. It is aimed to provide evening classes for adults, using stored solar energy to supply electricity. Programmes have already been developed for this purpose.

16.3.1. Focus on tourism and fishing

The development of tourism is the key factor in the plan, since it would provide a market for goods produced by the people of the island. It was envisaged that subsistence farming would become a subsidiary occupation and the emphasis would be changed to horticulture and the breeding of small stock largely by women. Vegetables grow well on broad strips of land around the larger and smaller swamps. The requirements of tourists could easily be met locally. More fruit trees would be planted so that mangos, oranges and cashew nuts could be sold to tourists and the plentiful *indigenous* fruits would also be collected from the forests for sale. Men would be encouraged to participate in horticulture.

It was recommended that larger stocks of poultry, including chickens, ducks, geese and turkeys, should be bred in the mutis and pigs should be kept. Goats would be phased out since they destroy the newly grown vegetation. The meat from local stock and fish from the Fisherman's Combine should find a ready market among the local restaurants for tourists. It was proposed that bee-keeping should become an important industry on the island, since the sale of honey is very profitable. The National Programme for Apiculture would train potential bee-keepers and supply them with hives. The honey would be harvested and the storage of honey for sale would be supervised by the National Programme of Apiculture so as to maintain a high standard. In addition people should be encouraged to sell local foods and fruit drinks, such as coconut milk, to tourists.

The Development Plan points out that fisheries should be expanded by increasing the number of boats with engines in order to fish in deeper water, further afield. Larger boats would be constructed on the island and provided with engines. Instruction in the use and repair of engines will be given. Nets would still be made locally. Italian and Norwegian Aid Agencies are already concerned with this field of development.

The following proposals to attract tourists were considered, and some are already being implemented. The present hotel in the District of Ridjene should be upgraded to 3-star status and should cater for 150 people. Two smaller hotels were proposed, one in the forest near the lighthouse in the north and the other across the Ponta Torres Strait on the Machangulo Peninsula. (The latter would be an incentive for the displaced persons to return home and sell their fish and produce.) A third type of hotel would have a rustic or traditional atmosphere, and be built in the same style as the local habitations in the mutis. A few bungalows of a similar construction have already been built, scattered around the island but not in areas where mutis and machambas exist. Camping sites with water and toilets are to be established on Inhaca Island as well as those currently in use on Portuguese Island and in Ridjene. They would be under the control of the rangers on the staff at the Marine Biological Station. In this way the interests of many types of tourists of various income levels would be catered for. They in turn would provide a viable market for local produce. It was agreed that the maximum tourist load at any time would not exceed 500 persons. A high standard of water supply, sanitation and mosquito control would be maintained by the staff.

16.3.2 Social welfare and administration

A programme for raising the standard of health on the island has been proposed. It is intended that existing wells should be upgraded, lined with cement and covered. The Hydrology Institute would survey the accessibility of fresh water and the extent of infiltration of sea water. New wells would be sunk, care being taken to site them at least 30 m away from latrines. Standard type earth closets would be built by the people outside each muti, at very low cost. Rubbish collection would be organised and biodegradable refuse converted into compost; the rest would be sold for recycling. In support of this modernisation, a study programme would be developed to educate people in the construction, use and repair of these facilities.

A second clinic has been built in the 'urban centre' at Ridjene incorporating a children's clinic and a male ward. An ambulance would be available to transport sick people to the clinic and standing arrangements made for conveying them to Maputo when necessary.

Links with the mainland at Maputo would be improved by means of a daily ferry service and regular air flights. Tourism would be confined very largely to the three peninsulas of the island and would not impinge on the mutis and machambas of the people in the central area (see Fig. 16.1). The traditional inheritance of land would not be changed. No foreigner would be allowed to occupy or buy land.

It is expected that the 'displaced persons' would return of their own accord, to their traditional land in Machangulo. The fishermen among them would be allowed to share the facilities of the Fishermen's Combine refrigeration plant.

Finally, the Nature Reserves will continue to protect the vegetation on the eastern and western dunes until the trees have recovered sufficiently for use. The staff of rangers employed by the University at the Marine Biological Station will enforce the conservation regulations both in the forests and along the coral reefs.

COPYRIGHT ACKNOWLEDGEMENT

The authors and Witwatersrand University Press wish to thank the copyright holders for permission to use material from the following publications:

ALEXANDER, R. MACNEILL. 1979. *Invertebrates*. Cambridge: Cambridge University Press; ALLANSON, B.R. 1958. *Siphonaria* species on the South African and Mozambican coasts. *Hydrobiologia* 12(2-3):140-50; ANNECKE, D.P. AND MORAN, B.C. 1982. *Insects, Mites and Cultivated Plants in South Africa*. Borough Green: Butterworths; BALINSKY, J.B. 1957. The Ophiuroidea of Inhaca Island. *Annals of the Natal Museum* 14(1):1-32; BARNARD, K.H. 1950. Descriptive catalogue of South African decapod Crustacea (crabs and shrimps). *Ann. South African Museum* 38:1-864; BARNARD, K.H. 1958. Further additions to the Crustacean fauna list of Portuguese East Africa. *Memorias Museu Alvaro de Castro* 4:2-23; BARRAS, R. 1963. The burrows of *Ocypode ceratophthalmus* (Pallas) (Crustacea, Ocypodidae) on a tidal, wave beach on Inhaca Island, Mozambique. *Journal of Animal Ecology* 32:73-85; BERJAK, P., CAMPBELL, G.K., HUCKERT, B.I. AND PAMMENTER, N.W. 1986. *In the Mangroves of Southern Africa* (2 edn.). Durban: Natal Wild Life Society of Southern Africa; BRANCH, G. AND BRANCH, M. 1981. *The Living Shores of Southern Africa*. Cape Town: Struik; CHAMBERLAIN, Y.M. 1958. *Dasyclados ramosus*: A new species from Inhaca Island and Peninsula, Portuguese East Africa. *South African Journal of Botany* 24(3):119-21; CHOLNOKY, B.J. 1960. Beitrage zur Kenntinis de Diatomeen-Flora von Natal. *Nova Hedwiga* 2(1&2); CROSNIER, A. 1962. Portunidae. *Faune de Madagascar* XVL:1-154; CROSNIER, A. 1965. Grapsidae and Ocypodidae. *Faune de Madagascar* XVIII:1-141; DAY, J.H. 1934. Polychaetes from South Africa, Angola, Moçambique and Madagascar. *J. Linn. Soc.* 39; DAY, J.H. 1951. Part 1: The intertidal and estuarine Polychaeta of Natal and Mozambique. *Ann. Natal Museum* XII(1):1-67; DAY, J.H. 1957. Part 4: New species and records from Natal and Mozambique. *Ann. Natal Museum* XIV(1):59-129; DAY, J.H. 1974. *A Guide to Marine Life on the South African Shores*. Cape Town: A.A. Balkema; DE FREITAS, A.J. 1984. The Penaeoidae of South East Africa. I. The study area and key to South East African species. *Investigational Report* 56:2-31. Durban: Oceanographic Research Institute; DE FREITAS, A.J. 1986. Selection of nursery areas by six Southeast African Penaeidae. *Estuarine, Coastal and Shelf Science* 23:901-08; DODGE, J.D. 1982. *Marine Dinoflagellates of the British Isles*. Norwich: HMSO Publications; DRENNAN, P.M., BERJAK, P., LAWTON, J.R. AND PAMMENTER, N.W. 1987. Ultrastructure of the salt glands of the mangrove tree *Avicennia marina* (Forsk.) Vierh. as indicated by the use of selective membrane staining. *Planta* 172:176-82; EDGAR, L.A. AND PICKETT-HEAPS, J.D. 1983. The mechanism of diatom locomotion. An ultrastructural study of the motility apparatus. *Proceedings of the Royal Society London B* 218:331-43; GIFFEN, M.H. 1963. Contributions on the diatom flora of South Africa. I. Diatoms of the estuaries of the Eastern Cape Province. *Hydrobiologia* 31(3-4):201-58; GORDON, H. 1958. Synchronous claw-waving of fiddler crabs. *Animal Behaviour* 6(3):238-42; GRIFFITHS, C.L. 1976. *Guide to the Benthic Marine Amphipods of Southern Africa*. Cape Town: S.A. Museum; HARTNOLL, R.G. 1973. Factors affecting the distribution and behaviour of the crab *Dotilla fenestrata* on East African shores. *Estuarine and Coastal Marine Science* 1:137-52; HORIGUCHI, T. AND PIENAAR, R.N. 1988. Ultrastructure of a new sand-dwelling dinoflagellate, *Scripsiella arenicola* sp. nov. *Journal of Phycology* 24:426-38; HORIGUCHI, T. AND PIENAAR, R.N. 1991.

Ultrastructure of a marine dinoflagellate *Peridium balticum* from South Africa with particular reference to its chrysophyte endosymbiont. *Botanica Marinara* 34:123-33; ICELY, J.D. AND JONES, D.A. 1978. Factors affecting the distribution of the genus *Uca* (Crustacea, Ocypodidae) on an East African shore. *Estuarine and Coastal Marine Science* 6:315-25; Instituto de Desenvolvimento Rural, Mozambique; ISAAC, F.M. 1958. Marine Angiosperms of the Kenya Coast. *Journal of the East Africa Natural History Society and National Museum* 27(1):28-47; ISAAC W.E. 1956. Marine Algae of Inhaca Island and Peninsula I. *South African Journal of Botany* 22(4):161-93; ISAAC, W.E. 1957. Some marine algae from Xai-Xai. *South African Journal of Botany* 23(3); ISAAC, W.E. AND CHAMBERLAIN, Y.M. 1958. Marine Algae of Inhaca Island and Peninsula II. *South African Journal of Botany* 24(3):124-58; JAASUND, E. 1976. Intertidal seaweeds of Tanzania. University of Trømso, Sweden; KALK, M. 1958. The fauna of intertidal rocks at Inhaca Island, Delagoa Bay. *Annals of the Natal Museum* 14:189-242; KALK, M. 1970. The organisation of a tunicate heart. *Tissue and Cell* 2:99-118; KENSLEY, B.F. 1972. *Shrimps and Prawns of Southern Africa*. Cape Town: South African Museum; KENSLEY, B.F. 1978. *A Guide to Isopods of Southern Africa*. Cape Town: South African Museum; KILBURN, R. AND RIPPEY, E. 1982. *Sea shells of Southern Africa* Johannesburg: Macmillan; LUBKE, R.A. AND VAN WYK, Y. 1988. Estuarine Swamps. In Lubke, R.A., Hess, F.W. and Bruton, N.N. (eds.) *A Field Guide to the Eastern Cape*. Grahamstown: Grahamstown Centre of the Wildlife Society of South Africa; MACNAE, W. 1968. A general account of the fauna and flora of mangrove swamps and forests in the Indo-West Pacific Region. *Advances in Marine Biology* 6:73-270; MACNAE, W. AND KALK, M. 1962. The ecology of the mangrove swamps at Inhaca Island, Moçambique; MCLAUGHLIN, P.A. 1980. Comparative Morphology of Recent Crustacea. San Francisco: W.H. Freeman; MILLAR, R.H. 1956. Ascidians from Mozambique. *Ann. Mag. Nat. Hist.* 11:913-22; MILLARD, N.A.H. 1975. Monograph on the Hydroidea of Southern Africa. *Annals of the South African Museum* 68:1-513; MOURA, A.R. 1965. Foraminiferos da Ilha da Inhaca. *Revista dos Estudos Universidade Gerais Universitarios de Moçambique* 11:1-74; NEWELL, R.C. 1979. *Biology of Intertidal Animals* (3 edn.). Kent, England: Marine Ecological Surveys Ltd; NICHOLAS, W.L. 1984. *Biology of Free-living Nematodes*. Oxford: Clarendon Press; PANTREATH, R.J. 1970. Feeding mechanisms and the functional morphology of podia and spines in some New Zealand ophiuroids. *Journal of Zoology* 162:395-429; PASSMORE, N.I. AND CARRUTHERS, V.C. 1979. *South African Frogs*. Johannesburg: Witwatersrand University Press; PRUDHOE, S. 1989. Polyclad turbellarians recorded from African waters. *Bulletin of the British Museum of Natural History (Zoology)* 55(1):47-96; SEAGRIEF, S.C. 1980. Seaweeds of Maputoland. In Bruton, R.N. and Cooper, K.H. (eds.), *Studies in the Ecology of Maputoland*. Grahamstown: Rhodes University and the Wildlife Society of Southern Africa (Natal Branch); SILVA, S.M.F. 1991. Flora de cianoficice bentonica da Ilha da Inhaca littoral de Moçambique. *Hoehnea* 18(1):107-25; SILVA, S.M.F. 1991. Cyanophyceae associated with mangrove trees at Inhaca Island, Mozambique. *Bothalia* 21(2):143-50; SILVA, S.M.F. AND CUAMBO, S.N.F. 1991. Contribuiacao ao conhecimento das cianoficeas filamentosus do plancton marinho da Ilha da Inhaca. *Hoehnea* 18(1):127-42; SIMONS, R.H. 1976. Seaweeds of Southern Africa. *Sea Fisheries Bulletin 7*. Cape Town: Department of Environmental Affairs; SKINNER, J.D. AND SMITHERS, R.H.N. 1990. *The Mammals of the Southern African Subregion*. Pretoria: University of Pretoria; SLEIGH, A. 1973. *The Biology of Protozoa*. London: Edward Arnold; SMITH, M.M. AND HEEMSTRA, P.C. 1986. *Smith's Fishes of Southern Africa*. Johannesburg: Macmillan, South Africa; STEBBINS, R.C. 1986. *A Field Guide to the Western Reptiles and Amphibians*. Copyright (c) 1966 by Robert Stebbins. Reprinted by permission of Houghton Mifflin Co. All rights reserved; STEBBINS, R.C. AND KALK, M. 1961. *Copeia* 1:21, fig. 1. Used with permission; STEBBINS, R.C, AND KALK, M. 1961. *Copeia* 1:23, fig. 2. Used with permission; THANDAR, A.S. 1987. The South African stichopid holothurians. *South African J. of Zoology* 24(4):278-86. 8.14d,f; THANDAR, A.S. 1987. The status of some nominal species of *Pseudocnella* and their ecological isolation. *South African J. of Zoology* 24(4):290-304; TIXIER-DURIVAULT, A. 1960. Les Octocoralliares de L'Ile Inhaca. *Bulletin du Museum National d'Histoire Naturelle* 32(4):359-67; TRUEMAN, E.R. 1975. *The Locomotion of Soft-bodied Animals*. Contemporary Biology Series. London: Edward Arnold; TRUEMAN, E.R. AND BROWN,

A.C. 1992. The burrowing habits of marine gastropods. *Advances in Marine Biology* 28:389-431; VAN DER HORST, C.J. 1940. Enteropneusts from Inyack Island. *Annals of the South African Museum* 32:293-380; VOGEL, F.C. 1984. Comparative functional morphology of the spoon-tipped setae on the second maxilliped in *Dotilla* (Decapoda, Brachyura, Ocypodidae). *Crustaceana* 17(3):1-12; WALENKAMP, J.H.C. 1990. Systematic and zoogeography of Asteroidea (Echinodermata) from Inhaca Island, Mozambique. *Zoologische Verhandelingen* 281:1-86; WAY, M.J. 1954. Life history and ecology of the ant, *Oecophylla longinoda* Latreille. *Bulletin of Entomological Research* 45(1):93-112; WEBB, M. 1957. On the lancelets of Southern Africa. *Ann. of the South African Museum* 43(5):249; WELLS, J.B.J. 1967. The littoral copepods (Crustacea) of Inhaca Island, Mozambique. *Transactions of the Royal Society of Edinburgh* 67(7):189-358; WICKSTEAD, J.H. 1965. *An Introduction to the Study of Tropical Plankton.* New York: Hutchinson.

AIDS TO THE IDENTIFICATION OF ORGANISMS AT INHACA ISLAND

A. Books with Keys to the Identification of some Common Organisms

Branch, G. and Branch, M. 1981. *The Living Shores of Southern Africa.* Cape Town: Struik.
Branch, G.M, Griffiths, C.L., Branch, M.L. and Beckley, L.E. 1994. *Two Oceans: A Guide to the Marine Life of Southern Africa.* Cape Town: David Philip
Day, J.H. 1974. *A Guide to the Marine Life on the South African Shores.* Cape Town: AA Balkema.

B. References with Detailed Species Descriptions

Balinsky, J.B. 1957. The Ophiuroidea of Inhaca Island. *Annals of the Natal Museum* 14(1):1-32.
Barnard, K.H. 1950. Descriptive Catalogue of Southern African Decapoda (crabs, shrimps and lobsters). *Annals of the South African Museum* 38:1-837.
—— 1950. Stomatopoda (mantid shrimps). *Annals of the South African Museum* 38:838-864.
—— 1954. South African Pycnogonida. *Annals of the South African Museum* 41:91-159.
—— 1955. Additions to the fauna list of South African Crustaceans. *Annals of the South African Museum* 43:1-122.
—— 1955. Additions to the Fauna list of Pycnogonida. *Annals of the South African Museum* 44:105-107.
—— 1958. Further additions to the Crustacean fauna list of Portuguese East Africa. *Memorias Museu Alvaro de Castro* 4:2-23.
—— 1962. New records of marine Crustacea. *Crustaceana* 3(3):239-245.
Berry, P.F. 1971. The spiny lobsters (Palinuridae) of the east coast of Southern Africa. Distribution and ecological notes. *Oceanographic Research Institute Investigational Report* 27:1-23.
Boshoff, P.H. 1981. An annotated checklist of Southern African Scleractinia. *Oceanographical Research Institute Investigational Report* 40:1-45.
Braak, L. 1991. *Field Guide to Insects of the Kruger National Park* (with a key by Brothers, D.J.). Cape Town: Struik.
Branch, B. *Field Guide to Snakes and other Reptiles of Southern Africa.* Cape Town: Struik.
Broadley, D.G. 1990. The Herpetofauna of the Islands off the coast of Southern Mozambique. *Arnoldia, Zimbabwe* 9(35):469-493.
Cholnoky, B.J. 1962. Eine Beitrag zu der Okologie des Diatomen in Englischer Protektorat, Swaziland. *Hydrobiologia* 20(4):309-355.
Chretiennot-Dinet, M.J. 1990. *Atlas du Phytoplancton Marin III. Chlorarachniophycees, Chlorophycees, Chrysophycees, Cryptphycees, Euglenophycees, Eustigmatophycees, Prasinophycees, Prymnesiophycees, Rhodophycees, Tribophycees.* Museum National d'Histoire Naturelle Centre, Paris.
Clarke, A.M. and Courtman Stock, J. 1976. *Echinoderms of Southern Africa* (excluding Holothuroidia). London: British Museum (Natural History).
Coates Palgrave, K. 1983. *Trees of Southern Africa.* Third updated impression. Cape Town: Struik.
Crosnier, A. 1962. Portunidae (swimming crabs). *Faune de Madagascar* 16:1-154. Office de la Recherche Scientifique et Technique Outra-Mer (ORSTOM) Paris.
—— 1965. Grapsidae et Ocypodidae. *Faune de Madagascar* 18:1-141. Office de la Recherche Scientifique et Technique Outra-Mer (ORSTOM) Paris.

Day, J.H. 1967. *A Monograph of the Polychaeta of Southern Africa 1. Errantia, 2. Sedentaria.* British Museum (Natural History) London. (Proofs of the revised edition available from the Zoology Department, University of Cape Town).

De Freitas, A.J. 1986. Selection of nursery areas by six Southeast African Penaeidae. *Estuarine, Coastal and Shelf Science* 23:901-908.

Flora of Southern Africa. 1966. Volumes 1-33. Botanical Research Institute Department of Agriculture, Pretoria.

Flora Zambesiaca. 1960 (Mozambique, Malawi, Zimbabwe and Botswana.) Management Committee, London.

Giffen, M.H. 1963. Contributions to the Diatom flora of South Africa. 1. Diatoms of the estuaries of the Eastern Cape Province. *Hydrobiologia* 31(3-4):201-258.

Griffiths, C.L. 1976. *Guide to the Benthic Marine Amphipods of Southern Africa.* 1-106. Cape Town: South African Museum.

Isaac, F.M. 1968. Marine Angiosperms on the Kenya Coast (with keys to species). *Journal of the East Africa Natural History Society and National Museum* 27(1):28-47.

Kensley, B.F. 1972. *Shrimps and Prawns of Southern Africa* 1-65. Cape Town: South African Museum.

——— 1978. *Guide to the Marine Isopods of Southern Africa.* 1-173. Cape Town: South Africa Museum.

——— 1978. On the Zoogeography of Southern African Decapod Crustacea with a Distribution Checklist of the species. *Smithsonian Contributions to Zoology* 338:1-51.

Kilburn, R. and Rippey, E. 1982. *Sea Shells of Southern Africa.* (Gastropods, shelled opisthobranchs, marine pulmonates, chitons, cephalopods, scaphopods, and bivalves). Johannesburg: Macmillan, South Africa.

Macnae, W. 1954. On four Saccoglossan Molluscs, new to South Africa. *Annals of the Natal Museum* 13(1):51-64.

——— 1957. The families Polyceridae and Goniodoridae (Mollusca, Nudibranchiata) in Southern Africa. *Transactions of the Royal Society of South Africa* 5(4):341-372.

——— 1962. Notaspidean opisthobranchiate molluscs from South Africa. *Annals of the Natal Museum* 15:167-181.

——— 1962. Tectibranch molluscs from Southern Africa. *Annals of the Natal Museum* 16:183-189.

Millar, R.H. 1956. Ascidians from Mozambique, East Africa. *Annals of the Magazine of Natural History* 12(9):913-932.

Millard, N.A.H. 1975. A Monograph on the Hydroidea of Southern Africa. *Annals of the South African Museum* 68:1-513.

Moura, A.R. 1965. Foraminifera da Ilha da Inhaca. *Revista dos Estudos gerais Universitarios de Mozambique.* 11:1-74.

Passmore, N.I. and Carruthers, V.C. 1979. *South African Frogs.* Johannesburg: Witwatersrand University Press.

Prudhoe, S. 1989. Polyclad turbellarians recorded from African waters. *Bulletin of the British Museum (Natural History) (Zoology)* 55(1):47-96.

Ricard, M. 1987. *Atlas du Phytoplancton Marin II: Diatomphycees.* Museum National d'Histoire Naturelle Centre, Paris.

Roberts' Birds of Southern Africa. 1978. Revised by McLachan and Liversidge, R. 1-848, and 1985, revised by Maclean, G.L. Cape Town: John Voelker Trust Bird Book Fund. (Authors of specific names given in the fourth edition, 1978; new numbers in the fifth edition, 1985.

Round, F.E. Crawford, R.M. and Mann, D.G. 1990. *The Diatoms: Biology and Morphology of the Genera.* Cambridge: Cambridge University Press.

Schelpe, C.L.E. and Dinoz, M.A. 1983. Pteridophytes. In Fernandes, A. and Merde, E.J. (eds) *Flora of Moçambique.* Lisbon.

Schleyer, M.H. (ed.) (in prep). Corals of the South-east Indian Ocean. *Oceanagraphic Research Institute Investigational Report.*

Scholtz, C.H. and Holm, E. 1985. *Insects of Southern Africa.* Durban: Butterworths.

Seagrief, S.C. 1980. Seaweeds of Maputaland, 4-41. In Bruton, R.N. and Cooper, K.H. (eds). *Studies on the Ecology of Maputaland*. Rhodes University and Natal Branch, Wild Life Society, Southern Africa.

Silva, S.M.F. 1991. Cyanophyceae associated with mangrove trees at Inhaca Island, Mozambique. *Bothalia* 21(2):143-150.

—— 1991. Flora de cianoficeas bentonicas da Ilha da Inhaca littoral de Mozambique, I. *Hoehnea* 18(1):107-125.

Silva, S.M.F. and Cuamba, N.J.B. 1991. Contribução ao conhecimento das cianoficeas filamentosas do plancton marinho da Ilha da Inhaca, Moçambique. *Hoehnea* 18(1):127-142.

Skinner, J.D. and Smithers, R.H.N. 1990. *The Mammals of the Southern African Subregion*. Pretoria: University of Pretoria.

Smith, M.M. and Heemstra, P.C. 1986. *Smith's Fishes of Southern Africa*. 1-1047. Johannesburg: Macmillan.

Sourna, A. 1986. *Atlas du Phytoplancton Marin*. Volume 1. Cyanophycees, Dictyophycees, Dinophycees and Raphidophycees. Museum National d'Histoire Naturelle Centre, Paris.

Stephen, A.C. and Robertson, J.D. 1952. A preliminary report on the Echiuridae and Sipunculidae of Zanzibar. *Proceedings of the Royal Society of Edinburgh B* 64(4):426-444.

Thandar, A.S. 1984. The Holothurian Fauna of Southern Africa. *Ph.D Thesis, University of Durban-Westville*, South Africa.

—— 1987. The status of some nominal species of *Cucumaria*, referable to a new genus *Pseudocnella* and their ecological isolation. *South African Journal of Zoology* 24(4):280-304.

—— 1989. The sclerodactylid holothurians of South Africa with the erection of a new subfamily and two new genera (Echinodermata, Holothuroidea). *South African Journal of Zoology* 24(4):290-304.

Thandar, A.S. and Rowe, F.W. 1989. New species and new records of apodous holothurians from Souther Africa. *Zoologica Scripta* 18(1):145-155.

Tixier-Durivault, A. 1960. Les Octocoralliares de L'Ile Inhaca. *Bulletin du Museum National d'Histoire Naturelle* 32(4):359-367.

Van Bruggen, A.C. 1978. Land Molluscs. 877-923. In Werger, M.J.A. (ed.), *Biogeography and Ecology of Southern Africa*. The Hague: Dr. W. Junk.

Van der Elst, R. 1991. *A Field Guide to South African Sea Fishes*. Cape Town: Struik.

Van der Horst, C.J. 1940. Enteropneusta from Inyack Island. *Annals of the South African Museum* 32(5):293-380.

Voss, G.L. 1961. South African Cephalopoda. *Transactions of the Royal Society of South Africa* 36:245-272.

Wallenkamp, J.H.C. 1990. Systematics and Zoogeography of Asteroidea from Inhaca Island, Mozambique. *Zoologische Verhandelingen Nationaal Natuurhistorisch Museum, Leiden* 261:1-87.

Webb, J.E. 1957. On the lancelets of South and East Africa. *Annals of the South African Museum* 43(5):240-270.

Wells, J.B.J. 1967. The littoral copepoda (Crustacea) of Inhaca Island, Mozambique. *Transaction of the Royal Society of Edinburgh* 68:189-358.

REFERENCES

Alberto, M.S. 1959. The Island of Inhaca, historical and demographic sketch. *South African Journal of Science* 55(7):163-65.

Alexander, R. MacNeill. 1979. *Invertebrates*. Cambridge: Cambridge University Press.

Allanson, B.R. 1958. *Siphonaria* species on the South African and Mozambican coasts. *Hydrobiologia* 12(2-3):140-50.

Balinsky, J.B. 1957. The ophiuroidea of Inhaca Island. *Annals of the Natal Museum* 14(1):1-32.

Bandeira, S.O. 1989. Ecology of the sea grass communities at Inhaca Island, Mozambique. Report at the Marine Sciences Conference, Tanzania.

Bandeira, S.O. 1992. The ethnobotany of non-medicinal plants of Inhaca Island, Mozambique. *Aefat Congress* XII.

Barnard, K.H. 1958. Further additions to the Crustacean fauna list of Portuguese East Africa. *Memorias Museu Alvaro de Castro* 4:2-23.

Barnes, R.B. 1974. *Invertebrate Zoology*. London: W.B. Saunders.

Barras, R. 1963. The burrows of *Ocypode ceratophthalmus* (Pallas) (Crustacea, Ocypodidae) on a tidal, wave beach on Inhaca Island, Mozambique. *Journal of Animal Ecology* 32:73-85.

Battistine, R. and Richard-Vindard, G. 1972. *Biogeography and Ecology of Madagascar*. The Hague: Dr. W. Junk.

Benayahu, O.Y. (in prep.). Alcyonaceae from Sodwana Bay, South Africa. In Schleyer, M.H. (ed.), Corals of the Southwest Indian Ocean. *Investigational Report*. Durban: Oceanographic Research Institute.

Berggen, M. 1990. *Dasella herdmaniae* (Lebour) (Decapoda: Natantia: Pontoniinae) from Mozambique and establishment of a new species, *Dasella brucei*. *Journal of Crustacean Biology* 10(3):554-59.

―――― 1991. *Athanopsis rubricinctura*, new species (Decapoda: Natantia: Alpheidae) a shrimp associated with an echiuroid at Inhaca Island, Mozambique. *Journal of Crustacean Biology* 11(1):166-78.

―――― 1992. *Naushonia lactoalbida*, new species (Decapoda: Thalassinidea: Laomediidae): a mud shrimp from Inhaca Island, Mozambique. *Journal of Crustacean Biology* 12(3):514-22.

Berjak, P., Campbell, G.K., Huckert, B.I. and Pammenter, N.W. 1986. *In the Mangroves of Southern Africa* (2 edn.). Durban: Wild Life Society of Southern Africa (Natal Branch).

Berry, P.F. 1971. The spiny lobsters *(Palinuridae)* of the east coast of Southern Africa: Distribution and ecological notes. *Investigational Report* 27:1-23. Durban: Oceanographic Research Institute.

Boltt, G. and Heeg, J. 1975. The osmoregulatory ability of three grapsid crab species in relation to their penetration of an estuarine system. *Zoologica Africana* 10(2):167-82.

Brady, J. 1979. Biological Clocks. *Studies in Biology* 104. London: Edward Arnold.

Branch, B. 1988. *Field Guide to Snakes and Other Reptiles of Southern Africa*. Cape Town: Struik.

Branch, G. and Branch, M. 1981. *The Living Shores of Southern Africa*. Cape Town: Struik.

Brehout, R.N. 1982. Ecology of Rocky Shores. *Studies in Biology* 139. London: Edward Arnold.

Bolwig, N. 1954. The influence of light and touch on the orientation and behaviour of *Gonodactylus glabrous (Stomatopoda)*. *British Journal of Animal Behaviour* 2:144-45.

Boney, A.D. 1975. Phytoplankton. *Studies in Biology* 52. London: Edward Arnold.

Boshoff, P.H. 1958. Development and Constitution of the Coral Reefs (at Inhaca Island). In Macnae, W. and Kalk, M. (eds.), *A Natural History of Inhaca Island, Mozambique*. Johannesburg: Witwatersrand University Press, pp. 49-56.

―― 1981. An annotated checklist of Southern African Scleractinia. *Investigational Report* 49:1-45. Durban: Oceanographic Research Institute.

Bosazza, V.L. 1956. The geology and development of the bays of the coastline of the Sul de Save of Mozambique. *Boletim Sociedade Estudos Mozambique* 98:19-28.

Broadley, D.G. 1990. The Herpetofauna of the islands off the coast of Southern Mozambique. *Arnoldia, Zimbabwe* 9(35):469-93.

Broekhuysen, C.J. 1940. A preliminary investigation of the importance of desiccation, temperature and salinity as factors controlling the vertical distribution of certain intertidal marine gastropods in False Bay, South Africa. *Trans. Royal Society of South Africa* 281:245-92.

Brooke, R.K. and Tuer, F.V. 1968. Additional records for (migrant) birds at Inhaca Island, Mozambique. *Ostrich* 39:266.

Brooke, R.K., Cooper, J., and Sinclair, J.C. 1981. Additional records of sea birds at Inhaca Island, Mozambique. *Cormorant* 9:30-40.

Brown, D.S. 1971. Ecology of gastropods in a South African mangrove swamp. *Proc. Malacol. Soc. London* 39:262.

Bruce, A.J. 1976. Shrimps and Prawns of Coral Reefs with Special Reference to Commensalism. In Jones, D.A. and Endean, R. (eds.), *Geology and Biology of Coral Reefs*, Vol. III. New York, Sydney: Academic Press, pp. 57-94.

Campbell, B.M., Attwell, C.A.M., Hatton, J.C., de Jager, P., Gabmiza, J., Lyman, T., Mizutani, F. and Wynter, P. 1988. Secondary dune succession on Inhaca Island, Mozambique. *Vegetatio* 78:3-11.

Campbell, B.M., Lyman, T. and Hatton, J.C. 1990. Small-scale patterning in the recruitment of forest species during recession in tropical forests on Inhaca Island, Mozambique. *Vegetatio* 87(1):51-57.

Cantrell, M.A. 1979. Invertebrate Communities in the Lake Chilwa Swamp in Years of High Level. In Kalk, M., McLachlan, A.J. and Howard-Williams, C. (eds.), *Lake Chilwa: Studies of Change in a Tropical Ecosystem*. The Hague: Dr W. Junk.

Cardosa, J.G.A. 1959. The settling of sand dunes on the Island of Inhaca. *South African Journal of Science* 55(7):166.

Chamberlain, Y.M. 1958. *Dasyclados ramosus*: A new species from Inhaca Island and Peninsula, Portuguese East Africa. *South African Journal of Botany* 24(3):119-21.

Chelazzi, G. and Vannini, M. 1980. Zonation of intertidal molluscs on rocky shores in Southern Somalia. *Estuarine and Marine Coastal Science* 10:569-84.

Cholnoky, B.J. 1960. The relationship between algae and the chemistry of natural waters. (Translated by H. Welsh). *South African Council for Scientific and Industrial Research, National Institute of Water Reseach*, Reprint 129.

―― 1962. Ein betrag zu der ökologie der diatomen in dern Englischen Protectorat, Swaziland. *Hydrobiologia* 20(4):309.

Clancey, P.A. 1971. A handlist of the birds of Southern Mozambique. *Memorias Instituto Cientifica Moçambique*, Series A. 10:145-303.

Clarke, A.M. and Courtman-Stock, J. *The Echinoderms of Southern Africa*. London: British Museum (Natural History).

Cockroft, V.G. and Forbes, A.T. 1981. Tidal activity in the mangrove snail, *Cerithidea decollata*. *South African Journal of Zoology* 16(1):5-9.

Connell, J.H. 1973. Population Ecology of Reef-Building Corals. In Jones, D.A. and Endean, R.A. (eds.), *Geology and Biology of Coral Reefs* Vol. II. New York, Sydney: Academic Press, pp. 205-44.

Connolly, M. 1925. The non-marine Mollusca of Portuguese East Africa. *Transactions of the Royal Society of South Africa* 12:105-220.

Costa, F. 1989. Marine Turtle Conservation at Nyaka (Inhaca) Island. Report to the Marine Sciences Workshop, Dar es Salaam.

―― 1990. Conservation and Management at Inhaca Island. Report of the Universidade Eduardo Mondlane, Maputo, Mozambique.

Crosnier, A. 1965. Grapsidae and Ocypodidae. *Fauna de Madagascar* 18:1-141. Paris: ORSTOM.

Darwin, C. 1874. *On the Structure and Distribution of Coral Reefs*. (2 edn.). London: Smith, Elder.

Day, J.H. 1951. The intertidal and estuarine Polychaeta of Natal and Mozambique. *Ann. Natal Museum* XII(1):1-67.
—— 1957. A new species and records from Natal and Mozambique. *Ann. Natal Museum* XIV(1):59-129.
—— 1967. *A Monograph of the Polychaetes of Southern Africa. 1.Errantia, 2. Sedentaria*. British Museum (Natural History) London.
—— 1974. The ecology of the Morrumbene estuary, Mozambique. *Transactions of the Royal Society of South Africa* 41(1):43-97.
—— 1974. *A Guide to the Marine Life on South African Shores*. Cape Town: A.A. Balkema.
—— 1981. *Estuarine Ecology: With Particular Reference to Southern Africa*. Cape Town: A.A. Balkema.
De Freitas, A.J. 1964. Breeding in *Periophthalmus kolbreuteri*. Unpublished report.
—— 1984. The Penaeidae of South-East Africa. I. The study area and key to South-East African species. *Investigational Report* 56:2-31. Durban: Oceanographic Research Institute.
—— 1986. Selection of nursery areas by six south-east African Penaeidae. *Estuarine, Coastal and Shelf Science* 23:901-08.
De Koning, J. 1993. Checklist of vernacular plant names in Mozambique. *Wageningen Agricultural University Papers*, Wageningen Publicatie en Documentatie Centrum.
De Morais, T. 1959. Parasitism on Inhaca Isalnd. *South African Journal of Science* 55(7):177.
Drennan, P.M. and Berjak, P. 1982. Physiology of salt excretion in the mangrove tree, *Avicennia marina* (Forsk.) *New Phytologist* 91:597-606.
—— 1984. The role of phenolics in ion movement in halophytes with particular reference to *Avicennia spp*. *Proceedings of the Electron Microscopy Society of South Africa* 14:41-42.
Drennan, P.M., Berjak, P., Lawton, J.R. and Pammenter, N.W. 1987. Ultrastucture of the salt glands of the mangrove tree *Avicennia marina* (Forsk.) Vierh. as indicated by the use of selective membrane staining. *Planta* 172:176-82.
Dring, M.J. 1982. *The Biology of Marine Plants*. Contemporary Biology Series. London: Edward Arnold.
Dye, A.H. and Lasiak, T.A. 1986. Microbenthos, meiobenthos and fiddler crabs: trophic interactions in a tropical mangrove sediment. *Marine Ecological Progress Series* 22:259-64.
—— 1987. Assimilation efficiencies of fiddler crabs and deposit-feeding gastropods from tropical mangrove sediments. *Comparative Biochemistry and Physiology* 87A (2):341-44.
Edgar, L.A. and Pickett-Heaps, J.D. 1983. The mechanism of diatom locomotion. An ultrastructural study of the motility apparatus. *Proceedings of the Royal Society London B* 218:331-43.
Edney, E.B. 1960 The water and heat relationships of fiddler crabs *(Uca spp.) Transactions of the Royal Society of Southern Africa* 36(2):71-91.
Eloff, F.C. and de Graaf, G. 1963. A note on the Golden Mole, bats and rodents of Inhaca Island. *South African Journal of Science* 59:88-89.
Eltringham, S.K. 1971. *Life in Mud and Sand*. Oxford: English Universities Press.
Endean, R.E. 1972. Destruction and Recovery of Coral Reef Communities. In Jones, D.A. and Endean, R.E. (eds.), *Geology and Biology of Coral Reefs* Vol. I. New York, Sydney: Academic Press, pp. 215-50.
Endean, R.E. 1957. The Cuvierian tubules of *Holothuria leucospilota*. *Quarterly Journal of Microscopical Science* 98(4):455-72.
Ferreira, G da Verga. 1964. Longicorneos de Moçambique. *Revista Entomologia Moçambique* 7:451-838.
Ferreira, M.C. and Fereira G. da Verga, 1952. Importance of the Island of Inhaca in the biological survey of Moçambique. *Boletim Sociedade Estudos Moçambique* 73:55-64.
Fingerman, M. 1968. Crustacean colour changes with emphasis on the fiddler crab. *Scientia* 103:571:1-16.
Fishelson, L. 1983. Population ecology and biology of *Dotilla sulcata* (Decapoda, Brachyura, Ocypodidae). *Crustaceana* 17(3):455-72.

Franca, M. de Lourdes. 1960. Sobre uma colecçâo malacologica recolhida na Ihla da Inhaca, Moçambique. *Memorias Junta Investigaçâo Ultramarino* 15:41-102.

Giffen, M.H. 1963. Contributions on the diatom flora of South Africa. I. Diatoms of the estuaries of the Eastern Cape Province. *Hydrobiologia* 31(3-4):201-58.

Goldman, B and Talbot, F.H. 1976. Aspects of the Ecology of Coral Fishes. In Jones, D.A. and Endean, R.E. (eds.), *Geology and Biology of Coral Reefs* Vol. III. New York, Sydney: Academic Press, pp. 125-54.

Gordon, H. 1955. Displacement activities in fiddler crabs. *Nature* 176:356-57.

—— 1958. Synchronous claw-waving of fiddler crabs. *Animal Behaviour* 6(3):238-42.

Goreau, T.F. and Goreau, N.I. 1959. The physiology of skeleton formation in corals. I. A method for measuring the rate of calcium deposition by corals under different conditions. *Biological Bulletin of the Marine Biological Laboratory, Woods Hole* 116:59-67.

Gove, D. and Cuambo, N.J. 1990. Seasonal variation of Inhaca plankton and some physical parameters of the sea. Report, Universidade Eduardo Mondlane, Maputo.

Griffiths, C.L. 1976. *Guide to the Benthic Marine Amphipods of Southern Africa.* Cape Town: S.A. Museum.

Grindley, J.R. 1963. A specimen of the asteroid *Acanthaster planci* (L) from the Mozambique coast. *Novitates Durban Museum, South Africa*, pp. 265-68.

Halle, F., Oldeman, R.A.A. and Tomlinson, P.B. 1978. *Tropical Trees and Forests - An Architectural Analysis.* New York, Berlin, Heidelberg: Springer-Verlag.

Hancock, F.D. 1973. Algal ecology of a stream polluted through gold mining on the Witwatersrand. *Hydrobiologia* 43(2):189-229.

Harte, J.M. and Pammenter, N.W. 1983. Leaf nutrient content in relation to senescence in the coastal dune pioneer, *Scaevola (thunbergi) plumeri. South African Journal of Science* 79:420-22.

Hartnoll, R.G. 1971. Swimming crabs. *Animal Behaviour* 19:34-50.

—— 1973. Factors affecting the distribution and behaviour of the crab *Dotilla fenestrata* on East African shores. *Estuarine and Coastal Marine Science* 1:137-52.

—— 1975. The Grapsidae and Ocypodidae of Tanzania. *Journal of Zoology* 177:305-28.

—— 1976. The ecology of some rocky shores in East Africa. *Estuarine and Coastal Marine Science* 4:1-21.

Hatton, J.C. and Couto, A.L. (in press). The effect of coastline configuration changes on mangrove community structure, Portuguese Island, Mozambique. *Hydrobiologia.*

—— (in press). The effect of coastline configuration changes on mangrove community structure, Portuguese Island, Mozambique. *Hydrobiologia.*

Hawkins, S.J. and Hartnoll, R.G. 1983. Grazing of intertidal algae by marine invertebrates. *Oceanography and Marine Biology Annual Review* 21.

Hill, B.J. 1971. Osmoregulation of an estuarine and a marine species of *Upogebia* (Anomura, Crustacea). *Zoologica Africana* 6(2):229-36.

Hobday, D.K. 1977. Late Quaternary sedimentary history of Inhaca Island, Mozambique. *Transactions of the Geological Society of South Africa* 80:183-91.

Horiguchi, T. and Pienaar, R.N. 1991. Ultrastructure of a marine dinoflagellate *Peridinium balticum* (Peridinales) from South Africa with particular reference to the Chrysophyte endosymbiont. *Botanica Marina* 14:123-35.

Horiguchi, T. and Pienaar, R.N. 1991. Ultrastructure of a new sand-dwelling dinoflagellate: *Scripsiella arenicola* sp.nov. *Journal of Phycology* 24:426-38.

Hughes, D.A. 1966. Behavioural and ecological investigation of the crab, *Ocypode ceratophthalmus* (Crustacea, Ocypodidae). *Journal of Zoology London* 150:129-43.

—— 1966. An investigation of the nursery areas and habitat preferences of juvenile penaeid prawns in Mozambique. *Journal of Applied Biology* 3:349-54.

—— 1971. On the mating and `copulation burrows' of crabs of the genus *Ocypode* (Decapoda, Brachyura). *Crustaceana* 24:72-5.

Hughes, G.R. 1974a. The Sea-Turtles of South-East Africa I. Status, morphology and distributions. *Investigational Report* 35:1-144. Durban: Oceanographic Research Institute.

────── 1974b. The Sea-Turtles of South-East Africa II. The Biology of the Tongaland Loggerhead Turtle *Caretta caretta* L. and the Green Turtle *Chelonia mydas* L. in the study region. *Investigational Report* 36:1-96. Durban: Oceanographic Research Institute.

Icely, J.D. and Jones, D.A. 1978. Factors affecting the distribution of the genus *Uca* (Crustacea, Ocypodidae) on an East African shore. *Estuarine and Coastal Marine Science* 6:315-25.

Isaac, F.M. 1958. Marine Angiosperms of the Kenya Coast. *Journal of the East Africa Natural History Society and National Museum* 27(1):28-47.

────── 1969. Floral structure and germination in *Cymodocea (Thalassodendron) ciliata*. *Phytomorphology* 19(1):44-51.

Isaac W.E. 1956. Marine Algae of Inhaca Island and Peninsula I. *South African Journal of Botany* 22(4):161-93.

────── 1958. Some marine algae from Xai-Xai. *South African Journal of Botany* 23(3):75-102.

Isaac, W.E. and Chamberlain, Y.M. 1958. Marine Algae of Inhaca Island and Peninsula II. *South African Journal of Botany* 24(3):124-58.

Isaac, W.E. and Isaac, F.M. 1967. A general account of the environment, flora and vegetation of the Kenya coast. A first list of marine Algae. *Journal of the East African Natural History Society and National Museum* 24(2):78-83.

────── 1968. A second list of marine algae from the Kenya coast. *Journal of the East African Natural History Society and National Museum* 27(1):1-6.

Jaasund, E. 1976. Intertidal seaweeds of Tanzania. University of Trømso, Sweden.

Jansen, P.C.M. and Mendes, O. 1983,1984 and 1985. *Plantas Medicinais seu uso tradiçional em Moçambique*. Parts 1 and 3. Maputo: Gabinete de Estudos de Medicina Tradicional. Part 2. Maputo: Instituto Naçional do Livro e do Disco.

Kalk, M. 1954. Marine biological research at Inhaca Island, Mozambique: An interim report. *South African Journal of Science* 55(7):107-15.

────── 1957. A Comparative Study of the Shores of Inhaca Island. Doctoral thesis, University of the Witwatersrand, Johannesburg.

────── 1958. The fauna of intertidal rocks at Inhaca Island, Delagoa Bay. *Annals of the Natal Museum* 14:189-242.

────── 1959a. The zoogeographical composition of the intertidal fauna of Inhaca Island, Mozambique. *South African Journal of Science* 55(7):178-80.

────── 1959b. A general survey of some shores in Northern Mozambique. *Revista de Biologia Lisbon* 2:1-25.

────── 1963a. Absorption of vanadium by tunicates. *Nature* 198:1010-11.

────── 1963b. Intracellular sites of activity in the histogenesis of tunicate vanadocytes. *Quarterly Journal of Microscopic Science* 104(4):483-93.

────── 1963c. Cytoplasmic transmission of vanadium in a tunicate oocyte. *Acta Embryologia Morphologia Experimentia* 6:289-303.

────── 1970. The organisation of a tunicate heart. *Tissue and Cell* 2:99-118.

Kensley, B.F. 1972. *Shrimps and Prawns of Southern Africa*. Cape Town: South African Museum.

Kilburn, R. and Rippey, E. 1982. *Sea shells of Southern Africa* Johannesburg: Macmillan.

King, P.E. 1973. *Pycnogonids*. London: Hutchinson Universities Library.

Lawson, G.W. 1969. Littoral ecology on rocky shores in East Africa (Kenya and Tanzania). *Transactions of the Royal Society of South Africa* 39:329-39.

Lubke, R.A. and van Wyk, Y. 1988. Estuarine Swamps. In Lubke, R.A., Hess, F.W. and Bruton, N.N. (eds.) *A Field Guide to the Eastern Cape*. Grahamstown: Grahamstown Centre of the Wildlife Society of South Africa.

Macnae, W. 1962. Tectibranch molluscs from Southern Africa. *Annals of the Natal Museum* 15(5):167-81.

────── 1968. A general account of the fauna and flora of mangrove swamps and forests in the Indo-West Pacific Region. *Advances in Marine Biology* 6:73-270.

Macnae, W. and Kalk, M. 1962. The fauna and flora of sand flats at Inhaca Island. *Journal of Animal Ecology* 31:93-128.

———1962. The ecology of mangrove swamps at Inhaca Island, Mozambique. *Journal of Ecology* 50:19-34.

Maitland, D.P. 1986. Crabs that breathe air with their legs: *Scopimera* and *Dotilla*. *Nature* 319:493-95.

Markus, E. and Macnae, W. 1954. Architomy in a species of *Convoluta*. *Nature* 173:130.

McCallan, E. 1964. Ecology of Sand Dunes with Special Reference to the Insect Communities. In Davies, D.H.S. (ed.), *Ecological Studies of Southern Africa*. The Hague, Dr W. Junk, pp. 174-85.

McGinitie, G.E. 1937. The natural history of *Callianassa californiensis* Dana. *The American Midland Naturalist* 18 (6):1031-37.

McIntyre, A.D. 1968. The meiofauna of some tropical beaches. *Journal of Zoology* 156:377-82.

McLachlan, A. 1977. Composition, distribution, abundance and biomass of the macrofauna and meiofauna of four sandy beaches. *Zoologica Africana* 12:279-306.

McRoy, C.P. and McMillan, C. 1977. Production Ecology and Physiology of Sea Grasses. In McRoy, C.P. and Helfferich, C. (eds.), *Sea Grass Ecosystems*. New York: Dekker, pp. 55-87.

Millar, R.H. 1956. Ascidians from Mozambique. *Ann. Mag. Nat. Hist.* 11:913-22.

Millard, N.A.H. 1975. Monograph on the Hydroidea of Southern Africa. *Annals of the South African Museum* 68:1-513.

Mogg, H.O.D. 1969. An Annotated Checklist of Flowering Plants and Ferns of Inhaca Island, Mozambique (with local names and uses of plants). In Macnae, W. and Kalk, M. *A Natural History of Inhaca Island, Mozambique* (2edn.). Johannesburg: Witwatersrand University Press.

———1963. A preliminary investigation of salinity in the zonation of species in salt marsh and mangrove swamp associations. *South African Journal of Science* 59:81-86.

Moll, E.J. and White, F. 1978. In Werger, M.J.A. (ed.), *Biogeography and Ecology of Southern Africa*. The Hague: Dr W. Junk.

Morton, J.E. 1979. *Molluscs*. London: Hutchinson University Library.

Moura, A.R. 1965. Foraminiferos da Ilha da Inhaca. *Revista dos Estudos Universidade Gerais Universitarios de Moçambique* 11:1-74.

Nestler, A., Paech, H.J. and Schmidt, W. 1984. Inhaca Island, Mozambique und die Entwicklung des Riffs vor Barreira Vermelha. *Petermanns Geographische Mitteilungen* DDR. 1:31-37.

Newell, R.C. 1979. Biology of Intertidal Animals (3 edn.). Kent, England: Marine Ecological Surveys Ltd.

Nicholas, W.L. 1984. *Biology of Free-living Nematodes*. Oxford: Clarendon Press.

Noel, A.R.A. 1959. The vegetation of the freshwater swamps of Inhaca Island. *South African Journal of Botany* 25:189-205.

Nolte, D.J. 1954. Polymorphic coloration in *Donax faba*. *Caryologia Supplement* pp. 889-90.

Pammenter, N.W. 1985. Photosynthesis and transpiration of the subtropical coastal sand dune pioneer *Scaevola plumieri* under controlled conditions. *South African Journal of Botany* 51:421-24.

Pantreath, R.J. 1970. Feeding mechanisms and the functional morphology of podia and spines in some New Zealand ophiuroids. *Journal of Zoology* 162:395-429.

Pardi, L. and Ercoli, A. 1986. Zonal recovery mechanisms in talitrid crustaceans. *Bolletino Zoologico* 53:139-160.

Passmore, N.I. and Carruthers, V.C. 1979. *South African Frogs*. Johannesburg: Witwatersrand University Press.

Perreira, M.de Cavalho. 1959. Culicidae of Mozambique. *Boletim Sociedade Estudos Moçambique* 17:112.

Petrarca, V., Carrara, G.C., Deco, M.A.Di. and Petrangeli, G. 1984. Observazioni citogenetiche e biometriche sui membri del complesso *Anopholes gambiae* in Mozambico. *Parasitologica* 26:247-69.

Pethick, J. 1984. An Introduction to Coastal Morphology. London: Edward Arnold.

Pichon, M. 1972. The Coral Reefs of Madagascar. In Battistine, R. and Richard-Vindard (eds.), *Biogeography and Ecology of Madagascar*. The Hague: Dr. W. Junk, pp. 367-410.

Pienaar, R.N. 1965. A first account of the freshwater diatoms of Inhaca Island. Appendix: Marine Diatoms on sandy shores. Doctoral thesis, University of the Witwatersrand, Johannesburg.

Pinhey, E.C.G. 1981. Checklist of the Odonata of Mozambique. *Occasional Papers of the Natural History Museum, Zimbabwe (B). Natural Science* 6:555-632.

Pinto, A.A. da Rosa. 1959. A contribution towards the study of the avifauna of the Island of Inhaca. *Boletim Sociedade Estudos Moçambique* 17(112):29-60.

Plano de Desenvolvimento Integrado Ilha da Inhaca, 1990. Projecto MOZ/86/026. Commissão Nacional do Plano Moçambique, Maputo.

Prudhoe, S. 1989. Polyclad turbellarians recorded from African waters. *Bulletin of the British Museum of Natural History (Zoology)* 55(1):47-96.

Rato, J.M. 1959. The Marine Biological Station of Inhaca. *South African Journal of Science* 55(7):107-15.

Riegl, B. (in prep.) Notes on the Hard Coral Genus *Acropora* Oken 1815 (Scleractinia, Asterocoeniia, Acroporidae) in South Africa. In Schleyer, M.H. (ed.), Corals of the Southeast Indian Ocean II. *Investigational Report.* Durban: Oceanographic Research Institute.

Robert's Birds of Southern Africa. 1978. Revised by McLachlan, G.R. and Liversidge, R. Cape Town: John Voelker Trust Bird Book Fund. (Authors of specific names given in the Fourth Edition, 1978; new numbers appear in the Fifth Edition, 1985: revised by Maclean, G.L.).

Salm, R. 1976. The dynamic management of the Ponta Torres coral reef. *Memorias Instituto Investigaçao Cientifica Moçambique* 12A.

Salmon M. and Zucker, N. 1988. Interpreting Differences in the Reproductive Behaviour of Fiddler Crabs (genus *Uca*). In Chelazzi, G and Vannini, M. (eds.) *Behaviour Adaptation to Intertidal Life.* NATO Advanced Science Institutes, Series A. New York and London: Plenum Press.

Schleyer, M.H. 1986. Decomposition in estuarine systems. *Journal of the Limnological Society of South Africa* 12(2):902-8.

Schleyer, M.H. (in prep.). A Checklist of Southern African Corals. In Schleyer, M.H. (ed.), Corals of the Southeast Indian Ocean. *Investigational Report.* Durban: Oceanographic Research Institute.

Seagrief, S.C. 1988. Marine Algae. In Lubke, R.A., Hess, F.W. and Bruton, R.N. (eds.), *A Field Guide to the Eastern Cape.* Grahamstown: Grahamstown Centre of the Wildlife Society of South Africa.

Silva, S.F.M. 1991. Flora de cianoficeas bentonicas da Ilha da Inhaca littoral de Moçambique. I. *Hoehnea* 18(1):107-25.

Silva, S.M.F. and Cuambo, N.J.R. 1991. Cyanophyceae associated with mangrove trees at Inhaca Island, Mozambique. *Bothalia* 21(2): 143-50.

―― 1991. Cyanophyceae in plankton at Inhaca Island. *Bothalia* 21(2):151-72.

Skinner, J.D. and Smithers, R.H.N. 1990. *The Mammals of the Southern African Subregion.* Pretoria: University of Pretoria.

Sleigh, A. 1973. *The Biology of Protozoa.* London: Edward Arnold.

Smith, D.A. 1975. Polymorphism and selective predation in *Donax faba* (Bivalvia, Tellinaceae). *Journal of Marine Biology and Ecology* 17:205-19.

Smith, J.L.B. 1958. The Marine Fishes of Inhaca Island. In Macnae, W. and Kalk, M. (eds.) *A Natural History of Inhaca Island, Mozambique* (1 edn.). Johannesburg: Witwatersrand University Press.

―― 1977. *The Sea Fishes of Southern Africa* (4 edn.). Johannesburg: Macmillan.

Smith, M.M. and Heemstra, P.C. 1986. *Smith's Fishes of Southern Africa.* Johannesburg: Macmillan, South Africa.

Stebbins, R.C, and Kalk, M. 1961. Observations on the natural history of the mudskipper, *Periophthalmus sobrinus. Copeia* 1:19-27.

Stephen, A.C. and Robertson, J.D. 1952. A preliminary report on the Echiuroidea and Sipunculidae of Zanzibar. *Proc. Royal Society Edinburgh, B.* 64(4):426-44.

Stephenson, T.A. and Stephenson, A. 1972. *Life Between Tidemarks on Rocky Shores.* San Francisco: W.H. Freeman.

Swart, H.J. 1958. An investigation of the Mycoflora in the soil of some mangrove swamps at Inhaca Island. I. *Acta Botanica Neerlandica* 7:741-68.

Tabela de Marés do Porto de Maputo, 1985. Instituto Hydrografico Português, Moçambique.

Talbot, F.A. 1965. A description of the coral reef structure at Tutia Reef, Tanganyika, East Africa and the fish fauna. *Proceedings of the Zoological Society, London* 145:431-71.

Taylor, W. and Lewis, J.B. 1970. Fauna and flora of the marine grass beds. *Journal of the East African Natural History Society and National Museum* 4:311-18.

Thandar, A.S. 1984. The holothurian fauna of Southern Africa. Doctoral thesis, University of Durban-Westville, Durban.

―――― 1987a. The Southern African stichopodid holothurians, with notes on changes in spicule composition with age in the endemic *Neostichopus granmatus* (H.L. Clarke). *South African Journal of Zoology* 22(4):278-86.

―――― 1987b. The status of some Southern African nominal species of Cucumarine (s.c.) referable to a new African genus and thir ecological isolation. *South African Journal of Zoology* 22(4):287-96.

―――― 1989. The sclerodactylid holothurians of Southern Africa, with the erection of one new subfamily and two genera (Echinodermata: Holothuriodea). *South African Journal of Zoology* 24(4):290-304.

Thompson, D. and Dunbar, R. 1988. Sex for dragons and damsels. *New Scientist* 160:45-48.

Tinley, K.L. 1985. *Coastal Dunes of South Africa.* National Programme for ecosystem research (South Africa), Committee for Nature Conservation Research. Pretoria: Council for Scientific and Industrial Research.

Tixier-Durivault, A. 1960. Les Octocoralliares de L'Ile Inhaca. *Bulletin du Museum National d'Histoire Naturelle* 32(4):359-67.

Tobias, P.V. 1978. A little-known chapter in the life of Eduardo Mondlane. *Acta Africana, Geneva - Africa* 16(1):119-24.

Trueman, E.R. 1975. *The Locomotion of Soft-bodied Animals.* Contemporary Biology Series. London: Edward Arnold.

Trueman, E.R. and Brown, A.C. 1987. Respiration in the foot of *Donax serra,* a burrowing bivalve mollusc. *Comparative Biochemistry and Physiology* 87A:1059-62.

―――― 1992. The burrowing habits of marine gastropods. *Advances in Marine Biology* 28:389-431.

Van Bruggen, A.C. 1966a. Studies on the land molluscs of Zululand with notes on the distribution of land molluscs in Southern Africa. *Zoologische Verhandelingen, Leiden* 193:1-116.

―――― 1966b. The terrestrial molluscs of the Kruger National Park: A contribution to the malacology of the Eastern Transvaal. *Annals of the Natal Museum* 18:315-399.

―――― 1978. Land Molluscs. In Werger, M.J.A. (ed.). *Biogeography and Ecology of Southern Africa.* The Hague: Dr W. Junk.

Van der Horst, C.J. 1940. Enteropneusts from Inyack Island. *Annals of the South African Museum* 32:293-380.

Vannini, M. 1975. Research on the coast of Somalia. The shores and dunes of Sar Uanle. 4. Orientation and anemotaxis in the land hermit crab, *Coenobita rugosus* Milne Edwards. 5. Description and rhythmicity of digging behaviour. *Monitore Zoologico Italiano* Supplemento 6:57-90 and 233-42.

―――― 1976. Field observations on the periodical transdunal migrations of the hermit crab, *Coenobita rugosus* Milne Edwards. *Monitore Zoologico Italiano,* Supplemento 7:145-95.

Vogel, F.C. 1984. Comparative functional morphology of the spoon-tipped setae on the second maxilliped in *Dotilla* (Decapoda, Brachyura, Ocypodidae). *Crustaceana* 17(3):1-12.

Walenkamp, J.H.C. 1990. Systematic and zoogeography of Asteroidea (Echinodermata) from Inhaca Island, Mozambique. *Zoologische Verhandelingen* 261:1-86.

Way, M.J. 1954. Life history and ecology of the ant, *Oecophylla longinoda* Latreille. *Bulletin of Entomological Research* 45(1):93-112.

Wells, J.B.J. 1967. The littoral copepods (Crustacea) of Inhaca Island, Mozambique. *Transactions of the Royal Society of Edinburgh* 67(7):189-358.

Wells, M.H. 1962. *Brain and Behaviour in Cephalopods*. London: Heinemann.
Wilson, E.O. 1971. *The Insect Societies*. Cambridge: Belknap.
Wilson, M. and Thompson, L. (eds.) 1969. *The Oxford History of South Africa* Vol. I. Oxford University Press. Oxford.
Wolfowitz, E. 1938. Noticia preliminar da flora da Ilha da Inhaca. *Moçambique Documento Trimestre* 13.
World Health Organisation Report. 1989. *Geographical Distribution of Arthropod-borne Diseases and their Principal Vectors*. Geneva.
Worth, C. de Sousa, J. and Weinbren, M.P. 1961. Studies on the life history of *Aedes pembaensis* (Theobald) (Diptera, Culicidae). *Bulletin of Entomological Research* 52:257-61.
Yonge, C.M. 1960. *Oysters*. London: Collins.
—— 1963. Rock-boring organisms: Mechanisms of tissue destruction. *American Association for the Advancement of Science* 75:1-24.
—— 1968. Living Corals. *Proceedings of the Royal Society B* 169:329-44.
Yonge, C.M. and Nicholls, A.G. 1931. Studies on the physiology of corals. IV. The structure, distribution and physiology of zooxanthellae. *Scientific Reports of the Great Barrier Reef Expedition* 1928-9. British Museum (Natural History) London 1:135-76.
Youthed, E.D. and Moran, V.C. 1969. Pit construction by Myrmeleontid larvae. *Journal of Insect Physiology* 15:1103-16.

Index

Acorn worms 73-75, 144
Agriculture 300
    fruit trees, preservation 302
Amphibians 318
    frogs 318, 320
    toads 319
Amphioxus *see Lancelets*
Amphipods
    in rock pools 202
    in the sea meadows 254
    in the supratidal fringe 152
Anemones 79, 227, 230, 239
    association with hermit crabs 231
    burrowing 227
    cerianthid 70, 260, 274
    colonial 227
    compound 192
    creeping sea 260
    green sand 263
    in the cliff pools 182
    on the sand-banks 274
    viviparity 182
Anthozoa 260
Ants
    Cocktail 345
    Tailor 344, *345*
Ascidians *see Tunicates*
Atmospheric gases 26 *et seq.*
    carbon dioxide 27
    nitrogen 27
    oxygen 26
Avicennia marina 90, 91, *91, 97*, 103
    biology 96
    excretory glands 97
    gaseous exchange 97
    root system 97
    salinity tolerance 96

Bacteria 94
Barnacles 253
    feeding habits 153
    in mangroves 115
    in shallow rock pools 184
    in the supratidal fringe 153, 182
    nauplius larvae 153
    on the shore rocks 157
    pelagic larvae 154
    reproduction 154

Barreira Vermelha 212
Basket star 232
Bats
    Angolan Free-tailed 327
    Little Free-tailed 327, *327*
    Wahlberg's Epauletted Fruit 327, *327*
Bay of Maputo 125, 126
Bees
    Carpenter 347, *347*
    Honey 348
    Stingless 348
Beetles
    Blister 337
    Dung 338
    False ground 337
    Flower chafers 338, *338*
    Fruit chafers 338, *338*
    Ground 337, *337*
    Leaf 339
    Longhorned 338, *338*
    Net-winged 339
    Rhinoceros 338
    Rove 337
    Staphylinid *87*, 88
    Tiger 337
    Weevils 339
    Wood-boring 339
Bernouilli effect 159, *159*
Biogeography 13
Birds
    Albatross, Shy 317
    Albatross, Wandering 317
    Albatross, Yellownosed 317
    Barbet, Blackcollared 312
    Batis, Chinspot 111
    Bee-eater, Little 313
    Bulbul, Blackeyed 310
    Chat, Stone 313
    Cormorant, Cape 316
    Cormorant, Reed 316
    Cormorant, Whitebreasted 316, 317
    Crow, Indian house 312
    Crow, Pied 312
    Cuckoo, Emerald 311, *312*
    Curlew 69, 111, 315
    Dikkop, Spotted 314
    Drongo, Forktailed 311
    Drongo, Squaretailed 311

Eagle, African Fish 314
Egret, Cattle 313, 315
Egret, Little 315
Flamingo, Greater 111, 127, *127*, 314
Flycatcher, Dusky 310
Flycatcher, Paradise 111
Francolin, Rednecked 314
Gannet, Cape 317
Greenshank 315
Gull, Greyheaded 316
Gull, Kelp 316
Gymnogene 314
Heron, Blackheaded 312, 315
Heron, Greenbacked 111
Heron, Grey 312, 315
Hoopoe 313
Hornbill, Trumpeter 313
Ibis, Hadeda 311
in mangroves 111
in sandy habitats 70
in the coastal bush 309
in the southern bay 127
Kingfisher, Brownhooded 311
Kingfisher, Greyhooded 311
Kingfisher, Mangrove 111, 311
Kingfisher, Pied 311
Kingfisher, Pygmy 311
Kingfisher, Striped 311
Korhaan, Blackbellied 314
Longclaw, Yellowthroated 313
Nightjar, Fierynecked 310
Owl, Barred 311
Owl, Spotted eagle 311
Pelican, Pinkbacked 316
Petrel, Pintado 317
Petrel, Softplumaged 317
Petrel, Whitechinned 317
Pigeon, Green 310
Pipit, Plainbacked 313
Plover, Crab 315
Plover, Grey 70, 315
Plover, Ringed 70, 315
Plover, Whitefronted 70, 315
Prion, Broadbilled 317
Quail, Blue 314
Robin, Natal 310
Sanderling 70, 315
Sandpiper, Common 315
Sandpiper, Curlew 70, 315
Sandpiper, Marsh 315
Shrike, Bush 311
Skua, Arctic 316
Skua, Pomarine 316
Skua, Subantarctic 316
Sparrow, House 312
Sparrow, Yellowthroated 312
Starling, Blackbellied 310
Stint, Little 315
Sunbird, Collared 311

Sunbird, Purplebanded 111
Swift, Palm 312
Terek 315
Tern, Blacknaped 316
Tern, Caspian 316
Tern, Common 316
Tern, Lesser crested 316
Tern, Sooty 316, 317
Tern, Swift 316
Trogon, Narina 310
Turnstone 70
Twinspot, Green 310
Wagtail, Cape 313
Warbler, Bleating 311
Weaver, Lessermasked 111
Weaver, Spottedbacked 311
Whimbrel 70, 111, 315
White-eye, Yellow 310
Bivalves
　among the sea grasses 77
　cockles 78
　clams 255
　Clam, giant 225
　estuarine bivalves 127
　false cockles 171
　fan shells 131, 146
　feeding habits 155, 255
　file shells 170
　in sandy habitats 53
　in the northern bay 146
　in the reef flats 169
　in the sea meadows 255
　in the southern bay 127, 131
　in the supratidal fringe 153, 154, 182
　mussels 88, 153, 157, 183
　Mussel, brown 193
　Mussel, wedge 53-55, 153
　Nestlers 171
　on the sand-banks 270
　on the shore rocks 157
　oysters 111, 154, 170, 182
　Oyster, pearl 170
　Oyster, thorny 170
　scallops 170
Brachiopoda 68, 69
Brittle stars
　feeding habits 68, 129, 232
　in rock pools 204, 206
　in sandy habitats 63, 67
　in the coral reefs 231
　in the reef flats 164
　in the sea meadows 252
　in the southern bay 129, 133, 138
　mobility 232
Bugs
　Cotton Stainers 334
　Stink 335, *335*
　Twig-wilter 334, *335*
　Water strider 335

Butterflies
    Blues 341, *343*
    Browns 342
    Commodores 342
    Distasteful 342, *343*
    Forest 342
    Monarchs 342, *343*
    Skippers 342
    Swallowtails 342, *343*
    Whites 341, *343*

Carbon dioxide 70
Cephalopoda 236, 258
Ceriantharia 70
Cerianthus 227
Chaetognatha 224
Chlorophyta 189
Ciliates, biology 81
Cirratulids 65, 258
Cirripede 254
Cladophorales 194
Clams
    anatomy 226
    Giant 223, 225
    symbiosis 226
Climate
    rainfall 22
    relative humidity 22
    temperature 20
    wind 22
Coelenterates 213, 259
    anemones 274
    hydroids 259
    in the sea meadows 259
    on the sand-banks 273
    sea fans 259
    sea feathers 273
    sea pens 274
Commensalism 36, 37, 128, 158, 230, 239
Competition 47
    between algae and animals 191
    in corals 216
    interspecific 230
Conservation 361, 364
Consumers 35
Copepods 223
    harpacticoids 83
    in sandy habitats 83
Coral polyps
    calcareous skeleton 213
    cnidocytes 213
    digestion 214
    feeding 214
    mesenteries 213
    metabolism 214
    nematocysts 213
    sensory organelles 214
    skeletal cups 213
    spirocysts 213
    toxins 214
Corals, reef building
    biology 213
    commensals 213
    feeding area 213
    nerve net 213
    rate of growth 213
    structure 213
    tentacle concerts 213
Coral reefs 211 *et seq.*
    Acropora 48, 198, 217, *218*, 228
    ahermatypic 136, 215, 216, 221, 226
    animal associates 211, 246
    Anomastrea 198, *198*
    biogeographical perspectives 245 *et seq.*
    biology of reef-building corals 213
    brain 219
    commensals 228
    common genera 216 *et seq.*
    communities in the ecosystem 211
    criteria for success 211
    dependence on light 211
    destruction 243
    detached 221
    distribution 212 *et seq.*
    ecosystems 246
    Favia 136, 219, *219*, 229
    Favites 219, *219*
    Fire coral 214
    fish compositon 243
    Fungia 145, 216, 220
    Galaxea 137, 220, *220*, 229
    Goniopora 136, *137*, 137, 219
    hermatypic 215, 216, 221
    hermatypic, taxonomic classification 216
    Herpetolitha 145, *145*, 221
    Heteropsammia 221
    hydroid 214
    in rock pools 198
    in sea meadows 211
    interspecific competition 216
    in the northern bay 145
    in the southern bay 136
    light penetration 211
    limits to development 245
    metabolism 214
    Montipora 136, 217, *218*
    mortality 245
    Mushroom 216, 220
    on sand-banks 211
    patterns of branching 215
    Pavona 218, 219
    phytoplankton 221 *et seq.*
    planula larvae 216
    Pocillopora 136, 198, 217, 218, *218*, 228, 229
    Portuguese Island 147
    recovery 243
    reef associates 225
    reef flats 211

relationship with algae 214
reproduction 216
Seriatopora 218, 228
Serpent stone 221
skeleton 215
soft 226, 259
special features 216 *et seq.*
Stag's horn 217
Stylophora 136, 198, 218, 228, *229*
Tubastrea 136, *136*, 221, 227
Turbinaria 185, 186, 196, 221, *220*
zooplankton 221 *et seq.*

Crabs
adaptations in mangroves 116
Box 267
Bubble 61
Burrowing 267
burrows 61
carnivorous 167
commensalism in coral reefs 228
courtship 141
Elbow 268
feeding habits 51, 228
Fiddler 101, 104, 106, 115, 116
Frog 268
Gall 228
Ghost 50, 85
Grapsid 103, 106, 200
heat tolerance 40
Hermit 134, 141, 254
in coral reefs 228
in mangroves 103, 106, 114, 115
in rock pools 200, 206
in sandy habitats 50, 61, 64, 85
in the reef flats 166
in the southern bay 127, 132, 134, 140
in the supratidal fringe 152
land 103
life-history 167
locomotion 51
Masked 268
myochordotonal organ 52
Ocypodid 64, 105
on the shore rocks 158
Pea 132
Porcellanid 200, 201, 258
respiration 51, 62, 104
Sentinel 64
Sesarmid 122
Shame-faced 267
Spider 167, 200, 253
Sponge 253
swimming 78, 106, 254
Xanthid 106, 200, 201, 230, 259

Crinoids
feeding habits 232
life history 233
structure 232

Crops 362

Crustaceans
amongst the sea grasses 78
amphipods 152, 202, 254
barnacles 115, 153, 157, 182, 184
copepods 83, 223
coral reef associates 230
crabs 78, 85, 106, 114, 127, 132, 134, 140, 152, 158, 166, 200, 206, 230, 267
in rock pools 184, 200, 206
in mangroves 101, 103, 106, 111, 115
in sandy habitats 50, 61, 64, 83
in the reef flats 166
in the sea meadows 253 *et seq.*
in the southern bay 127, 131, 132, 134, 140
in the supratidal fringe 152, 182
isopods 254
lobsters 231
on the shore rocks 157, 158
on the sand-banks 267
ostracods 224
prawns 78, 112
shrimps 78, 107, 111, 131, 168
zooplankton 223

Cushion stars
Golden 233
in the southern bay 138

Cyanophyta 180, 223

Demospongia 227
Detritus 35
Detrivores 35
Diatoms 80, 222, 353 *et seq.*
benthic 55
cell wall structure 222
epilithic 180, 183
freshwater 353
motility 56
in sandy habitats 55
in the southern bay 128

Dictyotales 195, *196*

Dinoflagellates 213, 214
in sandy habitats 56
symbiosis 222

Districts, administrative 361

Dolphin
Humpback 329
Indian Ocean Bottlenosed 329, *330*

Dragonflies, life cycle 332, *332*

Dugongs 329, *330*

Echinoderms
among the sea grasses 75
brittle stars 63, 67, 129, 133, 138, 164, 204, 206, 231, 252, 266
cushion stars 138
feather stars 232
heart urchins 266
in rock pools 204, 206
in the northern bay 145

in the reef flats 163
in the sea meadows 250
in the southern bay 129, 133, 137
on the sand-banks 264
on the shore rocks 157
sand-dollars 265
sea cucumber 75, 138, 157, 164, 204, 235, 252, 267
sea urchins 137, 164, 204, 235, 251
starfish 163, 233, 250, 264
Economy 2, 362
Employment 362
Enteropneusts *see Acorn worms*
Environment
    chemical factors 26 *et seq.*
    estuarine influences 128
    nutrient salt cycles 27
    organic compounds 27
    problems 363
    variations in salinity 27, 95
Erosion 14, 20
Evaporation 41

Feather star 230, 232
Fertility, soil 300
Fish
    Barracuda 208, *209*
    benthic feeders 239
    Blennies 205
    Blue emperor 243, *242*
    Bonita, striped 208, *209*
    Box fish 239, 240, *240*, 264
    Brindlebass 208
    Butterfly fish 178, *178*, 239, *240*
    Cardinal 239, *239*, 243
    Cardinal, striped 178, *178*
    carnivores 207, 240
    cartilaginous 142
    Catfish, eel-tail 146, *147*
    Clinids 205, *205*
    Cod, rock 208
    composition in coral reefs 243
    Cow fish 239, 240
    Crested weed fish 262
    Damsel fish 178, *178*, 239, 263, 264
    Devilfish 241
    diversity of species in coral reefs 236
    Dogfish 142
    Eels 240, 241
    Eels, conger 241, *242*
    Eels, moray 241, *242*
    Eels, serpent 241
    Eels, snake 262, *261*
    File fish 239, *240*
    Fire fish 241
    Fire fish, Devil's 177, *178*
    Flagtail 205
    Flat-head 264
    Fusilier, beautiful 147, *147*
    Garrupa 208, *209*
    Goat fish 147, *147*
    Gobies 114, 205, *205* 262
    Goby, candy-stick 262, *262*
    Goby, meander 115
    Goby, mud-skipping 109
    Grunter 142, *143*, 240, 264
    Half-beak fish 263, *263*
    herbivores 238
    Herring, wolf 146, *147*
    in mangroves 109, 114
    in rock pools 205
    in the northern bay 146
    in the open sea 207
    in the reef flats 177
    in the sea meadow 260
    in the southern bay 142
    Kingfish 143, 207, *209*
    Klipfish 205
    Labrids 239, 240
    Ladyfish 208
    Mackerel 208, *209*
    Marlin, black 208, *209*
    Milkfish 264
    mouth-brooding 237
    Mullet 142, *143*
    Mullet, red 147
    Needle fish 263
    Parrot fish 238, *238*
    phytoplankton feeding 238
    Pipefish, alligator 260
    Pipefish, blue-spotted 115
    Pipefish, ghost 261, *261*
    Pomfret 142, 143
    Porcupine fish 239, 240, *241*
    Puffers 239, 240, *241*, 264
    Pursemouth, blue-spotted 147, *147*
    Queenfish 143
    Rabbit fish 238, 260, 263, *263*
    Rainbow fish 239, 240, 264
    Ray, electric 142, *143*
    Razor fish 261, *261*
    reproductive adaptations 237
    reversal of sex 237
    Sand smelt 142, *143*
    Scorpion fish 177, 240, 241, *242*, 262, 264
    Sea goldies 238, *239*
    Sea horse 260, *261*
    Sharks 142
    Sharks, ragged-toothed 210, *209*
    Snappers 240, 241, *242*
    Soldier fish 240, 241, *242*, 243
    speciation 236
    Stonefish 177, *178*, 241
    Surgeon fish 238, *238*
    Tobies 239, 240, 264
    Trigger fish 239, *240*
    Wrasses 240, *242*, 263, 264
    Wrasse, bluestreak cleaner 205, *205*
    Wrasse, cocktail 264

    Wrasse, seagrass 264
    zooplankton feeders 238
Fishermen's Combine 362, 367
Flatworms
    dependence on algal symbionts 58
    in rock pools 200
    in sandy habitats 57, 86
    in the reef flats 173
    speciation 48
Flies
    Bee 340
    Blowflies 341
    Flesh 341
    Hover 340
    Long-legged 340
    Robber 340, *340*
    Tachinid 341
Food webs 88
Foraminifera 80, 225, 266
    life cycle 81
Forest
    canopy, closed 283
    primary evergreen dune 297
    regenerating, characteristics 297
    scrub, undisturbed 283
Freshwater swamps *see Swamps, freshwater*
Frogs
    Bushveld Rain 320, *320*
    Dwarf Puddle 319
    Grass 319, *319*
    Painted Reed 318, *319*
    Tinker Reed 318
Fucoxanthins 186
Fuel 362
Fungi, decompostion of mangrove detritus 94

Gastropods, marine
    among the sea grasses 76
    carnivorous snails 172
    cone shells 256
    cowries 172, 203, 256
    grazing snails 157
    in mangroves 107
    in rock pools 203
    in sandy habitats 66
    in the reef flats 171
    in the sea meadows 255
    in the southern bay 130
    in the supratidal fringe 154, 156, 183
    mangrove snails 107-08
    mitre 256
    on the sand-banks 268 *et seq.*
    scavenging whelks 66, 130
    triton 256
    tulip shell 271
    whelks 86, 87
Gastropods, non-marine 349 *et seq.*
Geomorphology 17
Gorgonids 227, 232, 259

Grasshoppers
    Elegant 334
    Toad 334

Halophytes 102
Heart urchins, on the sand-banks 266
Herbivores 33
History 1
Homoptera
    Cicadas 335
    Mealy bugs 336
    Plant lice 336
    Scale insects 336
    Spittle Bugs 335
Hotel 367
Hydroids
    in rock pools 199
    in the sea meadows 259

Inguane 361
Insects
    antlions 336
    ants 108, 344
    bees 109, 347
    beetles 88, 109, 336
    bugs 334
    butterflies 341
    crickets 333
    distribution of habitats 331
    dragonflies 332
    flies 340
    grasshoppers 333
    Homoptera 335
    in mangroves 108
    in sandy habitats 88
    lacewings 336
    locusts 333, *334*
    mantids 333
    mosquitoes 108, 339
    moths 343
    termites 332
    wasps 345
Instituto de Investigação Científica de Moçambique 5
Iridiocytes 236
Iridiophores 226
Isle of Goa 211
Isopods 192
    in the sea meadows 254

Kunwadlaniheni 92, 93

Labour, migratory 362
Lancelets, on the sand-banks 273
Land
    abandoned 300
    regeneration 303
Lizards
    Chameleon, Flap-necked 320, *321*
    Gecko, Cape Dwarf 320, *321*

Gecko, Tropical House 320
Leguaan, tree 324, *324*
Lizard, Yellow-throated plated *321*, 322
Skink, Giant legless 324, *324*
Skink, Golden blind legless 322
Skink, Mozambique 321, *321*
Skink, Mozambique dwarf burrowing 324
Skink, Wahlberg's snake-eyed 321
Skink, Zulu dwarf burrowing 324

Lobsters
coral reef associates 231

Machangulo Peninsula 125
Mammals
bats 327
dolphins 329
dugongs 329
marine 329
moles 327
rodents 328
terrestrial 326
whales 330
Management 364
Mangroves 90 *et seq.*
adaptations of crabs 116
adaptations of tree species 96
aeration 94
animal distribution 100
biological processes 94
birds 111
creek 93
crustacea 101, 103, 106, 111, 115
environmental variables 94
fern 92
goby 109, 114
halophytes 102 insects 108
interior of the swamp 106
landward fringe 101
meiofauna 108
molluscs 105, 107, 111, 115
organic content of soils 95, 96
preservation 123
salinity 94, 95
salt gradient 95
sand grain size 95
seaward fringe 115
tidal streams 111
tree species 90
types of associations 91
unzoned trees 91
zoned 92
Marine Biological Station 5, 212
Medical facilities 3
Meiofauna
in mangrove mud 108
in sandy habitats 82
Microflora, in sandy habitats 55
Mole, Yellow Golden 327, *328*
Molluscs, marine
among the sea grasses 76
bivalves 53, 77, 88, 111, 127, 131, 146, 153, 154, 168, 182, 255, 270
gastropods, marine 66, 107, 130, 156, 171, 183, 203, 236, 255, 268
in mangroves 105, 107, 111, 115
in rock pools 203
in sandy habitats 53, 86
in the coral reef 236
in the northern bay 146
in the reef flats 169
in the sea meadows 255 *et seq.*
in the southern bay 127, 130, 131, 139
in the supratidal fringe 149, 153, 154, 181, 182
octopus 236
on the sand-banks 268
on the shore rocks 157
periwinkles 105, 149, 181
predatory molluscs 236
sea slugs 130, 139, 203, 257, 270
squid 236, 257, 258
Molluscs, non-marine 349 *et seq.*
Moss, Margaret 5
Moths
Army worms 343
Bagworms 344
Boll worms 343
Cutworms 343, *343*
Mud-skippers *see Fish; Periophthalmus kalolo*
Mussels 193
adaptations 54
colour patterns 54
Date-stone 157
feeding habits 53
genetics 54
in sandy habitats 53
on the shore rocks 157
Wedge 53, 153

Nature reserves *see Reserves*
Nematodes 84
Nemertines 68, 173
Nhaquene 362
Niche 39
Nitrogen 27
Northern bay
bare flats 144
corals 145
echinoderms 145
fish 146
habitats 143
molluscs, marine 146
sand-bar boundary 148
sea meadows 145, 146
Nudibranchs 230, 257
in rock pools 203
in the southern bay 139
Nutrient salt cycles 27
Nyaka, Chief 1

Ocean currents 21
Oceanographic Research Institute of South Africa 7
Octocorallines 227, 274
Octopus 236
    feeding habits 236
    iridiocytes 236
Opisthobranchs 115
Osmoconformers 41
Osmotic stress 41
Ostracods 224
Oxygen 26
Oysters 111
    food resources 155
    in the reef flats 170
    in the supratidal fringe 154, 182
    reproduction 155
    Thorny 170

Pansy shells *see* Sand-dollars
Pennatulids 274
Periodicity 43
Periophthalmus kalolo 109, *110*
    anatomy 109
    feeding habits 110
    life history 111
    nests 110
    respiration 111
Periwinkles
    distribution on rocks 181, *181*
    feeding habits 151
    in mangroves 105
    in the supratidal fringe 149, 181
    locomotion 152
    migration 151
Phycocyanin 180, 187
Phycoerythrin 180, 187
Physical factors
    adaptation to abiotic variables 40
    environmental stresses 39
    periodicity 43
    tidal zones 38
Phytoplankton 221 *et seq.*
Plankton
    density 221
    numbers of phytoplankters 223
    permanent zooplankters 223
Plants, sand binding 105, 134, 136, 292
Platanna 318, *319*
Pneumatophores 91
Polychaetes 66, 258
    division of food resources 64
    Eunicids 63, 175, 200, 206, 258
    feeding habits 59, 61
    in rock pools 200
    in sandy habitats 63, 64
    in the reef flats 173, 174
    in the subtidal fringe 162
    in the southern bay 127, 134
    locomotion 272
    Maldanids 65, 127
    Nereids 175, 200, 206
    on the sand-banks 272
    Orbinids 65, 127
    reef-building 205
    reproduction 66
    Sabellariids 175, 206
    Serpulids 175, 200, 228
    Spionids 65, 127, 258
    Syllids 175, 200
    Terebellids 259
    Terebellids, feeding habits 59
    zooplankton 225
Ponta Punduine 126
Ponta Torres 125, 212
Population 4
Portuguese Island 15, 20, 213
    coral reefs 147
    mangroves 94
Prawns
    amongst the sea grasses 78, 80
    biology 113
    in mangroves 112
Primary producers 33
Problems, contemporary 19
Protists
    ciliates 81
    foraminiferans 80
Protochordates 273
Pseudo-faecal pellets 62
Pycnogonids 202, 206
Pyramidellids 270

Quintanilha, Aurelio 5

Radiolaria 225
Reef flats 162 *et seq.*
    crustaceans 166
    echinoderms 163
    fish 177
    molluscs 169
    sea weeds 176
Reproductive adaptations, in coral reef fish 237
Reproductive practices 44
Reptiles
    chameleons 320
    geckos 320
    lizards 321, 324
    snakes 319, 322, 325
    terrapins 319
    turtles 85, 325
Research 363
Reserves
    Barreira Vermelha Marine Reserve 366
    Nyakeni Forest Reserve 365, 366
    Portuguese Island 366
    Yingwani Forest Reserve 356
Respiratory organs, modifications 42
Rhizopoda 225

Ridjene 5, 361, 367
Rock pools
    algae 184
    corals 198
    crustaceans 184, 200, 206
    echinoderms 204, 206
    fish 205
    flatworms 200
    hydroids 199
    molluscs 203
    polychaetes 200
    seaweeds 207
    sponges 198
    tunicates 186, 207
Rocks, shore 157 *et seq.*
    crustacea 157, 158
    echinoderms 157
    molluscs 157
Rodents
    Mouse, house 328
    Mouse, multi-mammate 328, *328*
    Rat, black 328
    Rat, tree 328
    Rat, vlei 328
Ronga 1

Saco da Inhaca 125
Salinity 42
Sand banks 264 *et seq.*
    burrowing animals 247
    burrowing worms 272
    coelenterates 273
    crustacea 267
    echinoderms 264
    lancelets 273
    molluscs 268
Sand dollars 247
    biology 265
    feeding habits 266
    on the sand-banks 265
Sand dunes 19
Sand grains, microbiota 80
Sandy habitats
    birds 69
    charactersitics 49
    crustacea 50, 61, 64, 78, 83, 85
    east coast 50
    echinoderms 63, 67, 75
    insects 88
    meiofauna 82
    microflora 55
    molluscs 53, 66, 76, 86, 88
    north-west coast 50
    polychaetes 64
    reptiles 85
    sea grasses 71
    south-west coast 49
    west coast 49, 50
Schools 4

Scientific research 5
Sea
    currents 211
    temperature 211
Sea cucumbers 131
    among the sea grasses 75
    coral reef associates 230
    Cuvierian tubules 76
    distribution in coral reefs 235
    in rock pools 204
    in the reef flats 164
    in the sea meadows 252
    in the southern bay 138
    locomotion 75
    on the sand-banks 267
    on the shore rocks 157
    reproduction 76
Sea feathers 247
    on the sand-banks 273
Sea grasses 71 *et seq.*, 126
    associations 126, 248
    biology 248
    broad-leaved 72, 247
    distribution 247 *et seq.*
    flowers 249
    in the northern bay 146
    in the southern bay 126, 129, 131, 133, 135
    narrow-leaved 71, 247
    reproduction 249
    seaweeds 72
Sea hare 257
Sea meadows
    animals 250
    coelenterates 259
    crustaceans 253
    echinoderms 250
    fish 260
    in the northern bay 145
    in the southern bay 126 *et seq.*
    molluscs 255 *et seq.*
    polychaete worms 258
    seaweeds 250
    urochordates 260
Sea pens 247
Sea slugs
    in rock pools 203
    in the sea meadows 257
    in the southern bay 139
    see Nudibranchs, Tectibranchs
Sea urchins
    distribution in coral reefs 235
    embryological studies 252
    feeding habits 252
    in rock pools 204
    in the reef flats 164
    in the sea meadows 251
    in the southern bay 137
    Slate-pencil urchins 235
Seaweeds

articulated coralline 184, 187
blue-green 180, 223
encrusting coralline 188, 247, 250
epilithic 180-84, 188
epiphytic 116
flatworm associations 58
in coral reefs 214
in rock pools 184, 189 *et seq.*, 207
in sandy habitats 55, 72
in the reef flats 176
in the sea meadows 250
in the subtidal fringe 193
on the lower platform 189
red coralline 247, 250
symbiosis associations 35, 58, 214
unicellular 55
Sex, reversal of in fish 237
Shrimps
    alpheid 112, 220, 229, 262
    amongst the sea grasses 78
    cleaner shrimp 229, 230
    commensalism 37, 229
    in mangroves 107, 111
    in the reef flats 168
    in the southern bay 127, 131
    mantid 131, 169
    palaemonid 79 229, 230
    pistol 79, 127, 229
    scavenging 80
    thalassinids 169
    Zebra 169
Sichwane 92, *93*
Sipunculids 128, 173, 221, 228, 272
    feeding habits 60
Skototaxis 169
Snakes
    Adder, puff 322, *323*
    Boomslang 322, *323*
    Brown Water 319
    Cobra, Egyptian 322
    Cobra, Forest 322, *323*, 325
    Common Brown House 324
    Eastern Purple-glossed 325
    Fornasini's Blind 322
    Herald 324
    Mozambique Shovel-snout 325
    Natal Green 319
    Olive Grass 323
    Savanna Vine 323, *323*
    Variegated Slug-eater 325
Social welfare 368
Southern bay
    birds 127
    coral reef 136
    crustaceans 127, 131, 132, 134, 140
    diatoms 128
    echinoderms 129, 133, 137
    fish 142
    habitats 125

molluscs 127, 130, 131, 139
mouth 134
mud flats 127, 129
polychaetes 127
sandy areas 127
sea grass associations 126
sea grasses 129, 131, 133, 135
shelly banks 128
Speciation 48
    in coral reef fish 236
Specific Mate Recognition 237
Spiders 348
Sponges
    Boring 227
    Cup 227
    Encrusting 227
    feeding habits 159
    in the subtidal fringe 158
    in rock pools 198
    leuconoid 228
    reproduction 160
Squid 236
    biology 258
    in the sea meadows 257
Starfish 247, 250
    burrowing starfish 234
    juveniles 235
    larvae 235
    in the coral reef 233
    in the reef flats 163
    locomotion 251
    on the sand-banks 264
    respiratory exchange 251
    water vascular system 251
Subtidal fringe 158 *et seq.*
    polychaetes 162
    seaweeds 193
    sessile animals 247
    sponges 158
    tunicate 160
Supratidal fringe 149 *et seq.*, 180 *et seq.*
    crustacea 152, 182
    microflora 180
    molluscs 149, 153, 154, 181, 182
Swamps, freshwater 19, 95, 299
    changes in the diatom flora 353
    cultivation in 302
    emergent vegetation 299
    gastropods 351
    grass 300
Swamps, mangrove *see Mangroves*
Symbiosis
    algae 35
    in dinoflagellates 222
    in giant clams 226

Tectibranchs 257
    in the southern bay 130
    on the sand-banks 270

Temperature 21
    adaptations 41
    sea 211
    heat tolerance 40
Tengeni 92, *93*
Termites
    dry wood 332
    harvester 333
    subterranean wood-eating 333
Terrapin, Eastern Hinged 319
Thigmokinesis 169
Tidal current 247
Tides 22 *et seq.*, *23*
    currents 26
    modified local 24
    tidal zones 38
Toad, Guttural 319
Topography 13 *et seq.*
    north-east point 179
    shores 15
Tourism 367
Trophic relationships 33, 88
Tunicates 160, 193, 260
    cellulose synthesis 161
    heart 161
    in rock pools 186, 207
    in the sea meadows 260
    in the subtidal fringe 160
    planktonic 224
Turbellaria *see Flatworms*
Turtle
    Green 326, *326*
    Hawksbill 325
    Leatherback 86, *85*, 325, *326*
    Loggerhead 85, *85*, 325, *326*

Ulvales 207
United Nations Development Plan of 1990 7, 363
Universidade Eduardo Mondlane 6, 367
University of the Witwatersrand 5, 6
Urochordata 224
    in the sea meadows 260
    tunicates 260

Van der Horst, Professor 5
Vegetation
    adaptation of mangrove species 96
    bulrush 299
    bushland, evergreen 286, 289, 294
    Bush-tick berry 286
    Cashew nut 302
    Casuarina 307, 308
    Cat's whiskers 304
    climbers 289
    Cycad, Zululand 296
    Date palm, wild 285, 298
    Dog rose, African 304
    ebony 306
    Euphorbia, rubber hedge 285
    Euphorbia, tidal river 290
    Fern 284, 289, 296, 298, 300
    Fig, Hottentot's 293
    Fig, wild 297
    forest, undisturbed scrub 283
    fruit trees, conserved 304
    grass 300, 304, 307
    Grass, Bermuda 288
    Grass, imperial 288
    Grass, red-top 303
    Grass, strand 293
    Grewia, climbing 308
    hardwood 285
    herbs, halophytic 300
    Ironwood, white 296
    Jackal berry 298
    Jackal berry, Zulu 296
    Knobwood, dune 307
    Kooboo-berry 285
    mahogany 306
    Mahogany, pod 292
    Manga 284
    Mango 302
    Mangosteen, African 302
    Mangrove, black 98
    Mangrove, Indian 98
    Mangrove, red 98
    Mangrove, spring tide 100
    Mangrove, white 96
    Marula 302
    Milkweed 294
    Milkwood, red 286, 288
    Milkwood, white 289, 297
    Myrtle, dune 304
    Num-num 296
    Oak, wild silver 289
    Ochna 296
    oleander 296
    Olive, sand 308
    on the dunes 289, 292
    on the west coast 282
    Orchid, terrestrial 286, 289
    Orchid, tree 285
    pioneers 303
    Plane tree, Natal 296
    regeneration 303
    Rhus, Natal 286, 288
    secondary vegetation 304
    Sedge, maritime 288
    Soap-berry, dune 304
    Strychnos, coffee bean 285
    succulents 285
    Sweet goat's-foot 293
    Sweet thorn 286
    Teclea, Natal 296
    trees, fruit 302
    Water berry 289, 302,
    White pear 298
    Wildekornoelie 297

Xikweja 284
Venturi effect 159, *159*

Wasps
    Chalcidid 345
    Digger 346
    Hairy 346, *346*
    Mason 347, *347*
    Paper 347, *347*
    Parasitic 345
    Sand 346
    Spider-hunting 346, *346*
    Thread-waisted 346, *346*
    Velvet ants 345
Water table 19, 247
Wave action, environmental effects 179

Whales, Humpback 330, *330*

Xanthophyll 222
Xilthangalweni 93, *93*

Zoanthids 192, 227, *227*
Zones 39, 130, 182
Zooplankton 221 *et seq.*
    crustaceans 223
    foraminifera 225
    ostracods 224
    permanent zooplankters 223
    polychaete worms 225
    protozoans 225
    tunicates 224
Zooxanthellae 213, 226-228